Nutrients, Infectious and Inflammatory Diseases

Special Issue Editor
Helieh S. Oz

MDPI • Basel • Beijing • Wuhan • Barcelona • Belgrade

MDPI

Special Issue Editor
Helieh S. Oz
University of Kentucky Medical Center
USA

Editorial Office
MDPI AG
St. Alban-Anlage 66
Basel, Switzerland

This edition is a reprint of the Special Issue published online in the open access journal Nutrients (ISSN 2072-6643) in 2017 (available at: http://www.mdpi.com/journal/nutrients/special_issues/nutrients_infectious_infla mmatory_diseases).

For citation purposes, cite each article independently as indicated on the article page online and as indicated below:

Lastname, F.M.; Lastname, F.M. Article title. *Journal Name* **Year**. *Article number, page range.*

First Edition 2018

ISBN 978-3-03842-791-9 (Pbk)
ISBN 978-3-03842-792-6 (PDF)

Cover image courtesy of Dr. Helieh S. Oz

Table of Contents

About the Special Issue Editor

Helieh S. Oz has a DVM, MS (U. IL); PhD (U. MN) and clinical translational research certificate (U. KY Medical Center). Dr Oz is an active member of American Association of Gastroenterology (AGA) and AGA Fellow (AGAF) and Associate in Rome Foundation. Dr Oz is Immuno-Microbiologist with expertise in inflammatory and infectious diseases, pathogenesis, innate and mucosal immunity, micronutrient, animal models, pain related behavioral modifications and drugs discovery. Dr Oz has over 90 publications in peer-reviewed journals in the areas of chronic inflammatory disorders (pancreatitis, hepatitis, colitis, periodontitis), infectious parasitic (Chagas disease, toxoplasmosis, babesiosis) and microbial infectious diseases. Dr Oz has served as Lead Editor for special issues, book and book chapters and different editorial advisory board committees including Center of Excellence for Medical Research and Innovative Products, Walailak University Thailand. Dr Oz is an avid reviewer for several peer-reviewed journals and invited speaker for national and interntional conferences.

Preface to "Nutrients, Infectious and Inflammatory Diseases"

A balanced diet with sufficient essential nutritional elements is critical for maintaining a healthy body. Both nutritional excess and deficiency are associated with disease. For example, nutritional excess, particularly in refined carbohydrates and saturated fats, coupled with physical inactivity, can result in chronic inflammatory conditions such as obesity and cardiovascular disease. On the other hand, deficiencies in essential nutrients can lead to stunted growth, poor immune function and classical conditions such as scurvy, osteoporosis, depression and xeropththalmia. The gastrointestinal (GI) track takes in food and water, digests the food, extracts the nutrients and expels undigested/unabsorbed material as waste. Nutrients such as amino acids, oligosaccharides, and short-chain fatty acids have been recognized to be beneficial to the GI track and to human health in general, and they participate in shaping the immune system and in energy metabolism [1,2]. Short-chain fatty acids (acetate, propionate, and butyrate) are also produced naturally by the intestinal microbiome acting on prebiotics such as oligosaccharides and other indigestible fermentable fibers [3]. GI infection by microbial, viral or parasitic agents alters the gut microbiome and increases permeability to toxins. The microbiome is further altered by ingested antibiotics to treat bacterial infections. Microbial invasion stimulates inflammation, a defensive mechanism of the body's immune system. This helps clear the invading microorganisms. However, persistent and excessive inflammatory response is a significant risk factor for developing various chronic inflammatory conditions and cancer, and increases the risk of succumbing to infectious diseases, owing to T cell exhaustion [4]. The collection of articles in this issue have been compiled to help illuminate the contribution nutrients make to the prevention, treatment and taming of a range of inflammatory and infectious diseases.

Chronic inflammatory diseases affect millions of people globally. While inflammation contributes to the tissue healing process, chronic inflammation can lead to the loss of tissue function and organ failure. Chronic inflammation, which accompanies conditions such as chronic hepatitis, inflammatory bowel disease and neurodegenerative disorders, increases the risk of malignancy. Despite rapid advancement in diagnostics and the availability of therapeutic options, there remains no effective cure for patients who suffer from inflammatory diseases. Therefore, patients seek alternatives and complimentary agents as adjunct therapies to relieve symptoms and possibly prevent consequences of inflammation. Oz H.S. [5], has investigated the anti-inflammatory properties of green tea polyphenols (GrTPs) with potent antioxidant activities in a variety of settings. These include their ability to inhibit the I-κB kinase nuclear factor-kappa B (NF-κB) signaling pathway, induction of programmed cell death (caspases, Bcl-2), release of inflammatory cytokines and production of lipid mediators of the cyclooxygenase (Cox) system. The author reviews relevant investigations regarding protective and adverse effects as well as possible applications for GrTPs in the treatment of chronic and inflammatory complications. GrTPs also possess antimicrobial properties, including the ability to inhibit the growth of Mycobacterium Tuberculosis (TB) in macrophages [6] Worldwide, approximately 2–3 billion people are infected with TB and 5–15% of these individuals will develop some form of active TB. Using a structured questionnaire, Soh A.Z. et al. [7] investigated the effect of drinking black or green tea, or coffee on the risk of TB activation in a prospective population-based cohort involving 63,257 tea-drinking Singaporean Chinese. With a mean follow-up period of 16.8 years, the authors reported that drinking black or green tea was associated with a dose-dependent reduction in the risk of TB infection. This association was not evident with coffee or caffeine intake. The authors concluded that regular tea drinking is associated with reduced risk of active TB.

Inflammatory bowel disease (IBD), predominantly Crohn's disease and ulcerative colitis, affects 1.8 million people in the US and there is no available cure. Although environmental factors have been implicated in the etiology of IBD, a breakdown in immune homeostasis, caused by a dysregulated adaptive immune response against intestinal bacterial flora in genetically predisposed individuals, is believed to be a key pathogenetic factor [8,9]. Crohn's disease is frequently a progressive disease. Up to half of these patients require surgical interventions within about 10 years of diagnosis and over 75% of these operated patients require at least one further surgery in their lifetime. Currently, guidelines for nutrition in general surgery also apply to Crohn's patients. To minimize the risk of surgery, it is necessary to optimize the nutritional status for these patients. The systematic review by Grass F. et al. [10] investigated preoperative nutritional support in adult Crohn's patients between 1997 and 2017 and aimed to overview screening modalities, routes of administration and expected benefits in these patients. They selected 29 studies of which 14 were original and 15 were review articles. Malnutrition was found to be a major risk factor for postoperative complications, and both enteral and parenteral routes were efficient in decreasing postoperative morbidity. The authors recommended that the route of administration should be chosen based on disease presentation and the patients' condition. Further studies are needed to strengthen this evidence. Further, IBD patients may be at risk of experiencing vitamin B (Vit B) and folate deficiencies owing to malabsorption in the IBD-affected intestines. However, an association between IBD and serum folate and Vit B12 concentrations remains controversial. A multiple-database meta-analysis performed by Pan Y. et al. [11] to compare serum folate and Vit B12 concentrations in IBD compared to controls revealed that the average serum folate concentration in IBD patients was significantly lower than controls. Interestingly, this difference was only observed with ulcerative colitis patients, but not with Crohn's disease patients. No difference was detected for the mean serum Vit B12 levels. The authors concluded that folate deficiency might play a role in the development of IBD, although the data did not indicate causation. The authors suggested that folate and Vit B12 supplementation in IBD patients might improve their nutritional status and prevent other conditions.

Dietary considerations are important in IBD as nutritional factors may be directly involved in the pathogenesis and recurrence of IBD. Furthermore, dietary factors may also influence the treatment of IBD. Exclusive enteral nutrition (EEN) has been shown to have beneficial effects in various conditions including in Crohn's disease, and this has been recommended in Europe to be the first-line therapy for inducing remission in pediatric luminal Crohn's disease [12–14]. However, the mechanism of action is elusive. One possibility is that EEN could alter the microbiota in the patients. Gatti S. et al. [15] reviewed 14 different clinical trials, involving 216 Crohn's patients, which investigated the effect EEN had on the microbiota. Interestingly, patients on EEN had a profound decrease in microbiota diversity, which switched back to their natural status upon conclusion of EEN. The EEN appeared to cause metabolomic changes. Despite the interesting finding, inconsistencies were detected between studies in the effect EEN had on specific bacterial strains and this awaits further microbiological analysis using newer techniques such as next-generation DNA sequencing.

Probiotics and synbiotics are used to treat chronic inflammatory diseases such as IBD. The probiotic and synbiotic effects have been studied on chronic intestinal diseases in in vitro animal models and in humans in randomized clinical trials. Probiotic strains and their cell-free supernatants reduce the expression of pro-inflammatory cytokines via actions that are principally mediated by toll-like receptors. Probiotic administration improves clinical symptoms, histological alterations, and mucus production in most of the evaluated animal studies. Probiotic supplementation appears to be well tolerated, effective and safe in IBD patients. For instance, Bifidobacterium longum improved clinical symptoms in patients with mild-to-moderate active colitis. However, some results suggest that caution should be taken when administering these agents in the relapsed stages of IBD. In addition, no effects are reported on chronic enteropathies. Consequently, although probiotics are shown to provide benefits, Plaza-Díaz J [16] suggested that the risks and benefits should be carefully assessed before initiating therapy in these patients. Further studies are required to understand the exact mechanism by which probiotics and synbiotics affect these diseases.

Obesity has become a global problem. One of the emerging issues is the link between high pre-pregnancy body mass index (BMI) and an increased risk of adverse pregnancy outcomes. There have been limited studies regarding the relationship between pre-pregnancy BMI and the inflammatory potential of the diet during pregnancy. Shin D. et al. [17] included 630 pregnant women from the U.S. National Health and Nutrition Examination Survey (NHANES) with cross-sectional examinations from 2003 to 2012. Pre-pregnancy BMI was calculated based on self-reported weight and measured height. The authors reported that women with pre-pregnancy obesity (high-BMI) are more likely than those with normal weight to have a high dietary inflammatory index and elevated levels of C-reactive protein (CRP).

Micronutrient homeostasis is a key factor in maintaining a healthy immune system. Zinc (Zn) is an essential micronutrient that is involved in the regulation of the innate and adaptive immune responses. Malnutrition is the main cause of Zn deficiency and leads to cell-mediated immune dysfunctions and other manifestations [18]. Consequently, immune dysfunction leads to a worse outcome toward bacterial infection and sepsis. Zn is required for pathogen-eliminating signal pathways leading to neutrophil extracellular traps formation, as well as cell-mediated immunity over humoral immunity. Zn deficiency plays a role in inflammation damaging the host tissues. Zn is involved in the modulation of the proinflammatory response by targeting NF-κB, a transcription factor that is the master regulator of proinflammatory responses. It is also involved in controlling oxidative stress and regulating inflammatory cytokines. Zn is critical for sustaining proper immune function. Gammoh N.Z. and Rink L. [19] review the role of zinc and its deficiency during infections and inflammatory responses and modulation of the immune system.

Copper (Cu) is another essential trace element that is required for development. Infections alter Cu and Zn metabolism, deficiencies of which can increase infection risks. Wisniewska M. et al. [20] conducted a prospective, observational, case-control study in 21 infected and 23 control term and preterm newborns. Median concentration of Cu at birth (day 1) was 522.8 μg/L, and Zn was 1642.4 μg/L. Cu and Zn correlated positively with gestational age in control newborns. The authors concluded that infections affect trace element homeostasis in newborns; while serum concentration of Zn is reduced, Cu and CRP levels are increased. The Cu/Zn ratio may constitute a meaningful diagnostic biomarker for the early onset of infections. Selenium is a specific trace element and essential for normal metabolism. Minor selenium deficiency is accompanied by health defects, whereas severe deficiencies are associated with immunodeficiency, affecting both cell-mediated and humoral immune functions. Selenium supplementation improves immune function in depleted individuals. Pregnant women and infants are at risk of selenium deficiency, with negative effects on immune and brain function. Varsi K. et al. [21] investigated selenium levels in two different groups: (1) 158 healthy women who had never had any children; and (2) 140 women with singleton pregnancy who were followed from pregnancy week 18 through to 6 months postpartum. The prevalence of infant infection was reported by the mothers from 6 weeks to 6 months of age and the neurodevelopment of the infants were assessed using the Parental Questionnaire Ages and Stages at 6 months of age. The authors reported that low maternal selenium status (≤0.78 μmol/L) at week 36 of pregnancy was associated with an increased risk of infection in the babies during the first 6 weeks of age and low maternal selenium status (≤0.90 μmol/L) in pregnancy week 18 was associated with a lower psychomotor growth scores at 6 months of age. The authors recommended that maternal serum selenium should be greater than 0.90 μmol/L in pregnancy week 18 and 0.78 μmol/L in pregnancy week 36 to benefit the infants.

Chronic pancreatitis leads to pancreatic cancer, one of the most aggressive forms of cancer. Since pancreatic cancer cell lines express high levels of insulin-like growth factor (IGF-I and the IGF-I receptors that stimulate cell proliferation, angiogenesis and the invasiveness of cancer cells [22,23]) it is possible that IGF-I and related growth factors could play a role in promoting pancreatic cancer. Gong Y. et al. [24], in a meta-analysis study, investigated the association between the serum concentration of IGF-I and the risk of pancreatic cancer. Ten studies, published between 1997–2013, which met their inclusion criteria were selected from the Medline and EMBASE databases. The authors found no correlation between the serum concentrations of IGF-I and -II, IGFBP-1 and -3, and IGF-I/IGFBP-3 ratio with a risk of pancreatic cancer. Thus, serum IGF-I, IGF-II, IGFBP-1 and IGFBP-3, as well as the IGF-I/IGFBP-3 ratio, may not be associated with the development of pancreatic cancer. Further studies are required to confirm these findings.

Ginseng is an herbal supplement with a wide range of medicinal effects and few side effects. Ginsenoside (Rg1) is one of the major active ingredients of ginseng with some beneficial effects in neurodegenerative diseases such as Alzheimer's disease. Furthermore, Rg1 has anti-inflammatory effects. Given these actions, it is possible that Rg1 may have therapeutic value against osteoarthritis, a condition that is characterized by degenerative changes and inflammatory responses in chondrocytes. Cheng W. et al. [25] assessed the anti-inflammatory effects of Rg1 in human chondrocytes and whether Rg1 was able to reduce articular cartilage damage in a rat model of osteoarthritis. The authors reported that Rg1 suppressed IL-1β-induced inflammatory responses in human chondrocytes and reduced disease activity in the joints of the rats.

The microbiota affects the functioning of physiological, metabolic and immunological processes. For instance, the microbiota regulates the growth and function of immune cells in the intestines. Evidence indicates that alterations in the gut microbiota can influence infectious and inflammatory diseases. Bacteria that reside on the mucosal surface or within the mucosal layer interact with the host immune system. Thus, a healthy gut microbiota is essential for the development of mucosal immunity. In HIV patients and those with controlled viremia from using antiretroviral drugs, the gut microbiome is different to those uninfected HIV controls. Recent data suggest the patients with dysbiosis may have a breakdown in their gut immunological activity, causing systemic bacterial diffusion and inflammation. Treating GI tract disorders becomes a difficult task in HIV-infected patients and those on antiretroviral drugs. Therefore, trials are investigating the ability of probiotics to modulate epithelial barrier functions, microbiota composition, and microbial translocation. D'Angelo C. et al. [26] conducted a review on the use of probiotics to treat HIV-induced gastrointestinal tract disorders and in improving gut-associated lymphoid tissue (GALT) immunity.

The beneficial properties of pistachio nuts have been extensively reported for pistachio polyphenolic extracts from nuts, resins and leaves. These effects include antioxidant and anti-inflammatory, antipyretic, antibacterial and antiviral effects, and are used to treat infections, eczema, asthma, kidney stones, diarrhea/GI complications and abdominal pain. Paterniti I. et al. [27] investigated the anti-inflammatory and anti-oxidative stress properties of polyphenolic extracts from "raw pistachios" and "roasted, salted pistachio" utilizing the lipopolysaccharide (LPS)-stimulated monocyte/macrophage cell line J774 and a carrageenan-induced inflammatory paw edema in a rat model. The authors found anti-inflammatory and antioxidant properties of pistachio at lower doses than those reported before. The data support beneficial effects associated with the consumption of pistachios.

Pomegranate (Punicagranatum), a phytochemical-rich fruit, has been used for centuries to prevent and treat inflammatory conditions. Mandal A. et al. [28] previously reported that pomegranate extract inhibited dimethylbenz (a) anthracene (DMBA)-initiated rat mammary tumorigenesis by anti-proliferative actions and apoptosis. This is a continuation of their previous study to investigate the mechanism of the anti-inflammatory action of pomegranate extract in the same model. The authors demonstrated that pomegranate emulsion was able to prevent DMBA-evoked mammary carcinogenesis through anti-inflammatory mechanisms by inhibiting NF-κB while upregulating Nrf2 signalling.

L-Arginine is a non-essential amino acid and nitrogen carrier for the synthesis of urea, polyamines, proline, and other proteins. Arginine has immune-regulatory and anti-inflammatory properties. Consequently, it is frequently administered to critically ill patients with sepsis. Yeh C-L and her group [29] investigated the influence of intravenous arginine administration on altering circulating proangiogenic cells and remote lung injury in a mouse model of polymicrobial sepsis induced by cecal ligation and puncture. The authors demonstrated that arginine administration promoted the mobilization of circulating proangiogenic cells while down-regulating the sepsis-induced production of inflammatory cytokines and the expression of angiopoietin 1 and 2, and their receptor, Tie-2 mRNA in lungs. The authors suggested that more studies are required to determine whether their observations are involved in mediating the beneficial effects of arginine.

Vitamin D (Vit D), a fat-soluble steroid and a pro-hormone, is produced endogenously in the skin by a direct action of ultraviolet sunlight, and a portion delivered through dietary intake. Two forms of Vit D supplements available over-the-counter are ergocalciferol (Vit D2) and cholecalciferol (Vit D3). Vit D2 is commonly added to foods, whereas Vit D3 is mainly synthesized in the skin and is present in food animal products. Numerous studies have suggested the importance of Vit D in protecting against diseases, including obesity and malignancies. Vit D production and its receptors have been reported in multiple tissues, with a vital role in promoting the immune system. Rickets, a defect in bone growth in children due to Vit D deficiency, was first recognized in 1650. From the 1930s, Vit D2 has been added to milk in U.S. and Europe to eliminate rickets. Vit D deficiency is highly prevalent throughout the world, affecting host immunity and leading to an increased incidence and severity of several infectious diseases. Vit D deficiency may be caused by different drug interactions, including steroids, chemotherapies and a lack of sunlight exposure. However, high doses of Vit D supplements are linked with an increased risk of fractures, kidney stones and certain cancers. Gois P.H.F. et al. [30] reviewed the recent literature on the relationship between Vit D and the immune system; Vit D status and the risk of contracting infectious diseases such as TB, respiratory tract infections and fungal infections; Vit D status and sepsis; and disease progression and mortality in patients with human immunodeficiency virus-infected. In addition, the authors reviewed the outcomes of Vit D supplementation as a treatment/prophylaxis for the above mentioned diseases/conditions. Overall, there appears to be lack of agreement between the results of these investigations.

Dendritic cells (DC) are vital for antigen presentation and the initiation of an adaptive immune response against hostile infection as well as immune tolerance for harmless/beneficial microbiota. Food products can modulate the inflammatory status of intestinal DCs. Quercetin is a phytochemical (flavonoid), which can suppress the secretion of inflammatory cytokines, antigen presentation and DC migration towards the draining lymph nodes. De Santis S. et al. [31] recently identified that secretory leukocyte peptidase inhibitor (Slpi) expression is required for quercetin to inhibit the secretion of proinflammatory cytokines and chemokines. Quercetin-enriched food was found to be able to induce Slpi expression in the ileum, while few effects are detectable in the duodenum. Slpi-expressing cells are located at the tip of the intestinal villi, as quercetin exposure could be more efficient for DCs projecting "periscopes" in the intestinal lumen. The data suggest that quercetin could suppress inflammation in ileo-colonic tract.

Toxoplasmosis is a common foodborne and congenital disease caused by Toxoplasma gondii, an apicomplexan microorganism that infects muscle, neural tissue, and the brain in humans and animals [32]. Toxoplasmosis has been reported to be associated with behavioral and cognitive modulation, yet the mechanism of action is not well known. There is some evidence that suggests that Toxoplasma extracts folate from neurons. Since reduced folate availability is known to be associated with an increased risk of neurodevelopmental disorders, neurodegenerative diseases, and cognitive decline, Berrett A.N. et al. [33] analyzed data from the third National Health and Nutrition Examination Survey to determine the associations between Toxoplasma infection, multiple folate-cycle factors, and cognitive function in adults aged 20 to 59 years in United States. The data suggest that Toxoplasma infection affected levels of folate and/or Vit B-12 in the brain to alter cognitive functioning.

Millions of people are infected with Hepatitis C virus (HCV), which can lead to hepatocellular carcinoma. The mechanism by which HCV infection affects adipokines in the host remains unclear. Chang M-L et al. [34] presented a prospective trial with 450 patients with genotype 1 and 2 who had completed anti-HCV therapy. Patients surveys were used to assess patients pre-therapy and at 24 weeks post-therapy, as well as levels of adipokines including leptin, adiponectin and plasminogen activator inhibitor-1 (PAI-1). The authors reported specific associations between some of these parameters and suggested that HCV infection could mask other associations e.g., between increased adipokine levels and metabolic and hepatic profiles.

A major cause of admission into pediatric intensive care units (ICU) is severe inflammation/infections and sepsis. The metabolic response to stress and infection correlates with the severity of the insult and the energy requirement obtained from protein, fats and carbohydrates. The nutritional status in the admitted children tends to deteriorate during the course of the illness and this has a negative impact on clinical outcomes. Thus, it is important to accurately determine the energy requirements in the ICU in order to avoid under- or over-feeding. De Cosmi V. et al. [35] discussed the metabolic changes in critically ill children and the notion of personalized pediatric nutritional interventions. The authors report the major role of macronutrients, blood glucose levels and acute phase proteins by means of indirect calorimetric procedures. The authors concluded that personalized nutritional interventions in these patients are needed to use the glucose/fat balance to reduce catabolic consequences in critical stages and speed their recovery.

Parkinson's disease is a neurological disorder characterized by neuroinflammation and the loss of dopaminergic neurons within the midbrain. Due to insufficient therapies and the adverse effects of conventional drugs, there is an urge for the use of new, unconventional interventions for the treatment of Parkinson's. Atractylenolide-I (ATR-I) is a major bioactive ingredient isolated from the plant rhizomes of AtractylodesMacrocephala, also known as "Baizhu", a traditional Chinese medicine used for anti-gastrointestinal dysfunctions that has anti-oxidant and anti-cancer activity. More S and Choi D-K [36] investigated the anti-neuroinflammatory mechanisms of ATR-I in in vitro and in vivo models of Parkinson's disease. Intraperitoneal administration of ATR-I decreased microglial activation and protected dopaminergic neurons. In vitro, ATR-I inhibited NF-κB activation and enhanced the expression of hemoxygenase-1. The authors argue that ATR-I may be useful as a new therapeutic agent for Parkinson disease.

Neuromuscular diseases (NMDs) are a heterogeneous group of acquired or inherited syndromes. NMDs frequently accompany nutritional complications. Salera S. et al. [37] argued that with the prolongation of survival in patients with NMDs, it is important to consider nutritional issues in these patients. These include over-nutrition, glucose metabolism, mobility, respiratory and cardiologic functions. Hypo-nutrition affects muscles and ventilatory function, constipation and other GI disorders, chewing/swallowing difficulties that are risk factors for aspiration, which predisposes to infections, respiratory complications, osteoporosis and increased risk of fractures. Focusing on the care of children with Duchenne muscular dystrophy, the authors reported that appropriate nutritional care can improve the quality of life in these patients. Further studies are needed to investigate the relevance of over-nutrition and under-nutrition, GI, infections, dysphagia and reduced bone mass in the different types of neuromuscular diseases; information on appropriate percentiles of weight, height, body mass index and body composition are critical for the improved management of these patients.

Respiratory tract infections are the most common infections in children and adults. Recurrent respiratory tract infections are prevalent mostly in early childhood, causing high indirect and direct costs to the healthcare system. Recurrent respiratory tract infections are usually the consequence of immature immunity in children and immunosuppressed complications in adults with high exposure to various respiratory pathogens. Biologically active polysaccharides like β-glucans are extensively studied as natural immunomodulators, with anti-inflammatory and anti-infectious activities. Jesenak M. et al. [38] reviewed the use of β-glucans as a possible therapeutic and preventive approach in managing and preventing recurrent respiratory tract infections in children (β-glucans from PleurotusOstreatus), adults (yeast-derived β-glucans), and in elite athletes (β-glucans from PleurotusOstreatus or yeast).

Finally, a balanced and vigorous gut microbiota is necessary to support health and growth in the host. Overgrowth of gut microbes or pathogens can change the ecosystem and compromise gut integrity to initiate GI complications. So far there is no safe and effective modality against GI pathogenic coccidiosis. Antibiotic additives routinely fed to food animals to protect against infection will inevitably enter the food chain, contaminate food products and be passed on to the consumers. A century after the original discovery of poultry coccidiosis, Oz HS [39] introduces mechano-chemically-altered coccidial organisms with distinct ultrastructures, without abolishing their immunogenicity. These aberrant organisms were tolerated by cyclophosphamide immunodeficient animals yet were non-pathogenic, and provide novel immune protection in immuno-intact animals against pathogenic challenges, which included diarrhea, malnutrition and weight loss. This study warrants further investigations toward vaccine production. In conclusion, this Special Issue includes a collection of innovative articles on the basics, translational findings and clinical trials, as well as reviews of the relationship between infectious/inflammatory diseases and nutrients. Prospective clinical trials and the novel diagnostic, preventive and therapeutic modalities which are discussed can aid the development of nutritional strategies for the treatment as well as prevention of inflammation and infection. Finally, the original reviews are of particular interest in helping to advance our understanding of signaling pathways, and the molecular and biochemical mechanisms behind the effects of nutrients on inflammatory and infectious diseases. Different nutritional and dietary life styles, whether poor or lacking essential nutritional elements, as well as excess intake, can result in inflammatory complications and loss of function. Nutritional deficiency is linked with several infectious and inflammatory diseases as a cause or consequence. Studies indicate that nutrients, such as amino acids, oligosaccharides, and short-chain fatty acids, exert inhibitory and anti-inflammatory functions. Gastrointestinal (GI) infections alter the gut microbiome and increase permeability to toxins. Various invasions by microbial, viral and parasitic agents stimulate inflammation, a defensive mechanism of the body's immune system. Other stimuli include environmental stimuli, oxidative stress, aging and the physiological process. A long-lasting, persistent and excessive inflammatory response is a significant risk factor for developing various chronic inflammatory and infectious diseases. The following investigations may help to understand nutritional contributions to the prevention, treatment and taming of certain inflammatory and infectious diseases.

Conflicts of Interest: The author declares no conflict of interest.

Abbreviations

ATR-I	Atractylenolide-I
BMI	body mass index
Vit	D3 cholecalciferol
DC	Dendritic Cells'
Vit	D2 Ergocalciferol
EEN	Exclusive Enteral Nutrition
GI	gastrointestinal
HCV	Hepatitis C virus
IBD	Inflammatory bowel disease
IGF	insulin-like growth factor
ICU	Intensive care unit
LPS	lipopolysaccharide
NMDs	Neuromuscular diseases
PAI-1	plasminogen activator inhibitor-1
Vit	Vitamin
RTIs	Respiratory tract infections
RV	Rotaviruses
Toxoplasma	Toxoplasma gondii

References

1. Ruth, M.R.; Field, C.J. The immune modifying effects of amino acids on gut-associated lymphoid tissue. J. *Anim. Sci.Biotechnol.* **2013**, *4*, doi:10.1186/2049-1891-4-27.
2. Hardy, H.; Harris, J.; Lyon, E.; Beal, J.;Foey, A.D. Probiotics, prebiotics and immunomodulation of gut mucosal defences: Homeostasis and immunopathology. *Nutrients* **2013**, *5*, 1869–1912, doi:10.3390/nu5061869.
3. Soldavini, J.;Kaunitz, J.D. Pathobiology and potential therapeutic value of intestinal short-chain fatty acids in gut inflammation and obesity. *Dig. Dis. Sci.* **2013**, *58*, 2756–2766, doi:10.1007/s10620-013-2744-4.
4. Wherry, E.J.;Kurachi, M. Molecular and cellular insights into T cell exhaustion. *Nat. Rev. Immunol.* **2015**, *15*, 486–499, doi:10.1038/nri3862.
5. Oz, H.S. Chronic Inflammatory Diseases and Green Tea Polyphenols. *Nutrients* **2017**, *9*, 561, doi:10.3390/nu9060561.
6. Anand, P.K.; Kaul, D.; Sharma, M. Green tea polyphenol inhibits Mycobacterium tuberculosis survival within human macrophages. *Int. J.Biochem. Cell Biol.* **2006**, *38*, 600–609.
7. Soh, A.Z.; Pan, A.; Chee, C.B.C.; Wang, Y.-T.; Yuan, J.-M.; Koh, W.-P.Tea drinking and its association with active tuberculosis incidence among middle-aged and elderly adults: The Singapore Chinese health study.*Nutrients* **2017**, *9*, 544, doi:10.3390/nu9060544.
8. Globig, A.M.;Hennecke, N.; Martin, B.;Seidl, M.;Ruf, G.;Hasselblatt, P.;Thimme, R.;Bengsch, B. Comprehensive intestinal T helper cell profiling reveals specific accumulation of IFN-γ+IL-17+coproducing CD4+ T cells in active inflammatory bowel disease. Inflamm. *Bowel Dis.* **2014**, *20*, 2321–2329, doi:10.1097/MIB.0000000000000210.
9. Harbour, S.N.; Maynard, C.L.;Zindl, C.L.;Schoeb, T.R.; Weaver, C.T. Th17 cells give rise to Th1 cells that are required for the pathogenesis of colitis. *Proc. Natl. Acad. Sci.USA* **2015**, *112*, 7061–7066, doi:10.1073/pnas.1415675112.

10. Grass, F.;Pache, B.; Martin, D.;Hahnloser, D.;Demartines, N.;Hübner, M. Preoperative nutritional conditioning of Crohn's patients—systematic review of current evidence and practice. *Nutrients* **2017**, *9*, 562, doi:10.3390/nu9060562

11. Pan, Y.; Liu, Y.; Guo, H.; Jabir, M.S.; Liu, X.; Cui, W.; Li, D. Associations between folate and vitamin B12 levels and inflammatory bowel disease: A meta-analysis. *Nutrients* **2017**, *9*, 382, doi:10.3390/nu9040382.

12. Ruemmele, F.M.;Veres, G.;Kolho, K.L.; Griffiths, A.; Levine, A.; Escher, J.C.;Amil Dias, J.;Barabino, A.; Braegger, C.P.; Bronsky J.;et al. Consensus guidelines of ECCO/ESPGHAN on the medical management of pediatric Crohn's disease. *J. Crohns Colitis* **2014**, *8*, 1179–1207, doi:10.1016/j.crohns.2014.04.005.

13. Oz, H.S.; Chen, T.; Neuman, M. Nutritional intervention against systemic inflammatory syndrome. *J. Parent. Enter.Nutr.* **2009**, *33*, 380–389,doi:10.1177/01486071 08327194.

14. Oz, H.S.; Ray, M.; Chen, T.; McClain, C. Efficacy of a transforming growth factor beta 2 containing nutritional support formula in a murine model of inflammatory bowel disease. *J. Am. Coll.Nutr.* **2004**, *23*, 220–226.

15. Gatti, S.;Galeazzi, T.;Franceschini, E.;Annibali, R.; Albano, V.;Verma, A.K.; De Angelis, M.;Lionetti, M.E.;Catassi, C. Effects of the exclusive enteral nutrition on the microbiota profile of patients with Crohn's disease: A systematic review. *Nutrients* **2017**, *9*, 832, doi:10.3390/nu9080832.

16. Plaza-Díaz, J.; Ruiz-Ojeda, F.J.;Vilchez-Padial, L.M.; Gil, A. Evidence of the anti-inflammatory effects of probiotics and synbiotics in intestinal chronic diseases. *Nutrients* **2017**, *9*, 555, doi:10.3390/nu9060555.

17. Shin, D.;Hur, J.; Cho, E.-H.; Chung, H.-K.;Shivappa, N.; Wirth, M.D.; Hébert, J.R.; Lee, K.W. Pre-pregnancy body mass index is associated with dietary inflammatory index and C-reactive protein concentrations during pregnancy. *Nutrients* **2017**, *9*, 351, doi:10.3390/nu9040351.

18. Lisa, M.;Gaetkea, L.M.;Frederich, R.C.; Oz, H.S.; McClain, J.C. Decreased food intake rather than zinc deficiency is associated with changes in plasma leptin, metabolic rate, and activity levels in zinc deficient rats. *J.Nutr.Biochem.* **2002**, *13*, 237–244.

19. Gammoh, N.Z.; Rink, L. Zinc in infection and inflammation. *Nutrients* **2017**, *9*, 624, doi:10.3390/nu9060624.

20. Wisniewska, M.; Cremer, M.;Wiehe, L.; Becker, N.-P.; Rijntjes, E.;Martitz, J.;Renko, K.;Bührer, C.;Schomburg, L. Copper to zinc ratio as disease biomarker in neonates with early-onset congenital infections.*Nutrients* **2017**, *9*, 343, doi:10.3390/nu9040343.

21. Varsi, K.;Bolann, B.;Torsvik, I.;Eik, T.C.R.;Høl, P.J.;Bjørke-Monsen, A.-L. Impact of maternal selenium status on infant outcome during the first 6 months of life. *Nutrients* **2017**, *9*, 486, doi:10.3390/nu9050486.

22. Wang, N.;Rayes, R.; Elahi, S.; Lu, Y.; Hancock, M.; Massie, Rowe, G.;Aomari, Hossain, Durocher, Y.;et al. The IGF-trap: Novel inhibitor of carcinoma growth and metastasis. *Mol. Cancer Ther.* **2015**, *14*, 982–993, doi:10.1158/1535-7163.MCT-14-0751.

23. Bauer, T.W.; Liu, W.; Fan, F.; Camp, E.R.; Yang, A.;Somcio, R.J.;Bucana, C.D.; Callahan, J.; Parry, G.C.; Evans, D.B.;et al. Targeting of urokinase plasminogen activator receptor in human pancreatic carcinoma cells inhibits c-Met- and insulin-like growth factor-Ireceptor-mediated migration and invasion and orthotopic tumor growthin mice. *Cancer Res.* **2005**, *65*, 7775–7781.

24. Gong, Y.; Zhang, B.; Liao, Y.; Tang, Y.; Mai, C.; Chen, T.; Tang, H. Serum insulin-like growth factor axis and the risk of pancreatic cancer: Systematic review and meta-analysis. *Nutrients* **2017**, *9*, 394, doi:10.3390/nu9040394.

25. Cheng, W.; Jing, J.; Wang, Z.; Wu, D.; Huang, Y. Chondroprotective effects of ginsenoside Rg1 inhuman osteoarthritis chondrocytes and a rat model of anterior cruciate ligament transection. *Nutrients* **2017**, *9*, 263, doi:10.3390/nu9030263.

26. D'Angelo, C.;Reale, M.;Costantini, E. Microbiota and probiotics in health and HIV infection. *Nutrients* **2017**, *9*, 615, doi:10.3390/nu9060615.

27. Paterniti, I.;Impellizzeri, D.;Cordaro, M.;Siracusa, R.;Bisignano, C.;Gugliandolo, E.;Carughi, A.; Esposito, E.;Mandalari, G.;Cuzzocrea, S. The anti-inflammatory and antioxidant potential of pistachios (Pistaciavera L.) in vitro and in vivo. *Nutrients* **2017**, *9*, 915, doi:10.3390/nu9080915.

28. Mandal, A.; Bhatia, D.;Bishayee, A. Anti-inflammatory mechanism involved in pomegranate-mediated prevention of breast cancer: The role of NF-κB and Nrf2 signaling pathways. *Nutrients* **2017**, *9*, 436, doi:10.3390/nu9050436.

29. Yeh, C.-L.;Pai, M.-H.; Shih, Y.-M.; Shih, J.-M.;Yeh, S.-L. Intravenous arginine administration promotes proangiogenic cells mobilization and attenuates lung injury in mice with polymicrobial sepsis. *Nutrients* **2017**, *9*, 507, doi:10.3390/nu9050507.

30. Gois, P.H.F.; Ferreira, D.;Olenski, S.; Seguro, A.C. Vitamin D and infectious diseases: Simple bystander or contributing factor? *Nutrients* **2017**, *9*, 651, doi:10.3390/nu9070651.

31. De Santis, S.;Galleggiante, V.;Scandiffio, L.;Liso, M.;Sommella, E.;Sobolewski, A.;Spilotro, V.;Pinto, A.;Campiglia, P.;Serino, G.;et al. Secretory leukoprotease inhibitor (Slpi) expression is required for educating murine dendritic cells inflammatory response following quercetin exposure. *Nutrients* **2017**, *9*, 706, doi:10.3390/nu9070706.

32. Oz, H.S.Fetomaternaland pediatric toxoplasmosis. *J. Pediatr. Infect. Dis.* **2017**, *12*, 202–208, doi:10.1055/s-0037-1603942.

33. Berrett, A.N.; Gale, S.D.; Erickson, L.D.; Brown, B.L.; Hedges, D.W. Toxoplasma gondii moderates the association between multiple folate-cycle factors and cognitive function in U.S. adults. *Nutrients* **2017**, *9*, 564, doi:10.3390/nu9060564.

34. Chang, M.-L.; Chen, T.-S.; Hsu, C.-M.; Lin, C.-H.; Lin, C.-Y.;Kuo, C.-J.; Huang, S.-W.; Chen, C.-W.; Cheng, H.-T.;Yeh, C.-T.;et al. The evolving interplay among abundant adipokines in patients with hepatitis C during viral clearance. *Nutrients* **2017**, *9*, 570, doi:10.3390/nu9060570.

35. De Cosmi, V.; Milani, G.P.;Mazzocchi, A.;D'Oria, V.;Silano, M.;Calderini, E.;Agostoni, C. The metabolic response to stress and infection in 2 critically ill children: The opportunity of an individualized approach. *Nutrients* **2017**, *9*, 1032, doi:10.3390/nu9091032.

36. More, S.; Choi, D.-K. Neuroprotective role of atractylenolide-I in an in vitro and in vivo model of Parkinson's disease. *Nutrients* **2017**, *9*, 451, doi:10.3390/nu9050451.

37. Salera, S.;Menni, F.;Moggio, M.;Guez, S.;Sciacco, M.; Esposito, S. Nutritional challenges in Duchenne muscular dystrophy. Nutrients2017, 9, 594, doi:10.3390/nu9060594.

38. Jesenak, M.;Urbancikova, I.;Banovcin, P. Respiratory tract infections and the role of biologically active polysaccharides in their management and prevention. *Nutrients* **2017**, *9*, 779, doi:10.3390/nu9070779.

39. Oz, H.S. Induced aberrant organisms with novel ability to protect intestinal integrity from inflammation in an animal model. *Nutrients* **2017**, *9*, 864, doi:10.3390/nu9080864.

Helieh S. Oz

Special Issue Editor

nutrients

MDPI

Review

Chronic Inflammatory Diseases and Green Tea Polyphenols

Helieh S. Oz

Department of Physiology, Internal Medicine, College of Medicine, University of Kentucky Medical Center, Lexington, KY 40536-0298, USA; hoz2@email.uky.edu

Received: 22 April 2017; Accepted: 29 May 2017; Published: 1 June 2017

Abstract: Chronic inflammatory diseases affect millions of people globally and the incidence rate is on the rise. While inflammation contributes to the tissue healing process, chronic inflammation can lead to life-long debilitation and loss of tissue function and organ failure. Chronic inflammatory diseases include hepatic, gastrointestinal and neurodegenerative complications which can lead to malignancy. Despite the millennial advancements in diagnostic and therapeutic modalities, there remains no effective cure for patients who suffer from inflammatory diseases. Therefore, patients seek alternatives and complementary agents as adjunct therapies to relieve symptoms and possibly to prevent consequences of inflammation. It is well known that green tea polyphenols (GrTPs) are potent antioxidants with important roles in regulating vital signaling pathways. These comprise transcription nuclear factor-kappa B mediated I kappa B kinase complex pathways, programmed cell death pathways like caspases and B-cell lymphoma-2 and intervention with the surge of inflammatory markers like cytokines and production ofcyclooxygenase-2. This paper concisely reviews relevant investigations regarding protective effects of GrTPs and some reported adverse effects, as well as possible applications for GrTPs in the treatment of chronic and inflammatory complications.

Keywords: chronic inflammatory diseases; green tea polyphenols; (−)-Epigallocatechin-3-gallate (EGCG)

1. Introduction

Chronic inflammatory diseases affect millions of people and the incidence rate is on the rise. While inflammation contributes to the tissue healing process, chronic inflammation can lead to life-long debilitation and loss of tissue function and organ failure. Tumor necrosis factor α (TNFα) is a proinflammatory cytokine that promotes various chemokines and cytokines to commence acute and chronic stages of inflammation. TNFα is released chiefly by activated macrophages, astroglia, microglia, CD4+ lymphocytes, natural killer cells (NK), and neurons [1–3]. TNFα release is connected with inflammation and pain related sensation in patients with inflammatory diseases like hepatitis, inflammatory bowel disease, pancreatitis, and neuropathic complications [4]. TNFα is known to contribute to the progression of neuropathic pains [5]. Soluble TNF receptors (R1 and R2) are capable of neutralizing TNFα circulation to improve pain related responses to mechanical and thermal hypersensitivity or peripheral nerve injuries [6,7]. TNFα reveals vital functions in the pathogenesis of inflammatory diseases, as inhibition of TNFα ameliorates the duration of experimental pancreatitis [2,8,9]. Genetic manipulation such as TNFα receptor 1 (TNFR1) gene deletion and anti-TNF monoclonal antibodies' (e.g., etanercept) application improve acute inflammation in animal models [8]. Although current clinical applications to use these biological drugs may ease the inflammatory cascades and pain by reducing TNFα and other cytokines, the inflammation and pain are likely to re-surface in the patients who suffer from autoimmune diseases including arthritis and inflammatory bowel disease [7,10]. In addition, anti-TNFα monoclonal antibodies are economically unfeasible (expensive) and administration can cause potential complications [11] to provoke severe infectious diseases with viral (JC virus disease), fungal (aspergillosis) and microbial (tuberculosis) agents.

Inflammatory diseases include metabolic syndrome, nonalcoholic hepatitis, neurodegenerative diseases and gastrointestinal complications which can lead to malignancy. Indeed, the gut comprises major neuronal systems in the body and neurodegenerative disorders (e.g., Parkinson's and Alzheimer's) are commonly manifested with severe gastrointestinal complications. With the millennial advancements in diagnostic as well as anti-inflammatory therapeutic modalities (e.g., biological therapies, anti-TNFα monoclonal antibodies), there remains no effective cure available for those who suffer from chronic inflammatory diseases. Therefore, patients seek alternatives and complementary medications to relieve the symptoms and possibly to prevent consequences of inflammatory diseases. Yet, the safety and efficacy of these agents are not fully known as well as their possible interaction with the standard-of-care therapies. Thus, the consequences can become life-threatening, from uptake of contaminated toxic metals to complex interactions with conventional therapies [10].

Regular consumption of tea and plants rich in polyphenols are known for various health-promoting functions such as diuretic, anticarcinogenic, immunity-support, antimicrobial, and anti-inflammatory effects. Different relevant investigations have revealed green tea polyphenols as potent antioxidants [12,13] to play important roles in inactivation of several signaling pathways involved in inflammation. These include transcription nuclear factor-kappa B (NF-κB) mediated I kappa B kinase complex (IKK) pathways [14], TNFα [9,12,13], downregulation of cyclooxygenase (Cox)-2and B-cell lymphoma-2 (Bcl-2) activities [15], and upregulation of protective programmed cell death pathways [16,17]. Clinical trials and meta-analyses reveal constant consumption of diets rich in polyphenols to protect against chronic inflammatory diseases including cardiovascular [18] and neurodegenerative diseases [19] in humans. Consumption of a polyphenol-rich diet is linked with elevated plasma antioxidant [20], lowered markers for oxidative stress [21] and improved albuminuria in diabetic patients [22]. Additionally, polyphenols' use includes possible production of safe plastics, nanomaterials and storage for food products. Recent studies reveal an attractive field for the possible use of these natural extracts in biopolymer formulations [23]. Polyphenols with antioxidant and antimicrobial activities are candidates for use as active compounds in bio-additives for food packaging materials to reduce the oxidation and deterioration of food [23–26] and to prevent spoilage and contamination with infectious pathogens. For instance, dibutyltin dichloride, a stabilizer used in the production of polyvinyl chloride plastics, can cause severe inflammatory complications including pancreatitis in animal models [2]. Another organic synthetic compound, bisphenol A (BPA), is utilized to produce plastics for food packaging which has the potential to disrupt endocrine hormones and to increase the risk for type 2 diabetes [27]. Polyphenolic use, as a natural packaging product, is a novel venue to replace these synthetic compounds and additives with indications of severe side effects in consumers, including elevation of chronic inflammatory complications and malignancies [2,24–27].

This paper reviews an insight into some of the relevant investigations regarding protective effects of green tea polyphenols and their possible applications in the treatment of chronic and inflammatory diseases. Additionally, this paper briefly discusses some of the reported side effects due to consumption of these polyphenols.

2. Green Tea and Polyphenols

Leaves from *Camellia Sinensis* shrub form three types of teas including black tea, green tea and oolong tea, depending on their processing techniques. Green tea is prepared by steaming and drying the tealeaves (20% of total production). Tea extracts are commonly used as beverages, food additives or integrated into cosmetic and pharmaceutical formulations [20]. Tea extracts are rich in vitamins (B and C), minerals, polyphenols, caffeic acid, fertaric acid, tannins and volatiles. These polyphenols have been investigated for various biological and physiological activities. Tea extracts are used as coloring agents, antioxidants, and nutritional additives. Polyphenols have phenolic molecules and polymeric structures [28]. Polyphenols are ubiquitous secondary metabolites in tea and some other plants [29] and contribute to pigmentation in plant organs. Polyphenols play an important role as the mechanism of defense against environmental and biological stressors including in response to fungal and other pathogens' attacks [16]. Polyphenols have been shown to protect against oxidative damage by inhibiting the formation of free radicals and reactive oxygen species (ROS).

The purified green tea polyphenols contain >95% polyphenols when analyzed with high-performance liquid chromatography (HPLC). Pure GrTP extracts contain the following percentage composition of polyphenols (each catechin): (−)-epicatechin (EC) 35%, (−)-epigallocatechin (EGC) 15%, (−)-epicatechin-gallate (ECG) 4%, and (−)-epigallocatechin-3-gallate (EGCG) 38–40% [9]. The molecular structure of EGCG and EC, two of the most abundant GrTPs, are presented in the Figure 1.

The most prevalent individual polyphenolic constituent, EGCG (98% purity), is believed to account for several therapeutic effects of polyphenols. For instance, GrTPs are reported to attenuate inflammation in different inflammatory bowel disease models [9,13,17,30,31]. Following ingestion, GrTPs are extensively dispersed amongst the organs, including the hepatic system [32]. However, the anti-inflammatory effects of GrTPs are not limited to the scavenging of toxic oxidants, as GrTPs, specifically EGCG, can block the activation of the NF-κB and the release of proinflammatory TNFα in intestinal epithelia [14]. The ability of GrTPs to inhibit NF-κB activation and release of TNFα can be responsible for the anti-inflammatory effects of tea consumption.

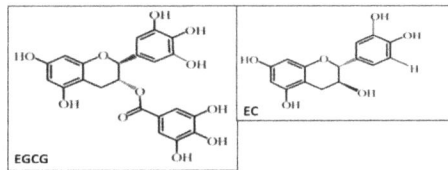

Figure 1. Molecular structure of (−)-epigallocatechin-3-gallate (EGCG) and (−)-epicatechin (EC).

3. Inflammatory Bowel Disease and Green Tea Polyphenols

Crohn's disease and ulcerative colitis are known chronic idiopathic inflammatory bowel diseases (IBD) mediated by immune dysregulation. In a normal gut, the number of intestinal epithelial cells (IEC) is tightly regulated to cover the surface of villi and crypts. IEC are generated by stem cells in the crypts which differentiate and migrate to the tip of villi then are "sloughed off" in approximately two to three days [15,33–35] and replaced with brand new epithelial cells. Therefore, the ratio of villus height in crypt stays in a constant state as it is regulated by the apoptosis (death) pathways [17]. Defective apoptosis impairs intestinal epithelial barrier function, activates immune system and macrophages, increases production of proinflammatory cytokines like TNFα and leads to IBD [17,35,36]. In Crohn's patients, the lamina propria lymphocytes (LPL) in intestinal mucosa are chronically activated [35] with increased expression of anti-apoptotic molecules [37]. Dysregulated apoptosis in IEC and activated LPL are key pathognomonic mechanisms for IBD. Further, ROS are increased in IBD patients and implicated as mediators of intestinal inflammation [9,35].

Despite available targeted therapies and advancements in the humanized monoclonal antibodies and complementary and alternative agents [9,35], the consequences are not yet fully explored. Sulfasalazine is a standard of therapy and commonly used in IBD patients. Yet, it has severe adverse effects, such as pulmonary fibrosis, infertility, and lack of response, which ultimately leads to intestinal resection in these patients. In a bold animal study, green tea polyphenols (GrTP, EGCG) were compared to sulfasalazine for their anti-inflammatory properties. Wild type mice were given dextran sodium sulfate (DSS) for a chemically induced ulcerative colitis model [13]. Interleukin-10 (IL-10) deficient mice spontaneously develop IBD when exposed to the normal gut microbiota from their control wild type background to provoke enterocolitis similar to Crohn's disease. Colitis and enterocolitis animals tolerated treatments with GrTP, EGCG, or sulfasalazine which was added into the diets. Treated animals similarly developed less severe symptoms compared to the sham-treated animals. The inflammatory markers (TNFα, IL-6, serum amyloid A) were significantly upregulated along with pathological symptoms but drastically decreased with GrTP, EGCG or sulfasalazine treatment.

While hepatic and colonic antioxidants (glutathione, cysteine) are depleted in IBD patients and colitic models [9,13], GrTP and EGCG significantly restored antioxidant concentrations and attenuated colitis symptoms similar to sulfasalazine administration [9]. In addition, GrTP decreased disease activity and inhibited inflammatory responses in interleukin-2-deficient (IL-2$^{-/-}$) mouse models for chronic inflammatory disease [31]. Colonic explants and LPL cultures from GrTP-treated mice had decreased spontaneous interferon-gamma and TNFα secretions [31]. In another study, lymphocytes from IBD patients and healthy subjects were chemically damaged (hydrogen peroxide) in vitro and then treated with epicatechin (0–0.1 mg/mL). A significant reduction in induced-DNA damage was discovered in lymphocytes from patients (48.6%) and normal controls (35.2%) when compared with lymphocytes from untreated subjects (both $p < 0.001$). Therefore, epicatechin significantly decreased oxidative stress in lymphocytes and supported beneficial effects of epicatechin inclusion in diets for IBD patients [38].

IKK mediates activation of NF-κB cascade in inflammatory responses to release the proinflammatory cytokine, TNFα. In vitro studies using GrTP and EGCG have shown a series of anti-inflammatory activities by inhibiting NF-κB through inactivation of I-κB kinase complex in intestinal epithelia cells. Pretreatment of intestinal cells with GrTPs (0.4 mg/mL) diminished TNFα-induced IKK and NF-κB activity [14]. The gallate group from polyphenols was required to block TNFα initiated IKK activation. When intestinal epithelial cells were transiently transfected with NF-κB-inducing kinase (NIK) for continued IKK activation, EGCG significantly decreased IKK activity in these NIK transfected epithelia [14]. Therefore, GrTP and specifically EGCG, but not other polyphenols (EC, EGC, ECG), were reported as effective inhibitors of IKK activity and as natural anti-inflammatory agents [14]. In a case-control study of 678 ulcerative colitis patients (2008–2013), increased risk for colitis was associated with serial factors as follow: (1) irregular meal times (OR: 2.287; 95% CI: 1.494–3.825); (2) consumption of fried (OR: 1.920; 95% CI: 1.253–3.254); (3) salty (OR: 1.465; 95% CI: 1.046–2.726) and frozen dinners (OR: 1.868; 95% CI: 1.392–2.854); (4) intestinal infectious diseases (1–2/year, OR: 1.836; 95% CI: 1.182–2.641); (5) frequent use of drugs such as antibiotics and NSAIDs (OR: 2.893; 95% CI: 1.619–5.312); (6) and high work stressors (OR: 1.732; 95% CI: 1.142–2.628, $p < 0.05$). In contrast, drinking tea (OR: 0.338, 95% CI: 0.275–0.488) and physical activities (1–2/week, OR: 0.655, 95% CI: 0.391–0.788; $\geq$3 times/week, OR: 0.461, 95% CI: 0.319–0.672, $p < 0.05$) were linked with significant protective effects in these patients [39]. These studies support possible beneficial effects of GrTP inclusion in diets for IBD patients.

4. Gastrointestinal Associated Malignancies and Green Tea Polyphenols

Cancer is one of the most prevalent causes of morbidity and mortality. Additionally, cancer patients are at risk for developing severe complications including diarrhea, nausea and abdominal pain during chemotherapy, mainly due to cytotoxic effects of anticancer drugs [17]. Studies suggest a protective effect of tea consumption on malignancies including those with gastrointestinal involvement mainly based on animal trials and in vitro studies. Yet, strong clinical trials and investigations are lacking to support the anticancer effects. GrTP is shown to inhibit carcinogen-induced gastrointestinal tumors in rodents [40–44] and in abnormal cell growth to induce apoptosis in various carcinoma cell lines [17,45,46]. GrTP and EGCG were proven to regulate apoptosis in the intestinal epithelia. In a normal gut, highly organized epithelial cells cover surfaces of villi and crypts. The number of IEC which blanket the normal gut is tightly regulated by programmed cell death (apoptosis) pathways. Dysregulated and defective apoptosis lead to IEC overgrowth and severe consequences including malignancy. Apoptosis "programmed cell death" is a process of self-destruction that can be initiated via extrinsic and intrinsic pathways. In the gut, extrinsic pathway leads to Fas-associated death domain proteins (FADD) and enzymatic activities of cysteine-aspartate specific proteases "caspasescascades" [47]. When IEC were treated with GrTP (0.4–0.8 mg/mL), they induced DNA fragmentation in a dose responsive fashion [17]. In higher concentrations (>0.8 mg/mL), GrTP caused a mixture of cytolysis and apoptosis. In addition, epithelial cells exposed to GrTP and EGCG, but not other polyphenols (i.e., EC, EGC), had increased caspase-3, caspase-8 and caspase-9 activities;

but caspase inhibitors could rescue cells from imminent apoptosis [17]. Furthermore, GrTP caused activation of Fas-associated proteins with FADD recruitment to Fas/CD95 domains. Indeed, GrTP blocked NF-κB activation, yet NF-κB inhibitor (MG132) only promoted cytolysis and not apoptosis. GrTP- and EGCG-induced apoptosis in intestinal epithelia and the activation of death pathways mediated by the caspase-8 through FADD dependent pathways [17] are presenting promising results as possible anticancer agents. In a meta-analysis of 6123 gastric cancer cases and 134,006 controls, green tea consumption had a minor inverse association with risk of gastric cancer (OR = 0.68, 95% CI = 0.49–0.92). A consumption of up to five cups per day of green tea was reported safe (OR = 0.99, 95% CI = 0.78–1.27) and to prevent gastric cancer [48]. A recent case-control survey trial studied the risk factor for gastric cancer and tea consumption. Similarly, this investigation indicated protective effects of regular tea consumption (OR 0.72; CI 95%) and in a large amount ($\geq$35 g/week) (OR 0.53; CI 95%) against gastric malignancy [49]. Future investigations may reveal use for GrTPs and EGCG as adjuvant therapies in malignancies.

5. Hepatic Complications and Green Tea Polyphenols

Acetaminophen [*N*-acetyl-p-aminophenol (APAP)] has been widely used as an over the counter anti-pyretic and analgesic drug since 1955. APAP overdose is a common cause of acute hepatic failure and mortality. APAP overdose is indicated in 50% of acute hepatic failures and approximately 20% of the liver transplant cases in the USA [50]. APAP-induced liver toxicity acts through many factors, including generation of ROS, glutathione (GSH) depletion, upregulation of apoptosis and Cox-2 generation, and inflammatory cytokines production. Generation of ROS causes tissue damage. GSH is a tripeptide and a powerful source of endogenous antioxidants which counteracts against the destructive deposits of free radicals. Yet, GSH sources in liver become depleted due to APAP to cause hepatotoxicity and inflammatory complications.

Current antidote practice is the use of the antioxidant Nacetylcystiene (NAC), which may not be always effective [50]. Therefore, more efficient compounds are urgently needed to protect against hepatic failure, death or liver transplants. In an investigation, mice were given a toxic dose of APAP (0.75 mg/g) by oral gavage. Animals developed profound up regulation of inflammatory markers, TNFα and Serum Amyloid A (SAA) release, as well as Cox-2 activities and Bcl-2 production. The inflammatory markers caused extensive centrilobular apoptosis, necrosis, and severe infiltration of leukocytes accompanied with generation of ROS and depletion of hepatic GSH concentration. GrTP supplementation in the diet prior to APAP injection significantly improved concentration of hepatic GSH, attenuated inflammatory markers, liver lesions and down regulated Cox-2 and Bcl-2 expression. In addition, GrTP normalized pathologically elevated hepatic enzyme activity of alanine aminotransferase (ALT) released by damaged hepatocytes, and protected against liver injury. Therefore, GrTP attenuated hepatotoxicity through normalizing antioxidants, inflammatory markers and Cox-2 and Bcl-2 activation [12,15], suggesting a potential for GrTP additives protecting against APAP toxicity.

As unhealthy diet and inactivity in urban areas are on the rise in recent decades, as are the health consequences including metabolic syndrome, the obesity epidemic and fatty liver inflammation to trigger further unforeseen socio-economic burdens. Looking for a magic wand would be a simple, effective, safe and feasible dietary compound to alleviate these public health loads. EGCG has been reported to show putative health effects including protection against inflammation and obesity. In a dose dependent study, inclusion of high dose (1.3 mg/g) EGCG in daily diets caused weight loss in DSS-treated BALB/c mice [9]. Recently, another study used very high doses of EGCG (3.2 mg/g) for three days in the same model (DSS-treated mice) and reported decreased colonic lipid peroxides and gut permeability and enhanced body weight loss [51]. Therefore, high dose EGCG might have different effects, including being responsible for lowering digestion of consumed protein and lipid and its possible application in obesity and weight loss patients. In addition, very high doses of EGCG require further safety trials.

Nonalcoholic fatty liver disease (NASH) is manifested with obesity and other complications with severe life threatening consequences. EGCG (0.05 mg/g/day) oral gavage regulated hepatic mitochondrial respiratory cascades and improved lipid metabolism, and insulin sensitivity in obese mice [52]. EGCG increased energy expenditure, and prevented oxidation of lipid substrates-stimulated by mitochondria and hepatic steatosis in this obesity model. EGCG is reported to specifically inhibit activated hepatic stellate cells by upregulating de novo biosynthesis of GSH [53]. Further, GrTP is reported to protect against NASH by decreasing hepatic steatosis and NF-κB activation in a model on a high fat diet given for eight weeks [54]. GrTP attenuated prostaglandin E2 (PGE2) accumulation and lipid peroxidation to reduce Cox-2 activity which was independent of arachidonic acid. GrTP protected against hepatic damage induced by a high fat diet in obese rats [54]. GrTP (10–20 mg/g) normalized liver malondialdehyde without affecting cytochrome P450 2E1 mRNA expression, and decreased upregulated hepatic Cox-2 activity and PGE2 elevated levels provoked by the high fat diet. In addition, GrTP attenuated increases in total hepatic short chain fatty acids without affecting the n-6/n-3 ratio and decreased total liver arachidonic acid [54]. Additionally, a double-blind, randomized clinical trial was reported in NASH patients with diagnostic ultrasonography symptoms and elevated hepatic enzymes, ALT >31 mg/dL and AST >41 mg/dL, respectively. Subjects were given green tea extract (500 mg tablet/day) or placebo for 90 days. Green tea significantly decreased hepatic enzymes; ALT and AST compared to placebo $p < 0.001$ [55]. In another randomized, double-blind trial, diabetic subjects with albuminuria received GrTP (containing 800mg of epigallocatechin-3-gallate) supplements for 12 consecutive weeks in addition to their standard of care therapy. Patients who received green tea polyphenol supplements showed significant improvements in urinary albumin-creatinine ratio (41% $p = 0.019$) compared to the placebo (standard therapy alone) group [22]. Further investigations may support the use of GrTPs in diabetic subjects and their hepatic complications including hepatotoxicity and fatty liver, as well as application in weight loss programs in obesity subjects.

6. Neurodegenerative Disorders and Green Tea Polyphenols

Neurodegeneration is associated with central nervous system (CNS) disorders, such as Parkinson's and Alzheimer's diseases, which are caused by multiple environmental and genetic factors [56]. Neurodegenerative disorders are commonly associated with severe gastrointestinal complications, as the gut is comprised of complex neuronal systems. Alzheimer's disease is a progressive neurodegenerative disorder characterized by amyloid β plaques formation, neurofibrillary tangles, microglial and astroglial activation leading to neuronal dysfunction and death. Microglia are primary immune cells which release proinflammatory cytokines (e.g., TNFα) and neurotoxins in the brain and contribute to neuroinflammation [1]. Neuroinflammation is a hallmark for Alzheimer's disease. Current therapies primarily focus on symptomatic improvement of cholinergic transmission. A mechanism by which to provide neuroprotection [19] and to prevent microglial activation may be useful in the treatment of Alzheimer's. EGCG was reported to protect neuronal cells from microglia-induced cytotoxicity and to suppress amyloid β-induced TNFα release [57]. Parkinson's disease, the second most prevalent neurodegenerative disease, is characterized by the loss of the neurotransmitter dopamine and neuronal degeneration in the substantia nigra. Studies revealed EGCG to improve dopaminaergic degeneration and may be beneficial for Parkinson's patients [58]. In a rat model for Parkinson's, green tea extract or EGCG reversed pathological and behavioral modifications, demonstrating neuroprotection by decreasing rotational and increased locomotor activities. Additionally, green tea extracts and EGCG improved cognitive dysfunction by antioxidant and anti-inflammatory properties [59]. In a double blinded, randomized trial, daily consumption of 2000 mg green tea powder (containing 220 mg of catechins) for 12 months did not significantly improve cognitive function in elderly Japanese (nursing home) participants [21]. However, levels of markers for oxidative stress, malondialdehyde-modified low-density lipoprotein, were significantly lower in the green tea group (OR −1.73, 95% CI, $p = 0.04$) compared to those in the placebo arm [21].

Table 1. Chronic inflammatory diseases and tea. Table summarizes applied investigations into green tea and polyphenols against different chronic inflammatory diseases using in in vitro, in vivo and in human trials. EGCG: epigallocatechin-3-gallate; GrTP: green tea polyphenols; IEC: intestinal epithelial cells; WT: wildtype mice, IL-/-: interleukin knockout mice; APAP: acetaminophen.

Applied Investigations	Tea Extract, GrTP, EGCG	In Vitro/Animal/Human Trial	References
Inflammatory Bowel Disease			
	GrTP	DSS-WT mouse model	Oz et al. [13]
	GrTP, EGCG	IL-10-/- spontaneous and DSS-WT	Oz et al. [9]
	GrTP	IL-2-/- spontaneous	Varilek et al. [31]
	Tea consumption	Patients	Niu J et al. [39]
		in vitro (patients with lymphocytes)	Najafzadeh et al. [38]
	GrTP, EGCG, EGC, ECG	in vitro IEC	Yang et al. [14]
GI malignancy/prevention	Tea extract	WT-mice, rats	Ju et al. [40], Metz et al. [41]; Issa et al. [42], Ohishi et al. [43]
	GrTP, EGCG	in vitrocell lines	Oz et al. [17], Isemura et al. [44], Wu [45], Basu et al. [46]
	Tea consumption	Human subjects	Zhou et al. [48]
Hepatic complications	GrTP	WT-mice and APAP toxicity	Oz et al. [12,15]
NASH	GrTP	Rat model	Chung et al. [54]
Diabetic	Tea extract	Patients	Borges et al. [22]
Metabolicweight loss,	EGCG	WT-mice	Oz et al. [9], Bitzer et al. [51], Santamarina et al. [52]
Fatty liver disease		WT-mice	Hirsch et al. [60]
Neurodegenerative Disorders			
Alzheimer's.	EGCG	in vitro neuronal cells	Cheng-Chung et al. [57]
Parkinson's disease	EGCG	Patients	Renaud et al. [58]
	EGCG	Rat model	Bitu et al. [59]
Cognitive function	Tea extract	Elderly	Ide et al. [21]
Diabetic retinopathy	Green tea	Human subjects	Ma et al. [61]
Retinalneurodegeneration	EGCG	Tat retina	Yang et al. [62]
Stroke	EGCG	WT-mice	Bai et al. [63], Zhang et al. [64]
	Tea consumption	Human subjects	Pang et al. [65]
Autism spectrum	Tea extract	WT-mice pups	Banji D et al. [66]
Tea and Side effects			
Weight loss	EGCG	Human subjects	Chow et al. [67]
Microbia, toxic metal contaminant	GrTP, EGCG	WT-mice	Oz et al. [9], Bitzer et al. [51]
Microbial contaminant/ provocation	Tea		Ting et al. [68]
Gastroesophageal reflux disease	Tea consumption	Human subjects	Lessa et al. [69], Evans et al. [70]
Iron deficiency	Tea consumption	Human subjects	Vossoughinia et al. [71]
	EGCG	WT-mice	Yeoh et al. [72]

Retinal neurodegeneration is a major cause of blindness specifically in the elderly population. Diabetic retinopathy is a recurrent complication of diabetes (type 1, type 2) which results in increased inflammation, oxidative stress, and vascular dysfunction. The inflammation and neurodegeneration may occur even before the development of clinical signs of diabetes. During the process of diabetes, the retina triggering proinflammatory signaling pathways becomes chronically activated, leading to retinal neurodegeneration and the loss of vision [61]. In a case-control clinical trial, 100 patients with diabetic retinopathy were recruited along with 100 age- and sex-matched diabetic controls without retinopathy in China. Diabetic retinopathy was confirmed from retinal photographs and the pattern of green tea consumption was collected using a face-to-face interview. The odds ratio for green tea consumption for diabetic retinopathy patients was 0.49 (95% CI: 0.26–0.90). When stratified by sex, the green tea consumption and protective effect of green tea on retinopathy was more significant in female (p = 0.01) than male participants (p = 0.63). When adjusted for age and sex, green tea consumption was reported to be significantly associated with reversed diabetic retinopathy (OR = 0.48; p = 0.04), high systolic blood pressure (OR = 1.02; p = 0.05), duration of diabetes (OR = 1.07; p = 0.02), and the presence of family history of diabetes (OR = 2.35; p = 0.04). Therefore, those diabetic patients who regularly consumed green tea (for at least one year) had a significant retinopathy risk reduction of about 50% compared with those who had not.

EGCG with potent antioxidants is reported to neuroprotect outer retinal degeneration after sodium iodate insult [62]. Indeed, EGCG has at least twice the antioxidant potential of vitamin E or C [59]. The retinal protection with orally administered EGCG was linked with reduced expression of superoxide dismutase, GSH peroxidase, caspase-3 and suppression of 8-iso-prostaglandin generation in the retina [73], suggesting a possible therapeutic/maintenance action of EGCG in these inflammatory neurodegenerative diseases.

Stroke is a major cerebrovascular disease which results in disability and mortality, thus far with inadequate neuroprotective and neurotherapeutic agents. Tissue plasminogen activator (t-PA) is the only United States Food and Drug Administration (FDA)-approved therapy against acute ischemic stroke. Yet, clinical outcomes of t-PA depend on its short therapeutic period and grave adverse effects, such as neurotoxicity and hemorrhagic transformation. Adjuvant therapies such as EGCG may reduce the side effects and improve the outcomes [64]. EGCG has anti-angiogenic properties and a possible preventive effect against ischemic stroke via the nuclear factor erythroid 2-related factor 2 (Nrf2) signaling pathway. EGCG therapy for the acute phase of ischemic stroke has been reported to promote angiogenesis in a mouse model of transient middle cerebral artery occlusion (MCAO), conceivably by upregulating the Nrf2 signaling pathway [65]. Additionally, EGCG was shown to augment proliferation and differentiation of neural progenitor cells (NPCs) isolated from the ipsilateral subventricular zone with subsequent spontaneous recovery after ischemic stroke [64]. In a meta-analysis, 259,267 individuals were included from nine different clinical trials [65]. The amount of green tea consumption had a negative correlation with intracerebral hemorrhage and cerebral infarction. The risk increased for intracerebral hemorrhage (OR = 1.24, 95% CI: 1.03–1.49) and cerebral infarction (OR = 1.15, 95% CI: 1.01–1.30) in those who did not consume green tea, compared to those consuming more than one cup of green tea per day. The risk reduced for myocardial infarction (OR = 0.81, 95% CI: 0.67–0.98) and stroke (OR = 0.64, 95% CI: 0.47–0.86) for one to three cups of green tea consumed per day, compared to those who drank less than one cup per day. Likewise, those drinking four or more cups per day had a reduced risk of myocardial infarction (OR = 0.68, 95% CI: 0.56–0.84) compared to those who drank less than one cup per day [65]. Taken together, green tea and EGCG may exert a beneficial effect on neurogenesis, stroke recovery and prevention.

Prenatal and postnatal contact with toxic elements can cause severe consequences in newborns. Postnatal exposure of two–week-old mice pups to a single dose of valproate (0.4 mg/g subcutaneous) provokes experimental autism spectrum and related neurobehavioral abnormalities. Valproate exposed pups were treated daily with green tea extract (0.075 or 0.3 mg/g) orally for about four weeks [66]. Extensive behavioral improvements (nociceptive response, locomotion, anxiety, motor co-ordination) were detected particularly in those pups treated with 0.3 mg/g green tea extract. These modifications were consistent with reduction in oxidative stress formation as well as neuronal cytoprotection. The antioxidant prosperity of green tea polyphenols suggests a possible application in autism spectrum patients. Table 1 summarizes applied investigations into green tea and polyphenols against different chronic inflammatory diseases.

7. Tea Polyphenols and Possible Side Effects

Tea, a popular beverage, has been consumed for many centuries. A preclinical trial described EGCG to have no detectable side effects at 800 mg/day in subjects [67]. However, some deleterious effects of tea and its GrTPs are as follows: tea is a known diuretic agent; overuse may result in dehydration. Prolonged GrTP supplementation may alter bile acid synthesis and increase hepatic oxidative stress with inflammatory hepatic injury, as reported in mice fed high cholesterol diets [60]. Weight loss may be considered a beneficial as well as a side effect of high dose GrTP (2.6 mg/g [9]) and EGCG consumption (1.3 mg/g [9], 3.2 mg/g [51]). Although tea has antimicrobial and antifungal properties, different toxic metals [68] and microbial contaminations such as *Clostridial* spp. have been isolated from unpasteurized tea [68].

Clostridium difficle (*C. diff*) is a facultative gram negative microbial which can cause recurrent and life threatening complications in about 0.2% of the population [69]. The recent increase in rates of recurrent *C. diff* was associated with tea consumption in the vulnerable group. A recent retrospective clinical trial in *C. diff* patients (Veteran Administration hospitals) with recurrent infection who drank tea showed the possible antimicrobial effects of tea. It was suggested that tea in the gut of these patients may reduce the normal microbiome and provoke overgrowth of the facultative pathogens [70]. However, the low number of participants in this trial requires further in-depth investigations to confirm these findings.

Gastroesophageal reflux disease (GERD) is a common chronic inflammatory disease characterized by persistent regurgitation and heartburn with increased prevalence in recent years. In a trial risk, factors were evaluated in 1685 participants. Of these, 420 (26%) suffered from GERD symptoms and the risk factors with significant effects coincided with the use of tea, coffee, smoking, NSAIDs and food indulgence [71]. The risk factors seemed to be similar to other previously reported trials, but the prevalence was remarkably higher among the studied group [71].

Finally, EGCG is counter-regulated by the presence of iron and lipocalin 2. EGCG prevents the peroxidase-catalyzed reaction by reverting the reactive peroxidase heme (compound I: oxoiron) back to its native inactive ferric state, possibly via the exchange of electrons [72]. Therefore, dietary oral intake of iron tablets can diminish EGCG, rendering it to become ineffective in inhibiting myeloperoxidase activity as an antioxidant to establish mucosal protection and anti-inflammatory effects of EGCG.

8. Conclusions

Chronic inflammatory diseases affect many humans worldwide, yet there is no available cure. Tea is one of the most consumed beverages globally and has been around for over 10,000 years. The polyphenols have shown varieties of possible applications, including increasing antioxidants (e.g., GSH, cysteine) depots in vital organs [9,13,35,53] and protecting against chronic inflammation in in vivo and in vitro models [31]. Studies revealed GrTPs attenuate inflammatory responses in signaling pathways, by downregulating IKK, NF-κB (0.4 mg/mL [14]), cytokines like TNFα, inflammatory markers [9,12–15,35,39], Cox-2 and Bcl-2, to protect against hepatic [12–15] and colonic [9,31,38,39] neurodegenerative complications [57–59,61–66,73,74] and various anti-malignancy effects [40–46,48,49]. Importantly, studies have revealed that GrTPs act dose dependently, as high doses (0.4–0.8 mg/mL) activate apoptotic pathways through caspases and DNA breakdown to provoke anti-malignant effects [17]. Also, GrTPs promote weight loss in high doses (EGCG 1.3mg/g/daily [9], 3.2 mg/g every three days [51]) which can be beneficial for the regulation of hepatic enzymes (10–20 mg/g [54]) and metabolism (500 mg tablet/day [55]), as well as in metabolic syndrome and obesity (EGCG 0.05

mg/g [52]); yet, this is a side effect in certain situations where weight loss is not favored. In addition, GrTPs and EGCG in very high doses (3.2 mg/g) require further safety trials. It is a common belief that constant consumption of tea provides anti-inflammatory and cardiovascular beneficial effects. Whether tea and its polyphenols provide preventive or therapeutic effects requires supporting clinical trials. In the era of antibiotic resistance and the hospital superbug epidemic, the use of GrTPs as natural antimicrobial and antifungal agents is an attractive area to be explored. However, GrTPs natural antimicrobial status that possibly alters the gut microbiome may be perceived as an adverse effect with can support facultative *C. diff* in certain vulnerable populations [70]. These warnings against GrTP utilization requires further in-depth trials before any recommendations can be implemented. Finally, GrTP has an attractive potential for use as a natural packaging material [23–26] to replace synthetic compounds with possible severe side effects by promoting chronic inflammation and malignancies in consumers [2,24,27].

Acknowledgments: National Institutes of Health NCCAM-AT1490 (HO).

Conflicts of Interest: Authors declare no conflict of interest.

References

1. Gregersen, R.; Lambertsen, K.; Finsen, B. Microglia and macrophages are the major source of tumor necrosis factor in permanent middle cerebral artery occlusion in mice. *J. Cereb. Blood Flow Metab.* **2000**, *20*, 53–65. [CrossRef] [PubMed]
2. Oz, H.S. Multiorgan chronic inflammatory hepatobiliary pancreatic murine model deficient in tumor necrosis factor receptors 1 and 2. *World J. Gastroenterol.* **2016**, *22*, 4988–4998. Available online: http://www.wjgnet.com/1007-9327/full/v22/i21/4988.htm (accessed on 7 June 2016). [CrossRef] [PubMed]
3. Westlund, K.N.; Zhang, L.; Ma, F.; Oz, H.S. Chronic inflammation and pain in a tumor necrosis factor receptor (TNFR) (p55/p75-/-) dual deficient murine model. *Transl. Res.* **2012**, *160*, 84–94. [CrossRef] [PubMed]
4. Uçeyler, N.; Schäfers, M.; Sommer, C. Mode of action of cytokines on nociceptive neurons. *Exp. Brain Res.* **2009**, *196*, 67–78. [CrossRef] [PubMed]
5. Marchand, F.; Perretti, M.; McMahon, S.B. Role of the immune system in chronic pain. *Nat. Rev. Neurosci.* **2005**, *6*, 521–532. [CrossRef] [PubMed]
6. Sommer, C.; Schmidt, C.; George, A. Hyperalgesia in experimental neuropathy is dependent on the TNF receptor 1. *Exp. Neurol.* **1998**, *151*, 138–142. [CrossRef] [PubMed]
7. Ma, F.; Zhang, L.; Oz, H.S.; Mashni, M.; Westlund, K.N. Dysregulated TNFα promotes cytokine proteome profile increases and bilateral orofacial hypersensitivity. *Neuroscience* **2015**, *300*, 493–507. [CrossRef] [PubMed]
8. Malleo, G.; Mazzon, E.; Genovese, T.; Di Paola, R.; Muià, C.; Centorrino, T.; Siriwardena, A.K.; Cuzzocrea, S. Etanercept attenuates the development of cerulein-induced acute pancreatitis in mice: A comparison with TNF-alpha genetic deletion. *Shock* **2007**, *27*, 542–551. [CrossRef] [PubMed]
9. Oz, H.S.; Chen, T.; de Villiers, W.J. Green Tea Polyphenols and Sulfasalazine have Parallel Anti-Inflammatory Properties in Colitis Models. *Front Immunol.* **2013**, *4*, 132. [CrossRef] [PubMed]
10. Stub, T.; Quandt, S.A.; Arcury, T.A.; Sandberg, J.C.; Kristoffersen, A.E.; Musial, F.; Salamonsen, A. Perception of risk and communication among conventional and complementary health care providers involving cancer patients' use of complementary therapies: A literature review. *BMC Complement. Altern. Med.* **2016**, *16*, 353. [CrossRef] [PubMed]
11. Saraceno, R.; Chimenti, S. How to manage infections in the era of biologics? *Dermatol. Ther.* **2008**, *21*, 180–186. [CrossRef] [PubMed]
12. Oz, H.S.; McClain, C.J.; Nagasawa, H.T.; Ray, M.B.; de Villiers, W.J.; Chen, T.S. Diverse antioxidants protect against acetaminophen hepatotoxicity. *J. Biochem. Mol. Toxicol.* **2004**, *18*, 361–368. [CrossRef] [PubMed]
13. Oz, H.S.; Chen, T.S.; McClain, C.J.; de Villiers, W.J. Antioxidants as novel therapy in a murine model of colitis. *J. Nutr. Biochem.* **2005**, *16*, 297–304. [CrossRef] [PubMed]
14. Yang, F.; Oz, H.S.; Barve, S.; de Villiers, W.J.; McClain, C.J.; Varilek, G.W. The green tea polyphenol (−)-epigallocatechin-3-gallate blocks nuclear factor-kappa B activation by inhibiting I kappa B kinase activity in the intestinal epithelial cell line IEC-6. *Mol. Pharmacol.* **2001**, *60*, 528–533. [PubMed]

15. Oz, H.S.; Chen, T.S. Green-tea polyphenols downregulate cyclooxygenase and Bcl-2 activity in acetaminophen-induced hepatotoxicity. *Dig. Dis. Sci.* **2008**, *53*, 2980–2988. [CrossRef] [PubMed]
16. Beckman, C.H. Phenolic-storing cells: Keys to programmed cell death and periderm formation in wilt disease resistance and in general defense responses in plants? *Physiol. Mol. Plant Pathol.* **2000**, *57*, 101–110. [CrossRef]
17. Oz, H.S.; Ebersole, J. Green tea polyphenols mediate apoptosis in Intestinal Epithelial Cells. *J. Cancer Ther.* **2010**, *1*, 105–113. [CrossRef] [PubMed]
18. Ohmori, R.; Kondo, K.; Momiyama, Y. Antioxidant beverages: Green tea intake and coronary artery disease. *Clin. Med. Insights Cardiol.* **2014**, *8*, 7–11. [CrossRef] [PubMed]
19. Venigalla, M.; Sonego, S.; Gyengesi, E.; Sharman, M.J.; Münch, G. Novel promising therapeutics against chronic neuroinflammation and neurodegeneration in Alzheimer's disease. *Neurochem. Int.* **2016**, *95*, 63–74. [CrossRef] [PubMed]
20. Munin, A.; Edwards-Lévy, F. Encapsulation of natural polyphenolic compounds; a review. *Pharmaceutics* **2011**, *3*, 793–829. [CrossRef] [PubMed]
21. Ide, K.; Yamada, H.; Takuma, N.; Kawasaki, Y.; Harada, S.; Nakase, J.; Ukawa, Y.; Sagesaka, Y.M. Effects of green tea consumption on cognitive dysfunction in an elderly population: A randomized placebo-controlled study. *Nutr. J.* **2016**, *15*, 49. [CrossRef] [PubMed]
22. Borges, C.M.; Papadimitriou, A.; Duarte, D.A.; Lopes de Faria, J.M.; Lopes de Faria, J.B. The use of green tea polyphenols for treating residual albuminuria in diabetic nephropathy: A double-blind randomised clinical trial. *Sci. Rep.* **2016**, *6*, 28282. [CrossRef] [PubMed]
23. Del Mar Castro-López, M.; López-Vilariño, J.M.; González-Rodríguez, M.V. Analytical determination of flavonoids aimed to analysis of natural samples and active packaging applications. *Food Chem.* **2014**, *150*, 119–127. [CrossRef] [PubMed]
24. Beltran, A.; Valente, A.J.; Jiménez, A.; Garrigós, M.C. Characterization of Poly (ε-caprolactone)-Based Nanocomposites Containing Hydroxytyrosol for Active Food Packaging. *J. Agric. Food Chem.* **2014**, *62*, 2244–2252. [CrossRef] [PubMed]
25. Ramos, M.; Jiménez, A.; Peltzer, M.; Garrigós, M.C. Development of novel nano-biocomposite antioxidant films based on poly (lactic acid) and thymol for active packaging. *Food Chem.* **2014**, *162*, 149–155. [CrossRef] [PubMed]
26. Perazzo, K.K.; Conceição, A.C.; dos Santos, J.C.; Assis Dde, J.; Souza, C.O.; Druzian, J.I. Properties and antioxidant action of actives cassava starch films incorporated with green tea and palm oil extracts. *PLoS ONE* **2014**, *9*, e105199. [CrossRef] [PubMed]
27. Provvisiero, D.P.; Pivonello, C.; Muscogiuri, G.; Negri, M.; de Angelis, C.; Simeoli, C.; Pivonello, R.; Colao, A. Influence of Bisphenol A on Type 2 Diabetes Mellitus. *Int. J. Environ. Res. Public Health* **2016**, *13*, E989. [CrossRef] [PubMed]
28. Dreosti, I.E. Antioxidant polyphenols in tea, cocoa, and wine. *Nutrition* **2000**, *16*, 692–694. [CrossRef]
29. Vickery, M.L.; Vickery, B. *Secondary Plant Metabolism*; Macmillan Press: London, UK, 1981; p. 335.
30. Farzaei, M.H.; Rahimi, R.; Abdollahi, M. The role of dietary polyphenols in the management of inflammatory bowel disease. *Curr. Pharm. Biotechnol.* **2015**, *16*, 196–210. [CrossRef] [PubMed]
31. Varilek, G.W.; Yang, F.; Lee, E.Y.; deVilliers, W.J.; Zhong, J.; Oz, H.S.; Westberry, K.F.; McClain, C.J. Green tea polyphenol extract attenuates inflammation in interleukin-2-deficient mice, a model of autoimmunity. *J. Nutr.* **2001**, *131*, 2034–2039. [PubMed]
32. Suganuma, M.; Okabe, S.; Oniyama, M.; Tada, Y.; Ito, H.; Fujiki, H. Wide distribution of [3H] (−)-epigallocatechin gallate, a cancer preventive tea polyphenol, in mouse tissue. *Carcinogenesis* **1998**, *10*, 1771–1776. [CrossRef]
33. Hendry, J.H.; Potten, C.S. Cryptogenic cells and proliferative cells in intestinal epithelium. *Int. J. Radiat. Biol. Relat. Stud. Phys. Chem. Med.* **1974**, *25*, 583–588. [CrossRef] [PubMed]
34. Hermos, J.A.; Mathan, M.; Trier, J.S. DNA synthesis and proliferation by villous epithelial cells in fetal rats. *J. Cell Biol.* **1971**, *50*, 255–258. [CrossRef] [PubMed]
35. Oz, H.S.; Ebersole, J.L. Application of prodrugs to inflammatory diseases of the gut. *Molecules* **2008**, *13*, 452–474. [CrossRef] [PubMed]
36. Strober, W.; Fuss, I.J.; Blumberg, R.S. The immunology of mucosal models of inflammation. *Annu. Rev. Immunol.* **2002**, *20*, 495–549. [CrossRef] [PubMed]

37. Doering, J.; Begue, B.; Lentze, M.J.; Rieux-Laucat, F.; Goulet, O.; Schmitz, J.; Cerf-Bensussan, N.; Ruemmele, F.M. Induction of T lymphocyte apoptosis by sulphasalazine in patients with Crohn's disease. *Gut* **2004**, *53*, 1632–1638. [CrossRef] [PubMed]

38. Najafzadeh, M.; Reynolds, P.D.; Baumgartner, A.; Anderson, D. Flavonoids inhibit the genotoxicity of hydrogen peroxide (H_2O_2) and of the food mutagen 2-amino-3-methylimadazo[4,5-f]-quinoline (IQ) in lymphocytes from patients with inflammatory bowel disease (IBD). *Mutagenesis* **2009**, *24*, 405–411. [CrossRef] [PubMed]

39. Niu, J.; Miao, J.; Tang, Y.; Nan, Q.; Liu, Y.; Yang, G.; Dong, X.; Huang, Q.; Xia, S.; Wang, K.; et al. Identification of Environmental Factors Associated with Inflammatory Bowel Disease in a Southwestern Highland Region of China: A Nested Case-Control Study. *PLoS ONE* **2016**, *11*, e0153524. [CrossRef] [PubMed]

40. Ju, J.; Liu, Y.; Hong, J.; Huang, M.T.; Conney, A.H.; Yang, C.S. Effects of green tea and high-fat diet on arachidonic acid metabolism and aberrant crypt foci formation in an azoxymethane-induced colon carcinogenesis mouse model. *Nutr. Cancer* **2003**, *46*, 172–178. [CrossRef] [PubMed]

41. Metz, N.; Lobstein, A.; Schneider, Y.; Gossé, F.; Schleiffer, R.; Anton, R.; Raul, F. Suppression of azoxymethane-induced preneoplastic lesions and inhibition of cyclooxygenase-2 activity in the colonic mucosa of rats drinking a crude green tea extract. *Nutr. Cancer* **2000**, *38*, 60–64. [CrossRef] [PubMed]

42. Issa, A.Y.; Volate, S.R.; Muga, S.J.; Nitcheva, D.; Smith, T.; Wargovich, M.J. Green tea selectively targets initial stages of intestinal carcinogenesis in the AOM-ApcMin mouse model. *Carcinogenesis* **2007**, *28*, 1978–1984. [CrossRef] [PubMed]

43. Ohishi, T.; Kishimoto, Y.; Miura, N.; Shiota, G.; Kohri, T.; Hara, Y.; Hasegawa, J.; Isemura, M. Synergistic effects of (−)-epigallocatechin gallate with sulindac against colon carcinogenesis of rats treated with azoxymethane. *Cancer Lett.* **2002**, *177*, 49–56. [CrossRef]

44. Isemura, M.; Saeki, K.; Kimura, T.; Hayakawa, S.; Minami, T.; Sazuka, M. Tea catechins and related polyphenols as anti-cancer agents. *Biofactors* **2000**, *13*, 81–85. [CrossRef] [PubMed]

45. Wu, P.P.; Kuo, S.C.; Huang, W.W.; Yang, J.S.; Lai, K.C.; Chen, H.J.; Lin, K.L.; Chiu, Y.J.; Huang, L.J.; Chung, J.G. (−)-Epigallocatechingallate induced apoptosis in human adrenal cancer NCI-H295 cells through caspase-dependent and caspase-independent pathway. *Anticancer Res.* **2009**, *29*, 1435–1442. [PubMed]

46. Basu, A.; Haldar, S. Combinatorial effect of epigallocatechin-3-gallate and TRAIL on pancreatic cancer cell death. *Int. J. Oncol.* **2009**, *34*, 281–286. [CrossRef] [PubMed]

47. Mercier, I.; Vuolo, M.; Jasmin, J.F.; Medina, C.M.; Williams, M.; Mariadason, J.M.; Qian, H.; Xue, X.; Pestell, R.G.; Lisanti, M.P.; et al. ARC (apoptosis repressor with caspase recruitment domain) is a novel marker of human colon cancer. *Cell Cycle* **2008**, *7*, 1640–1647. [CrossRef] [PubMed]

48. Zhou, Y.; Li, N.; Zhuang, W.; Liu, G.; Wu, T.; Yao, X.; Du, L.; Wei, M.; Wu, X. Green tea and gastric cancer risk: Meta-analysis of epidemiologic studies. *Asia Pac. J. Clin. Nutr.* **2008**, *17*, 159–165. [PubMed]

49. Wang, Y.; Duan, H.; Yang, H. A case-control study of stomach cancer in relation to Camellia sinensis in China. *Surg. Oncol.* **2015**, *24*, 67–70. [CrossRef] [PubMed]

50. Yoon, E.; Babar, A.; Choudhary, M.; Kutner, M.; Pyrsopoulos, N. Acetaminophen-Induced Hepatotoxicity: A Comprehensive Update. *J. Clin. Transl. Hepatol.* **2016**, *4*, 131–142. [CrossRef] [PubMed]

51. Bitzer, Z.T.; Elias, R.J.; Vijay-Kumar, M.; Lambert, J.D. (−)-Epigallocatechin-3-gallate decreases colonic inflammation and permeability in a mouse model of colitis, but reduces macronutrient digestion and exacerbates weight loss. *Mol. Nutr. Food Res.* **2016**. [CrossRef] [PubMed]

52. Santamarina, A.B.; Carvalho-Silva, M.; Gomes, L.M.; Okuda, M.H.; Santana, A.A.; Streck, E.L.; Seelaender, M.; do Nascimento, C.M.; Ribeiro, E.B.; Lira, F.S.; et al. Decaffeinated green tea extract rich in epigallocatechin-3-gallate prevents fatty liver disease by increased activities of mitochondrial respiratory chain complexes in diet-induced obesity mice. *J. Nutr. Biochem.* **2015**, *26*, 1348–1356. [CrossRef] [PubMed]

53. Fu, Y.; Zheng, S.; Lu, S.C.; Chen, A. Epigallocatechin-3-gallate inhibits growth of activated hepatic stellate cells by enhancing the capacity of glutathione synthesis. *Mol. Pharmacol.* **2008**, *73*, 1465–1473. [CrossRef] [PubMed]

54. Chung, M.Y.; Mah, E.; Masterjohn, C.; Noh, S.K.; Park, H.J.; Clark, R.M.; Park, Y.K.; Lee, J.Y.; Bruno, R.S. Green Tea Lowers Hepatic Cox-2 and Prostaglandin E2 in Rats with Dietary Fat-Induced Nonalcoholic Steatohepatitis. *J. Med. Food* **2015**, *18*, 648–655. [CrossRef] [PubMed]

55. Pezeshki, A.; Safi, S.; Feizi, A.; Askari, G.; Karami, F. The Effect of Green Tea Extract Supplementation on Liver Enzymes in Patients with Nonalcoholic Fatty Liver Disease. *Int. J. Prev. Med.* **2016**, *7*, 28. [CrossRef] [PubMed]

56. Schwartz, M.; Deczkowska, A. Neurological Disease as a Failure of Brain-Immune Crosstalk: The Multiple Faces of Neuroinflammation. *Trends Immunol.* **2016**, *37*, 668–679. [CrossRef] [PubMed]

57. Cheng-Chung Wei, J.; Huang, H.C.; Chen, W.J.; Huang, C.N.; Peng, C.H.; Lin, C.L. Epigallocatechin gallate attenuates amyloid β-induced inflammation and neurotoxicity in EOC 13.31 microglia. *Eur. J. Pharmacol.* **2016**, *770*, 16–24. [CrossRef] [PubMed]

58. Renaud, J.; Nabavi, S.F.; Daglia, M.; Nabavi, S.M.; Martinoli, M.G. Epigallocatechin-3-Gallate, a Promising Molecule for Parkinson's Disease? *Rejuvenation Res.* **2015**, *18*, 257–269. [CrossRef] [PubMed]

59. Bitu Pinto, N.; da Silva Alexandre, B.; Neves, K.R.; Silva, A.H.; Leal, L.K.; Viana, G.S. Neuroprotective Properties of the Standardized Extract from Camellia sinensis (Green Tea) and Its Main Bioactive Components, Epicatechin and Epigallocatechin Gallate, in the 6-OHDA Model of Parkinson's Disease. *Evid. Based Complement. Altern. Med.* **2015**, *2015*, 161092. [CrossRef] [PubMed]

60. Hirsch, N.; Konstantinov, A.; Anavi, S.; Aronis, A.; Hagay, Z.; Madar, Z.; Tirosh, O. Prolonged Feeding with Green Tea Polyphenols Exacerbates Cholesterol-induced Fatty Liver Disease in Mice. *Mol. Nutr. Food Res.* **2016**. [CrossRef] [PubMed]

61. Arroba, A.I.; Valverde, A.M. Modulation of microglia in the retina: New insights into diabetic retinopathy. *Acta Diabetol.* **2017**, *54*, 527–533. [CrossRef] [PubMed]

62. Yang, Y.; Qin, Y.J.; Yip, Y.W.; Chan, K.P.; Chu, K.O.; Chu, W.K.; Ng, T.K.; Pang, C.P.; Chan, S.O. Green tea catechins are potent anti-oxidants that ameliorate sodium iodate-induced retinal degeneration in rats. *Sci. Rep.* **2016**, *6*, 29546. [CrossRef] [PubMed]

63. Bai, Q.; Lyu, Z.; Yang, X.; Pan, Z.; Lou, J.; Dong, T. Epigallocatechin-3-gallate promotes angiogenesis via up-regulation of Nrf2 signaling pathway in a mouse model of ischemic stroke. *Behav. Brain Res.* **2017**, *321*, 79–86. [CrossRef] [PubMed]

64. Zhang, J.C.; Xu, H.; Yuan, Y.; Chen, J.Y.; Zhang, Y.J.; Lin, Y.; Yuan, S.Y. Delayed Treatment with Green Tea Polyphenol EGCG Promotes Neurogenesis After Ischemic Stroke in Adult Mice. *Mol. Neurobiol.* **2017**, *54*, 3652–3664. [CrossRef] [PubMed]

65. Pang, J.; Zhang, Z.; Zheng, T.Z.; Bassig, B.A.; Mao, C.; Liu, X.; Zhu, Y.; Shi, K.; Ge, J.; Yang, Y.J.; et al. Green tea consumption and risk of cardiovascular and ischemic related diseases: A meta-analysis. *Int. J. Cardiol.* **2016**, *202*, 967–974. [CrossRef] [PubMed]

66. Banji, D.; Banji, O.J.; Abbagoni, S.; Hayath, M.S.; Kambam, S.; Chiluka, V.L. Amelioration of behavioral aberrations and oxidative markers by green tea extract in valproate induced autism in animals. *Brain Res.* **2011**, *1410*, 141–151. [CrossRef] [PubMed]

67. Chow, H.H.; Cai, Y.; Hakim, I.A.; Crowell, J.A.; Shahi, F.; Brooks, C.A.; Dorr, R.T.; Hara, Y.; Alberts, D.S. Pharmacokinetics and Safety of Green Tea Polyphenols after Multiple-Dose Administration of Epigallocatechin Gallate and Polyphenon E in Healthy Individuals. *Clin. Cancer Res.* **2003**, *9*, 3312–3319. [PubMed]

68. Ting, A.; Chow, Y.; Tan, W. Microbial and heavy metal contamination in commonly consumed traditional Chinese herbal medicines. *J. Tradit. Chin. Med.* **2013**, *33*, 119–124. [CrossRef]

69. Lessa, F.C.; Mu, Y.; Bamberg, W.M. Burden of *Clostridium difficile* infection in the United States. *N. Engl. J. Med.* **2015**, *372*, 825–834. [CrossRef] [PubMed]

70. Evans, M.O., II; Starley, B.; Galagan, J.C.; Yabes, J.M.; Evans, S.; Salama, J.J. Tea and Recurrent *Clostridium difficile* Infection. *Gastroenterol. Res. Pract.* **2016**, *2016*. [CrossRef]

71. Vossoughinia, H.; Salari, M.; Mokhtari Amirmajdi, E.; Saadatnia, H.; Abedini, S.; Shariati, A.; Shariati, M.; Khosravi Khorashad, A. An epidemiological study of gastroesophageal reflux disease and related risk factors in urban population of Mashhad, Iran. *Iran. Red Crescent Med. J.* **2014**, *16*, e15832. [CrossRef] [PubMed]

72. Yeoh, B.S.; Aguilera Olvera, R.; Singh, V.; Xiao, X.; Kennett, M.J.; Joe, B.; Lambert, J.D.; Vijay-Kumar, M. Epigallocatechin-3-Gallate Inhibition of Myeloperoxidase and Its Counter-Regulation by Dietary Iron and Lipocalin 2 in Murine Model of Gut Inflammation. *Am. J. Pathol.* **2016**, *186*, 912–926. [CrossRef] [PubMed]

73. Salah, N.; Miller, N.J.; Paganga, G.; Tijburg, L.; Bolwell, G.P.; Rice-Evans, C. Polyphenolicflavanols as scavengers of aqueous phase radicals and as chain-breaking antioxidants. *Arch. Biochem. Biophys.* **1995**, *322*, 339–346. [CrossRef] [PubMed]
74. Ma, Q.; Chen, D.; Sun, H.P.; Yan, N.; Xu, Y.; Pan, C.W. Regular Chinese Green Tea Consumption is Protective for Diabetic Retinopathy: A Clinic-Based Case-Control Study. *J. Diabetes Res.* **2015**, *2015*, 231570. [CrossRef] [PubMed]

nutrients

MDPI

Article

Tea Drinking and Its Association with Active Tuberculosis Incidence among Middle-Aged and Elderly Adults: The Singapore Chinese Health Study

Avril Zixin Soh [1], An Pan [2], Cynthia Bin Eng Chee [3], Yee-Tang Wang [3], Jian-Min Yuan [4] and Woon-Puay Koh [1,5,*]

[1]   Saw Swee Hock School of Public Health, National University of Singapore, Singapore 117549, Singapore; avril.soh@u.nus.edu
[2]   Department of Epidemiology and Biostatistics, Ministry of Education Key Laboratory of Environment and Health, and State Key Laboratory of Environmental Health (incubation), School of Public Health, Tongji Medical College, Huazhong University of Science and Technology, Wuhan 430030, China; panan@hust.edu.cn
[3]   Singapore Tuberculosis Control Unit, Tan Tock Seng Hospital, Singapore 308089, Singapore; cynthia_chee@ttsh.com.sg (C.B.E.C.); yee_tang_wang@ttsh.com.sg (Y.-T.W.)
[4]   Division of Cancer Control and Population Sciences, University of Pittsburgh Cancer Institute, and Department of Epidemiology, University of Pittsburgh Graduate School of Public Health, Pittsburgh, PA 15261, USA; yuanj@upmc.edu
[5]   Office of Clinical Sciences, Duke-NUS Medical School, 8 College Road, Singapore 169857, Singapore
*   Correspondence: woonpuay.koh@duke-nus.edu.sg; Tel.: +65-6601-3147

Received: 18 January 2017; Accepted: 23 May 2017; Published: 25 May 2017

Abstract: Experimental studies showed that tea polyphenols may inhibit growth of *Mycobacterium tuberculosis*. However, no prospective epidemiologic study has investigated tea drinking and the risk of active tuberculosis. We investigated this association in the Singapore Chinese Health Study, a prospective population-based cohort of 63,257 Chinese aged 45–74 years recruited between 1993 and 1998 in Singapore. Information on habitual drinking of tea (including black and green tea) and coffee was collected via structured questionnaires. Incident cases of active tuberculosis were identified via linkage with the nationwide tuberculosis registry up to 31 December 2014. Cox proportional hazard models were used to estimate the relation of tea and coffee consumption with tuberculosis risk. Over a mean 16.8 years of follow-up, we identified 1249 incident cases of active tuberculosis. Drinking either black or green tea was associated with a dose-dependent reduction in tuberculosis risk. Compared to non-drinkers, the hazard ratio (HR) (95% confidence interval (CI)) was 1.01 (0.85–1.21) in monthly tea drinkers, 0.84 (0.73–0.98) in weekly drinkers, and 0.82 (0.71–0.96) in daily drinkers (*p* for trend = 0.003). Coffee or caffeine intake was not significantly associated with tuberculosis risk. In conclusion, regular tea drinking was associated with a reduced risk of active tuberculosis.

Keywords: tea; tuberculosis; epidemiology

---

1. Introduction

Tuberculosis is caused by infection with *Mycobacterium tuberculosis* (Mtb), and approximately 5–15% of the estimated 2–3 billion infected individuals in the world will develop active tuberculosis during their lifetime [1]. With the very large reservoir of latently-infected individuals, understanding the factors affecting reactivation of latent tuberculosis infection is imperative in preventing new cases of active tuberculosis.

During latent tuberculosis infection, a continuum of immunologic responses are encompassed within the dynamic balance between the host and pathogen, and active disease occurs when bacterial replication exceeds host protective responses [2,3]. One of the host-defense mechanisms during tuberculosis infection include the production of reactive intermediates against Mtb [4], and this generation of free radicals in excess of the host antioxidant capacity leads to oxidative stress [5]. Furthermore, progressive oxidative stress during experimental tuberculosis in guinea pigs has been shown to be partially restored with antioxidant treatment, suggesting that the therapeutic strategies that reduce oxidant-mediated tissue damage may be beneficial as an adjunct therapy in the treatment and prevention of tuberculosis in humans [6].

Tea is a widely consumed beverage worldwide. Both black tea and green tea exhibit antioxidant properties [7,8], and many experimental studies have shown tea polyphenols to be beneficial against many diseases by ameliorating levels of oxidative stress [9–12]. Existing literature investigating the effect of tea polyphenols on tuberculosis has mainly focused on green tea catechins. Green tea extract has been reported to reduce oxidative stress associated with tuberculosis in infected mice in an experimental study [13], and in tuberculosis patients in a clinical study [14]. Epigallocatechin gallate (EGCG), the major component of green tea catechins, has also been shown experimentally to inhibit mycobacterial survival [15–17]. Epidemiological evidence to support the role of tea drinking in the development of tuberculosis is scarce; only one recent case-control study has reported tea drinking to be inversely associated with prevalent tuberculosis [18]. However, the temporal association between tea drinking and tuberculosis remains unclear.

In this study, we investigated the prospective relation between tea drinking and risk of developing active tuberculosis in a population-based cohort in Singapore. Participants of this cohort went through periods when tuberculosis was highly prevalent in the country a few decades ago, and those who acquired latent tuberculosis infection in those early years would be at risk of disease reactivation at advanced age [19].

2. Materials and Methods

2.1. Study Population

A total of 63,257 Chinese adults (27,959 men and 35,298 women) were enrolled in the Singapore Chinese Health Study between 1993 and 1998 [20], and the inclusion criteria was based on the participant's age, dialect group, and residency status. The recruited study participants were aged 45–74 years at recruitment, and were restricted to the two major dialect groups in Singapore: the Hokkiens who came from Fujian Province, and the Cantonese who came from Guangdong province in China. The cohort included citizens or permanent residents of Singapore living in government-built housing estates, where 86% of the Singapore population resided during the period of recruitment [20]. Recruitment was initiated using posted letters to invite residents from public housing estates to take part in the study. Interviewers went door-to-door 5–7 days later to recruit participants for the study if they met the inclusion criteria based on their ethnicity, age, dialect group, and residency status. Approximately 85% of the eligible subjects invited agreed to participate [21]. This study was approved by the Institutional Review Board at the National University of Singapore in September 2011 (approval number NUS-1396), and all participants gave informed consent.

2.2. Assessment of Tea Intake and Other Covariates at Baseline

At recruitment, a face-to-face interview was conducted using a structured questionnaire, and information collected included participant demographics, height, weight, lifetime use of tobacco, alcohol consumption, and history of physician-diagnosed medical conditions, such as diabetes and cancer. Body mass index (BMI) of each participant was calculated with the use of the formula: weight (kg)/height (m)^2. A 165-item semi-quantitative food-frequency questionnaire (FFQ) specifically developed and validated for this study population was used to assess the participant's usual diet over the past year [20]. For the intake frequency of black tea, green tea, and coffee, participants were asked to choose from nine predefined categories (never or hardly ever, 1–3 cups/month, 1 cup/week, 2–3 cups/week, 4–6 cups/week, 1 cup/day, 2–3 cups/day, 4–5 cups/day, and 6 or more cups/day). Oolong, a semi-fermented tea, was grouped together with green tea in the FFQ since it was drunk interchangeably with green tea in our study population. Caffeine intake was estimated from the participant's reported intake of tea and coffee. The main sources of caffeine in this study population include coffee (82%), black tea (13%), and green tea (<5%) [22].

2.3. Ascertainment of Tuberculosis Cases

Cases of active tuberculosis were identified via linkage with the National Tuberculosis Notification Registry, which was started in 1957 [23]. Notification of tuberculosis cases is compulsory under the Infectious Diseases Act in Singapore [24], and all doctors are mandated by law to notify all suspected and confirmed cases of tuberculosis to the Ministry of Health within 72 h. Most of the tuberculosis cases in Singapore are diagnosed by passive case-finding when patients present with symptoms, such as persistent cough, blood-stained sputum, fever and chills, and night sweats. A case of active tuberculosis is diagnosed by positive sputum smear and confirmed by culture tests. The principal sources of notification are the restructured public hospitals and the Tuberculosis Control Unit, accounting for 56% and 34% of notifications, respectively. In addition to notification by doctors, all culture-positive tuberculosis patients in Singapore are also captured comprehensively in the National Tuberculosis Notification Registry via electronic linkage with the two mycobacterial laboratories in Singapore [23].

The cohort was also actively followed by regular linkage to the Singapore Registry of Births and Deaths to update vital status of the cohort members. As of 31 December 2014, only 52 participants were known to be lost to follow-up due to migration out of Singapore or other reasons.

2.4. Statistical Analysis

We excluded participants with a history of active tuberculosis before recruitment (n = 3012), identified through linkage with the National Tuberculosis Notification Registry. The final analysis included 60,245 participants (Supplementary Figure S1). The baseline characteristics of the participants by their frequency of tea intake were compared using analysis of variance (ANOVA) [25] for continuous variables and chi-square test [26] for categorical variables. For each participant, person-years were calculated from the date of recruitment to date of diagnosis of tuberculosis, death, lost-to-follow-up, or 31 December 2014, whichever occurred earlier. Participants were categorized based on their intake frequency of tea and coffee into non-drinkers (less than monthly), monthly, weekly, or daily drinkers. Cox proportional hazard regression models [27] were used to assess the associations between intake frequency of tea and coffee, and quartile intake of caffeine and tuberculosis risk. The strength of an association was measured by the hazard ratio (HR) and its corresponding 95% confidence interval (CI). p-values for trend were computed via the likelihood ratio test by using the ordinal values of intake categories for tea and coffee, or quartile intake of caffeine as continuous variables in the Cox regression models. There was no violation of Cox proportional hazard assumptions for our variables of interest.

The model was first adjusted for age at recruitment (years), year of recruitment (1993–1995, 1995–1998), gender, and dialect group (Hokkien, Cantonese). Additionally, we adjusted for the level of education (no formal education, primary school, secondary school, or higher), BMI (kg/m^2, continuous), baseline history of diabetes (yes, no), smoking status and intensity (never, former 1–12 cig/day, former 13–22 cig/day, former 23+ cig/day, current 1–12 cig/day, current 13–22 cig/day, current 23+ cig/day), and alcohol consumption (none, monthly, weekly, daily). These factors have been shown to affect tuberculosis risk either in the literature [28] or in our cohort (Supplementary Table S3) and could be potential confounders. Finally, we adjusted for the intake of green tea, black tea, and coffee concurrently.

We also explored the interaction between tea drinking and other established factors of tuberculosis, such as age, gender, BMI categories, smoking status, alcohol consumption, and baseline history of diabetes. The classification of BMI levels was based on categories recommended for potential public health action points for Asian populations by the World Health Organization [29]. The heterogeneity of the tea-tuberculosis associations by different factors was tested using an interaction term (product between tea drinking categories and factor of interest) in the Cox model. Finally, to overcome the possibility of reverse-causality bias, we repeated our analysis by excluding tuberculosis cases diagnosed within two years post-enrollment and the corresponding observed person-years.

All statistical analyses were conducted using SAS version 9.3 (SAS Institute, Cary, NC, USA) statistical software package. Two-sided *p*-value < 0.05 was considered statistically significant.

3. Results

In this cohort, 41.3% of participants were non-drinkers of tea (defined as drinking less than 1 cup per month), and 22.3% were daily tea drinkers (Table 1). Among daily tea drinkers, 40.0% drank only black tea, 46.4% drank only green tea, and the remaining 13.6% were not exclusive in the type of tea they consumed. Among daily black tea drinkers, 19.1% also drank green tea daily, and among daily green tea drinkers, 17.0% also drank black tea daily (Supplementary Table S1). Compared with non-drinkers, daily tea drinkers were more likely to be men, former or current smokers, have higher level of education, consume alcohol and report history of diabetes at baseline (Table 1). Compared to daily green tea drinkers, daily black tea drinkers were younger, had a lower prevalence of diabetes at baseline, and were more likely to be men, Hokkiens, and current smokers (Supplementary Table S1). Tuberculosis cases were also more likely to be older, men, smokers, have a history of diabetes at baseline, lower BMI, received lower level of education, and have a higher intake frequency of alcohol compared to participants who did not develop active tuberculosis (Supplementary Table S2).

Table 1. Baseline characteristics of participants according to frequency of tea consumption [1].

Characteristics [2]	Intake Frequency of Black or Green Tea			
	None	Monthly	Weekly	Daily
No. of participants (%)	24,859 (41.3)	7275 (12.1)	14,705 (24.4)	13,406 (22.3)
Age at interview, years	57.0 ± 8.1	56.3 ± 8.0	55.6 ± 7.9	56.1 ± 7.9
Body mass index, kg/m^2	23.0 ± 3.2	23.2 ± 3.3	23.3 ± 3.3	23.5 ± 3.3
Men	8554 (34.4)	2766 (38.0)	6969 (47.4)	7625 (56.9)
Dialect				
Cantonese	10,641 (42.8)	3544 (48.7)	6892 (46.9)	6900 (51.5)
Hokkien	14,218 (57.2)	3731 (51.3)	7813 (53.1)	6506 (48.5)
Level of education				
No formal education	8796 (35.4)	2092 (28.8)	3261 (22.2)	2461 (18.4)
Primary school (1–6 years)	10,623 (42.7)	3280 (45.1)	6558 (44.6)	6055 (45.2)
Secondary school and above	5440 (21.9)	1903 (26.2)	4886 (33.2)	4890 (36.5)
Smoking status				
Never	18,046 (72.6)	5298 (72.8)	10,410 (70.8)	8678 (64.7)
Former	2253 (9.1)	739 (10.2)	1621 (11.0)	1824 (13.6)
Current	4560 (18.3)	1238 (17.0)	2674 (18.2)	2904 (21.7)
Alcohol intake				
None	21,288 (85.6)	5894 (81.0)	11,570 (78.7)	10,298 (76.8)
Monthly	1239 (5.0)	654 (9.0)	1291 (8.8)	1145 (8.5)
Weekly	1507 (6.1)	531 (7.3)	1398 (9.5)	1400 (10.4)
Daily	825 (3.3)	196 (2.7)	446 (3.0)	563 (4.2)
Baseline history of diabetes	2080 (8.4)	648 (8.9)	1329 (9.0)	1344 (10.0)

[1] Data shown are *n* (%) for categorical variables and mean ± SD for continuous variables; [2] All *p*-values for differences in baseline characteristics of participants according to frequency of tea consumption by ANOVA (continuous variables) or chi-square test (categorical variables) were <0.001.

Over a mean 16.8 ± 5.2 years of follow-up, we identified 1249 incident cases of active tuberculosis. The incidence rates of tuberculosis within this cohort, adjusted to the age structure of the whole cohort, were 224 per 100,000 person-years in men and 55 per 100,000 person-years in women. For tuberculosis cases, the mean duration from time of recruitment to tuberculosis diagnosis was 8.9 ± 5.6 years, and mean age at diagnosis was 68.7 ± 9.1 years. We observed an inverse association between the consumption of both black and green tea with tuberculosis risk in a dose-dependent manner (Table 2). Compared to non-drinkers of the particular type of tea, a reduced risk of tuberculosis was observed among participants who had a daily consumption of black tea (HR 0.79; 95% CI 0.65–0.95; *p* for trend = 0.02) or green tea (HR 0.84; 95% CI 0.70–1.00; *p* for trend = 0.03). Compared to non-drinkers of tea, drinking either black or green tea daily reduced risk of tuberculosis (HR 0.82; 95% CI 0.71–0.96; *p* for trend = 0.003). There was no significant association between coffee and caffeine intake and tuberculosis risk (Table 2).

Table 2. Intake of tea, coffee, and caffeine in relation to risk of tuberculosis.

Beverage	Intake Frequency				*p* for Trend
	None	Monthly	Weekly	Daily	
Black tea					
Person-years	646,006	78,649	174,715	113,319	
Cases	816	99	202	132	
HR (95% CI) [1]	1.00	1.01 (0.82–1.25)	0.84 (0.72–0.98)	0.75 (0.63–0.91)	<0.001
HR (95% CI) [2]	1.00	1.06 (0.86–1.31)	0.92 (0.78–1.07)	0.79 (0.66–0.95)	0.02
HR (95% CI) [3]	1.00	1.10 (0.89–1.36)	0.94 (0.80–1.10)	0.79 (0.65–0.95)	0.02
Green tea					
Person-years	595,265	118,101	175,055	124,268	
Cases	764	130	200	155	
HR (95% CI) [1]	1.00	0.86 (0.71–1.03)	0.83 (0.71–0.97)	0.78 (0.66–0.93)	<0.001
HR (95% CI) [2]	1.00	0.90 (0.75–1.09)	0.89 (0.76–1.04)	0.84 (0.70–1.00)	0.02
HR (95% CI) [3]	1.00	0.89 (0.74–1.08)	0.90 (0.77–1.06)	0.84 (0.70–1.00)	0.03
Black or green tea					
Person-years	415,819	122,737	248,903	225,232	
Cases	531	155	277	286	
HR (95% CI) [1]	1.00	0.98 (0.82–1.17)	0.78 (0.67–0.90)	0.77 (0.66–0.89)	<0.001
HR (95% CI) [2]	1.00	1.01 (0.85–1.21)	0.84 (0.73–0.98)	0.82 (0.71–0.96)	0.003
Coffee					
Person-years	187,395	19,633	93,401	712,261	
Cases	219	16	90	924	
HR (95% CI) [1]	1.00	0.70 (0.42–1.16)	0.81 (0.63–1.03)	1.10 (0.95–1.28)	0.08
HR (95% CI) [2]	1.00	0.72 (0.43–1.19)	0.84 (0.65–1.07)	0.97 (0.83–1.12)	0.89
HR (95% CI) [3]	1.00	0.71 (0.43–1.18)	0.82 (0.64–1.06)	0.92 (0.79–1.08)	0.55
	Quartile Intake				
	Q1	Q2	Q3	Q4	
Caffeine (mg/day)					
Person-years	252,489	254,103	259,394	246,704	
Cases	205	305	316	373	
HR (95% CI) [1]	1.00	1.20 (1.02–1.42)	1.06 (0.90–1.25)	1.20 (1.02–1.41)	0.10
HR (95% CI) [2]	1.00	1.11 (0.94–1.32)	0.95 (0.80–1.12)	0.98 (0.83–1.16)	0.40

[1] HR = hazard ratio, CI = confidence interval. Model 1 was adjusted for age at recruitment (years), year of recruitment (1993–1995, 1996–1998), gender, and dialect group (Hokkien, Cantonese); [2] Further adjusted for education level (no formal education, primary school, secondary school or higher), body mass index (kg/m^2, continuous), baseline history of diabetes (yes, no), smoking status and intensity (never, former 1–12 cig/day, former 13–22 cig/day, former 23+ cig/day, current 1–12 cig/day, current 13–22 cig/day, current 23+ cig/day), alcohol intake (none, monthly, weekly, daily); [3] Further adjusted for intake of black and green tea and coffee (none, monthly, weekly, daily).

We did not observe any statistically significant modification of the tea-tuberculosis association with age, gender, smoking status, or baseline history of diabetes (*p* for interaction > 0.20; data not shown). A significant interaction between tea drinking and BMI categories was observed. Daily tea drinking was associated with reduced tuberculosis risk among leaner participants with BMI < 23 kg/m^2 (HR 0.66; 95% CI 0.54–0.81), but was not among participants with higher BMI (HR 1.09; 95% CI 0.87–1.36; *p* for interaction = 0.004). There was also significant interaction between tea drinking and alcohol consumption. A greater reduction in tuberculosis risk with daily tea drinking was observed among weekly/daily alcohol drinkers (HR 0.54; 95% CI 0.38–0.77) compared to participants with less regular alcohol consumption (HR 0.90; 95% CI 0.77–1.06; *p* for interaction = 0.01) (Table 3).

Table 3. Interaction between tea drinking and stratifying variables in relation to risk of tuberculosis.

Stratifying Variables	Intake Frequency of Black or Green Tea				p for Trend	p for Interaction
	None	Monthly	Weekly	Daily		
Body mass index						0.004
<23 kg/m^2						
Cases	350	96	168	146		
HR (95% CI) [1]	1.00	0.99 (0.79–1.25)	0.79 (0.66–0.96)	0.66 (0.54–0.81)	<0.001	
≥23 kg/m^2						
Cases	181	59	109	140		
HR (95% CI) [1]	1.00	1.05 (0.78–1.41)	0.92 (0.72–1.17)	1.09 (0.87–1.36)	0.70	
Alcohol intake						0.01
None/monthly						
Cases	432	137	225	242		
HR (95% CI) [1]	1.00	1.10 (0.90–1.33)	0.86 (0.73–1.02)	0.90 (0.77–1.06)	0.08	
Weekly/daily						
Cases	99	18	52	44		
HR (95% CI) [1]	1.00	0.62 (0.38–1.03)	0.71 (0.50–0.99)	0.54 (0.38–0.77)	<0.001	

[1] HR = hazard ratio, CI = confidence interval. Model was adjusted for age at recruitment (years), year of recruitment (1993–1995, 1996–1998), gender, dialect group (Hokkien, Cantonese), education level (no formal education, primary school, secondary school or higher), body mass index (kg/m^2, continuous), baseline history of diabetes (yes, no), smoking status and intensity (never, former 1–12 cig/day, former 13–22 cig/day, former 23+ cig/day, current 1–12 cig/day, current 13–22 cig/day, current 23+ cig/day), alcohol intake (none, monthly, weekly, daily).

The association between tea drinking and tuberculosis risk remained essentially the same after exclusion of participants diagnosed with tuberculosis within the first two years after their recruitment (data not shown). We also examined the associations among participants who were exclusive drinkers of either black or green tea. Compared to non-drinkers, participants who only drank black or green tea daily had similarly reduced risk of tuberculosis; HR (95% CI) were 0.79 (0.63–0.98) for daily drinkers of black tea only and 0.87 (0.71–1.06) for daily drinkers of green tea only. Comparatively, daily tea drinkers who drank both types of tea had the lowest reduced risk of 0.73 (95% CI: 0.56–0.95). We have also additionally adjusted for dietary factors such as fruits and vegetables intake in the model, and the results remained essentially unchanged (data not shown).

4. Discussion

To our best knowledge, this is the first prospective study investigating the relation between tea drinking and tuberculosis risk, and we found that drinking tea (either black or green tea) was inversely associated with risk of developing active tuberculosis in a dose-dependent manner. The reduction in tuberculosis risk with daily tea drinking was greater in leaner participants (BMI < 23 kg/m^2) relative to their overweight/obese counterparts, as well as in weekly/daily alcohol drinkers compared to participants who drank alcohol less frequently. No association was observed with coffee or caffeine consumption.

Epidemiologic evidence on the relation between tea drinking and tuberculosis risk is limited. Only one recent case-control study reported a possible inverse association with consumption of ≥1 cup/week of black tea (odds ratio 0.68; 95% CI 0.52–0.90) or green tea (odds ratio 0.53; 95% CI 0.35–0.82) [18], which was consistent with the findings of the present study.

Both black and green teas are produced from the leaves of *Camellia sinensis*, and they vary in their polyphenol content due to different manufacturing processes. Green tea is made by drying fresh tea leaves without prior fermentation, and this preserves the naturally occurring polyphenols, such as tea catechins [30]. The production of black tea, on the other hand, involves the crushing of fresh tea leaves, leading to fermentation and oxidation of tea catechins into other polyphenols, such as theaflavins [30]. To date, experimental investigations of tea polyphenols on tuberculosis have only involved green tea catechins. EGCG, the most abundant catechin present in green tea, has been shown to have anti-mycobacterial effects by inhibiting Mtb enoyl-acyl reductase (InhA), an enzyme involved in the production of functional mycolic acids [16]. Structural damage in the mycobacterial cell wall due to impaired production of mycolic acids has been postulated as a mechanism behind the anti-mycobacterial function of EGCG [15]. In experimental studies, mice infected with Mtb had significantly increased levels of oxidative stress during early stages of tuberculosis, and oral administration of green tea extract led to the reversion of oxidative stress parameters to near normal levels [13]. Pre-treatment of macrophages with 60 µg/mL ($\approx$131 µmol/L) of EGCG has been shown to lead to down-regulation of the expression of the host molecule tryptophan-aspartate containing coat protein (TACO), and resultant inhibition of mycobacterium survival within macrophages [17]. The highest tea consumers in our cohort reported intake of six or more cups of black or green tea per day, and the consumption of six cups of black or green tea has been shown to increase blood catechin levels with an average maximum change from baseline by 0.10 µmol/L and 0.46 µmol/L respectively [31]. While the increase in blood catechin levels with tea consumption in human is much lower than what has been achieved in experimental settings using animal models, the increased levels of catechins with tea consumption has been linked to increased plasma antioxidant activity and possible protection against diseases [31–33]. Even though the effect of black tea polyphenols on tuberculosis infection or disease has not been investigated in experimental studies, black tea polyphenols, such as theaflavins have been shown to exhibit similar antioxidant potency as green tea catechins [7,34]. These experimental findings suggest the potential of tea polyphenols against development of active tuberculosis, and support the observations from epidemiological studies.

We attempted to understand the biological mechanism of tea in reducing tuberculosis risk by examining how other established risk factors of tuberculosis may modify this effect. We found daily tea drinking to have a more prominent effect in leaner individuals compared to obese/overweight individuals. Leanness has been thought to increase tuberculosis risk due to lower protein and energy intake among lean individuals [35]. However, this cannot explain the greater risk reduction with tea drinking among leaner participants in our study as tea does not contribute significantly to protein or energy intake. Furthermore, the estimates for the tea-tuberculosis association did not change when we included protein and total energy intake in the model (data not shown). Similarly, we found that daily tea drinking conferred a greater protective effect in weekly/daily alcohol drinkers compared to less regular alcohol drinkers. Although alcohol consumption can impair the immune system and increase tuberculosis risk [36–39], the meta-analysis by Lönnroth et al. concluded that a substantial increase in risk of developing active tuberculosis was only observed in individuals who consumed more than 40 g/day of alcohol, and/or had an alcohol use disorder, but not in people who drank less than 40 g/day [38]. Only 0.91% of the participants in our cohort reported an average alcohol intake of more than 40 g/day. On top of these plausible mechanisms, leanness and alcohol consumption are also factors associated with increased levels of oxidative stress [40,41]. Hence tea, a beverage rich in antioxidants, may confer a protective effect against tuberculosis, in part by ameliorating the increased levels of oxidative stress associated with leanness or alcohol consumption in affected individuals. Further studies are needed to evaluate possible anti-mycobacterial effects of polyphenols from black and green teas, and validate the effect of tea drinking in different subpopulations at risk of tuberculosis.

Even though caffeine has been reported to have immunomodulatory effects [42], we did not observe any associations between caffeine intake and tuberculosis risk in our study. A significant relation between coffee and tuberculosis risk has never been reported, and we did not find any significant association with coffee in this study. The results provide indirect evidence that caffeine in tea is unlikely responsible for the inverse association between tea and tuberculosis.

Even though Singapore currently has an intermediate tuberculosis incidence rate of about 40 per 100,000 population, our study cohort consist of older residents of the country who were likely to be exposed to the bacteria during the 1960s, a period where incidence of tuberculosis was as high as 300 per 100,000 population [43]. Hence, many of our participants could have acquired latent infection in those early years where tuberculosis was far more rampant [19], and this makes our cohort suitable to study risk factors associated with tuberculosis reactivation. The observation of lower BMI, greater proportion of men, smokers and prevalent diabetes among tuberculosis cases compared to participants who remained free of tuberculosis in our cohort was also consistent with findings from the local population and other study populations [28,43]. Another strength of this study is the variability in intake frequencies of the two different types of tea in the population, which allowed us to simultaneously examine the relation of both black and green tea with tuberculosis risk in the same analysis. The prospective population-based design of our study also ensured minimal recall bias in exposure data since they were obtained years before tuberculosis diagnosis. By linking up with the national tuberculosis registry, in which notification of tuberculosis cases in the country is mandated by law, we were able to attain near-complete ascertainment of all diagnosed active tuberculosis cases in our cohort.

Limitations of our study include using baseline intake of tea in our analysis, and changes to the frequency of tea consumption following the baseline interview were not accounted for. However, among 39,528 participants contacted for a follow-up interview between 2006 and 2010, an average of 12.7 years after the baseline interview, 85.6% of them retained their status as daily or non-daily drinkers of black tea, and 85.2% of them retained their status as daily or non-daily drinkers of green tea. This suggests the stability of tea drinking behaviour in our study population. Nevertheless, we acknowledge that any change in the habit of tea-drinking after recruitment could lead to potential non-differential misclassification of tea drinking, and underestimation of the true association. We were unable to determine whether green tea and semi-fermented oolong tea could have a different effect on tuberculosis risk as these two types of tea were included as a single inquiry in our FFQ. However, since black and green teas were both found to be associated with reduced active tuberculosis risk to a similar extent, we believe that drinking oolong tea would also have the same effect, and the grouping of oolong together with green tea is unlikely to influence the results of green tea. Our study is also limited to the analysis of late-life tuberculosis as our cohort consists of only middle-aged and elderly adults. Even though we did not adjust for the use of systemic immunosuppressant, a risk factor for tuberculosis, the use of systemic immunosuppressant that is sufficient to affect tuberculosis risk is unlikely to be high in a population-based cohort. We also lacked information on the participant's human immunodeficiency virus status, which is also a risk factor for active tuberculosis. However, Singapore has relatively low human immunodeficiency virus infection rate in the general population [43]. We also lacked information on whether the participants had been infected with Mtb previously, and there could also have been undiagnosed cases of active tuberculosis that were not captured in the national tuberculosis notification registry. However, such cases are not expected to be high as medical and healthcare services in Singapore are generally affordable and efficient, and there is comprehensive capture of active tuberculosis cases mandated by the Singapore Infectious Diseases Act. Finally, as in any observational studies, causation could not be established, and residual confounding is still possible despite having adjusted for a number of confounding factors in the statistical models.

5. Conclusions

In conclusion, we observed an inverse association between tea consumption and tuberculosis risk in a prospective cohort of Chinese adults in Singapore. The protective association with tea drinking was more prominent among lean individuals and among weekly/daily alcohol drinkers. Since these individuals could be more susceptible to oxidative stress, it is conceivable that the antioxidant properties of tea polyphenols could be the biological driver underlying our observations, but this theory needs to be validated in future studies. Since both black and green teas are widely consumed worldwide, our findings have significant clinical and public health implications and tea drinking could potentially be used as a prophylactic measure against tuberculosis in susceptible populations. However, further research is essential to identify the bioactive compounds in black and green teas and to determine the mechanistic ways for the anti-mycobacterial effects in these compounds.

Supplementary Materials: The following are available online at www.mdpi.com/2072-6643/9/5/544/s1, Figure S1: Flow diagram of participants included in the final analysis, Table S1: Baseline characteristics of daily tea drinkers by different types of tea, Table S2: Baseline characteristics of participants who developed tuberculosis (TB) and those who remained free of TB, Table S3: Baseline factors in relation to risk of active tuberculosis.

Nutrients **2017**, 9, 544

Acknowledgments: We thank Siew-Hong Low of the National University of Singapore for supervising the field work of the Singapore Chinese Health Study, and Renwei Wang for the maintenance of the cohort study database. We also thank Jeffrey Cutter of the Ministry of Health and Kyi-Win Khin Mar of the National Tuberculosis Notification Registry in Singapore for assistance with the identification of tuberculosis cases in the cohort. This work was supported by The United States National Cancer Institute, National Institutes of Health (grant numbers UM1 CA182876 and R01 CA144034). W-PK is supported by the National Medical Research Council, Singapore (NMRC/CSA/0055/2013).

Author Contributions: A.Z.S. and W.-P.K. contributed to the conception and design of the study and data analysis; C.B.E.C., Y.-T.W., J.-M.Y. and W.-P.K. participated in the acquisition of data; A.Z.S., A.P., C.B.E.C., Y.-T.W., J.-M.Y. and W.-P.K. were involved in the interpretation of the data and drafting of intellectual content; and W.-P.K. was responsible for the integrity of the work as a whole. All authors read and approved the final manuscript.

Conflicts of Interest: The authors declare no conflict of interest. The founding sponsors had no role in the design of the study; in the collection, analyses, or interpretation of data; in the writing of the manuscript, and in the decision to publish the results.

References

1. World Health Organization. *Global Tuberculosis Report 2015*; World Health Organization: Geneva, Switzerland, 2015.
2. Getahun, H.; Matteelli, A.; Chaisson, R.E.; Raviglione, M. Latent Mycobacterium tuberculosis infection. *N. Engl. J. Med.* **2015**, 372, 2127–2135. [CrossRef] [PubMed]
3. Achkar, J.M.; Jenny-Avital, E.R. Incipient and subclinical tuberculosis: Defining early disease states in the context of host immune response. *J. Infect. Dis.* **2011**, 204 (Suppl. 4), S1179–S1186. [CrossRef] [PubMed]
4. Chan, E.D.; Chan, J.; Schluger, N.W. What is the role of nitric oxide in murine and human host defense against tuberculosis? Current knowledge. *Am. J. Respir. Cell Mol. Biol.* **2001**, 25, 606–612. [CrossRef] [PubMed]
5. Kwiatkowska, S.; Piasecka, G.; Zieba, M.; Piotrowski, W.; Nowak, D. Increased serum concentrations of conjugated diens and malondialdehyde in patients with pulmonary tuberculosis. *Respir. Med.* **1999**, 93, 272–276. [CrossRef]
6. Palanisamy, G.S.; Kirk, N.M.; Ackart, D.F.; Shanley, C.A.; Orme, I.M.; Basaraba, R.J. Evidence for oxidative stress and defective antioxidant response in guinea pigs with tuberculosis. *PLoS ONE* **2011**, 6, e26254. [CrossRef] [PubMed]
7. Leung, L.K.; Su, Y.; Chen, R.; Zhang, Z.; Huang, Y.; Chen, Z.Y. Theaflavins in black tea and catechins in green tea are equally effective antioxidants. *J. Nutr.* **2001**, 131, 2248–2251. [PubMed]
8. Leenen, R.; Roodenburg, A.J.; Tijburg, L.B.; Wiseman, S.A. A single dose of tea with or without milk increases plasma antioxidant activity in humans. *Eur. J. Clin. Nutr.* **2000**, 54, 87–92. [CrossRef] [PubMed]
9. Tipoe, G.L.; Leung, T.M.; Hung, M.W.; Fung, M.L. Green tea polyphenols as an anti-oxidant and anti-inflammatory agent for cardiovascular protection. *Cardiovasc. Hematol. Disord. Drug Targets* **2007**, 7, 135–144. [CrossRef] [PubMed]
10. Beltz, L.A.; Bayer, D.K.; Moss, A.L.; Simet, I.M. Mechanisms of cancer prevention by green and black tea polyphenols. *Anticancer Agents Med. Chem.* **2006**, 6, 389–406. [CrossRef] [PubMed]
11. Wang, Y.; Wang, B.; Du, F.; Su, X.; Sun, G.; Zhou, G.; Bian, X.; Liu, N. Epigallocatechin-3-Gallate Attenuates Oxidative Stress and Inflammation in Obstructive Nephropathy via NF-kappaB and Nrf2/HO-1 Signalling Pathway Regulation. *Basic Clin. Pharmacol. Toxicol.* **2015**, 117, 164–172. [CrossRef] [PubMed]
12. Ahmed, S.; Rahman, A.; Hasnain, A.; Lalonde, M.; Goldberg, V.M.; Haqqi, T.M. Green tea polyphenol epigallocatechin-3-gallate inhibits the IL-1 beta-induced activity and expression of cyclooxygenase-2 and nitric oxide synthase-2 in human chondrocytes. *Free Radic. Biol. Med.* **2002**, 33, 1097–1105. [CrossRef]
13. Guleria, R.S.; Jain, A.; Tiwari, V.; Misra, M.K. Protective effect of green tea extract against the erythrocytic oxidative stress injury during mycobacterium tuberculosis infection in mice. *Mol. Cell. Biochem.* **2002**, 236, 173–181. [CrossRef] [PubMed]
14. Agarwal, A.; Prasad, R.; Jain, A. Effect of green tea extract (catechins) in reducing oxidative stress seen in patients of pulmonary tuberculosis on DOTS Cat I regimen. *Phytomedicine* **2010**, 17, 23–27. [CrossRef] [PubMed]

15. Sun, T.; Qin, B.; Gao, M.; Yin, Y.; Wang, C.; Zang, S.; Li, X.; Zhang, C.; Xin, Y.; Jiang, T. Effects of epigallocatechin gallate on the cell-wall structure of Mycobacterial smegmatis mc155. *Nat. Prod. Res.* **2014**, *29*, 2122–2124.

16. Sharma, S.K.; Kumar, G.; Kapoor, M.; Surolia, A. Combined effect of epigallocatechin gallate and triclosan on enoyl-ACP reductase of Mycobacterium tuberculosis. *Biochem. Biophys. Res. Commun.* **2008**, *368*, 12–17. [CrossRef] [PubMed]

17. Anand, P.K.; Kaul, D.; Sharma, M. Green tea polyphenol inhibits Mycobacterium tuberculosis survival within human macrophages. *Int. J. Biochem. Cell Biol.* **2006**, *38*, 600–609. [CrossRef] [PubMed]

18. Chen, M.; Deng, J.; Li, W.; Lin, D.; Su, C.; Wang, M.; Li, X.; Abuaku, B.K.; Tan, H.; Wen, S.W. Impact of tea drinking upon tuberculosis: A neglected issue. *BMC Public Health* **2015**, *15*, 515. [CrossRef] [PubMed]

19. Enhancing Public Health Measures Against Tuberculosis. Available online: https://www.moh.gov.sg/content/moh_web/home/pressRoom/pressRoomItemRelease/2008/enhancing_public_health_measures_against_tuberculosis.html (accessed on 13 August 2015).

20. Hankin, J.H.; Stram, D.O.; Arakawa, K.; Park, S.; Low, S.H.; Lee, H.P.; Yu, M.C. Singapore Chinese Health Study: Development, validation, and calibration of the quantitative food frequency questionnaire. *Nutr. Cancer* **2001**, *39*, 187–195. [CrossRef] [PubMed]

21. Koh, W.P.; Yuan, J.M.; Sun, C.L.; Lee, H.P.; Yu, M.C. Middle-aged and older Chinese men and women in Singapore who smoke have less healthy diets and lifestyles than nonsmokers. *J. Nutr.* **2005**, *135*, 2473–2477. [PubMed]

22. Goh, G.B.; Chow, W.C.; Wang, R.; Yuan, J.M.; Koh, W.P. Coffee, alcohol and other beverages in relation to cirrhosis mortality: The Singapore Chinese Health Study. *Hepatology* **2014**, *60*, 661–669. [CrossRef] [PubMed]

23. Chee, C.B.; James, L. The Singapore Tuberculosis Elimination Programme: The first five years. *Bull. World Health Organ.* **2003**, *81*, 217–221. [PubMed]

24. Infectious Diseases Act, The Statutes of the Republic of Singapore (Cap 137, 2003 Rev Ed). Available online: http://statutes.agc.gov.sg/aol/download/0/0/pdf/binaryFile/pdfFile.pdf?CompId:4c22780c-89cc-4820-acf3-38140427b099 (accessed on 25 May 2017).

25. Bewick, V.; Cheek, L.; Ball, J. Statistics review 9: One-way analysis of variance. *Crit. Care* **2004**, *8*, 130–136. [CrossRef] [PubMed]

26. Cochran, W.G. The χ^2 test of goodness of fit. *Ann. Math. Stat.* **1952**, *23*, 315–345. [CrossRef]

27. Cox, D.R. Regression Models and Life-Tables. *J. R. Stat. Soc. Ser. B Methodol.* **1972**, *34*, 187–220.

28. Patra, J.; Jha, P.; Rehm, J.; Suraweera, W. Tobacco smoking, alcohol drinking, diabetes, low body mass index and the risk of self-reported symptoms of active tuberculosis: Individual participant data (IPD) meta-analyses of 72,684 individuals in 14 high tuberculosis burden countries. *PLoS ONE* **2014**, *9*, e96433. [CrossRef] [PubMed]

29. World Health Organization. Appropriate body-mass index for Asian populations and its implications for policy and intervention strategies. *Lancet* **2004**, *363*, 157–163.

30. Yang, C.S.; Landau, J.M. Effects of tea consumption on nutrition and health. *J. Nutr.* **2000**, *130*, 2409–2412. [PubMed]

31. Van het Hof, K.H.; Kivits, G.A.; Weststrate, J.A.; Tijburg, L.B. Bioavailability of catechins from tea: The effect of milk. *Eur. J. Clin. Nutr.* **1998**, *52*, 356–359. [CrossRef] [PubMed]

32. Yashin, A.; Nemzer, B.; Yashin, Y. Bioavailability of Tea Components. *J. Food Res.* **2012**, *1*, 281–290. [CrossRef]

33. Serafini, M.; Ghiselli, A.; Ferro Luzzi, A. In vivo antioxidant effect of green and black tea in man. *Eur. J. Clin. Nutr.* **1996**, *50*, 28–32. [PubMed]

34. Sarkar, A.; Bhaduri, A. Black tea is a powerful chemopreventor of reactive oxygen and nitrogen species: Comparison with its individual catechin constituents and green tea. *Biochem. Biophys. Res. Commun.* **2001**, *284*, 173–178. [CrossRef] [PubMed]

35. Hanrahan, C.F.; Golub, J.E.; Mohapi, L.; Tshabangu, N.; Modisenyane, T.; Chaisson, R.E.; Gray, G.E.; McIntyre, J.A.; Martinson, N.A. Body mass index and risk of tuberculosis and death. *AIDS* **2010**, *24*, 1501–1508. [CrossRef] [PubMed]

36. Szabo, G. Alcohol's contribution to compromised immunity. *Alcohol Health Res. World* **1997**, *21*, 30–41. [PubMed]

37. Mason, C.M.; Dobard, E.; Zhang, P.; Nelson, S. Alcohol Exacerbates Murine Pulmonary Tuberculosis. *Infect. Immun.* **2004**, *72*, 2556–2563. [CrossRef] [PubMed]

38. Lönnroth, K.; Williams, B.G.; Stadlin, S.; Jaramillo, E.; Dye, C. Alcohol use as a risk factor for tuberculosis—A systematic review. *BMC Public Health* **2008**, *8*, 289. [CrossRef] [PubMed]

39. Rehm, J.; Samokhvalov, A.V.; Neuman, M.G.; Room, R.; Parry, C.; Lonnroth, K.; Patra, J.; Poznyak, V.; Popova, S. The association between alcohol use, alcohol use disorders and tuberculosis (TB). A systematic review. *BMC Public Health* **2009**, *9*, 450.

40. Das, S.K.; Vasudevan, D.M. Alcohol-induced oxidative stress. *Life Sci.* **2007**, *81*, 177–187. [CrossRef] [PubMed]

41. Loft, S.; Vistisen, K.; Ewertz, M.; Tjonneland, A.; Overvad, K.; Poulsen, H.E. Oxidative DNA damage estimated by 8-hydroxydeoxyguanosine excretion in humans: Influence of smoking, gender and body mass index. *Carcinogenesis* **1992**, *13*, 2241–2247. [CrossRef] [PubMed]

42. Horrigan, L.A.; Kelly, J.P.; Connor, T.J. Immunomodulatory effects of caffeine: Friend or foe? *Pharmacol. Ther.* **2006**, *111*, 877–892. [CrossRef] [PubMed]

43. Ministry of Health. *Communicable Diseases Surveillance in Singapore 2014*; Ministry of Health Communicable Diseases Division: Singapore, 2014; pp. 120–145.

nutrients

MDPI

Article

Preoperative Nutritional Conditioning of Crohn's Patients—Systematic Review of Current Evidence and Practice

Fabian Grass, Basile Pache, David Martin, Dieter Hahnloser, Nicolas Demartines and Martin Hübner *

Department of Visceral Surgery, Lausanne University Hospital CHUV, 1011 Lausanne, Switzerland;
fabian.grass@chuv.ch (F.G.); basile.pache@chuv.ch (B.P.); david.martin@chuv.ch (D.M.);
dieter.hahnloser@chuv.ch (D.H.); demartines@chuv.ch (N.D.)
* Correspondence: martin.hubner@chuv.ch; Tel.: +41-21-314-2400; Fax: +41-21-314-2411

Received: 7 April 2017; Accepted: 30 May 2017; Published: 1 June 2017

Abstract: Crohn's disease is an incurable and frequently progressive entity with major impact on affected patients. Up to half of patients require surgery in the first 10 years after diagnosis and over 75% of operated patients require at least one further surgery within lifetime. In order to minimize surgical risk, modifiable risk factors such as nutritional status need to be optimized. This systematic review on preoperative nutritional support in adult Crohn's patients between 1997 and 2017 aimed to provide an overview on target populations, screening modalities, routes of administration, and expected benefits. Pertinent study characteristics (prospective vs. retrospective, sample size, control group, limitations) were defined a priori. Twenty-nine studies were retained, of which 14 original studies (9 retrospective, 4 prospective, and 1 randomized controlled trial) and 15 reviews. Study heterogeneity was high regarding nutritional regimens and outcome, and meta-analysis could not be performed. Most studies were conducted without matched control group and thus provide modest level of evidence. Consistently, malnutrition was found to be a major risk factor for postoperative complications, and both enteral and parenteral routes were efficient in decreasing postoperative morbidity. Current guidelines for nutrition in general surgery apply also to Crohn's patients. The route of administration should be chosen according to disease presentation and patients' condition. Further studies are needed to strengthen the evidence.

Keywords: Crohn's; inflammatory bowel disease; nutrition; supplement; surgery; preoperative; complications

1. Introduction

Nutritional support strategies in malnourished patients became widely accepted tool to decrease postoperative morbidity in major gastrointestinal surgery [1,2]. Up to 85% of patients with Crohn's disease awaiting surgery are malnourished as a consequence of active and disabling disease [3], impeding proper dietary intake and resorption [4–6]. Since up to 70% of Crohn's patients at some point requires surgery [7,8], a bowel-sparing attitude is mandatory in order to prevent short bowel syndromes, malnutrition, and anemia [9,10]. Despite advances in non-surgical management of acute flares through multimodal concepts including biologics, antibiotics, and nutritional support, surgery remains a last treatment option for medically exhausted cases [11–13]. Interestingly, it should be emphasized that incidence of surgical procedures for Crohn's disease did not decrease after introduction of infliximab [14]. Postoperative complication rates, including most feared intra-abdominal septic complications, reach 30% [15,16]. Nutritional guidelines for Crohn's patients have been published by the European Society for Clinical nutrition and metabolism (ESPEN) in 2006 [17] and 2016 [4]. Most recommendations are based on consensus among experts or extrapolated from the general surgical population.

The aim of this present study was to systematically review scientific evidence over the last 20 years in an attempt to provide evidence-based recommendations.

2. Materials and Methods

2.1. Data sources and Search Strategies

Main electronic databases including Medline (searched through PubMed), Embase, the Cochrane Database of Systematic Review and the Cochrane Central Register of Controlled trials were systematically searched. For searching PubMed, the medical subject heading (MeSH) were "(Crohn OR Crohn's Disease OR inflammatory bowel disease) AND (nutrition OR conditioning OR nutritional support) AND (preoperative OR perioperative)".

Electronic links were searched for related articles and cross-referencing of selected articles was performed by two authors (FG, BP). The trial registry http://clinicaltrials.gov was screened for relevant unpublished prospective trials. The search was limited to studies published between 1 January 1997 and 28 February 2017 for the following reasons: First, to provide evidence over the last 20 years, considering 10 years before the implementation of guidelines [17] and 10 years thereafter. Second, U.S. Food and Drug Administration approval of infliximab was granted in 1997 [14]. All kinds of original scientific reports were considered, and reviews and book chapters were also included. No language restrictions were applied for the original search, but studies providing only English abstracts were consequently excluded.

2.2. Study Selection (Inclusion and Exclusion Criteria)

Original studies and reviews reporting on pre- or perioperative nutritional support in Crohn's disease were included. Excluded from the analysis were reports on general nutritional support in Crohn's disease (including postoperatively) or in pediatric populations. The included manuscripts were divided in original scientific reports and reviews.

2.3. Data Extraction and Quality Assessment

Pertinent study characteristics (prospective vs. retrospective, sample size, control group, matching) were defined a priori and each manuscript was assessed for potential sources of bias. Two authors independently performed the literature search. The search terms were firstly identified in the title, and secondly in the abstract or medical subject heading. All studies of interest were obtained as full text articles and scrutinized thoroughly. Three authors made the final decision on inclusion of a study.

Relevant data were extracted and documented in a database developed ad hoc for all publications. The following items were recorded for each study when available: authors, title, year of publication, disease presentation/surgical indication, details on nutritional regimen (type/formula/duration/timing), and potential limitations of original studies. Postoperative outcomes of interest were complications (overall, infectious/septic, non-infectious), recurrence rates, and changes in different nutritional parameters if available. Data are presented in accordance to the PRISMA statement [18] (Figure 1).

Based on the findings of this study, an algorithm was created for practical guidance.

Identification

| 189 records identified through Medline database search | 2 records identified through other sources and cross-referencing |

Screening

| 191 records screened (title) | → | 121 records excluded based on title |

| 70 records screened (abstract) | → | 35 records excluded:
- other outcomes/endpoints: 13
- no Crohn's disease: 6
- no surgery: 5
- *post*operative nutrition: 4
- pediatric population: 3
- other: 4 |

Eligibility

| 35 full text articles assessed for eligibility | — | 6 full-text articles excluded:
- data not within the scope ($n = 4$)
- language restriction ($n = 1$)
- duplicate data (=1) |

Inclusion

| 29 studies (14 original, 15 reviews) |

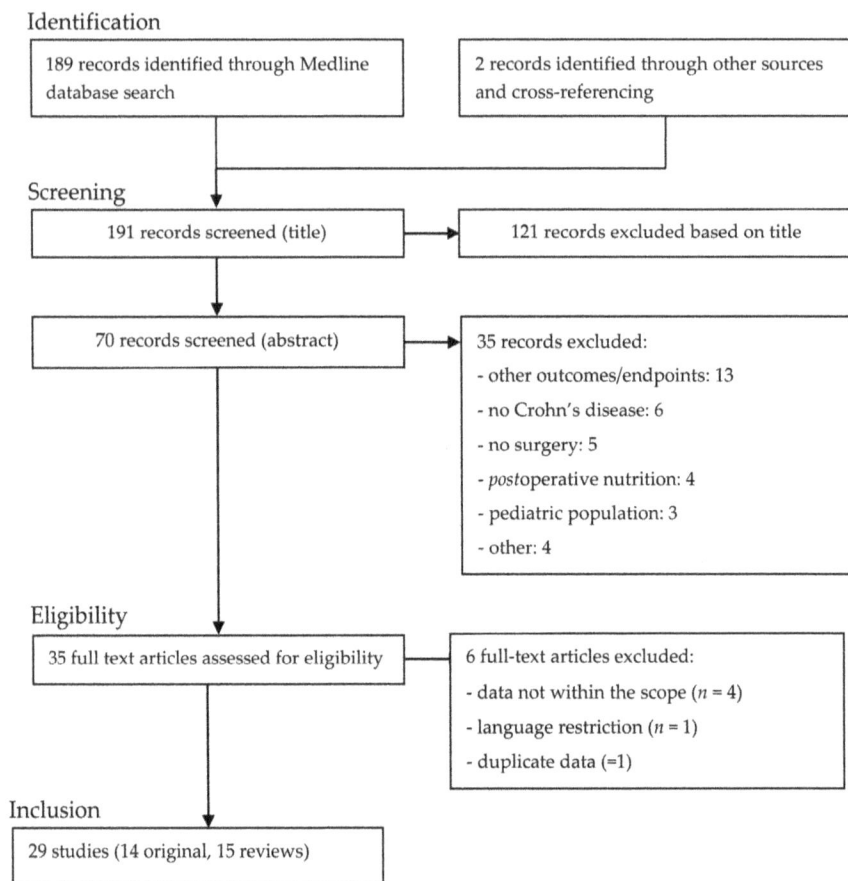

Figure 1. The selection process adhere to the guidelines outlined in the PRISMA statement [18].

2.4. Data Analysis

Meta-analysis of results was not feasible due to limited and heterogeneous original data. Instead, tables were created with descriptive statistics to display the most relevant findings of each original study and review to give a comprehensive overview of the most relevant results.

3. Results

Electronic search of PubMed yielded 189 studies. By cross-referencing and through other data sources, two further studies [16,19] were identified matching the inclusion criteria. Of these 191 studies, 121 were excluded based on the title, and 35 further studies based on the abstract with reasons for exclusion displayed in Figure 1, remaining 35 full text articles assessed. Six were subsequently excluded for the following reasons: incomplete data or not within the scope of the present analysis ($n = 4$) [20–23], language restriction ($n = 1$) [24], and risk of double publication ($n = 1$) [25] (Figure 1). For final analysis, 14 original studies [26–39] and 15 reviews [4,5,17,19,40–50] were retained (Tables 1–3). All studies except one were indexed in PubMed. The one not indexed was a conference paper [51] from Cochrane Database with no detailed data thus was excluded for final analysis. Two studies were found on http://clinicaltrials.gov, one completed study from a Chinese group (NCT01540942) and one Canadian study which was not yet recruiting by 8 March 2017 (NCT02985489).

3.1. Methodological Assessment of Inlcuded Studies

3.1.1. Study Design and Quality of Original Studies

Overall, 2141 patients with Crohn's disease were reported in study and control groups, respectively (Table 1). Nine studies (64%) were retrospective [26,27,29,30,32–34,36,37] totaling 1783 patients. The remaining 358 patients were studied in 4 (29%) prospective [28,35,38,39] and 1 randomized controlled trial (RCT) [31]. In the only RCT, patients were not randomized to a nutritional regimen but to different endpoint measures, and thus no nutritional control group was available. Nutritional control groups were available in nine studies (64%) [26,28–30,32,33,35,37,38]. Only three studies (21%) [26,29,35] matched study and control groups, however, matching criteria were inconsistent among these studies.

Table 1. Original studies on preoperative nutritional support in Crohn's disease patients.

Author	Year	Design	Control	Matching	N	Limitations
Heerasing [26]	2017	Retrospective	Yes	Yes	114	Incomplete matching for disease severity
Guo [27]	2017	Retrospective	No	na	118	No outcome data other than SSI
Beaupel [28]	2017	Prospective	Yes	No	56	Comparison high-risk to low-risk patients
Wang [29]	2016	Retrospective	Yes	Yes	81	Potential selection bias for study group
Zhang [30]	2015	Retrospective	Yes	No	64	Comparison high-risk to low-risk patients
Zhu [31]	2015	RCT	No	na	108	No nutritional control group
Li [32]	2015	Retrospective	Yes	No	708	Potential selection bias, <10% laparoscopy
Li [33]	2014	Retrospective	Yes	No	123	No dietary information of control group
Bellolio [34]	2013	Retrospective	No	No	434	No nutritional control group
Jacobson [35]	2012	Prospective	Yes	Yes	120	No matching for disease severity
Zerbib [36]	2010	Retrospective	No	Na	78	Heterogeneous study groups
Grivceva [37]	2008	Retrospective	Yes	No	63	Composition of diets not specified
Yao [38]	2005	Prospective	Yes	No	32	Small sample size
Smedh [39]	2002	Prospective	No	na	42	Small sample size, no nutritional control group

Abbreviations: RCT—randomized controlled trial, SSI—surgical site infection, N—number of included patients, na—not available

3.1.2. Patients, Disease Presentation and Nutritional Details

In 12 studies (86%) nutritional support was administered preoperative only, ranging from two weeks up to three months. The remaining two studies [31,38] administered nutritional support pre- and postoperatively. Nutritional regimens were heterogeneous among included studies. Details on formulas are displayed in Table 2. In six studies (43%) [30,31,34,36,37,39], nutritional formulas were not or incompletely described. Six studies [26,28,29,32,33,39] evaluated the impact of exclusive enteral nutrition (EEN), four studies [34,35,37,38] reported on total parenteral nutrition (TPN) and four studies [27,30,31,36] combined different ways of nutritional support. Most studies (71%) [26,28,30–35,37,38] reported on severely sick and malnourished patients with either obstructing or fistulizing disease (Table 2). The remaining studies [27,29,36,39] optimized their cohorts or reported on low-risk patients.

Table 2. Nutritional details and outcome of original studies.

Author	Disease	Type/Formula	Timing	Duration	Groups/Cohort	Main results (Nutritional Group)
Heerasing [26]	P/F	EEN [1]	Pre	6 w (mean)	EEN pre-treatment group vs. straight to surgery group	Nine-fold decreased infectious complications, shorter operating time
Guo [27]	F	PN+EN [2]	Pre	3 m	Preop optimized cohort (nutritional support, steroid weaning, abscess drainage, antibiotics)	EEN <3 m retained as independent risk factor for SSI
Beaupel [28]	P/F	ANS-TGF-b2 (EEN)	Pre	3 w (median)	Supplemented high-risk (steroids, malnutrition) vs. non supplemented low-risk patients	Similar overall and infectious complications
Wang [29]	FS	EEN [2]	Pre	4 w	Low-risk patients (no immunosuppression, no inflammation) in both groups (EEN vs. non-EEN)	Decreased overall and infectious complications, less recurrence at 6 m
Zhang [30]	F/O	TPN or PN or EN (na)	Pre	3 w (median)	Fortified nutrition support group (lower BMI, higher CDAI) vs. non-supplemented control group	Similar postoperative septic complications (3 m)
Zhu [31]	F/P	EEN [2] +/-PN +/-TPN (na)	Pre Post	4 w 4 w	Supplementation in all patients, randomization and blinding for two endpoints: ROI and IOM	Similar complications (4 w) in ROI group = better endpoint than IOM, less complications than historical controls
Li [32]	R/F/O/P	EEN [2]	Pre	4 w	Immunosuppressants-treated EEN patients vs. different non-supplemented control groups	Decreased overall and infectious complications (30 days) in EEN-group
Li [33]	F	EEN [3]	Pre	3 m	EEN group vs. normal diet group, abscess-drainage in all patients	Decrease of intra-abdominal septic complications at 3 m
Bellolio [34]	P/N-P	TPN (na)	Pre	na	TPN for bowel rest in patients with penetrating disease vs. few TPN in non-penetrating disease	Similar complication rates in both groups, beneficial effect of TPN and bowel rest
Jacobson [35]	O	TPN [52]	Pre	46 days (mean)	Matched cohort of preoperative TPN vs. straight to surgery group	Clinical remission achieved, postoperative complications (30 days), decreased
Zerbib [36]	F/P	EN [4]/TPN (na)	Pre	2 w/3 w	Preop optimized cohort (nutritional support, steroid weaning, abscess drainage, antibiotics)	Low postoperative morbidity (30 days) and stoma rate within a standardized pathway
Griveeva [37]	FS	TPN (na)	Pre	12 days (mean)	PN group (with lower BMI and higher CDAI) vs. non-supplemented control group	Improvement of BMI/CDAI, no difference in outcome
Yao [38]	O	TPN [5]	Peri	3 w	Severely malnourished cohort (BMI <15), TPN group vs. non-supplemented control group	TPN ameliorates immunity, reverses malnutrition (BMI), facilitates recovery
Smedh [39]	F/FS	EEN (na)	Pre	3–6 w	Preoperative optimized cohort (EEN in 50% of patients, steroid weaning, abscess drainage)	Few postop complications (30 days) compared to historical control groups

[1] Modulen IBD (Nestle, Vevey, Switzerland), [2] Peptisorb Liquid, Enteral Nutrition Suspension; Nutricia Company, Amsterdam, the Netherlands, [3] Peptison Liquid, Nutricia Company, (Shanghai, China), [4] elemental diet >30 kcal/kg ideal body weight/day, [5] nitrogen 0.2 g/kg/day, 30 kcal/kg/day, fat 40%, glucose 60%. Abbreviations: P—Penetrating, F—Fistulizing, FS—Fibrous Stenosis, O—Obstructing, R—Refractory Disease, EEN—Exclusive Enteral Nutrition, EN—Enteral Nutrition, PN—Parenteral Nutrition, TPN—Total Parenteral Nutrition, na—not available, w—weeks, m—months, d—days, Preop—Preoperative, Postop—Postoperative, BMI—Body Mass Index, CDAI—Crohn's Disease Activity Index, ROI—reduction of inflammation, IOM—improvement of malnutrition.

3.2. Outcome

Five studies [26,29,32,33,35] showed significantly better results in terms of overall and infectious complications in groups undergoing preoperative nutritional therapy compared to control groups (Table 2). Among these studies, four [26,29,32,33] used EEN formulas and one [35] TPN. In the study of Heerasing et al. [26], 25% of patients could avoid surgery due to EEN induced remission and were bridged to alternative immunosuppressant therapy, but follow-up was limited to one year. Further effects of EEN were a significant decrease in CRP levels, surgical complications (8% vs. 32%) and infectious complications (abscess, collection, or leak, 3% vs. 20%) [4]. Wang et al. [29] showed an effect of EEN on different nutritional parameters (significant improvement of BMI, anemia and CRP levels), significantly lower infectious (21% vs. 44%) and non-infectious (26% vs. 51%) complication rates and less recurrence at six months (7% vs. 26%). Lower incidences of total (19% vs. 29%) and specific infectious complications (wound infection, abscess, and leak) were observed in the study of Li et al. [32] when comparing the steroid-weaned EEN group with the steroid-weaned control group. Further, supplemented patients needed less emergency surgeries compared to the different control groups. Li et al. [33] demonstrated a significant improvement of albumin and CRP levels after EEN therapy and at three months postoperatively, intra-abdominal septic complications were significantly lower (4% vs. 18%). In the cohort of Jacobson [35], patients pre-treated with TPN showed clinical remission and improved nutritional status (albumin, weight) at the time of surgery and no serious early (30 days) postoperative complications were observed in these 15 consecutive patients, contrarily to 28% in the matched control group.

Three studies [28,30,37] could not demonstrate differences in outcome due to unequal nutritional baseline conditions between nutritional and control groups, as detailed in Table 2. Three studies [31,36,39] without control groups compared their results to historical controls of other authors and found overall complication rates of 14% [39] and 18% [31]. However, these two studies were designed to compare nutritional endpoints [31] and anastomotic techniques [39] rather than focusing on impact of nutritional support. The preoperatively optimized cohort of Zerbib et al. [36] with 64% of patients receiving nutritional support (in combination with bowel rest, weaning of steroids, abscess drainage, and antibiotics) presented an overall morbidity of 18% and a low rate of fecal diversion. Another study without a control group by Guo et al. [27] identified EEN of <3 months, preoperative anemia and bacteria in fistula tract as independent risk factors for surgical site infection (31%), while preoperative abscess drainage represented a protective factor. Another study did not compare nutritional regimens [34], but reported TPN in combination with antibiotics, drainage, and postponed surgery in patients with penetrating disease which led to similar complication rates compared to patients with non-perforating disease (13% vs. 11%). The cohort of Yao et al. [38] was severely malnourished and half of patients were supplement by TPN one week before surgery and continued two weeks postoperatively. IgM levels decreased and BMI increased significantly in the study group, while no changes were observed in the control group. No difference was found regarding overall postoperative complications between the two groups (27% each), but a six month follow-up showed that the rate of resuming work was higher in the study group [38].

3.3. Reviews

Fifteen reviews, guidelines or book chapters were retained [4,5,17,19,40–50] (Table 3). Most of them (80%) [5,19,40–44,46–50] were narrative, and only two (13%) [4,17] did perform systematic search to provide official guidelines by the ESPEN society. The more recent recommendations advocate:

-   enteral nutrition always preferred over parenteral nutrition in malnourished patients (weight loss >10–15% within six months, BMI <18.5 kg/m^2, albumin <30 g/L)
-   postpone surgery for 7–14 days if possible
-   parenteral nutrition should be used as supplementary to enteral nutrition if >60% energy needs cannot be met via the enteral route

These recommendations are congruent and 1:1 extrapolated from guidelines on enteral nutrition [53,54].

All reviews are consistent among each other regarding conclusions and agree on the importance of perioperative nutritional support. Further, they provide recommendations in line with current guidelines: if compared, enteral nutrition should be the preferred route of administration. All reviews underline the importance of a multimodal approach (preoperative optimization). Evidence-based recommendations however are scarce, since no solid evidence is available, and all authors agree that more high quality studies are needed to establish solid recommendations. Further, the impact of specific components of nutritional supplements should be studied to provide further evidence-based formulas [5,41].

Table 3. Reviews on preoperative nutritional support in Crohn's disease patients.

Author	Year	Design	Aim/Conclusions
Forbes [4]	2016	Guidelines	64 recommendations to guide nutritional support in IBD patients.
Nguyen [5]	2016	N. Review	Preoperative optimization by enteral and parenteral nutrition mandatory. Timing, route of administration, type, duration debated.
Nickerson [40]	2016	N. Review	Perioperative optimization imperative for favorable postoperative outcome.
Schwartz [41]	2016	N. Review	Evidence in favour of PN, but larger trials needed.
Montgomery [42]	2015	N. Review	Recommendations for nutritional assessment and preoperative optimization.
Horisberger [43]	2015	Book chapter	Preoperative protein supplements (at least one week) beneficial.
Crowell [44]	2015	N. Review	Preoperative optimization (nutritional support, abscess drainage) prevent septic complications and early recurrence.
Spinelli [45]	2014	N. Review	Preoperative optimization crucial for surgical outcome, preoperative enteral nutrition for at least 10–14 days to prefer over TPN.
Triantafillidis [19]	2014	N. Review	Indications for TPN are the same as in every major surgical patient.
Sharma [46]	2013	N. Review	Enteral support (immunonutrition and elemental diet) preferred over TPN.
Iesalnieks [47]	2012	N. Review	Preoperative enteral nutrition might be beneficial, more evidence needed.
Wagner [48]	2011	N. Review	EN preferred, preoperative and postoperative PN remain alternatives. Consider immunonutrition, fish oils, and probiotics.
Efron [49]	2007	N. Review	Perioperative TPN might be beneficial, more high quality studies needed.
Lochs [17]	2006	Guidelines	No specifics for Crohn's patients, perioperative nutrition as in general GI surgery.
Husain [50]	1998	N. Review	Nutrition has a critical benefit in postoperative Crohn's disease.

Abbreviations: N—Narrative, Postop—postoperative, PN—Parenteral Nutrition, TPN—Total Parenteral Nutrition, EN—Enteral Nutrition, IBD—Inflammatory Bowel Disease, GI—gastrointestinal.

4. Discussion

This systematic review scrutinized available evidence over the last 20 years to provide evidence-based guidelines for perioperative nutritional support in patients suffering from Crohn's disease. Fourteen original studies evaluated nutritional support in mostly severely ill patients, and a large heterogeneity was observed among studies regarding type, formula, and timing of nutrition. Only few prospective studies were available, and a randomized controlled study comparing different nutritional strategies was not to date. Hence, comparison between studies is delicate, and conclusions should be drawn cautiously. Even though nutritional support strategies were different, all studies presented encouraging results and emphasized the importance of nutritional support within a multimodal preoperative optimization concept. Some general principles in patients suffering from Crohn's disease must be discussed, including particularities of Crohn's patients, screening modalities, and current guidelines, which are discussed and compared to the evidence provided by this systematic review.

4.1. Particularities in Surgery for Crohn's Disease

Patients suffering from Crohn's disease are a particular subset of patients in many ways. At the time of surgery, most patients are treated by immunomodulating drugs, present with intra-abdominal infections, and are anemic and malnourished [55]. In a recent meta-analysis, steroid use, low albumin level, preoperative surgical history, and preoperative abscess were retained as risk factors for adverse surgical outcome [56]. Besides steroids and thiopurines, biologics such as anti-TNF provide new treatment options for disease control [57]. However, the influence of these drugs on postoperative outcome is matter of debate. While Fumery et al. described an increased risk of complications [15], a recent meta-analysis did not find any association between immunomodulating therapy and postoperative outcome [58]. Malnutrition on the other hand is common (up to 85%) among Crohn's patients awaiting surgery and is a well-known risk factor for adverse postoperative outcome in surgical patients in general [59]. For Crohn's patients needing surgery, anastomotic dehiscence, intraabdominal abscess, and fistula, regrouped as intraabdominal septic complications, represent most feared complications [60,61]. Intraabdominal septic complications hinder the postoperative course in up to 20% of patients with potentially severe consequences [15,62] and either reoperation or percutaneous drainage is needed in most cases. Hence, efforts to improve modifiable risk factors before surgery are of utmost importance.

4.2. Guidelines for Perioperative Nutrition and Preoperative Optimization

By the time of official ESPEN guidelines publication in 2006 [17], specific data on the effect of perioperative nutrition was lacking. Considerable evidence on nutritional support in general gastrointestinal surgery and in critically ill patients led by extrapolation to the recommendation to treat Crohn's patients accordingly [17]. This message was reinforced 10 years later by revised guidelines [4]. Hence, the ESPEN guidelines on enteral nutrition for surgery [53], published in 2006 and 2017 [54], do apply for Crohn's patients if they tolerate nutritional supplements to meet their metabolic needs. Most recommendations concerning enteral nutrition were elaborated on firm evidence and are hence highly recommended [53,54]. Whenever possible, the route of administration should be enteral, which is also advocated for patients with Crohn's disease [17]. This was previously emphasized by a review on nutritional support strategies in Crohn's disease [63]. Guidelines for parenteral nutrition [64] in the perioperative phase do likewise apply for Crohn's patients if metabolic needs are not met by enteral nutrition alone or if disease presentation at the time of scheduled surgery impedes enteral nutrition (e.g., intestinal obstruction or high output fistula). Nutritional guidelines merge with enhanced recovery after surgery (ERAS) guidelines [65], which are beneficial for surgical patients regarding outcome, length of stay, and costs. Recent reports suggested that enhanced recovery combined with minimally invasive techniques may lead to further improvements in surgical outcomes of Crohn's patients [66,67]. Whenever possible, elective surgical patients should be treated according to the ERAS protocol: avoidance of long term fasting, integration of nutritional strategies into the overall management of the patient, metabolic homeostasis, and early mobilization [4].

Nutritional strategies need to be part of a concept called preoperative optimization, including weaning of steroids if possible, drainage of percutaneous abscesses if applicable, and intravenous antibiotics if indicated [68]. Several of the studies retained for the present analysis presented promising data within such a multimodal approach [27,36,39,40,42]. Thus, surgery has to be delayed if possible in order to ensure best conditions. In case of emergency, EN or PN should start postoperatively [4].

4.3. Nutritional Screening

Several original studies [30,37,38] retained for the present analysis reported on nutritional screening tools or markers to provide nutritional support, especially by emphasizing the importance of body mass index (BMI). They all identified BMI as a follow-up tool of nutritional status during parenteral nutritional therapy. Recent guidelines advocate BMI <18.5 kg/m^2, weight loss >10–15% within six months and serum albumin <30 g/L as best reflectors of severe undernutrition in Crohn's disease [4]. Concerning screening, ESPEN guidelines for Inflammatory Bowel Disease [4] recommend that Crohn's patients should be screened for malnutrition as patients undergoing general surgery [53,54] through validated screening tools. Particularly recommended are the Nutritional Risk Score (NRS) [69] and the Malnutrition Universal Screening Tool (MUST) [70]. Patients with a NRS ≥3 are considered to be at risk for gastrointestinal surgery [71].

4.4. Further Nutritional Strategies for Perioperative Support in Crohn's Patients

Two concepts need special consideration as result of this systematic review: EEN and TPN. Interestingly, most included studies reported on either EEN or TPN or a combination of both. EEN, either in elemental or polymeric form, has a direct anti-inflammatory effect [72], promotes mucosal healing [73], modifies intestinal microflora [74], and might decrease the antigenic load through bowel rest. EEN can induce clinical remission in pediatric and adult patients [75,76], as observed by Heerasing et al. [26] in 25% of patients awaiting surgery. In a recent review analyzing EEN in non-surgical Crohn's patients [77], EEN has been associated with remission rates of up to 80%. Wang et al. [29] observed decreased recurrence rates at six months in the EEN group, however, clinical recurrence was similar two years after surgery in both groups. Li et al. [32] and Smedh et al. [39] further presented interesting data on EEN allowing subsequent steroid-weaning, contributing to lower complication rates in EEN-groups in these studies. Disease presentations were severe in all studies with EEN [26,28,32,33,39] but one, [29], and might thus be particularly useful in this context (Table 2).

TPN was mainly used for penetrating disease in the study of Bellolio et al. [34], and Yao et al. [38] and Jacobson [35] treated patients with obstructing disease to observe improved immunity and clinical remission. Concerning formulas and timing, data was heterogeneous. Hence, no solid conclusions can be drawn. As a consequence, guidelines on parenteral nutrition [64] should be used for guidance. Schwartz [41] emphasized the need for larger prospective trials to strengthen the evidence. With this respect and due to lacking data, parenteral nutrition should be reserved for patients who are unable to cover their energetic needs by enteral nutrition.

Further considerations regarding routes of administration and associated potential complications have been published before [63].

4.5. Particularities in Perioperative Nutrition for Crohn's Disease

Despite the particularities of Crohn's disease and potential clinical discrepancies with the general surgical population including disease flares at time of surgery, exhaustive immunomodulating and medical treatment, and unfavorable baseline conditions, guidelines on enteral and parenteral nutrition including screening modalities, nutritional support strategies, and nutritional follow-up can be extrapolated to Crohn's patients. However, severe malnutrition in high-risk patients or inability to cover energy needs in patients with obstructing or fistulizing disease might impede conventional nutritional support (including oral nutritional supplements and immunonutrition) [4]. In these circumstances, specific nutritional support strategies including EEN or TPN have to be discussed. The following algorithm gives an overview on treatment suggestions considering available guidelines and the evidence of this systematic review (Figure 2).

Several limitations of the present study need to be mentioned. Due to heterogeneity of data and modest study quality of original studies regarding nutritional treatment strategies, solid conclusions cannot be drawn, and further high-quality evidence will be needed. The suggested treatment algorithm (Figure 2) should thus rather help in decision-making than provide formal recommendations.

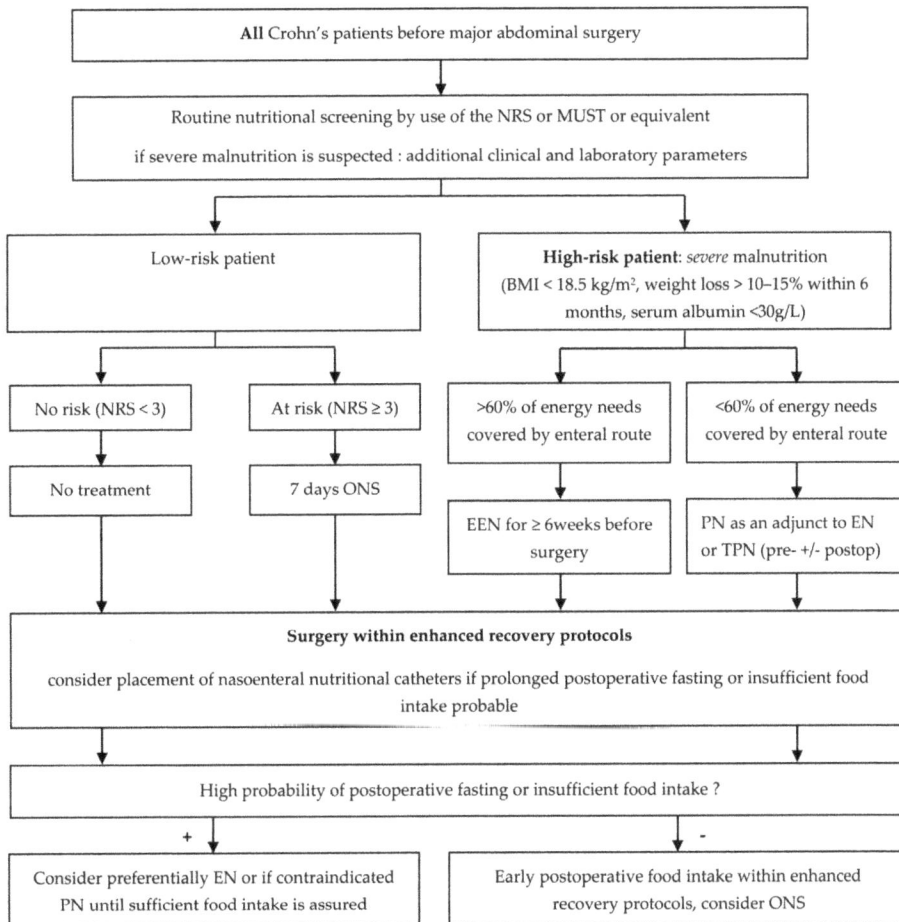

Figure 2. Nutritional treatment algorithm for preoperative nutritional screening and perioperative nutrition in digestive surgery in Crohn's patients. Abbreviations: NRS—Nutritional Risk Score, MUST—Malnutrition Universal Screening Tool; EEN—exclusive enteral nutrition; PN—parenteral nutrition; EN—enteral nutrition; TPN—total parenteral nutrition; preop—preoperative; postop—postoperative.

5. Conclusions

Perioperative nutrition in Crohn's patients awaiting surgery should be considered as a mandatory adjunct within preoperative optimization strategies. Guidelines including enteral nutrition and perioperative care for general surgery do also apply to Crohn's patients. Encouraging data for exclusive enteral or total parenteral nutrition, especially regarding induction of surgery-preventing disease remission and decreased recurrence, call for further high-quality studies.

Acknowledgments: No funding sources to declare.

Author Contributions: F.G. and B.P. independently performed the literature search. F.G., B.P., and M.H. conceived and designed the study and made the final decision on inclusion of a study; F.G., B.P., and D.M. analyzed the data; F.G., B.P., M.H., D.M., D.H., and N.D. wrote the paper; All authors approved the final manuscript. F.G. and B.P. share first authorship.

Conflicts of Interest: The authors declare no conflict of interest.

References

1. Hegazi, R.A.; Hustead, D.S.; Evans, D.C. Preoperative standard oral nutrition supplements vs immunonutrition: Results of a systematic review and meta-analysis. *J. Am. Coll. Surg.* **2014**, *219*, 1078–1087. [CrossRef] [PubMed]
2. Mazaki, T.; Ishii, Y.; Murai, I. Immunoenhancing enteral and parenteral nutrition for gastrointestinal surgery: A multiple-treatments meta-analysis. *Ann. Surg.* **2015**, *261*, 662–669. [CrossRef] [PubMed]
3. Goh, J.; O'Morain, C.A. Review article: Nutrition and adult inflammatory bowel disease. *Aliment. Pharmacol. Ther.* **2003**, *17*, 307–320. [CrossRef] [PubMed]
4. Forbes, A.; Escher, J.; Hebuterne, X.; Klek, S.; Krznaric, Z.; Schneider, S.; Shamir, R.; Stardelova, K.; Wierdsma, N.; Wiskin, A.E.; et al. ESPEN guideline: Clinical nutrition in inflammatory bowel disease. *Clin. Nutr.* **2016**, *36*, 321–347. [CrossRef] [PubMed]
5. Nguyen, D.L.; Limketkai, B.; Medici, V.; Saire Mendoza, M.; Palmer, L.; Bechtold, M. Nutritional Strategies in the Management of Adult Patients with Inflammatory Bowel Disease: Dietary Considerations from Active Disease to Disease Remission. *Curr. Gastroenterol. Rep.* **2016**, *18*, 55. [CrossRef] [PubMed]
6. Massironi, S.; Rossi, R.E.; Cavalcoli, F.A.; Della Valle, S.; Fraquelli, M.; Conte, D. Nutritional deficiencies in inflammatory bowel disease: Therapeutic approaches. *Clin. Nutr.* **2013**, *32*, 904–910. [CrossRef] [PubMed]
7. Bednarz, W.; Czopnik, P.; Wojtczak, B.; Olewinski, R.; Domoslawski, P.; Spodzieja, J. Analysis of results of surgical treatment in Crohn's disease. *Hepatogastroenterology* **2008**, *55*, 998–1001. [PubMed]
8. Frolkis, A.D.; Dykeman, J.; Negron, M.E.; Debruyn, J.; Jette, N.; Fiest, K.M.; Frolkis, T.; Barkema, H.W.; Rioux, K.P.; Panaccione, R.; et al. Risk of surgery for inflammatory bowel diseases has decreased over time: A systematic review and meta-analysis of population-based studies. *Gastroenterology* **2013**, *145*, 996–1006. [CrossRef] [PubMed]
9. Toh, J.W.; Stewart, P.; Rickard, M.J.; Leong, R.; Wang, N.; Young, C.J. Indications and surgical options for small bowel, large bowel and perianal Crohn's disease. *World J. Gastroenterol.* **2016**, *22*, 8892–8904. [CrossRef] [PubMed]
10. Maguire, L.H.; Alavi, K.; Sudan, R.; Wise, P.E.; Kaiser, A.M.; Bordeianou, L. Surgical Considerations in the Treatment of Small Bowel Crohn's Disease. *J. Gastrointest. Surg.* **2017**, *21*, 398–411. [CrossRef] [PubMed]
11. Gionchetti, P.; Dignass, A.; Danese, S.; Magro Dias, F.J.; Rogler, G.; Lakatos, P.L.; Adamina, M.; Ardizzone, S.; Buskens, C.J.; Sebastian, S.; et al. 3rd European Evidence-based Consensus on the Diagnosis and Management of Crohn's Disease 2016: Part 2: Surgical Management and Special Situations. *J. Crohn Colitis* **2017**, *11*, 135–149. [CrossRef] [PubMed]
12. Bailey, E.H.; Glasgow, S.C. Challenges in the Medical and Surgical Management of Chronic Inflammatory Bowel Disease. *Surg. Clin. N. Am.* **2015**, *95*, 1233–1244. [CrossRef] [PubMed]
13. Colombel, J.F.; Reinisch, W.; Mantzaris, G.J.; Kornbluth, A.; Rutgeerts, P.; Tang, K.L.; Oortwijn, A.; Bevelander, G.S.; Cornillie, F.J.; Sandborn, W.J. Randomised clinical trial: Deep remission in biologic and immunomodulator naive patients with Crohn's disease—A SONIC post hoc analysis. *Aliment. Pharmacol. Ther.* **2015**, *41*, 734–746. [CrossRef] [PubMed]
14. Jones, D.W.; Finlayson, S.R. Trends in surgery for Crohn's disease in the era of infliximab. *Ann. Surg.* **2010**, *252*, 307–312. [CrossRef] [PubMed]
15. Fumery, M.; Seksik, P.; Auzolle, C.; Munoz-Bongrand, N.; Gornet, J.M.; Boschetti, G.; Cotte, E.; Buisson, A.; Dubois, A.; Pariente, B.; et al. Postoperative Complications after Ileocecal Resection in Crohn's Disease: A Prospective Study From the REMIND Group. *Am. J. Gastroenterol.* **2017**, *112*, 337–345. [CrossRef] [PubMed]
16. El-Hussuna, A.; Iesalnieks, I.; Horesh, N.; Hadi, S.; Dreznik, Y.; Zmora, O. The effect of pre-operative optimization on post-operative outcome in Crohn's disease resections. *Int. J. Colorectal. Dis.* **2017**, *32*, 49–56. [CrossRef] [PubMed]
17. Lochs, H.; Dejong, C.; Hammarqvist, F.; Hebuterne, X.; Leon-Sanz, M.; Schutz, T.; van Gemert, W.; van Gossum, A.; Valentini, L.; DGEM (German Society for Nutritional Medicine); et al. ESPEN Guidelines on Enteral Nutrition: Gastroenterology. *Clin. Nutr.* **2006**, *25*, 260–274. [CrossRef] [PubMed]
18. Moher, D.; Liberati, A.; Tetzlaff, J.; Altman, D.G.; Group, P. Preferred reporting items for systematic reviews and meta-analyses: The PRISMA statement. *BMJ* **2009**, *339*, b2535. [CrossRef] [PubMed]
19. Triantafillidis, J.K.; Papalois, A.E. The role of total parenteral nutrition in inflammatory bowel disease: Current aspects. *Scand. J. Gastroenterol.* **2014**, *49*, 3–14. [CrossRef] [PubMed]

20. Zhang, T.; Cao, L.; Cao, T.; Yang, J.; Gong, J.; Zhu, W.; Li, N.; Li, J. Prevalence of Sarcopenia and Its Impact on Postoperative Outcome in Patients With Crohn's Disease Undergoing Bowel Resection. *JPEN J. Parenter. Enter. Nutr.* **2015**, *41*, 592–600. [CrossRef] [PubMed]

21. Bautista, M.C.; Otterson, M.F.; Zadvornova, Y.; Naik, A.S.; Stein, D.J.; Venu, N.; Perera, L.P. Surgical outcomes in the elderly with inflammatory bowel disease are similar to those in the younger population. *Dig. Dis. Sci.* **2013**, *58*, 2955–2962. [CrossRef] [PubMed]

22. Maeda, K.; Nagahara, H.; Shibutani, M.; Otani, H.; Sakurai, K.; Toyokawa, T.; Tanaka, H.; Kubo, N.; Muguruma, K.; Kamata, N.; et al. A preoperative low nutritional prognostic index correlates with the incidence of incisional surgical site infections after bowel resection in patients with Crohn's disease. *Surg. Today* **2015**, *45*, 1366–1372. [CrossRef] [PubMed]

23. Yamamoto, T.; Keighley, M.R. Factors affecting the incidence of postoperative septic complications and recurrence after strictureplasty for jejunoileal Crohn's disease. *Am. J. Surg.* **1999**, *178*, 240–245. [CrossRef]

24. Niu, L.Y.; Gong, J.F.; Wei, X.W.; Zhu, W.M.; Li, N.; Li, J.S. Effects of perioperative combined nutritional support in Crohn disease. *Zhonghua Wai Ke Za Zhi* **2009**, *47*, 275–278. [PubMed]

25. Yao, G.X.; Wang, X.R.; Jiang, Z.M.; Zhang, S.Y.; Ma, E.L.; Ni, A.P. The role of nutritional status on serum immunoglobulins, body weight and postoperative infectious-related complications in patients with Crohn's disease receiving perioperative parenteral nutrition. *Zhongguo Yi Xue Ke Xue Yuan Xue Bao* **2002**, *24*, 181–184. [PubMed]

26. Heerasing, N.; Thompson, B.; Hendy, P.; Heap, G.A.; Walker, G.; Bethune, R.; Mansfield, S.; Calvert, C.; Kennedy, N.A.; Ahmad, T.; et al. Exclusive enteral nutrition provides an effective bridge to safer interval elective surgery for adults with Crohn's disease. *Aliment. Pharmacol. Ther.* **2017**, *45*, 660–669. [CrossRef] [PubMed]

27. Guo, K.; Ren, J.; Li, G.; Hu, Q.; Wu, X.; Wang, Z.; Wang, G.; Gu, G.; Ren, H.; Hong, Z.; et al. Risk factors of surgical site infections in patients with Crohn's disease complicated with gastrointestinal fistula. *Int. J. Colorectal. Dis.* **2017**, *32*, 635–643. [CrossRef] [PubMed]

28. Beaupel, N.; Brouquet, A.; Abdalla, S.; Carbonnel, F.; Penna, C.; Benoist, S. Preoperative oral polymeric diet enriched with transforming growth factor-beta 2 (Modulen) could decrease postoperative morbidity after surgery for complicated ileocolonic Crohn's disease. *Scand. J. Gastroenterol.* **2017**, *52*, 5–10. [CrossRef] [PubMed]

29. Wang, H.; Zuo, L.; Zhao, J.; Dong, J.; Li, Y.; Gu, L.; Gong, J.; Liu, Q.; Zhu, W. Impact of Preoperative Exclusive Enteral Nutrition on Postoperative Complications and Recurrence After Bowel Resection in Patients with Active Crohn's Disease. *World J. Surg.* **2016**, *40*, 1993–2000. [CrossRef] [PubMed]

30. Zhang, M.; Gao, X.; Chen, Y.; Zhi, M.; Chen, H.; Tang, J.; Su, M.; Yao, J.; Yang, Q.; Chen, J.; et al. Body Mass Index Is a Marker of Nutrition Preparation Sufficiency Before Surgery for Crohn's Disease From the Perspective of Intra-Abdominal Septic Complications: A Retrospective Cohort Study. *Medicine* **2015**, *94*, e1455. [CrossRef] [PubMed]

31. Zhu, W.; Guo, Z.; Zuo, L.; Gong, J.; Li, Y.; Gu, L.; Cao, L.; Li, N.; Li, J. CONSORT: Different End-Points of Preoperative Nutrition and Outcome of Bowel Resection of Crohn Disease: A Randomized Clinical Trial. *Medicine* **2015**, *94*, e1175. [CrossRef] [PubMed]

32. Li, Y.; Zuo, L.; Zhu, W.; Gong, J.; Zhang, W.; Gu, L.; Guo, Z.; Cao, L.; Li, N.; Li, J. Role of exclusive enteral nutrition in the preoperative optimization of patients with Crohn's disease following immunosuppressive therapy. *Medicine* **2015**, *94*, e478. [CrossRef] [PubMed]

33. Li, G.; Ren, J.; Wang, G.; Hu, D.; Gu, G.; Liu, S.; Ren, H.; Wu, X.; Li, J. Preoperative exclusive enteral nutrition reduces the postoperative septic complications of fistulizing Crohn's disease. *Eur. J. Clin. Nutr.* **2014**, *68*, 441–446. [CrossRef] [PubMed]

34. Bellolio, F.; Cohen, Z.; Macrae, H.M.; O'Connor, B.I.; Huang, H.; Victor, J.C.; McLeod, R.S. Outcomes following surgery for perforating Crohn's disease. *Br. J. Surg.* **2013**, *100*, 1344–1348. [CrossRef] [PubMed]

35. Jacobson, S. Early postoperative complications in patients with Crohn's disease given and not given preoperative total parenteral nutrition. *Scand. J. Gastroenterol.* **2012**, *47*, 170–177. [CrossRef] [PubMed]

36. Zerbib, P.; Koriche, D.; Truant, S.; Bouras, A.F.; Vernier-Massouille, G.; Seguy, D.; Pruvot, F.R.; Cortot, A.; Colombel, J.F. Pre-operative management is associated with low rate of post-operative morbidity in penetrating Crohn's disease. *Aliment. Pharmacol. Ther.* **2010**, *32*, 459–465. [CrossRef] [PubMed]

37. Grivceva Stardelova, K.; Misevska, P.; Zdravkovska, M.; Trajkov, D.; Serafimoski, V. Total parenteral nutrition in treatment of patients with inflammatory bowel disease. *Prilozi* **2008**, *29*, 21–43. [PubMed]

38. Yao, G.X.; Wang, X.R.; Jiang, Z.M.; Zhang, S.Y.; Ni, A.P. Role of perioperative parenteral nutrition in severely malnourished patients with Crohn's disease. *World J. Gastroenterol.* **2005**, *11*, 5732–5734. [CrossRef] [PubMed]

39. Smedh, K.; Andersson, M.; Johansson, H.; Hagberg, T. Preoperative management is more important than choice of sutured or stapled anastomosis in Crohn's disease. *Eur. J. Surg.* **2002**, *168*, 154–157. [CrossRef] [PubMed]

40. Nickerson, T.P.; Merchea, A. Perioperative Considerations in Crohn Disease and Ulcerative Colitis. *Clin. Colon. Rectal. Surg.* **2016**, *29*, 80–84. [CrossRef] [PubMed]

41. Schwartz, E. Perioperative Parenteral Nutrition in Adults With Inflammatory Bowel Disease: A Review of the Literature. *Nutr. Clin. Pract.* **2016**, *31*, 159–170. [CrossRef] [PubMed]

42. Montgomery, S.C.; Williams, C.M.; Maxwell, P.J.T. Nutritional Support of Patient with Inflammatory Bowel Disease. *Surg. Clin. N. Am.* **2015**, *95*, 1271–1279. [CrossRef] [PubMed]

43. Horisberger, K.; Kienle, P. Surgery in Crohn's disease. *Chirurg* **2015**, *86*, 1083–1094. [CrossRef] [PubMed]

44. Crowell, K.T.; Messaris, E. Risk factors and implications of anastomotic complications after surgery for Crohn's disease. *World J. Gastrointest. Surg.* **2015**, *7*, 237–242. [CrossRef] [PubMed]

45. Spinelli, A.; Allocca, M.; Jovani, M.; Danese, S. Review article: Optimal preparation for surgery in Crohn's disease. *Aliment. Pharmacol. Ther.* **2014**, *40*, 1009–1022. [CrossRef] [PubMed]

46. Sharma, A.; Chinn, B.T. Preoperative optimization of crohn disease. *Clin. Colon. Rectal. Surg.* **2013**, *26*, 75–79. [CrossRef] [PubMed]

47. Iesalnieks, I.; Dederichs, F.; Kilger, A.; Schlitt, H.J.; Agha, A. Postoperative morbidity after bowel resections in patients with Crohn's disease: Risk, management strategies, prevention. *Z. Gastroenterol.* **2012**, *50*, 595–600. [PubMed]

48. Wagner, I.J.; Rombeau, J.L. Nutritional support of surgical patients with inflammatory bowel disease. *Surg. Clin. N. Am.* **2011**, *91*, 787–803. [CrossRef] [PubMed]

49. Efron, J.E.; Young-Fadok, T.M. Preoperative optimization of Crohn's disease. *Clin. Colon. Rectal Surg.* **2007**, *20*, 303–308. [CrossRef] [PubMed]

50. Husain, A.; Korzenik, J.R. Nutritional issues and therapy in inflammatory bowel disease. *Semin. Gastrointest. Dis.* **1998**, *9*, 21–30. [PubMed]

51. LeCouteur, J.; Patel, K.; O'Sullivan, S.; Ferreira, C.; Williams, A.B.; Darakhshan, A.A.; Irving, P.M.; Sanderson, J.; McCarthy, M.; Dunn, J.; et al. Preoperative parenteral nutrition in crohn's disease patients requiring abdominal surgery. *Gut* **2015**, *64* (Suppl. 1). [CrossRef]

52. Jacobson, S.; Wester, P.O. Balance study of twenty trace elements during total parenteral nutrition in man. *Br. J. Nutr.* **1977**, *37*, 107–126. [CrossRef] [PubMed]

53. Weimann, A.; Braga, M.; Harsanyi, L.; Laviano, A.; Ljungqvist, O.; Soeters, P.; DGEM (German Society for Nutritional Medicine); Jauch, K.W.; Kemen, M.; Hiesmayr, J.M.; et al. ESPEN Guidelines on Enteral Nutrition: Surgery including organ transplantation. *Clin. Nutr.* **2006**, *25*, 224–244. [CrossRef] [PubMed]

54. Weimann, A.; Braga, M.; Carli, F.; Higashiguchi, T.; Hubner, M.; Klek, S.; Laviano, A.; Ljungqvist, O.; Lobo, D.N.; Martindale, R.; et al. ESPEN guideline: Clinical nutrition in surgery. *Clin. Nutr.* **2017**, *36*, 623–650. [CrossRef] [PubMed]

55. Vatn, M.H.; Sandvik, A.K. Inflammatory bowel disease. *Scand. J. Gastroenterol.* **2015**, *50*, 748–762. [CrossRef] [PubMed]

56. Huang, W.; Tang, Y.; Nong, L.; Sun, Y. Risk factors for postoperative intra-abdominal septic complications after surgery in Crohn's disease: A meta-analysis of observational studies. *J. Crohn Colitis* **2015**, *9*, 293–301. [CrossRef] [PubMed]

57. Abraham, B.P.; Ahmed, T.; Ali, T. Inflammatory Bowel Disease: Pathophysiology and Current Therapeutic Approaches. *Handb. Exp. Pharmacol.* **2017**, *239*, 115–146. [PubMed]

58. Ahmed Ali, U.; Martin, S.T.; Rao, A.D.; Kiran, R.P. Impact of preoperative immunosuppressive agents on postoperative outcomes in Crohn's disease. *Dis. Colon Rectum* **2014**, *57*, 663–674. [PubMed]

59. Kondrup, J.; Allison, S.P.; Elia, M.; Vellas, B.; Plauth, M. Educational, Clinical Practice Committee ESoP, Enteral N: ESPEN guidelines for nutrition screening 2002. *Clin. Nutr.* **2003**, *22*, 415–421. [CrossRef]

60. Yamamoto, T.; Allan, R.N.; Keighley, M.R. Risk factors for intra-abdominal sepsis after surgery in Crohn's disease. *Dis. Colon Rectum* **2000**, *43*, 1141–1145. [CrossRef] [PubMed]

61. Yamamoto, T. Positive histological margins are risk factors for intra-abdominal septic complications after ileocolic resection for Crohn's disease? *Dis. Colon Rectum* **2013**, *56*, e50. [CrossRef] [PubMed]

62. Zuo, L.; Li, Y.; Wang, H.; Zhu, W.; Zhang, W.; Gong, J.; Li, N.; Li, J. A Practical Predictive Index for Intra-abdominal Septic Complications After Primary Anastomosis for Crohn's Disease: Change in C-Reactive Protein Level Before Surgery. *Dis. Colon Rectum* **2015**, *58*, 775–781. [CrossRef] [PubMed]

63. Altomare, R.; Damiano, G.; Abruzzo, A.; Palumbo, V.D.; Tomasello, G.; Buscemi, S.; Lo Monte, A.I. Enteral nutrition support to treat malnutrition in inflammatory bowel disease. *Nutrients* **2015**, *7*, 2125–2133. [CrossRef] [PubMed]

64. Braga, M.; Ljungqvist, O.; Soeters, P.; Fearon, K.; Weimann, A.; Bozzetti, F. Espen: ESPEN Guidelines on Parenteral Nutrition: Surgery. *Clin. Nutr.* **2009**, *28*, 378–386. [CrossRef] [PubMed]

65. Gustafsson, U.O.; Scott, M.J.; Schwenk, W.; Demartines, N.; Roulin, D.; Francis, N.; McNaught, C.E.; Macfie, J.; Liberman, A.S.; Soop, M.; et al. Guidelines for perioperative care in elective colonic surgery: Enhanced Recovery After Surgery (ERAS((R))) Society recommendations. *World J. Surg.* **2013**, *37*, 259–284. [CrossRef] [PubMed]

66. Spinelli, A.; Bazzi, P.; Sacchi, M.; Danese, S.; Fiorino, G.; Malesci, A.; Gentilini, L.; Poggioli, G.; Montorsi, M. Short-term outcomes of laparoscopy combined with enhanced recovery pathway after ileocecal resection for Crohn's disease: A case-matched analysis. *J. Gastrointest. Surg.* **2013**, *17*, 126–132. [CrossRef] [PubMed]

67. Gong, J.; Gu, L.; Li, Y.; Cao, L.; Xie, Z.; Guo, D.; Zhang, T.; Yang, J.; Zhu, W.; Li, N.; et al. Outcomes of laparoscopy combined with enhanced recovery pathway for Crohn's disease: A case-matched analysis. *Zhonghua Wei Chang Wai Ke Za Zhi* **2015**, *18*, 16–20. (In Chinese). [PubMed]

68. Patel, K.V.; Darakhshan, A.A.; Griffin, N.; Williams, A.B.; Sanderson, J.D.; Irving, P.M. Patient optimization for surgery relating to Crohn's disease. *Nat. Rev. Gastroenterol. Hepatol.* **2016**, *13*, 707–719. [CrossRef] [PubMed]

69. Kondrup, J.; Rasmussen, H.H.; Hamberg, O.; Stanga, Z. Ad Hoc EWG: Nutritional risk screening (NRS 2002): A new method based on an analysis of controlled clinical trials. *Clin. Nutr.* **2003**, *22*, 321–336. [CrossRef]

70. Sandhu, A.; Mosli, M.; Yan, B.; Wu, T.; Gregor, J.; Chande, N.; Ponich, T.; Beaton, M.; Rahman, A. Self-Screening for Malnutrition Risk in Outpatient Inflammatory Bowel Disease Patients Using the Malnutrition Universal Screening Tool (MUST). *JPEN J. Parenter. Enter. Nutr.* **2016**, *40*, 507–510. [CrossRef] [PubMed]

71. Schiesser, M.; Muller, S.; Kirchhoff, P.; Breitenstein, S.; Schafer, M.; Clavien, P.A. Assessment of a novel screening score for nutritional risk in predicting complications in gastro-intestinal surgery. *Clin. Nutr.* **2008**, *27*, 565–570. [CrossRef] [PubMed]

72. Yamamoto, T.; Nakahigashi, M.; Umegae, S.; Kitagawa, T.; Matsumoto, K. Impact of elemental diet on mucosal inflammation in patients with active Crohn's disease: Cytokine production and endoscopic and histological findings. *Inflamm. Bowel Dis.* **2005**, *11*, 580–588. [CrossRef] [PubMed]

73. Berni Canani, R.; Terrin, G.; Borrelli, O.; Romano, M.T.; Manguso, F.; Coruzzo, A.; D'Armiento, F.; Romeo, E.F.; Cucchiara, S. Short- and long-term therapeutic efficacy of nutritional therapy and corticosteroids in paediatric Crohn's disease. *Dig. Liver. Dis.* **2006**, *38*, 381–387. [CrossRef] [PubMed]

74. Lionetti, P.; Callegari, M.L.; Ferrari, S.; Cavicchi, M.C.; Pozzi, E.; de Martino, M.; Morelli, L. Enteral nutrition and microflora in pediatric Crohn's disease. *JPEN J. Parenter. Enter. Nutr.* **2005**, *29*, S173–S178. [CrossRef] [PubMed]

75. Heuschkel, R.B. Enteral nutrition in children with Crohn's disease. *J. Pediatr. Gastroenterol. Nutr.* **2000**, *31*, 575. [CrossRef] [PubMed]

76. Yamamoto, T.; Nakahigashi, M.; Saniabadi, A.R.; Iwata, T.; Maruyama, Y.; Umegae, S.; Matsumoto, K. Impacts of long-term enteral nutrition on clinical and endoscopic disease activities and mucosal cytokines during remission in patients with Crohn's disease: A prospective study. *Inflamm. Bowel Dis.* **2007**, *13*, 1493–1501. [CrossRef] [PubMed]

77. MacLellan, A.; Moore-Connors, J.; Grant, S.; Cahill, L.; Langille, M.G.I.; Van Limbergen, J. The Impact of Exclusive Enteral Nutrition (EEN) on the Gut Microbiome in Crohn's Disease: A Review. *Nutrients* **2017**, *9*. [CrossRef] [PubMed]

nutrients   MDPI

Article

Associations between Folate and Vitamin B12 Levels and Inflammatory Bowel Disease: A Meta-Analysis

Yun Pan [1], Ya Liu [1], Haizhuo Guo [2], Majid Sakhi Jabir [3], Xuanchen Liu [1], Weiwei Cui [1,*] and Dong Li [4,5,*]

1   Department of Nutrition and Food Hygiene, School of Public Health, Jilin University, 1163 Xinmin Avenue, Changchun 130021, China; yunpan14@mails.jlu.edu.cn (Y.P.); liuya@jlu.edu.cn (Y.L.); liuxc15@mails.jlu.edu.cn (X.L.)
2   Department of Radiology, The Second Part of the First Hospital, Jilin University, Changchun 130031, China; doyouknow_0330@163.com
3   Department of Biotechnology, University of Technology, Baghdad 00964, Iraq; msj_iraq@yahoo.com
4   Department of Immunology, College of Basic Medical Sciences, Jilin University, 126 Xinmin Avenue, Changchun 130021, China
5   Department of Hepatology, The First Hospital, Jilin University, Changchun 130021, China
*   Correspondence: cuiweiwei@jlu.edu.cn (W.C.); lidong1@jlu.edu.cn (D.L.); Tel.: +86-431-8561-9455 (W.C.); +86-431-8561-9476 (D.L.)

Received: 3 February 2017; Accepted: 12 April 2017; Published: 13 April 2017

Abstract: Background: Inflammatory bowel disease (IBD) patients may be at risk of vitamin B12 and folate insufficiencies, as these micronutrients are absorbed in the small intestine, which is affected by IBD. However, a consensus has not been reached on the association between IBD and serum folate and vitamin B12 concentrations. Methods: In this study, a comprehensive search of multiple databases was performed to identify studies focused on the association between IBD and serum folate and vitamin B12 concentrations. Studies that compared serum folate and vitamin B12 concentrations between IBD and control patients were selected for inclusion in the meta-analysis. Results: The main outcome was the mean difference in serum folate and vitamin B12 concentrations between IBD and control patients. Our findings indicated that the average serum folate concentration in IBD patients was significantly lower than that in control patients, whereas the mean serum vitamin B12 concentration did not differ between IBD patients and controls. In addition, the average serum folate concentration in patients with ulcerative colitis (UC) but not Crohn's disease (CD) was significantly lower than that in controls. This meta-analysis identified a significant relationship between low serum folate concentration and IBD. Conclusions: Our findings suggest IBD may be linked with folate deficiency, although the results do not indicate causation. Thus, providing supplements of folate and vitamin B12 to IBD patients may improve their nutritional status and prevent other diseases.

Keywords: folate; vitamin B12; inflammatory bowel disease; meta-analysis; nutrition

1. Introduction

Inflammatory bowel disease (IBD) is characterized by chronic and typically recurrent intestinal inflammation, and it includes Crohn's disease (CD) and ulcerative colitis (UC). Although the exact aetiology and pathogenesis of IBD is still largely unknown, it is considered to be related to individual immunity, an inherited predisposition, environmental factors and the interactions between the mucosal immune system and intestinal antigenic material (e.g., commensal bacteria) [1]. Abnormal immune responses are believed to be the direct cause of intestinal damage [2]. IBD can lead to many clinical symptoms, including impaired nutrient absorption, which can influence the absorption of folate and vitamin B12. Furthermore, many studies have indicated that serum folate and vitamin B12 concentrations influence the development of IBD. Folate is involved in the methylation of DNA and may produce epigenetic changes that affect the interaction between the gut microbiota and systemic immune responses [3]. The gut microbiota [4] and epigenetic changes [5] may be involved in the pathogenesis of IBD. Vitamin B12 acts as a coenzyme in various biochemical reactions, including DNA synthesis and folate metabolism [6]. Deficiencies in vitamin B12 and folate can lead to macrocytic anaemia, hyperhomocysteinemia, and neurologic and psychiatric disorders [7–9]. Compared with healthy subjects, IBD patients are at increased risk of hyperhomocysteinemia [10], and folic acid and vitamin B12 may play pivotal roles in homocysteine metabolic reactions [11]. Moreover, folate and vitamin B12 deficiencies may cause increased homocysteine levels, a risk factor for thrombosis [8,12–14].

Many studies have reported that serum folic and vitamin B12 concentrations differ between IBD patients and healthy individuals. However, the results are not consistent, and differences have been observed between patients with CD and UC. Whether serum folic acid and vitamin B12 concentrations are lower in IBD patients than in non-IBD patients is still largely unknown. Thus, a more comprehensive evaluation of the association between IBD and serum folic acid and vitamin B12 concentrations is needed. In this study, we conducted a meta-analysis to analyse the relationships between the serum concentrations of folic acid and vitamin B12 in IBD patients and healthy controls to provide additional insights into treating and rehabilitating IBD and maintaining a healthy nutritional status.

2. Materials and Methods

2.1. Sources and Methods of Data Retrieval

We performed a comprehensive literature search that included studies from 1970 to December 2016; the electronic databases included PubMed, Medline, Web of Science, and Google Scholar. The searches were conducted to identify all published studies that reported data on the mean differences and standard deviations of serum folate and vitamin B12 concentrations in IBD patients and healthy controls. The following terms were used for the literature search: folic acid, vitamin B9, vitamin M, folvite, folate, vitamin B12, cyanocobalamin, cobalamins, inflammatory bowel disease, Crohn's disease, ulcerative colitis. The term 'OR' was used as the set operator to combine different sets of results. The serum concentrations of folate and vitamin B12 and IBD were determined and used in a meta-analysis to understand how serum folate and vitamin B12 concentrations differ in IBD patients relative to healthy controls. Age, location, detection methods and other confounding factors were also considered.

2.2. Inclusion Criteria

The articles that were included in this meta-analysis matched the following five criteria: (1) inflammatory bowel disease patients were clinically diagnosed; (2) studies included a case group and a control group; (3) the folate and vitamin B12 values were presented as the mean ± standard deviation (SD); and (4) the patients and controls had not previously received folate and vitamin B12 supplementation; and (5) we excluded studies that did not provide initial data, animal studies, in vitro studies, reviews and conference papers. Three investigators independently reviewed and extracted all of the potentially eligible studies and discussed the inconsistencies until a consensus was reached (Figure 1). Additionally, the Newcastle-Ottawa Scale was used for assessing the quality of studies included in this meta-analysis (Table 1).

2.3. Data Abstraction

We reviewed all of the relevant studies and extracted the following data: (1) lead author, publication year, sample size, mean age of the patients and controls, and gender of the patients and controls; (2) serum folate and vitamin B12 concentrations of the patients and controls; (3) folate and vitamin B12 detection methods; and (4) the diagnosis of the patients and the number of patients and controls.

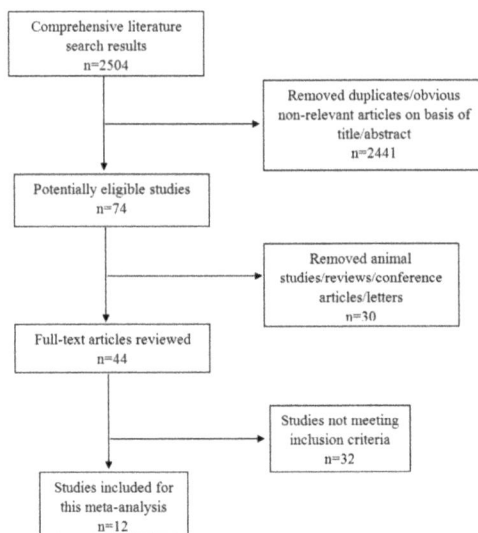

Figure 1. Flow diagram of the literature search.

2.4. Statistical Analysis

All statistical analysis was conducted using the statistical software Stata (version 12.0, StataCorp LLC, College Station, TX, USA). The mean difference, standard deviation and standard error of the serum folate and vitamin B12 concentrations in the IBD and control group were used for the meta-analysis. Units that were not unified were transformed into unified units. We combined the standardized mean difference (SMD) for studies that reported mean and standard deviation values for serum folate and vitamin B12 concentrations in IBD patients and controls. An inverse variance weighted random effect model was used to determine the SMD and 95% confidence intervals (CIs) and measure the different concentrations of folate and vitamin B12 in the patients and controls, and the results were used to evaluate the differences in serum concentrations of folate and vitamin B12 between the IBD patients and normal controls. In order to avoid double counting, both controls in two studies that included both UC and CD patients [15,16] were split approximately evenly into 2 control groups with the means and standard deviations left unchanged before entered into the meta-analysis [17,18].

We used Cochran's Q statistic and the I^2 statistic to assess the statistical heterogeneity in the meta-analysis [19]. If the data were homogeneous ($p > 0.05$), a fixed effect model meta-analysis was performed; if the data were heterogeneous ($p \leq 0.05$), a random effects model meta-analysis was performed. Heterogeneity was considered significant at $p < 0.05$ in the Q test, and the I^2 value was used to evaluate the degree of heterogeneity. We defined low, medium and high heterogeneity at I^2 values of 25%, 50%, and 75%, respectively [20]. Sensitivity analysis was used for analyzing heterogeneity. Subgroup analyses were performed for the type of disease, region of study, and method of detecting the vitamins. In addition, we performed a meta-regression analysis based on the folate and vitamin B12 measurement methods, the year of publication, the sample size, the quality of the study and the average age of the patients. We used a funnel plot to detect publication bias concerning this meta-analysis, with the symmetry of the funnel plot used to determine whether publication bias occurred. Furthermore, a formal statistical assessment of the funnel plot asymmetry was performed with Egger's regression asymmetry test [21].

Table 1. Studies showing the serum folate and vitamin B12 concentrations in IBD patients and controls.

Author	Region	Score	Year	n IBD/CD	n UC	Assay Method	Age IBD/CD/UC/C	Gender (M/F) IBD/CD/UC	Gender (M/F) C	IBD/CD/UC FA (ng/mL)	IBD/CD/UC B12 (pg/mL)	Controls FA (ng/mL)	Controls B12 (pg/mL)	p FA	p B12
Jiang et al. [22]	China	8	2010	252	654	chemiluminescence	-/-/45 ± 14/46 ± 17	147/105	279/374	4.97 ± 2.73*	437.53 ± 174.12*	6.74 ± 3.41	572.77 ± 175.33	<0.001	<0.001
Lambert et al. [23]	France	6	1996	21	20	radioimmunoassay	-	11/10	8/12	2.8 ± 2.8#	190 ± 146#	4.55 ± 3.26	207 ± 75	NS	NS
Kuroki et al. [24]	Japan	8	1993	24	24	radioimmunoassay	-/26.4 ± 12.4/-/27.0 ± 4.2	17/7	14/10	-	641 ± 186#	-	682 ± 488	<0.001	<0.001
Fernández-Miranda et al. [25]	Spain	7	2005	52	186	radioimmunoassay	41.7 ± 11.9/-/-/ 41.9 ± 10.1	23/29	71/115	7.6 ± 4.1	499 ± 287	8.9 ± 3.7	603 ± 231	<0.05	<0.05
Erzin et al. [26]	Turkey	7	2008	105	85	ELISA	38.69 ± 12.13/-/-/ 37.61 ± 10.05	-	-	3.72 ± 1.44	600.14 ± 145.30	4.96 ± 1.19	700.32 ± 141.58	<0.001	<0.001
Alkhouri et al. [27]	U.S.	8	2013	61	61	-	12.3 ± 3.9/12.1 ± 4.1/12.3 ± 3.5/12.1 ± 3.6	40/21	30/31	20.1 ± 6.5 20.3 ± 7# 20.9 ± 5.9*	775 ± 441 781 ± 372# 906 ± 669*	20.4 ± 5.5	727 ± 346		
Yakut et al. [15]	Turkey	6	2010	138	53	chemiluminescence	-	64/74	19/34	7.7 ± 5.3# 8.6 ± 8.3*	281 ± 166# 348 ± 218*	9.9 ± 3.3	342 ± 179	NS	NS
Akbulut et al. [28]	Turkey	9	2010	55	45	chemiluminescence	-/-/47.4 ± 13.80/ 46.4 ± 13.89	38/17	31/14	5.1 ± 2.19*	250.4 ± 82.49*	6.3 ± 0.87	327 ± 73.9	<0.001	<0.001
Kallel et al. [29]	Tunisia	6	2011	89	103	specific immunochemical methods	35.3 ± 12.6/-/-/ 36.5 ± 9.26	47/42	50/53	8.54 ± 3.04	295 ± 180	8.1 ± 3.11	378 ± 170	NS	<0.001
Geerting et al. [30]	Netherlands	7	2000	69	69	radioimmunoassay	-/30.1 ± 10.2/37.8 ± 14.7/35.4 ± 13.7	33/36	23/46	4.72 ± 4.02* 5.03 ± 3.80*	304.96 ± 82.27* 364.59 ± 119.14#	5.47 ± 2.47* 5.78 ± 2.96#	357.82 ± 121.58* 365.95 ± 119.54#	NS	0.05
Chen et al. [31]	China	6	2011	112	110	ELISA	-/39.4 ± 11.7/ 40.3 ± 10.8	58/54	56/54	3.37 ± 0.86*	147.25 ± 43.67*	4.03 ± 0.54	152.67 ± 45.17	0.005	0.004
Koutroubakis et al. [16]	Greece	6	2000	108	74	IMx assay	-	66/42	-	6.34 ± 3.05# 7.02 ± 3.13*	666.5 ± 366.8# 478.8 ± 257.7*	8.78 ± 3.07	377.5 ± 155.6	<0.05	<0.05

* Folate and Vitamin B12 levels reported in Crohn's disease. # Folate and Vitamin B12 levels reported in ulcerative colitis. - Not applicable. P: patients vs. controls; IBD: Inflammatory Bowel Disease; CD: Crohn's Disease; UC: Ulcerative Colitis; C: Controls; FA: Folate Acid; B12: Vitamin B12. SD: standard deviation. IMx assay: a multi-functional immune assay system developed by Abbott Diagnostics, Abbott Park, IL.

3. Results

Our search identified 2504 related references; however, only 12 papers met our inclusion criteria. The 12 studies included 2570 individuals in total, with 1086 IBD patients and 1484 controls [15,16,22–31]. The detailed results are expressed in Table 1. Among the 12 studies, the folate concentration in the study by Kuroki et al. [24] was excluded because it was detected in red blood cells. Folate and vitamin B12 were detected via chemiluminescence immunoassays in three studies [15,22,28], radioimmunoassays in four studies [23–25,30], ELISA in two studies [26,31], a specific immunochemical method in one study [29], and the IMx assay in one study [16]. Three studies did not include the mean age of patients and controls [15,16,23], and one study included the median age of patients and controls [23]. Of the 12 included studies, six studies were conducted in Asia [15,22,24,26,28,31], four in Europe [16,23,25,30], one in America [27], and one in Africa [29]. Seven studies investigated the association between UC and serum folate and vitamin B12 concentrations [15,16,22,27,28,30,31]; six studies investigated the association between CD and serum folate and vitamin B12 [15,16,23,27,29,30]; and two studies investigated IBD patients who were not classified by the type of disease [25,26]. The patients' basic characteristics are presented in Table 1. The participants in one study [27] were children, and those in the other studies were adults. The data included in this analysis was defined as continuous variables.

We conducted a meta-analysis of the serum folic acid and vitamin B12 concentrations in 1086 IBD patients and 1484 controls. The average serum folate concentration in the IBD patients was 0.46 nmol/L lower than that in the controls (SMD = -0.46 ng/mL, 95% CI = -0.64, -0.27 ng/mL, $I^2 = 74.7\%$, $p = 0.000$; Figure 2). An analysis of the results of these studies indicated that IBD patients did not have significantly lower serum vitamin B12 concentrations than the healthy controls (SMD = -0.20 ng/mL, 95% CI = -0.46, 0.05 ng/mL, $I^2 = 87.6\%$, $p = 0.123$; Figure 3). Because of the heterogeneity of the results, we applied a random effects model. In addition, publication bias was not observed in the serum folate concentrations (Egger's test: coefficient = 0.70, $p = 0.627$). However, publication bias was observed in the serum vitamin B12 concentration (Egger's test: coefficient = 4.09, $p = 0.021$). A sensitivity analysis revealed that the studies by Koutroubakis et al. [16] influenced the results; however, when we excluded the study by Koutroubakis et al. on vitamin B12, our final results indicated that the serum vitamin B12 concentrations for IBD patients were 0.35 ng/mL lower than the controls (overall effect size = -0.35 ng/mL, 95% CI = -0.56, -0.13, $I^2 = 79.3\%$, $p = 0.001$).

Significant heterogeneity was observed for the studies (folate: $I^2 = 74.7\%$, $p < 0.001$; vitamin B12: $I^2 = 87.6\%$, $p < 0.001$).

The included articles were divided into three groups by the type of disease as follows: only patients with UC (UC group) [15,16,22,27,28,30,31], only patients with CD (CD group) [15,16,23,27,29,30] and patients from studies [25,26] that did not indicate the type of IBD (IBD group). For the UC group (680 UC patients, 1151 controls), the average serum folate concentration was 0.50 ng/mL lower than that of the controls (SMD = -0.50 ng/mL, 95% CI = -0.71, -0.28 ng/mL, $I^2 = 63.0\%$, $p = 0.000$). For the CD group (247 CD patients, 367 controls), the average serum folate concentration was not statistically significant differences with the controls (SMD = -0.30 ng/mL, 95% CI = -0.63, 0.04 ng/mL, $I^2 = 69.0\%$, $p = 0.080$). For the IBD group, statistically significant differences with the controls were observed (SMD = -0.64 ng/mL, 95% CI = -1.21, -0.06 ng/mL, $I^2 = 85.9\%$, $p = 0.030$) (Figure 4).

The meta-analysis of 12 studies that included the mean serum vitamin B12 concentrations showed an overall non-significant summary effect between the UC patients (SMD = -0.17 pg/mL, 95% CI = -0.59, 0.24 pg/mL, $I^2 = 90.3\%$, $p = 0.406$) and CD patients (SMD = -0.05 pg/mL, 95% CI = -0.46, 0.37 pg/mL, $I^2 = 82.7\%$, $p = 0.829$). However, IBD patients had significantly lower concentrations than healthy controls (SMD = -0.57 pg/mL, 95% CI = -0.83, -0.30 pg/mL, $I^2 = 35.5\%$, $p = 0.000$) (Figure 5).

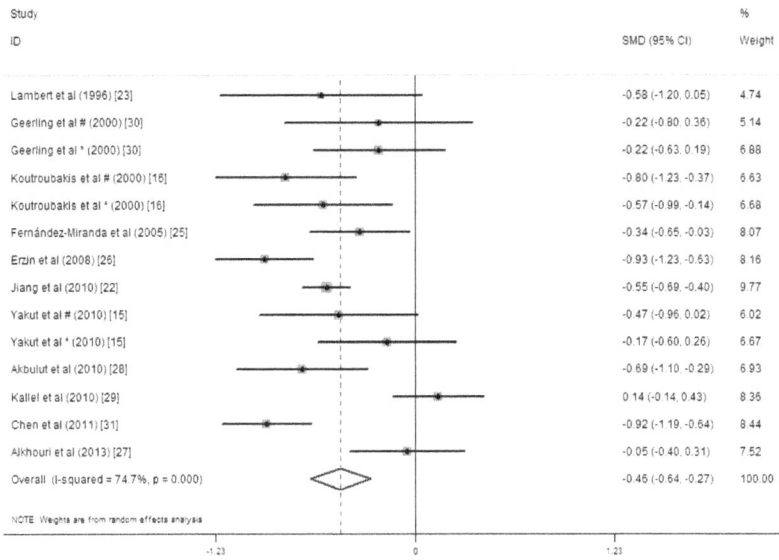

Figure 2. Forest plot of the serum folate concentrations in the inflammatory bowel disease (IBD) patients vs. controls; standardized mean differences with the 95% confidence interval and weight percentage are shown. * Reported in ulcerative colitis; # reported in Crohn's disease.

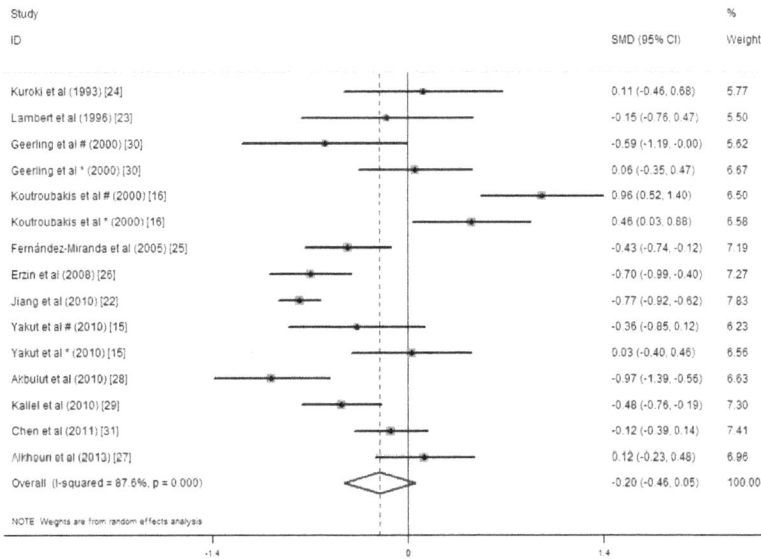

Figure 3. Forest plot of the serum concentrations of vitamin B12 in the IBD patients vs. controls; standardized mean differences with the 95% confidence interval and weight percentage are shown. * Reported in ulcerative colitis; # reported in Crohn's disease.

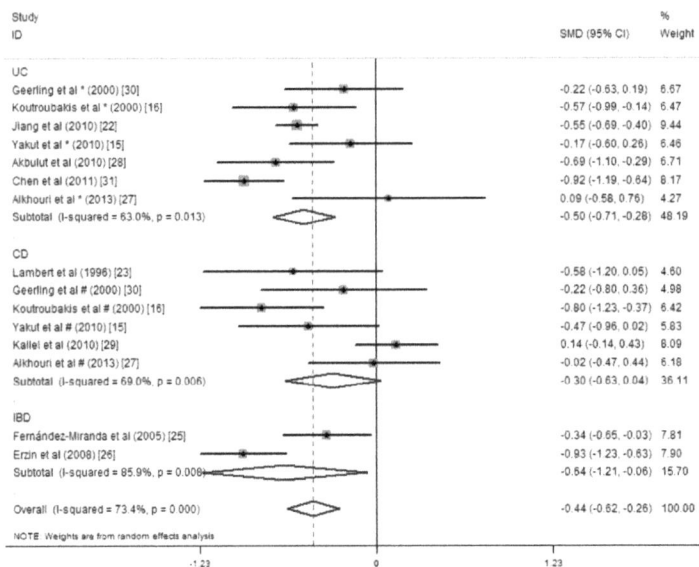

Figure 4. Forest plot of the serum concentrations of folate in the ulcerative colitis (UC), Crohn's disease (CD) and IBD patients vs. controls; standardized mean differences with the 95% confidence interval and weight percentage are shown. Subtotals are for the UC, CD and IBD patients.

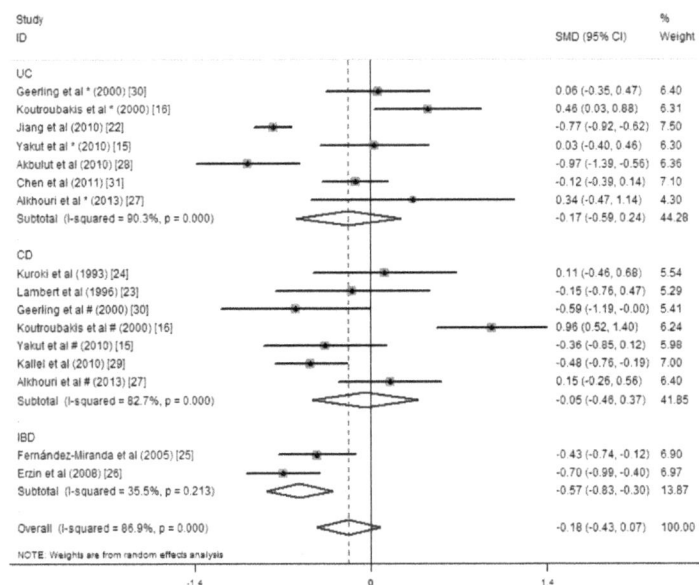

Figure 5. Forest plot of the serum concentrations of vitamin B12 in the UC, CD and IBD patients vs. controls; standardized mean differences with the 95% confidence interval and weight percentage are shown. Subtotals are for the UC, CD and IBD patients.

A subgroup analysis was performed according to the study area, and the studies were divided into three areas: Asia, Europe, and other (America and Africa). The studies in Asia showed that the serum folate concentrations of the IBD patients were significantly lower than those of the healthy controls (SMD = −0.65 ng/mL, 95% CI = −0.86, −0.44 ng/mL, I^2 = 64.3%, p = 0.000). The studies in Europe showed that the serum folate concentrations in the IBD patients were 0.44 ng/mL lower than those of the healthy controls (SMD = −0.44 ng/mL, 95% CI = −0.62, −0.26 ng/mL, I^2 = 3.1%, p = 0.000). The studies in the other areas did not show differences in the serum folate concentrations between the IBD patients and the controls (Figure 6).

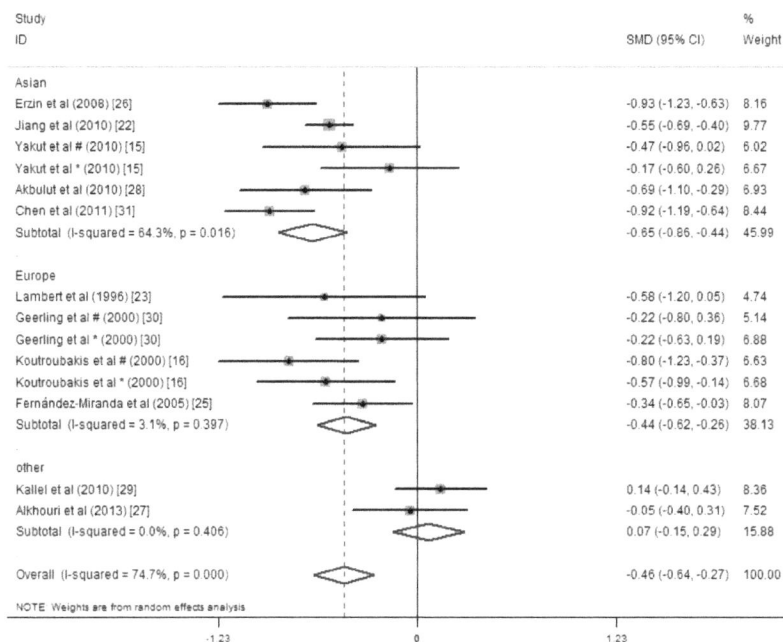

Figure 6. Forest plot of the serum concentrations of folate in the IBD patients vs. controls; standardized mean differences with the 95% confidence interval and weight percentage are shown. Subtotals are for the studies from Asia, Europe, and others. * Reported in ulcerative colitis; # reported in Crohn's disease.

In the subgroup meta-analysis on the mean serum vitamin B12 concentrations, the studies in Asia showed a significant summary effect, and the serum vitamin B12 concentrations of the IBD patients were 0.43 pg/mL lower than that of the controls (SMD = −0.43 ng/mL, 95% CI = −0.73, −0.13 ng/mL, I^2 = 83.4%, p = 0.005). However, significant differences in serum vitamin B12 concentrations were not detected between the IBD patients and controls in the studies from Europe or the other areas (Figure 7).

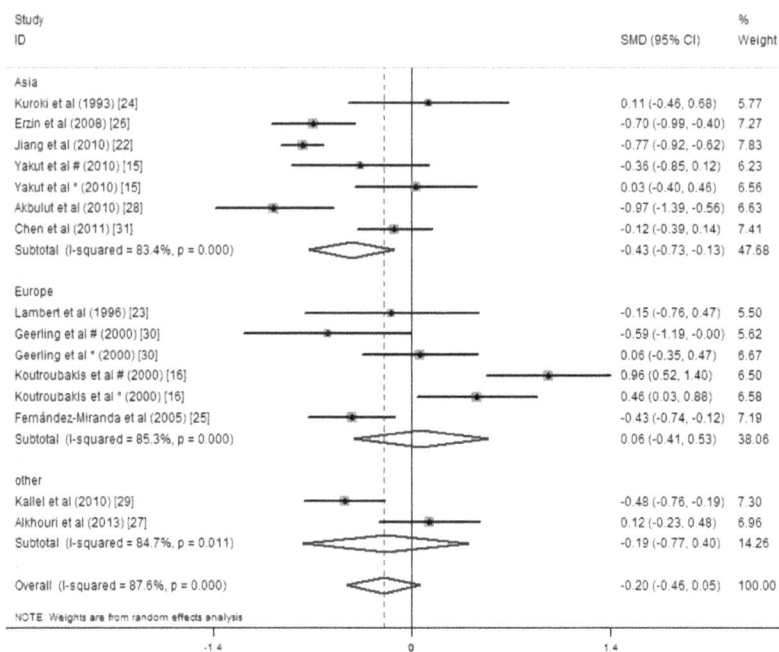

Figure 7. Forest plot of the serum concentrations of vitamin B12 in the IBD patients vs. controls; standardized mean differences with the 95% confidence interval and weight percentage are shown. Subtotals are for the studies from Asia, Europe, and others. * Reported in ulcerative colitis; # reported in Crohn's disease.

The main methods of detecting folic acid and vitamin B12 in the included studies were chemiluminescence immunoassays, radioimmunoassays and ELISA. Alkhouri et al. did not specify the detection method. Kallel et al.'s study used specific immunochemical methods [29], and Koutroubakis et al.'s study used the IMx assay [16]. We performed subgroup analysis according to the three main methods of detecting folate, and the results indicated that the heterogeneity among the three groups was reduced, and the subtotal I^2 values of the chemiluminescence immunoassay, radioimmunoassay and ELISA methods were 11.3%, 0% and 0%, respectively. Studies that applied these three detection methods showed lower average serum folate concentrations in the IBD patients relative to the controls (Figure 8). Thus, the detection methods may introduce heterogeneity into the serum folate concentration results in the different studies. A similar subgroup analysis was performed for the serum vitamin B12 concentrations. Significant differences were only observed among the studies that used chemiluminescence immunoassays, and they indicated that the serum vitamin B12 concentrations of the IBD patients were 0.54 pg/mL lower than that of the controls (SMD = -0.54 pg/mL, 95% CI = -0.94, -0.15 pg/mL, $I^2 = 80.6\%$, $p = 0.010$). However, in studies that detected serum vitamin B12 concentrations via radioimmunoassays and ELISA, statistically significant differences were not observed between the IBD patients and the controls (Figure 9).

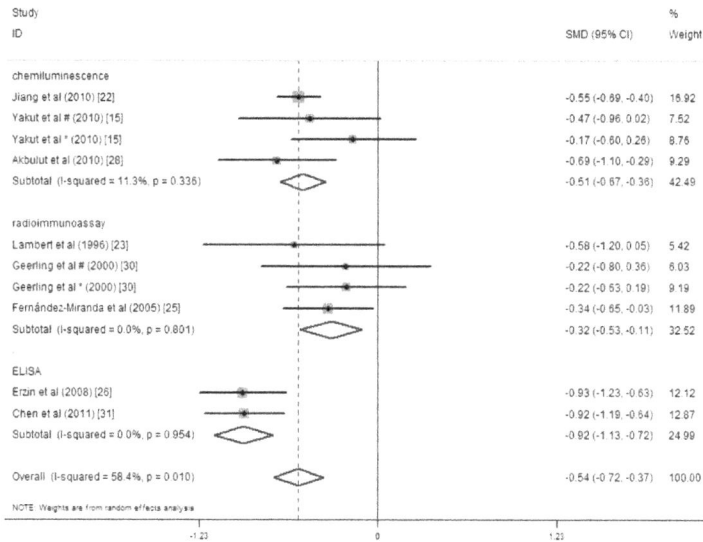

Figure 8. Forest plot of the serum concentrations of folate in the IBD patients vs. controls; standardized mean differences with a 95% confidence interval and weight percentage are shown. Subtotals are for the three detection methods. * Reported in ulcerative colitis; [#] reported in Crohn's disease.

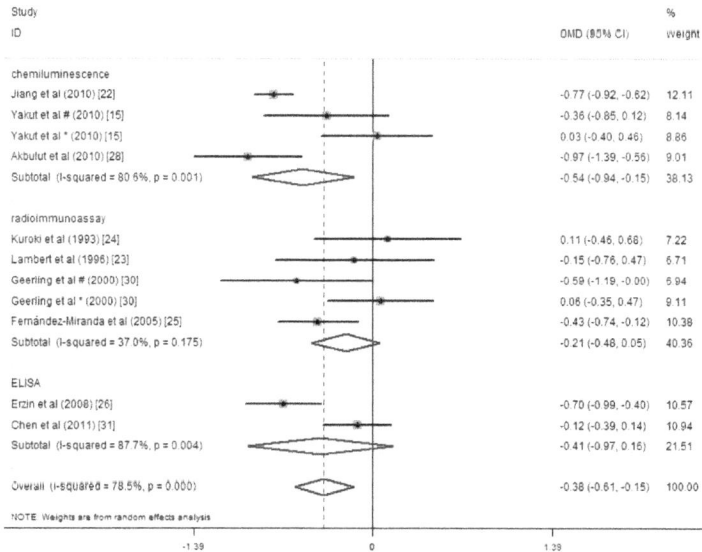

Figure 9. Forest plot of the serum concentrations of vitamin B12 in IBD patients vs. controls; standardized mean differences with a 95% confidence interval and weight percentage are shown. Subtotals are for the three detection methods. * Reported in ulcerative colitis; [#] reported in Crohn's disease.

A meta-regression analysis was performed because of the significant heterogeneity observed between studies. The meta-regression analysis evaluated the folate and vitamin B12 detection methods, the year of publication, the size of the samples, the quality of the study and the average age of the patients. The results showed that the detection methods impacted the serum folic acid concentrations and the total effect size ($p < 0.05$; Table 2). For serum vitamin B12 concentrations, the mean age of the IBD patients was the source of heterogeneity, and the difference in the results was statistically significant ($p < 0.05$; Table 3). The meta-regression analysis based on the size of the sample and the quality of study could not explain the between-study heterogeneity (Tables 2 and 3).

Table 2. Results of the folate regression analysis.

Variables	n	I^2	Adj R^2	exp(b)	Std. Err.	t	p	95% CI
				0.81	0.10	−1.60	0.154 *	0.61, 1.10
Testing method	10	0.00%	100.00%	0.55	0.08	−4.00	0.005 #	0.38, 0.78
				0.73	0.08	−2.95	0.021 &	0.56, 0.94
Year of publication	14	76.64%	−8.34%	1.007	0.02	0.35	0.729	0.97, 1.05
Sample size	14	76.12%	−11.49%	1.00	0.00	−0.35	0.729	0.99, 1.00
Quality of studies	14	76.51%	−10.63%	0.99	0.10	−0.05	0.959	0.79, 1.25
Average age of patients	9	80.56%	18.27%	0.98	0.01	−1.58	0.159	0.95, 1.01

n: number; I^2: percentage of total variation across studies; Adj R^2: adjusted R^2, the proportion of between-study variance; exp(b): the exponentiation of the B coefficient; Std. Err: standard error; t: t statistic, the coefficient divided by its standard error; p: an independent variable would be significant (<0.05) or not significant (≥0.05) in the model; CI: confidence interval; * chemiluminescence; # radioassay; & ELISA.

Table 3. Results of the vitamin B12 regression analysis.

Variables	n	I^2	Adj R^2	exp(b)	Std. Err.	t	p	95% CI
				0.86	0.27	−0.47	0.654 *	0.43, 1.76
Testing method	11	73.32%	−1.84%	1.22	0.37	0.66	0.530 #	0.60, 2.47
				0.67	0.16	−1.64	0.139 &	0.38, 1.18
Year of publication	15	84.87%	16.62%	0.96	0.02	−1.74	0.105	0.92, 1.01
Sample size	15	81.47%	12.78%	1.00	0.00	−1.58	0.139	0.99, 1.00
Quality of studies	15	82.38%	20.13%	0.78	0.10	−1.92	0.078	0.60, 1.03
Average age of patients	10	65.85%	55.42%	0.97	0.01	−2.73	0.026	0.95, 1.00

n: number; I^2: percentage of total variation across studies; Adj R^2: adjusted R^2, the proportion of between-study variance; exp(b): the exponentiation of the B coefficient; Std. Err: standard error; t: t statistic, the coefficient divided by its standard error; p: an independent variable would be significant (<0.05) or not significant (≥0.05) in the model; CI: confidence interval; * chemiluminescence; # radioassay; & ELISA.

4. Discussion

The results from this meta-analysis showed that the serum folate concentrations of patients with IBD were lower than that in normal controls, and the difference was significant. However, significant differences in serum vitamin B12 concentrations were not observed between the IBD patients and healthy controls. Serum folate concentrations may be reduced because of inadequate dietary intake [32], increased utilization, or drug effects, mainly salicylazosulfapyridine (SASP) [14]. Moreover, Burr et al. [33] showed that folic acid can be used as a supplement for the prevention of colorectal cancer in IBD patients. Considering the results of this study, IBD patients should be supplemented with folic acid. Research has shown that folate deficiency is more common than vitamin B12 deficiency [34]. Folate and vitamin B12 are important water-soluble vitamins for humans. Folate mainly assimilates in the duodenum and the proximal jejunum and is found in a variety of foods. Vitamin B12 is a component of coenzymes, and it is essential for cell biosynthesis and metabolism in vivo. Vitamin B12 is mainly absorbed in the terminal ileum, and the main source of vitamin B12 for humans is animal products. IBD may involve the small intestine; thus, the serum concentrations of folate and vitamin B12 may be lower in patients with IBD despite their presence in a wide range of foods.

Although our findings indicated that the serum concentrations of vitamin B12 were not significantly different between IBD patients and controls, studies have shown that a resection of more than 50–60 cm of the ileum frequently produces vitamin B12 malabsorption [35]. The reason for this difference might be due to 50 patients among 1086 patients included in this meta-analysis underwent surgery, which is less than 5 percent. In CD patients, prior intestinal surgery was an independent risk factor for low serum concentrations of vitamin B12 [15]. A meta-analysis by Battat et al. on vitamin B12 deficiency in IBD patients indicated that the only factor that predisposed CD patients to vitamin B12 deficiency was ileal resections greater than 20 cm [36]. Therefore, IBD patients, especially patients who have had ileal surgery, are recommended to take vitamin B12 supplements.

In the subgroup analysis based on the types of diseases, the serum folate concentrations in UC patients but not CD patients were lower than those in the healthy controls. However, serum folate concentrations in patients with unspecified IBD were lower than that in controls. The subgroup analysis by area showed lower serum concentrations of folate and vitamin B12 in the Asian study populations, which may be related to the typical Asian diet, which mostly includes plant-based foods and less meat than is typical in European diets, because vitamin B12 widely occurs in animal-based food products. Therefore, we suggest these patients may require moderate vitamin B12 supplements.

When we combined the results of all of the studies, a large degree of heterogeneity was observed, so we performed analysis to identify the source of heterogeneity. Firstly, the subgroup analysis of the detection methods showed that the serum folate concentrations were lower in the IBD patients than in the healthy control patients, regardless of the techniques that were used (chemiluminescence immunoassay, radioimmunoassay or ELISA), and the heterogeneity was lower when grouping by detection methods compare to by disease types. Therefore, we concluded that different test methods for folic acid may be one of the sources of heterogeneity in the associated studies. Secondly, the results of the sensitivity analysis showed that the studies by Koutroubakis et al. introduced a large amount of heterogeneity [16]. This might be due to inclusion of three control groups (healthy blood donors, visitors to the gynaecology/obstetrics and orthopaedics wards and normal hospital personnel) by investigators (Koutroubakis et al. [16]), which might have introduced a number of uncontrollable confounding factors. Thirdly, the results of meta-regression analysis showed that the method of detection explained the heterogeneity among the studies that examined serum folate concentrations; the average age of the patient was a heterogeneous factor among the studies that examined the serum vitamin B12 concentrations. Moreover, IBD patients often present morphological and functional disorders of the liver [23], and patients with hepatic dysfunction may have normal or high serum concentrations of vitamin B12 rather than increased body stores of vitamin B12 [37]. We did not analyse the disease severity because of the different ratings criteria for severity. In addition, certain studies did not specify or excluded severe patients, which led to incomplete data; also, disease severity may be not associated with lower serum vitamin concentrations [24].

We still need to acknowledge that this review has several limitations, which means that our results should be interpreted with caution. We only included studies that fit our inclusion criteria described in Section 2.2. Several important studies were excluded due to variety of reasons. These include: Ward et al. only reported the prevalence of vitamin B12 deficiency, but did not show the mean or standard deviation of serum concentrations of vitamin B12 [38]. We excluded this study based on our inclusion criteria (3). Jayaprakash et al. reported their data without a healthy control group, and some of the patients they included had been receiving vitamin B12 supplementation [39]. We excluded this study based on our inclusion criteria (2) and (4). The data included in the analysis were from observational studies; thus, potentially confounding factors may have been included in the baseline characteristics of the selected population. The serum concentrations of folate and vitamin B12 could be affected by ethnicity, gender and complications. Moreover, case-controlled studies and cross-sectional studies lack sufficient proof of the causal link between serum concentrations of folic acid and vitamin B12 and IBD status. Also, observational studies alone to prove cause–effect relationship between certain nutrition factors and certain diseases are very difficult unless the study is as large as the China–Cornell–Oxford Project in the late 20th century [40]. Future prospective studies are needed to identify the causal factors linking serum folate and vitamin B12 concentrations and IBD.

5. Conclusions

In conclusion, the serum folate concentrations in patients with IBD were lower than those in healthy controls, and low serum concentrations of folic acid may be an important risk factor for IBD patients. In addition, the serum vitamin B12 concentrations were lower in Asian patients. We recommend routine screening of patients with IBD for vitamin B12 and folate deficiency to determine whether supplements of folic acid or vitamin B12 should be administered to ensure the nutritional status of IBD patients and to improve the health of patients.

Acknowledgments: This work was supported by the National Natural Science Foundation of China (No. 81501423) and the Norman Bethune Program of Jilin University (No. 2015223).

Author Contributions: W.C., Y.P. and D.L. contributed to the data collection, analysis and manuscript preparation; H.G., M.S.J. and X.L. contributed to the data collection and analysis; and W.C., Y.L. and D.L. developed the study protocol, secured funds, supervised the study and guided the manuscript preparation.

Conflicts of Interest: The authors declare that there are no conflicts of interest regarding the publication of this paper.

References

1. Leone, V.; Chang, E.B.; Devkota, S. Diet, microbes, and host genetics: The perfect storm in inflammatory bowel diseases. *J. Gastroenterol.* **2013**, *48*, 315–321. [CrossRef] [PubMed]
2. Kaser, A.; Zeissig, S.; Blumberg, R.S. Inflammatory bowel disease. *Annu. Rev. Immunol.* **2010**, *28*, 573–621. [CrossRef] [PubMed]
3. Leddin, D.; Tamim, H.; Levy, A.R. Is folate involved in the pathogenesis of inflammatory bowel disease? *Med. Hypotheses* **2013**, *81*, 940–941. [CrossRef] [PubMed]
4. Shanahan, F. The microbiota in inflammatory bowel disease: Friend, bystander, and sometime-villain. *Nutr. Rev.* **2012**, *70* (Suppl. 1), S31–S37. [CrossRef] [PubMed]
5. Kellermayer, R. Epigenetics and the developmental origins of inflammatory bowel diseases. *Can. J. Gastroenterol.* **2012**, *26*, 909–915. [CrossRef] [PubMed]
6. Bermejo, F.; Algaba, A.; Guerra, I.; Chaparro, M.; De-La-Poza, G.; Valer, P.; Piqueras, B.; Bermejo, A.; Garcia-Alonso, J.; Perez, M.J.; et al. Should we monitor vitamin b12 and folate levels in crohn's disease patients? *Scand. J. Gastroenterol.* **2013**, *48*, 1272–1277. [CrossRef] [PubMed]
7. Owczarek, D.; Rodacki, T.; Domagala-Rodacka, R.; Cibor, D.; Mach, T. Diet and nutritional factors in inflammatory bowel diseases. *World J. Gastroenterol.* **2016**, *22*, 895–905. [CrossRef] [PubMed]
8. Lachner, C.; Steinle, N.I.; Regenold, W.T. The neuropsychiatry of vitamin B12 deficiency in elderly patients. *J. Neuropsychiatry Clin. Neurosci.* **2012**, *24*, 5–15. [CrossRef] [PubMed]
9. Cario, H.; Smith, D.E.; Blom, H.; Blau, N.; Bode, H.; Holzmann, K.; Pannicke, U.; Hopfner, K.P.; Rump, E.M.; Ayric, Z.; et al. Dihydrofolate reductase deficiency due to a homozygous DHFR mutation causes megaloblastic anemia and cerebral folate deficiency leading to severe neurologic disease. *Am. J. Hum. Genet.* **2011**, *88*, 226–231. [CrossRef] [PubMed]
10. Oussalah, A.; Gueant, J.L.; Peyrin-Biroulet, L. Meta-analysis: Hyperhomocysteinaemia in inflammatory bowel diseases. *Aliment. Pharmacol. Ther.* **2011**, *34*, 1173–1184. [CrossRef] [PubMed]
11. Stipanuk, M.H. Sulfur amino acid metabolism: Pathways for production and removal of homocysteine and cysteine. *Annu. Rev. Nutr.* **2004**, *24*, 539–577. [CrossRef] [PubMed]
12. Hoffbrand, V.; Provan, D. Abc of clinical haematology. Macrocytic anaemias. *Br. Med. J.* **1997**, *314*, 430–433. [CrossRef]
13. Vasilopoulos, S.; Saiean, K.; Emmons, J.; Berger, W.L.; Abu-Hajir, M.; Seetharam, B.; Binion, D.G. Terminal ileum resection is associated with higher plasma homocysteine levels in crohn's disease. *J. Clin. Gastroenterol.* **2001**, *33*, 132–136. [CrossRef] [PubMed]
14. Chowers, Y.; Sela, B.A.; Holland, R.; Fidder, H.; Simoni, F.B.; Bar-Meir, S. Increased levels of homocysteine in patients with Crohn's disease are related to folate levels. *Am. J. Gastroenterol.* **2000**, *95*, 3498–3502. [CrossRef] [PubMed]
15. Yakut, M.; Ustun, Y.; Kabacam, G.; Soykan, I. Serum vitamin B12 and folate status in patients with inflammatory bowel diseases. *Eur. J. Intern. Med.* **2010**, *21*, 320–323. [CrossRef] [PubMed]

16. Koutroubakis, I.E.; Dilaveraki, E.; Vlachonikolis, I.G.; Vardas, E.; Vrentzos, G.; Ganotakis, E.; Mouzas, I.A.; Gravanis, A.; Emmanouel, D.; Kouroumalis, E.A. Hyperhomocysteinemia in Greek patients with inflammatory bowel disease. *Dig. Dis. Sci.* **2000**, *45*, 2347–2351. [CrossRef] [PubMed]

17. Deeks, J.; Altman, D. Chapter 16: Special topics in statistics. In *Cochrane Handbook for Systematic Reviews of Interventions*; Version 5.1.0; The Cochrane Collaboration: London, UK, 2011.

18. Senn, S. Overstating the evidence: Double counting in meta-analysis and related problems. *BMC Med. Res. Methodol.* **2009**, *9*, 10. [CrossRef] [PubMed]

19. Cochran, W.G. The combination of estimates from different experiments. *Biometrics* **1954**, *10*, 101–129. [CrossRef]

20. Higgins, J.P.; Thompson, S.G.; Deeks, J.J.; Altman, D.G. Measuring inconsistency in meta-analyses. *Br. Med. J.* **2003**, *327*, 557–560. [CrossRef] [PubMed]

21. Egger, M.; Davey Smith, G.; Schneider, M.; Minder, C. Bias in meta-analysis detected by a simple, graphical test. *Br. Med. J.* **1997**, *315*, 629–634. [CrossRef]

22. Jiang, Y.; Zhao, J.; Jiang, T.; Ge, L.; Zhou, F.; Chen, Z.; Lei, Y.; Huang, S.; Xia, B. Genetic polymorphism of methylenetetrahydrofolate reductase g1793a, hyperhomocysteinemia, and folate deficiency correlate with ulcerative colitis in central China. *J. Gastroenterol. Hepatol.* **2010**, *25*, 1157–1161. [CrossRef] [PubMed]

23. Lambert, D.; Benhayoun, S.; Adjalla, C.; Gelot, M.A.; Renkes, P.; Felden, F.; Gerard, P.; Belleville, F.; Gaucher, P.; Gueant, J.L.; et al. Crohn's disease and vitamin B12 metabolism. *Dig. Dis. Sci.* **1996**, *41*, 1417–1422. [CrossRef] [PubMed]

24. Kuroki, F.; Iida, M.; Tominaga, M.; Matsumoto, T.; Hirakawa, K.; Sugiyama, S.; Fujishima, M. Multiple vitamin status in Crohn's disease. Correlation with disease activity. *Dig. Dis. Sci.* **1993**, *38*, 1614–1618. [CrossRef] [PubMed]

25. Fernandez-Miranda, C.; Martinez Prieto, M.; CasisHerce, B.; Sanchez Gomez, F.; Gomez Gonzalez, P.; Martinez Lopez, J.; Saenz-Lopez Perez, S.; Gomez de la Camara, A. Hyperhomocysteinemia and methylenetetrahydrofolate reductase 677c→t and 1298a→c mutations in patients with inflammatory bowel disease. *Rev. Esp. Enferm. Dig.* **2005**, *97*, 497–504. [CrossRef] [PubMed]

26. Erzin, Y.; Uzun, H.; Celik, A.F.; Aydin, S.; Dirican, A.; Uzunismail, H. Hyperhomocysteinemia in inflammatory bowel disease patients without past intestinal resections: Correlations with cobalamin, pyridoxine, folate concentrations, acute phase reactants, disease activity, and prior thromboembolic complications. *J. Clin. Gastroenterol.* **2008**, *42*, 481–486. [CrossRef] [PubMed]

27. Alkhouri, R.H.; Hashmi, H.; Baker, R.D.; Gelfond, D.; Baker, S.S. Vitamin and mineral status in patients with inflammatory bowel disease. *J. Pediatr. Gastroenterol. Nutr.* **2013**, *56*, 89–92. [CrossRef] [PubMed]

28. Akbulut, S.; Altiparmak, E.; Topal, F.; Ozaslan, E.; Kucukazman, M.; Yonem, O. Increased levels of homocysteine in patients with ulcerative colitis. *World J. Gastroenterol.* **2010**, *16*, 2411–2416. [CrossRef] [PubMed]

29. Kallel, L.; Feki, M.; Sekri, W.; Segheir, L.; Fekih, M.; Boubaker, J.; Kaabachi, N.; Filali, A. Prevalence and risk factors of hyperhomocysteinemia in Tunisian patients with Crohn's disease. *J. Crohns Colitis* **2011**, *5*, 110–114. [CrossRef] [PubMed]

30. Geerling, B.J.; Badart-Smook, A.; Stockbrugger, R.W.; Brummer, R.J. Comprehensive nutritional status in recently diagnosed patients with inflammatory bowel disease compared with population controls. *Eur. J. Clin. Nutr.* **2000**, *54*, 514–521. [CrossRef] [PubMed]

31. Chen, M.L.; Mei, Q.; Xu, J.M.; Hu, N.Z.; Lu, C.X.; Fang, H.M. Significance of plasmichomocysteine, folate and vitamin B(12) in ulcerative colitis. *Zhonghua Wei Chang Wai Ke Za Zhi* **2011**, *14*, 185–187 (in Chinese). [PubMed]

32. Zezos, P.; Papaioannou, G.; Nikolaidis, N.; Vasiliadis, T.; Giouleme, O.; Evgenidis, N. Hyperhomocysteinemia in ulcerative colitis is related to folate levels. *World J. Gastroenterol.* **2005**, *11*, 6038–6042. [CrossRef] [PubMed]

33. Burr, N.E.; Hull, M.A.; Subramanian, V. Folic acid supplementation may reduce colorectal cancer risk in patients with inflammatory bowel disease: A systematic review and meta-analysis. *J. Clin. Gastroenterol.* **2017**, *51*, 247–253. [CrossRef] [PubMed]

34. Hwang, C.; Ross, V.; Mahadevan, U. Micronutrient deficiencies in inflammatory bowel disease: From A to zinc. *Inflamm. Bowel Dis.* **2012**, *18*, 1961–1981. [CrossRef] [PubMed]

35. Bermejo, F.; Algaba, A.; Gisbert, J.P.; Nogueiras, A.R.; Poza, G.; Chaparro, M.; Valer, P.; Piqueras, B.M.; Villa, J.C.; Bermejo, A. Prospective controlled analysis of vitamin B12 and folate deficiency in Crohn's disease. *Gastroenterology* **2011**, *140*, S-434.

36. Battat, R.; Kopylov, U.; Szilagyi, A.; Saxena, A.; Rosenblatt, D.S.; Warner, M.; Bessissow, T.; Seidman, E.; Bitton, A. Vitamin B12 deficiency in inflammatory bowel disease: Prevalence, risk factors, evaluation, and management. *Inflamm. Bowel Dis.* **2014**, *20*, 1120–1128. [CrossRef] [PubMed]

37. Jacobson, S. Serum concentration of cobalamines during total parenteral nutrition in Crohn's disease. *JPEN J. Parenter. Enter. Nutr.* **1986**, *10*, 223–226. [CrossRef] [PubMed]

38. Ward, M.; Kariyawasam, V.; Mogan, S.; Patel, K.; Pantelidou, M.; Sobczyńska-Malefora, A.; Porté, F.; Griffin, N.; Anderson, S.; Sanderson, J.; et al. Prevalence and Risk Factors for Functional Vitamin B12 Deficiency in Patients with Crohn's Disease. *Inflamm. Bowel Dis.* **2015**, *21*, 2839–2847. [CrossRef] [PubMed]

39. Jayaprakash, A.; Creed, T.; Stewart, L.; Colton, B.; Mountford, R.; Standen, G.; Probert, C. Should we monitor vitamin B12 levels in patients who have had end-ileostomy for inflammatory bowel disease? *Int. J. Colorectal Dis.* **2004**, *19*, 316–318. [CrossRef] [PubMed]

40. Campbell, T.C.; Campbell, T.M. *The China Study*, 1st ed.; BenBella Books: Dallas, TX, USA, 2006.

nutrients

MDPI

Review

Effects of the Exclusive Enteral Nutrition on the Microbiota Profile of Patients with Crohn's Disease: A Systematic Review

Simona Gatti [1,*], Tiziana Galeazzi [2], Elisa Franceschini [1], Roberta Annibali [1], Veronica Albano [1], Anil Kumar Verma [2], Maria De Angelis [3], Maria Elena Lionetti [1] and Carlo Catassi [1]

[1] Department of Pediatrics, Università Politecnica delle Marche, 60123 Ancona, Italy; elisa.franceschini3@gmail.com (E.F.); robertannibali@hotmail.it (R.A.); vero.albano@ospedaliriuniti.marche.it (V.A.); mariaelenalionetti@gmail.com (M.E.L.); c.catassi@univpm.it (C.C.)
[2] Laboratory of Metabolic Diseases, Università Politecnica delle Marche, 60123 Ancona, Italy; t.galeazzi@univpm.it (T.G.); anilkrvermaa@gmail.com (A.K.V.)
[3] Department of Soil, Plant and Food Sciences, University of Bari A. Moro, 70126 Bari, Italy; maria.deangelis@uniba.it
*   Correspondence: simona.gatti@hotmail.it; Tel.: +39-071-5962-114

Received: 14 April 2017; Accepted: 28 July 2017; Published: 4 August 2017

Abstract: The mechanisms behind the efficacy of exclusive enteral nutrition (EEN) in Crohn's disease (CD) remain poorly understood, despite the high rate of treatment response. Evidence accumulated in the last 20 years suggests that a positive shift of the disrupted microbiota is one of the treatment effects. The purpose of this study was to critically review and summarize data reporting the microbiological effects of EEN in patients with CD. Fourteen studies were considered in the review, overall involving 216 CD patients on EEN. The studies were heterogeneous in methods of microbiota analysis and exclusion criteria. The most frequently reported effect of EEN was a reduction in microbiota diversity, reversible when patients returned to a normal diet. The effect of EEN on specific bacteria was very variable in the different studies, partially due to methodological limitations of the mentioned studies. The EEN seem to induce some metabolomic changes, which are different in long-term responder patients compared to patients that relapse earlier. Bacterial changes can be relevant to explaining the efficacy of EEN; however, microbiological data obtained from rigorously performed studies and derived from last generation techniques are largely inconsistent.

Keywords: microbiota; metabolome; exclusive enteral nutrition; Crohn's disease; IBD; next generation sequencing; 16S rRNA; *Faecalibacterium prausntzii*

1. Introduction

Inflammatory Bowel Disease (IBD), including Crohn's disease (CD), Ulcerative Colitis (UC) and Unclassified IBD (IBD-U), are chronic relapsing inflammatory disorders of the gastrointestinal tract resulting in significant morbidity. Despite the intense research on the underlying mechanisms, the etiology is still unclear. It has been proposed that IBD could be the consequence of an altered balance between the gastrointestinal (GI) microbiota and an inappropriate immune response to the GI bacterial community in subjects with a genetic predisposition [1,2]. The composition, the metabolic functions, and the roles of GI microbiota in IBD have been deeply investigated in the last 20 years, and the results have been recently reviewed and summarized [1,3–5]. Data suggest that bacterial dysbiosis is involved in pathogenesis, recurrence, and complications of IBD (especially increased abundance of Enterobacteriaceae, including adherent invasive *Escherichia coli* (AIEC) and a decreased abundance of Clostridiales, including *Faecalibacterium*) [5]. Additionally, changes in the microbiota composition (particularly at the mucosal level) potentially have prognostic and therapeutic implications [5]. At diagnosis, there is a window of opportunity to investigate the modifications of microbiota longitudinally, and studies conducted in untreated patients have shown interesting differences in children and adults with IBD [6].

Diet has a prominent role in IBD, and dietary factors have been implicated in the pathogenesis, recurrence, and treatment of IBD [7,8]. Exclusive enteral nutrition (EEN) is based on the administration of a liquid formula as the unique source of nutrition for a limited period of time (six to eight weeks). The EEN formulas differ in their protein and fat content and can be classified as elemental (monomeric), semi-elemental (oligomeric), or orpolymeric. Elemental formulas contain individual amino acids and glucose polymers and are generally low in fat. Semi-elemental formulas contain peptides of varying chain length, simple sugars, glucose polymers, or starch and fat, primarily as medium chain triglycerides (MCTs). Polymeric formulas contain intact proteins, complex carbohydrates, and mainly long-chain triglycerides (LCTs). A particular polymeric formula, enriched with an anti-inflammatory molecule (the transforming growth factor-beta 2, TGF-β2) has been designed specifically for patients with CD. European guidelines currently recommend EEN as the first-line therapy for inducing remission in pediatric luminal CD [9]. Such recommendations are based on a large amount of data showing rates of EEN-induced clinical remission up to 85% to 90%, and a rate of mucosal healing in 19% to 75% of patients [10–12]. The efficacy of EEN in CD is not correlated to the type of formula used (polymeric versus elemental) [13,14] or to the route of administration (oral versus naso-gastric tube) [15]. The rates of response to EEN are very high if compared to other treatment options in CD, while it is not useful in UC [16]. However, it is still unclear how this treatment works. Anti-inflammatory effects, avoiding specific dietetic components, restoration of the epithelial barrier, and alterations of the intestinal microbiota have been proposed and investigated as possible effects of EEN. The knowledge of these mechanisms can be important to understand the pathogenic events involved in IBD. Studies detailing changes in microbiota composition induced by EEN offer a unique chance to explore the isolated effect of a specific dietetic intervention in the IBD microbiota. Such studies require naïve treatment subjects (at diagnosis) and rigorous methodological criteria (i.e., exclusion subjects on antibiotics and probiotics and using new generation techniques of microbiota analysis).

Based on this background, we aimed to systematically review the existing literature on 'the effects induced by EEN on intestinal microbiota' in order to highlight the current knowledge on this topic, to reveal possible limitations of current studies, and to point out research gaps on this important issue.

2. The Normal Intestinal Microbiota: Techniques of Analysis

The human intestinal microbiota is represented by a complex system of bacteria, archaea, viruses, and fungi. In healthy subjects, the bacteria composing the intestinal microbiota primarily belong to two phyla that together represent 90% of the microbiota; the Firmicutes and the Bacteroidetes. Less represented phyla include Proteobacteria, Actinobacteria (including Bifidobacterium), Fusobacteria, Cyanobacteria, and Verrucomicrobia [17,18]. The characterization of the intestinal microbiota was originally based on microbiological techniques, with some limitations in the ability to culture a high proportion of intestinal bacteria [19]. More recently, culture efforts (culturomics) and new techniques in routine bacteriology have increased the number of identified species [20]. The introduction of molecular methods has further expanded the possibility of bacterial identification. More traditional methods, including electrophoresis and real-time quantitative polymerase chain reaction (PCR), are mainly based on the analysis of the 16S ribosomal RNA (rRNA) gene. This gene is ubiquitarious in all prokaryotes and contains both highly conserved and species–specific regions. The conserved regions work as amplification targets for PCR primers to extract one or more variable regions of the 16S rRNA gene. By using next-generation sequencing (NGS) technologies, the variable regions are replicated, producing thousands of sequenced fragments (reads), which are mapped to known 16S rRNA gene sequences of different microbial taxa such as genera or phyla or clustered according to their sequence similarity into "operational taxonomic units" (OTUs) [21]. The shotgun metagenomics sequencing approach has been developed in order to avoid some limitations of the more traditional techniques. Here, instead of targeting a specific genomic locus for amplification, all extracted DNA is broken-up into tiny fragments that are independently sequenced. These reads will be in part sampled from taxonomically informative genomic loci (e.g., 16S), while others will be sampled from coding sequences that provide insight into the biological functions encoded in the genome. As a result, metagenomic data provides the opportunity to simultaneously explore both the taxonomical and functional aspects of a microbial community [22].

3. Materials and Methods

All studies (including cross-sectional, cohort, case-control, case reports, and series) evaluating the effect induced by the EEN or partial enteral nutrition (PEN) on microbiota composition or metabolomic profile in patients with CD, were considered eligible. Abstracts presented at meetings and articles not published in English were excluded. Studies conducted in children and adults were both considered. No publication date or publication status was imposed. Our search was applied to the Medline database using PubMed by combining key words for Crohn's disease or Inflammatory Bowel Disease AND search terms for Enteral Nutrition, Exclusive Enteral Nutrition, Partial Enteral Nutrition, Exclusive Diet, or Polimeric Diet AND keywords for Microbiota, Microbiome, Metabolome, or Bacterial composition. All studies were published between 1950 (start of Medline) and January 2017. From each study, the following information were extracted: (1) Characteristics of the participants (number, age, disease's status, method of diagnosis, concomitant therapies); (2) Type of microbiota analysis (biological sample and technique of analysis); (3) Nutritional treatment (type of formula, duration of treatment, exclusivity); and (4) Outcome of nutritional treatment and correlations with microbiota changes.

4. Results

4.1. Studies Characteristics

Fourteen studies were identified (one case-control report and 13 prospective studies, 12 of them including a control group) [23–36]. All the studies were considered relevant for the purpose of this review. The main characteristics and outcomes of the included studies are presented in Table 1.

Two studies were conducted in adults [25,26] and 12 in children [23,24,27–36]. Overall 462 subjects were enrolled, 314 of them having CD. Out the 216 enrolled CD subjects that received EEN, longitudinal microbiological data (including at least two consecutive fecal or biopsy samples during an EEN course) were available only for 150 patients. All the studies included subjects with active CD (at diagnosis or during a flare-up). In five studies, the use of antibiotics (in a period of time variable between one week and three months prior to enrollment) was considered an exclusion criteria [23,24,29,30,32]; one study excluded only patients on antibiotic or probiotic treatment during the study period [26]; and another study excluded subjects who had received probiotics in the two weeks before the start of the study [31]. The remaining eight studies did not indicate the use of antibiotics or probiotics as an exclusion criterion.

Table 1. Characteristics and main outcomes of the 14 studies included in the review.

Author, Year	Type of Study	Groups of Subjects (Number, Characteristics)	Exclusion Criteria (ATBs, Probiotics, Other Drugs)	Biological Sample (Type and Number of Samples)	Type of Formula and Duration of Treatment	Microbiota Analysis	Outcomes
Lionetti P et al., 2005 [23]	Prospective, controlled	nine active Crohn's Disease (CD) adolescents (nine to 17 years), five controls (10–15 years)	Antibiotics or colon cleansing in the previous week	Fecal samples, multiple samples during the exclusive enteral nutrition (EEN) course and during partial enteral nutrition (total number not indicated)	Polymeric formula enriched with TGF-β2 for eight weeks	Temperature gradient gel electrophoresis (TGGE) analysis of 16S rRNA	TGGE profile varied greatly between subjects and required time to achieve stability of the band profile in each subject during exclusive and partial EN (no statistical analysis available).
Leach ST et al., 2008 [24]	Prospective, controlled	six CD children at diagnosis, mean age 10.2 years (2.5–13.5); seven controls, mean age 5.9 years (2.1–12)	Antibiotics or antiinflammatory agents in the previous four weeks. Severe CD requiring surgery or intensive medical treatment.	Fecal samples collected prior to endoscopy and at one, two, four, six, eight, 16 and 26 weeks after the start of EEN	eight weeks of EEN (formula not specified)	PCR amplification of the bacterial 16S rRNA gene followed by denaturating gel electrophoresis (DGGE)	CD children had a greater degree of change in the bacterial composition during EEN compared to controls on a normal diet ($p < 0.05$). The greatest change was seen in Ruminococcaceae ($p < 0.001$) and the least in the Bacteroides-Prevotella group ($p < 0.01$).
Jia W et al., 2010 [25]	Prospective, controlled	20 CD, 21 Irritable bowel syndrome (IBS), 14 Ulcerative colitis (UC), and 18 controls	Not indicated	Fecal samples collected before and after two weeks of EEN treatment	two weeks of elemental formula	PCR amplification of Faecalibacterium prausnitzii DNA (A2-165 and M21/2 subgroup)	Levels of F. prausnitzii A2-165 decreased significantly ($p = 0.0046$) after treatment compared to baseline and to other groups. Levels of F. prausnitzii M21/2 decreased without statistical significance ($p = 0.61$).
Shiga H et al., 2012 [26]	Prospective, controlled	33 active CD (median age: 30 years, 15–47), 17 controls	No antibiotic or probiotics during the study period.	Fecal samples at baseline, after 38 days for the EEN group, 35 days for the total parenteral nutrition (TPN) group, six weeks for the controls. In 12 healthy controls, a second fecal sample was collected after six weeks.	eight patients: eight weeks of elemental formula; nine patients on total parenteral nutrition	Terminal restriction fragment length polymorphism analysis of bacterial 16S rRNA to evaluate the whole microbiota. Specific quantitative PCR to determine predominant bacterial groups.	Number of bacterial species was reduced by EEN in CD ($p = 0.672$); the ratios of bifidobacteria and Bacteroides fragilis were reduced ($p = 0.664$ and 0.034, respectively), and Enterococcus was increased ($p = 0.788$).
Tjellstrom B et al., 2012 [27]	Prospective controlled	18 active CD children, median age 13.5 years (10–17); 12 healthy controls, median age 14.5 years (14–15.5)	Not indicated	Fecal samples (eight patients collected at the start and finish of EEN)	six weeks of polymeric formula	Determination of the fecal pattern of short-chain-fatty acids (SCFAs) using gas-liquid chromatogrphy	Concentration of fecal acetic acid was reduced by EEN ($p < 0.05$), and butyric and valeric acids were increased ($p = $ ns). 79% of CD showed response to EEN, showing a fecal pattern of SCFAs similar to healthy children.
D'Argenio et al., 2013 [28]	Case report, controlled	one active CD patient (14 years), one control with gut polyp (15 years)	Not indicated	Ileum samples	eight weeks of polymeric formula	16S rRNA next-generation sequencing	Bacterial diversity was reduced in CD patient at baseline compared to control and increased after EEN ($p < 0.05$). Composition changed after therapy (Bacteroides increased and Proteobacteria decreased) reaching a distribution similar to the healthy control (statistical significance not indicated).

Table 1. *Cont.*

Author, Year	Type of Study	Groups of Subjects (Number; Characteristics)	Exclusion Criteria (ATBs, Probiotics, Other Drugs)	Biological Sample (Type and Number of Samples)	Type of Formula and Duration of Treatment	Microbiota Analysis	Outcomes
Gerasimidis K et al., 2014 [29]	Prospective, controlled	15 active CD children (median age: 12.7 years), 11 newly diagnosed and four started a second EEN course; 21 healthy controls (median age: 9.9 years)	Antibiotics in the previous 3 months	68 fecal samples from CD subjects (baseline, 15 to 30 days on EEN, at EEN end, and two to four months after EEN). 40 samples from controls (two samples for each)	eight weeks of polymeric formula TGF-β2 enriched	16S rRNA amplification and quantification with real-time quantitative PCR. Measurement of SCFAs by gas cromatography. Measurement of D and L-lactate by enzymatic commercial assay. Measurement of fecal sulfide by a spetrophometric method.	After EEN, the global bacterial diversity abundance decreased ($p = 0.037$) and returned to normal on a free diet ($p = 0.041$). During EEN, concentrations of *F. prausnitzii* ($p = 0.002$) and *Bifidobacterium* genus decreased ($p = 0.053$) and re-increased on a normal diet ($p = 0.006$ for Faecalibacterium, $p =$ ns for Bacteroides/Prevotella), but remained lower compared to healthy subjects.
Quince C et al., 2015 [30]	Prospective controlled	23 active CD (age: 6.9–14.7 years) 21 controls (age: 4.6–16.9 years)	Antibiotics in the previous three months	78 fecal samples from CD patients (baseline, during EEN: 16th to 32th and 54th day and 63 days after EEN) 39 fecal samples from controls (collected at least two months apart)	eight weeks of polymeric formula TGF-β2 enriched	Sequencing of 16S rRNA gene performed on the MiSeq platform. Shotgun metagenome sequencing was performed for 69 samples with the Nextera XT Prep Kit and the Illumina dual barcoding Nextera XT Index kit. Shotgun metagenomics reads were used also for assignment to functional models through alignment to Kyoto Encyclopedia of Genes and Genomes (KEGG).	A decrease in species was evident after 15 days of EEN ($p = 0.037$). Diversity returned to baseline when patients were back to a normal diet but remained lower (at any time) compared to controls. At the community level, EEN made the CD microbial even more dissimilar to that of healthy controls. 34 genera significantly were reduced over the EEN course (including *F. prausnitzii*); only *Lactococcus* increased with EEN.
Lewis JD et al., 2015 [31]	Prospective controlled	90 active CD children (age: 10.1–15.5 years): 52 anti-TNF 21 EEN, 16 PEN 26 Healthy controls (age: 7.9–19.9 years, data collected from a previous study)	Probiotics in the previous two weeks, children with an ostomy	366 Fecal samples collected at baseline, one to four and eight weeks into therapy	EEN: 90% of calories from a not specified dietary formula; PEN: 53% of calories from formula	Bacterial DNA sequenced using the Illumina HiSeq method.	Microbiota composition changed within one week of EEN, moving farther from the centroid of healthy controls (overall $p = 0.05$, among responders $p = 0.02$, among non responders $p = 0.14$). Abundance of six genera changed after one week ($p = 0.05$). An opposite pattern was seen in anti-TNF treated patients (microbiota composition became similar to healthy controls in one week) and in PEN treated patients. At the end of the eight weeks both EEN and anti-TNF responders had a microbiota composition similar to healthy controls.
Kaakoush NO et al., 2015 [32]	Prospective, controlled	five newly diagnosed CD children, five healthy controls	Antibiotics or antinflammatory agents in the previous four weeks	39 fecal samples collected at baseline (at diagnosis, prior to bowel cleansing for endoscopy) and then at one, two, four, eight, 12, 16, and 26 weeks after diagnosis.	eight to 12 weeks of a polymeric formula	16S rRNA gene and whole-genome high throughout sequencing	The number of OTUs decreased during EEN in responder patients (no statistical analysis indicated).

Table 1. *Cont.*

Author, Year	Type of Study	Groups of Subjects (Number, Characteristics)	Exclusion Criteria (ATBs, Probiotics, Other Drugs)	Biological Sample (Type and Number of Samples)	Type of Formula and Duration of Treatment	Microbiota Analysis	Outcomes
Schwerd T et al., 2016 [33]	Prospective	15 CD children, 12 newly diagnosed (mean age: 13.5 years, SD: 2.2 years)	Not indicated	24 fecal samples collected from eight CD subjects at baseline, week two, and at cessation of EEN	Polymeric formula TGF-β2 enriched in 14 patients, elemental formula in one patient	High-throughput 16S rRNA gene sequencing	Altered fecal bacteria composition was seen after two weeks of EEN (bacterial profiles clustered from pre-EEN). EEN decreased the abundance of phylum Bacteroidetes ($p = 0.039$) and increased the abundance of Firmicutes ($p = 0.027$).
Guinet-Charpentier C et al., 2016 [34]	Prospective, controlled	34 CD children (median age 14.8 years, range 6.5–21); four children on EEN, eight on PEN, 22 on other treatments	Not indicated	Fecal samples collected at baseline, two weeks, and six weeks after EEN (three patients)	Polymeric formula TGF-β2 enriched	MiSeq sequencing of the 16S rRNA gene	A decrease in genera from the Proteobacteria phylum (particularly *Sutterella*) was observed ($p < 0.05$), whereas *Alistipes* ($p < 0.05$) and *Bifidobacterium* ($p < 0.1$) increased during EEN.
Dunn KA et al., 2016 [35]	Prospective, controlled	10 children with active CD (age 10 to 16 years) on EEN for 12 weeks; five controls (CD relatives) on normal diet	Use of other medications (including antibiotics) was not an exclusion criteria	19 Fecal samples from CD patients collected at baseline (at least 48 h after bowel preparation) and week 12	12 weeks of EEN by NG tube	High-throughput sequencing of the 16S rRNA gene targeting the V6-V8 region performed on the Illumina MiSeq platform	Species diversity (Chao-1 index) decreased among sustained remission (SR) samples, whereas it increased among the non-SR samples over the course. Taxonomic composition changed over the course of EEN treatment (no specific statistic measure is indicated).
Dunn KA et al., 2016 [36]	Prospective, controlled	15 CD patients (aged 10–16 years), five controls (age nine to 14 years, relatives of CD patients)	Use of other medications (including antibiotics) was not an exclusion criteria	33 CD patient samples (15 at baseline and 18 at various times at or after the end of EEN treatment (week 12). Five samples from healthy controls	12 weeks of polymeric formula	Metagenomic data obtained by next-generation sequencing (NGS) (Illumina MiSeq). Sequences were compared to 28 complete microbial genomes annotated with KEGG.	Eight KEGG pathways differed significantly between baseline CD patients and controls ($p < 0.05$). SR patients had greater similarity to controls than NSR patients in all cases.

4.2. Methodologies of Microbiota Determination

The earliest study was conducted separating PCR-amplified fragments of 16S rDNA in polyacrylamide gels containing a gradient of denaturing agents (DGGE) or in a temperature gradient (TGGE) [23]. With this method, heteroduplexes of different amplicons (with different G/C contents) were dissociated at different positions in the denaturing or temperature gradient, resulting in a hold of migration. The result was a pattern of bands, which is characteristic of the bacterial community present in the sample. A combination of electrophoresis and real time PCR was used in the studies by Leach et al. and Gerasimidis et al., where bacterial 16S rRNA genes were amplified from stools using primers for different groups of bacteria, allowing specific analysis of each bacteria group [24,29]. Real time quantitative PCR was used in two further studies [25,29]. The most recent studies were performed using next generation sequencing (NGS) techniques [28,30–35]. Microbiota was extracted from fecal samples in all but one single case report, where ileal histological specimens were analyzed instead [28].

4.3. Overall Effects Induced by EEN on Microbial Composition, Diversity and Abundance

A significant decrease in the bacterial diversity during EEN treatment was observed in different studies using different techniques. Gerasimidis et al. demonstrated a decrease in the bacterial diversity richness (calculated as the total number of bands on the TTGE images) after 30 days of EEN ($p = 0.041$), persisting at the end of EEN ($p = 0.037$) [29]. In the study by Quince et al. diversity in species (represented by the Shannon diversity index) decreased already after two weeks of EEN ($p = 0.037$) [30]. In the same study, at the species level, EEN increased the microbiological distance between CD patients and healthy controls. Lewis et al. reported the same effect from increasing the distance from the healthy subjects microbiota composition with EEN [31]. A reduction in the number of operational taxonomic units (OTUs) was reported also by Kaakoush and collaborators [32]. Quince et al. quantified such reduction, demonstrating a drop of 0.6 points in the Shannon diversity index calculated at the genus level every 10 days of EEN [30]. This corresponds approximately to a reduction in genus diversity of 20% after one month of EEN. Conversely, in the small study by Schwerd et al., the Shannon diversity index was not affected by the nutritional treatment [33]. The authors attributed this unexpected result to the very stringent parameters used to filter OTUs. Also in the study by Shiga et al., species diversity was not reduced by EEN, in contrast with the effect induced by total parenteral nutrition (TPN) [26]. The reported reduction in microbial diversity seemed to correlate well with the achievement of clinical remission induced by EEN [32] and to reverse when patients returned to a normal diet [29,30,32].

4.4. Effects on Specific Bacterial Species or Strains

The effect of the EEN on specific bacteria species varied largely between the different studies. Quince et al. described a statistically significant reduction in 33 genera over the EEN course (including genera that were already less abundant in CD at baseline compared to controls, including *Faecalibacteria*, *Bifidobacteria*, and *Ruminococcaceae*) [30]. In the same report, only *Lactococcus* increased significantly with EEN ($p = 0.017$). A diet-induced reduction in the abundance of *Faecalibacterium prausnitzii* (belonging to Firmicutes phylum), a bacterium previously considered protective towards the development and flare-up of CD [37,38], was reported both by Jia et al. in adults (a significant decrease was observed for the A2-165 subgroup, $p = 0.0046$) and by Gerasimidis et al. ($p = 0.023$) and Quince et al. (decrease of -0.0144 log genera abundance, $p = 0.068$) in CD children [25,29,30]. However, in the paper by Kaakoush et al., the authors observed that different OTUs classified under *Faecalibacterium* responded differently to EEN therapy [32], and, in contrast, Schwerd et al. described an increase in the relative abundance of Firmicutes ($p = 0.0227$), particularly in members of the family Christensenellaceae, with no mention of changes in *F. prausnitzii* [33]. Among the Bacteroidetes, a significant reduction in the relative sequence abundance of the phylum ($p = 0.039$) was reported by Schwerd et al. [33], and a decrease in *Bacteroides fragilis* ($p = 0.034$) was described by Shiga et al. [26].

A trend for decreasing concentrations of *Bifidobacterium* genus, belonging to Actinobacteria phylum (at 60 days of EEN $p = 0.120$), was reported by Gerasimidis et al., with a subsequent increase when patients were back to a normal free diet ($p = 0.031$) [29]. In the recent report by Guinet-Charpentier et al., a decrease in genera from the Proteobacteria phylum (particularly *Sutterella*, $p < 0.05$) was observed, whereas *Alistipes* ($p < 0.05$) and *Bifidobacterium* ($p < 0.1$) increased over the course of EEN [34]. Table 2 summarizes specific bacterial changes induced by EEN.

Table 2. Specific bacterial changes induced by exclusive enteral nutrition (EEN).

Increased during EEN	Decreased during EEN
Firmicutes	
Relative abundance of Firmicutes ($p = 0.227$) [33]	Levels of A2-165 *Faecalibacterium prausnitzii* ($p = 0.0046$) [25]
	Levels of M21/2 *Faecalibacterium prausnitzii* ($p = 0.61$) [25]
	Concentration (log10 16S Ribosomal RNA Gene Copy Number/g of dry stool) of *Faecalibacterium prausnitzii* ($p = 0.002$) [29]
	Relative abundance of *Faecalibacterium* ($p = 0.068$) [30]
Relative abundance of *Lactococcus* ($p = 0.017$) [30]	Relative abundance of *Dialister* ($p = 0.04$) [30]
Relative abundance of Christensenellaceae ($p = 0.0237$) [33]	Relative abundance of Ruminococcacae ($p = 0.04$) [30]
	Relative abundance of *Subdoligranulum* ($p = 0.023$) [30]
Bacteroidetes	
Concentration of *Bacteroides* ($n = 1$, p not reported) [28]	Concentration (log 10 cells per g of faeces) of *Bacteroides fragilis* ($p = 0.034$) [26]
Abundance of *Alistipes* ($p < 0.05$) [34]	Concentration of *Bacteroides/Prevotella* ($p = 0.053$) [29]
	Concentration of *Prevotella* ($p = 0.27$) [30]
	Relative abundance of Bacteroidetes ($p = 0.039$) [33]
Proteobacteria	
	Concentration of Proteobacteria ($n = 1$, p not reported) [28]
	Concentration of *Sutterella* ($p < 0.05$) [34]
Actinobacteria	
Abundance of *Bifidobacterium* ($p < 0.1$) [34]	Abundance of Bifidobacteriaceae genus ($p = 0.005$) [30]
	Concentration of *Bifidobacteria* ($p = 0.003$) [29]

4.5. EEN Induced Microbiota Changes and Relation to Disease Activity and Remission

The dietary treatment, in line with the literature, led to a high remission rate in CD subjects (up to 90%). The decline in specific bacteria, particularly the levels of *F. prausnitzii* [25,29] and *Bacteroides-Prevotella* [29], was found to correlate with the achievement of clinical remission. In the recent study by Guinet-Charpentier et al., patients responding to EEN and in clinical remission showed a reduction in *Dialister*, *Blautia*, unclassified Ruminococcaceae, and *Coprococcus* compared with patients in remission with other treatments such as anti-TNF and partial enteral nutrition (the clustering based on microbial distribution was significant with Monte Carlo, p value $= 0.029$, based on 10,000 replicates) [34].

In the recent study by Dunn and collaborators, interesting differences emerged between patients that achieved and maintained remission at week 24 after EEN (sustained remission-SR) and patients who did not achieve or maintain remission (non-SR) [35]. Species richness (estimated according to the Chao-1 index) was higher among the SR group (no statistical significance was shown in the text). Over the EEN course, species diversity tended to decrease in the SR group and to increase in the non-SR subjects. Furthermore, the taxonomic composition of the SR group was much more similar to healthy controls at the principle coordinate analysis than the non-SR group (no statistical calculation was made). *Akkermansia muciniphila* and *Bacteroides* were particularly prevalent in the SR group, whereas the non-SR group observed a prevalence of Proteobacteria. The authors reported that their 'proposed microbiological model' (based on the microbiota composition of pre-EEN samples) can predict sustained response to EEN with an accuracy of 80% [35]. The same group also looked at differences in the abundance of specific metabolic pathways in SR patients, compared to controls and non-SR patients. The authors initially identified eight metabolic pathways that differed significantly between CD patients and controls (p value < 0.1); the comparison of these eight pathways showed that SR patients were much more similar to controls compared to non-SR patients in the abundance of these pathways (non-SR patients differed significantly in the abundance of seven of the eight pathways, while SR-patients differed significantly from the controls in the abundance of two pathways only, p value < 0.05, after correction according to the method of Benjamini and Hochberg). However no significant difference was detected between SR and non-SR samples, possibly due to the small number of samples (10 in total) [36].

Finally, Kaakoush et al. reported re-colonization with specific microbial taxa belonging to six Firmicutes families and that increases in OTUs were correlated with disease recurrence) [32].

4.6. Effects on Metabolic Pathways

Short chain fatty acids (SCFAs) are produced in the intestinal lumen by the anaerobic fermentation of non-digestible dietary residues and endogenous epithelial-derived mucus. They are readily absorbed and used as energy sources by colonocytes and also by other tissues [39]. The fecal pool of SCFAs is correlated both to diet and to the microbiota composition and abundance. The three major luminal SCFAs are acetate (acetic acid), propionate (propionic acid), and butyrate (butyric acid). Butyrate is the most extensively studied, and its inflammatory properties are well documented [40]. The anti-inflammatory capacity of acetate is less documented, and interestingly it is also commonly used to induce inflammatory colitis in animal models [41,42], also suggesting pro-inflammatory properties.

Two studies specifically investigated the effects of EEN on the bacterial metabolism. Tjielstrom et al. [27] showed that EEN induced a significant reduction (p < 0.05) in the fecal median concentration of the acetic-acid. A parallel increase in anti-inflammatory SCFAs (butyric and valeric acids) was described in the same paper [27], although this change did not reach statistical significance. The same effects were not encountered in children with perianal CD. In the study by Gerasimidis et al., fecal pH and total sulfide increased, while butyric acid decreased during EEN, and these changes reverted to baseline on a free diet [29]. The metagenomics approach used by Quince et al. revealed that the effect of the EEN was associated with a reduction in the expression of genes involved in biotin and thiamine biosynthesis and a parallel increased expression of genes involved in spermidine/putrescine biosynthesis or in the shikimate pathway [30]. Changes in genes involved in the byosinthesis of vitamin B complex can potentially reflect a decrease in bacteria that express genes encoding these vitamins (e.g., bifidobacteria, *Lactobacillus reuteri*, and *E. coli*) or in the production of short or medium chain fatty acids that require these vitamins, or, alternatively, the supplementation of vitamin B complex by EEN may reduce bacterial production. The shikimate pathway is an alternative metabolic route for the biosynthesis of aromatic amino acids (phenylalanine, tyrosine, and tryptophan) employed by microorganisms and plants but not by animals and humans. The biosynthesis and transport of spermidine/putrescine and the shikimate pathway, implicated in the essential aminoacids synthesis, are both fundamental for cell growth and the overexpression of these genes can potentially indicate tissue regeneration [43].

In the recent study by Dunn et al., eight metabolic pathways were initially identified as substantially different in CD patients compared to controls, including pathways involved in xenobiotic and environmental pollutant degradation, succinate metabolism, bacterial HtpG, fatty acid metabolism, and the nucleotide-binding oligomerization domain (NOD)-like receptor signaling pathway. The abundance of pathways implicated in the degradation of environmental pollutants and xenobiotic degradation showed an increase in non-SR patients, whereas pathways involved in NOD-like receptor signaling were reduced in non-SR patients [36].

5. Discussion

The modification of the intestinal microbiota, commonly reported both in GI and non-GI diseases and widely described in IBD, is one principle potential mechanism behind the efficacy of EEN. In the last 15 years, some studies have tried to elucidate this aspect (which may further contribute to a general comprehension of IBD pathogenesis). Our review aimed to describe the literature on this specific topic, revealing a low number of underpowered and methodologically heterogeneous studies. However, the most recent trials, performed following strict inclusion criteria and using new generation techniques, led to intriguing results.

The origin of the microbial sample is a critical issue in studies describing intestinal bacterial composition. In fact, fecal samples have a big advantage in the low cost and non-invasiveness of the sampling, but the composition of the fecal bacterial population can be extremely different from the mucosal (ileal or colonic) one [6,44,45]. Only the well-detailed case-report by D'Argenio et al. [28] evaluated the intestinal microbiota based on histological samples. However, the importance of the sampling site is minimized by looking at the longitudinal changes induced by the EEN treatment. Both antibiotics and probiotics have been demonstrated to significantly affect the bacterial microbiota community in the GI tract [46,47]. Furthermore, bowel preparation can affect the composition and the diversity of both fecal and luminal microbiota [48]. As previously stated, only six of the considered studies in this review clearly indicated the use of antibiotics as an exclusion criteria, and, in two studies, patients on probiotics were excluded; additionally, bowel preparation was reported only in two studies [23,35]. We should consider these methodological pitfalls as possible limitations when extending the results of this review.

Overall, the more recent and well-documented studies seem to indicate a paradoxical effect of EEN on the intestinal microbiota. In fact, a reduction in microbial diversity and/or richness induced by EEN was found in three recent and well-designed studies [29,30,32] that all together examined 43 CD patients on EEN. The diversity at the end of EEN was found to be significantly reduced in comparison with both the microbial diversity of pre-treatment samples and the microbial diversity of healthy controls. This effect reverses following the resumption of a normal diet [29,30]. The transitory effect of EEN could explain the reason why this dietetic approach has a very high efficacy as an induction treatment and is able to induce mucosal healing but is less effective in maintenance of remission. These results are in accordance with animal studies [49], demonstrating a decrease in the diversity of bacterial species in interleukin-10 (IL 10) deficient mice fed with an elemental diet compared with mice fed with the regular diet. Interestingly, although limited by the paucity of data and the early techniques, in healthy subjects enteral nutrition also seems to induce a reduction of the total intestinal bacterial count [50]. The same effect has been described in pigs fed with EEN compared to those receiving TPN [51]. Possible explanations are the absence of dietary fiber, with consequent reductions in the exogenous carbohydrate available for fermentation, and/or an increase in intestinal transit time, which may independently reduce microbial mass [52].

The effects of the therapeutic formula on specific bacterial strains were very variable and not consistent between studies. This is not surprising considering both the small sample sizes and the variable methods used in different studies and the huge degree of inter-individual variation in microbiota composition normally encountered in different people. More frequently, studies reported a decrease in the abundance of Firmicutes following a course of EEN [25,29,30], thus increasing the difference of the IBD microbiota subjects from the healthy subjects' bacterial composition. In fact, in patients with IBD, a reduced representation of Firmicutes and a parallel increase in Proteobacteria are generally reported [53–55]. It is plausible, however, that this effect simply reflects a general depletion of all the species already present in the gut at the start of the intervention rather than being a specific effect induced by EEN. In a recent review on the therapeutic utility of EEN in CD, Cuiv and their co-authors suggest that the effect of EEN on intestinal bacteria is mainly based on the limitation of growth and metabolic activity rather than the selection of a specific microbiological pattern [56]. The same authors proposed that the bowel rest induced by EEN may induce mucosal healing by limiting the activity of potentially pathogenic microbes and by enhancing repairing mechanisms (autophagy) [56]. The reduction at the end of EEN of *F. prausnitzii*, a butyrate producing bacteria considered to be protective towards CD development [37,38], reported by some studies [25,29,30] is in agreement with this hypothesis. In fact this reduction could simply be the expression of the general effect induced by EEN. Interestingly, relevant changes in abundance of the AIEC, another bacteria described as potentially pathogenic in IBD [57–59], were not reported in any of the studies. This lack of consistency in pointing out possible community differences might indicate that efforts should be probably directed towards identifying other potentially more relevant discrepancies; for example, at a metabolic or functional level.

6. Conclusions

Currently there is limited data on the microbiological effect induced by EEN in subjects with CD. However, the most recent and well-documented results suggest a paradoxic effect of EEN, consisting in a reduction of bacterial diversity and richness. Long-term, multicenter, and rigorous studies based on NGS technology comparing the effect of different treatments are expected to clarify the relevance of microbiological changes induced by the EEN.

Author Contributions: All the authors conceived this work. S.G. performed the literature search and the data extraction and interpreted the data. S.G., C.C., and M.D.A. drafted the manuscript. All the authors revised and approved the final version of the paper.

Conflicts of Interest: The authors declare no conflict of interest.

References

1. Orel, R.; Kamhi Trop, T. Intestinal microbiota, probiotics and prebiotics in inflammatory bowel disease. *World J. Gastroenterol.* **2014**, *20*, 11505–11524. [CrossRef] [PubMed]
2. Serban, D.E. The gut microbiota in the metagenomics era: Sometimes a friend, sometimes a foe. *Roum. Arch. Microbiol. Immunol.* **2011**, *70*, 134–140. [PubMed]
3. Hold, G.L.; Smith, M.; Grange, C.; Watt, E.R.; El-Omar, E.M.; Mukhopadhya, I. Role of the gut microbiota in inflammatory bowel disease pathogenesis: What have we learnt in the past 10 years? *World J. Gastroenterol.* **2014**, *20*, 1192–1210. [CrossRef] [PubMed]
4. De Cruz, P.; Prideaux, L.; Wagner, J.; Ng, S.C.; McSweeney, C.; Kirkwood, C.; Morrison, M.; Kamm, M.A. Characterization of the gastrointestinal microbiota in health and inflammatory bowel disease. *Inflamm. Bowel Dis.* **2012**, *18*, 372–390. [CrossRef] [PubMed]
5. Serban, D.E. Microbiota in Inflammatory Bowel disease pathogenesis and therapy: Is it all about diet? *Nutr. Clin. Pract.* **2015**, *30*, 760–779. [CrossRef] [PubMed]
6. Gevers, D.; Kugathasan, S.; Denson, L.A.; Vázquez-Baeza, Y.; Van Treuren, W.; Ren, B.; Schwager, E.; Knights, D.; Song, S.J.; Yassour, M.; et al. The treatment-naive microbiome in new-onset Crohn's disease. *Cell Host Microbe* **2014**, *15*, 382–392. [CrossRef] [PubMed]
7. Lee, D.; Albenberg, L.; Compher, C.; Baldassano, R.; Piccoli, D.; Lewis, J.D.; Wu, G.D. Diet in the pathogenesis and treatment of inflammatory bowel diseases. *Gastroenterology* **2015**, *148*, 1087–1106. [CrossRef] [PubMed]
8. Penagini, F.; Dilillo, D.; Borsani, B.; Cococcioni, L.; Galli, E.; Bedogni, G.; Zuin, G.; Zuccotti, G V. Nutrition in Pediatric Inflammatory Bowel Disease: From Etiology to Treatment. A Systematic Review. *Nutrients* **2016**. [CrossRef] [PubMed]
9. Ruemmele, F.M.; Veres, G.; Kolho, K.L.; Griffiths, A.; Levine, A.; Escher, J.C.; Amil Dias, J.; Barabino, A.; Braegger, C.P.; Bronsky, J.; et al. Consensus guidelines of ECCO/ESPGHAN on the medical management of pediatric Crohn's disease. *J. Crohns Colitis* **2014**, *8*, 1179–1207. [CrossRef] [PubMed]
10. Grover, Z.; Muir, R.; Lewindon, P. Exclusive enteral nutrition induces early clinical, mucosal and transmural remission in paediatric Crohn's disease. *J. Gastroenterol.* **2014**, *49*, 638–645. [CrossRef] [PubMed]
11. Borrelli, O.; Cordischi, L.; Cirulli, M.; Paganelli, M.; Labalestra, V.; Uccini, S.; Russo, P.M.; Cucchiara, S. Polymeric diet alone versus corticosteroids in the treatment of active pediatric Crohn's disease: A randomized controlled open-label trial. *Clin. Gastroenterol. Hepatol.* **2006**, *4*, 744–753. [CrossRef] [PubMed]

12. Berni Canani, R.; Terrin, G.; Borrelli, O.; Romano, M.T.; Manguso, F.; Coruzzo, A.; D'Armiento, F.; Romeo, E.F.; Cucchiara, S. Short- and long-term therapeutic efficacy of nutritional therapy and corticosteroids in paediatric Crohn's disease. *Dig. Liver Dis.* **2006**, *38*, 381–387. [CrossRef] [PubMed]

13. Ludvigsson, J.F.; Krantz, M.; Bodin, L.; Stenhammar, L.; Lindquist, B. Elemental versus polymeric enteral nutrition in paediatric Crohn's disease: A multicentre randomized controlled trial. *Acta Paediatr.* **2004**, *93*, 327–335. [CrossRef] [PubMed]

14. Verma, S.; Brown, S.; Kirkwood, B.; Giaffer, M.H. Polymeric versus elemental diet as primary treatment in active Crohn's disease: A randomized, double-blind trial. *Am. J. Gastroenterol.* **2000**, *95*, 735–739. [CrossRef] [PubMed]

15. Rubio, A.; Pigneur, B.; Garnier-Lengline, H.; Talbotec, C.; Schmitz, J.; Canioni, D.; Goulet, O.; Ruemmele, F.M. The efficacy of exclusive nutritional therapy in paediatric Crohn's disease, comparing fractionated oral vs. continuous enteral feeding. *Aliment. Pharmacol. Ther.* **2011**, *33*, 1332–1339. [CrossRef] [PubMed]

16. Ruemmele, F.M.; Pigneur, B.; Garnier-Lengliné, H. Enteral nutrition as treatment option for Crohn's disease: In kids only? *Nestlé Nutr. Inst. Workshop Ser.* **2014**, *79*, 115–123. [CrossRef] [PubMed]

17. Human Microbiome Project Consortium. Structure, function and diversity of the healthy human microbiome. *Nature* **2012**, *486*, 207–214.

18. Lozupone, C.A.; Stombaugh, J.I.; Gordon, J.I.; Jansson, J.K.; Knight, R. Diversity, stability and resilience of the human gut microbiota. *Nature* **2012**, *489*, 220–230. [CrossRef] [PubMed]

19. Shah, R.; Kellermayer, R. Associations of therapeutic enteral nutrition. *Nutrients* **2014**, *21*, 5298–5311. [CrossRef] [PubMed]

20. Lagier, J.C.; Hugon, P.; Khelaifia, S.; Fournier, P.E.; La Scola, B.; Raoult, D. The rebirth of culture in microbiology through the example of culturomics to study human gut microbiota. *Clin. Microbiol. Rev.* **2015**, *28*, 237–264. [CrossRef] [PubMed]

21. Schmidt, T.S.B.; Matias Rodrigues, J.F.; von Mering, C. Ecological consistency of SSU rRNA-based operational taxonomic units at a global scale. *PLoS Comput. Biol.* **2014**, *10*. [CrossRef] [PubMed]

22. Sharpton, T.J. An introduction to the analysis of shotgun metagenomic data. *Front. Plant. Sci.* **2014**. [CrossRef] [PubMed]

23. Lionetti, P.; Callegari, M.L.; Ferrari, S.; Cavicchi, M.C.; Pozzi, E.; de Martino, M.; Morelli, L. Enteral nutrition and microflora in pediatric Crohn's disease. *J. Parenter. Enter. Nutr.* **2005**, *29*, S173–S175. [CrossRef] [PubMed]

24. Leach, S.T.; Mitchell, H.M.; Eng, W.R.; Zhang, L.; Day, A.S. Sustained modulation of intestinal bacteria by exclusive enteral nutrition used to treat children with Crohn's disease. *Aliment. Pharmacol. Ther.* **2008**, *28*, 724–733. [CrossRef] [PubMed]

25. Jia, W.; Whitehead, R.N.; Griffiths, L.; Dawson, C.; Waring, R.H.; Ramsden, D.B.; Hunter, J.O.; Cole, J.A. Is the abundance of *Faecalibacterium prausnitzii* relevant to Crohn's disease? *FEMS Microbiol. Lett.* **2010**, *310*, 138–144. [CrossRef] [PubMed]

26. Shiga, H.; Kajiura, T.; Shinozaki, J.; Takagi, S.; Kinouchi, Y.; Takahashi, S.; Negoro, K.; Endo, K.; Kakuta, Y.; Suzuki, M.; et al. Changes of faecal microbiota in patients with Crohn's disease treated with an elemental diet and total parenteral nutrition. *Dig. Liver. Dis.* **2012**, *44*, 736–742. [CrossRef] [PubMed]

27. Tjellstrom, B.; Hogberg, L.; Stenhammar, L.; Magnusson, K.E.; Midtvedt, T.; Norin, E.; Sundqvist, T. Effect of exclusive enteral nutrition on gut microflora function in children with Crohn's disease. *Scand. J. Gastroenterol.* **2012**, *47*, 1454–1459. [CrossRef] [PubMed]

28. D'Argenio, V.; Precone, V.; Casaburi, G.; Miele, E.; Martinelli, M.; Staiano, A.; Salvatore, F.; Sacchetti, L. An altered gut microbiome profile in a child affected by Crohn's disease normalized after nutritional therapy. *Am. J. Gastroenterol.* **2013**, *108*, 851–852. [CrossRef] [PubMed]

29. Gerasimidis, K.; Bertz, M.; Hanske, L.; Junick, J.; Biskou, O.; Aguilera, M.; Garrick, V.; Russell, R.K.; Blaut, M.; McGrogan, P.; et al. Decline in presumptively protective gut bacterial species and metabolites are paradoxically associated with disease improvement in pediatric Crohn's disease during enteral nutrition. *Inflamm. Bowel. Dis.* **2014**, *20*, 861–871. [CrossRef] [PubMed]

30. Quince, C.; Zeeshan Ijaz, U.; Loman, N.; Eren, A.M.; Saulnier, D.; Russell, J.; Haig, S.J.; Calus, S.T.; Quick, J.; Barclay, A.; et al. Extensive modulation of the fecal metagenome in children with Crohn's disease during Exclusive Enteral Nutrition. *Am. J. Gastroenterol.* **2015**, *110*, 1718–1729. [CrossRef] [PubMed]

31. Lewis, J.D.; Chen, E.Z.; Baldassano, R.N.; Otley, A.R.; Griffiths, A.M.; Lee, D.; Bittinger, K.; Bailey, A.; Friedman, E.S.; Hoffmann, C.; et al. Inflammation, antibiotics, and diet as environmental stressors of the gut microbiome in pediatric Crohn's disease. *Cell Host Microbe* **2015**, *18*, 489–500. [CrossRef] [PubMed]
32. Kaakoush, N.O.; Day, A.S.; Leach, S.T.; Lemberg, D.A.; Nielsen, S.; Mitchell, H.M. Effect of exclusive enteral nutrition on the microbiota of children with newly diagnosed Crohn's disease. *Clin. Transl. Gastroenterol.* **2015**, *6*, 1–11. [CrossRef] [PubMed]
33. Schwerd, T.; Frivolt, K.; Clavel, T.; Lagkouvardos, I.; Katona, G.; Mayr, D.; Uhlig, H.H.; Haller, D.; Koletzko, S.; Bufler, P. Exclusive enteral nutrition in active pediatric Crohn disease: Effects on intestinal microbiota and immune regulation. *J. Allergy Clin. Immunol.* **2016**, *138*, 592–596. [CrossRef] [PubMed]
34. Guinet-Charpentier, C.; Lepage, P.; Morali, A.; Chamaillard, M.; Peyrin-Biroulet, L. Effects of enteral polymeric diet on gut microbiota in children with Crohn's disease. *Gut* **2017**, *66*, 194–195. [CrossRef] [PubMed]
35. Dunn, K.A.; Moore-Connors, J.; MacIntyre, B.; Stadnyk, A.; Thomas, N.A.; Noble, A.; Mahdi, G.; Rashid, M.; Otley, A.R.; Bielawski, J.P.; et al. The gut microbiome of pediatric Crohn's disease patients differs from healthy controls in genes that can influence the balance between a healthy and dysregulated immune response. *Inflamm. Bowel Dis.* **2016**, *22*, 2607–2618. [CrossRef] [PubMed]
36. Dunn, K.A.; Moore-Connors, J.; MacIntyre, B.; Stadnyk, A.W.; Thomas, N.A.; Noble, A.; Mahdi, G.; Rashid, M.; Otley, A.R.; Bielawski, J.P.; et al. Early changes in microbial community structure are associated with sustained remission after nutritional treatment of pediatric Crohn's disease. *Inflamm. Bowel Dis.* **2016**, *22*, 2853–2862. [CrossRef] [PubMed]
37. Sokol, H.; Pigneur, B.; Watterlot, L.; Lakhdari, O.; Bermúdez-Humarán, L.G.; Gratadoux, J.J.; Blugeon, S.; Bridonneau, C.; Furet, J.P.; Corthier, G.; et al. *Faecalibacterium prausnitzii* is an anti-inflammatory commensal bacterium identified by gut microbiota analysis of Crohn disease patients. *Proc. Natl. Acad. Sci. USA* **2008**, *105*, 16731–16736. [CrossRef] [PubMed]
38. Sokol, H.; Seksik, P.; Furet, J.P.; Firmesse, O.; Nion-Larmurier, I.; Beaugerie, L.; Cosnes, J.; Corthier, G.; Marteau, P.; Doré, J. Low counts of *Faecalibacterium prausnitzii* in colitis microbiota. *Inflamm. Bowel Dis.* **2009**, *15*, 1183–1189. [CrossRef] [PubMed]
39. McNeil, N.I. The contribution of the large intestine to energy supplies in man. *Am. J. Clin. Nutr.* **1984**, *39*, 338–342. [PubMed]
40. Vinolo, M.A.R.; Rodrigues, H.G.; Nachbar, R.T.; Curi, R. Regulation of Inflammation by Short Chain Fatty Acids. *Nutrients* **2011**, *3*, 858–876. [CrossRef] [PubMed]
41. Elson, C.O.; Sartor, R.B.; Tennyson, G.S.; Riddell, R.H. Experimental models of Inflammatory bowel disease. *Gastroenterology* **1995**, *109*, 1344–1367. [CrossRef]
42. Fabia, R.; Willén, R.; Ar'Rajab, A.; Andersson, R.; Ahrén, B.; Bengmark, S. Acetic acid-induced colitis in the rat: A reproducible experimental model for acute ulcerative colitis. *Eur. Surg. Res.* **1992**, *24*, 211–225. [CrossRef] [PubMed]
43. Slezak, K.; Hanske, L.; Loh, G.; Blaut, M. Increased bacterial putrescine has no impact on gut morphology and physiology in gnotobiotic adolescent mice. *Benef. Microbes* **2013**, *4*, 253–266. [CrossRef] [PubMed]
44. Stearns, J.C.; Lynch, M.D.; Senadheera, D.B.; Tenenbaum, H.C.; Goldberg, M.B.; Cvitkovitch, D.G.; Croitoru, K.; Moreno-Hagelsieb, G.; Neufeld, J.D. Bacterial biogeography of the human digestive tract. *Sci. Rep.* **2011**. [CrossRef] [PubMed]
45. Lavelle, A.; Lennon, G.; O'Sullivan, O.; Docherty, N.; Balfe, A.; Maguire, A.; Mulcahy, H.E.; Doherty, G.; O'Donoghue, D.; Hyland, J.; et al. Spatial variation of the colonic microbiota in patients with ulcerative colitis and control volunteers. *Gut* **2015**, *64*, 1553–1561. [CrossRef] [PubMed]
46. Dethlefsen, L.; Relman, D.A. Incomplete recovery and individualized responses of the human distal gut microbiota to repeated antibiotic perturbation. *Proc. Natl. Acad. Sci. USA* **2011**, *108*, 4554–4561. [CrossRef] [PubMed]
47. Yang, B.; Xiao, L.; Liu, S.; Liu, X.; Luo, Y.; Ji, Q.; Yang, P.; Liu, Z. Exploration of the effect of probiotics supplementation on intestinal microbiota of food allergic mice. *Am. J. Transl. Res.* **2017**, *9*, 376–385. [PubMed]
48. Shobar, R.M.; Velineni, S.; Keshavarzian, A.; Swanson, G.; DeMeo, M.T.; Melson, J.E.; Losurdo, J.; Engen, P.A.; Sun, Y.; Koenig, L.; Mutlu, E.A. The Effects of Bowel Preparation on Microbiota-Related Metrics Differ in Health and in Inflammatory Bowel Disease and for the Mucosal and Luminal Microbiota Compartments. *Clin. Transl. Gastroenterol.* **2016**, *7*, e143. [CrossRef] [PubMed]

49. Kajiura, T.; Takeda, T.; Sakata, S.; Sakamoto, M.; Hashimoto, M.; Suzuki, H.; Suzuki, M.; Benno, Y. Change of intestinal microbiota with elemental diet and its impact on therapeutic effects in a murine model of chronic colitis. *Dig. Dis. Sci.* **2009**, *54*, 1892–1900. [CrossRef] [PubMed]

50. Whelan, K.; Judd, P.A.; Preedy, V.R.; Simmering, R.; Jann, A.; Taylor, M.A. Fructooligosaccharides and fiber partially prevent the alterations in fecal microbiota and short-chain fatty acid concentrations caused by standard enteral formula in healthy humans. *J. Nutr.* **2005**, *135*, 1896–1902. [PubMed]

51. Harvey, R.B.; Andrews, K.; Droleskey, R.E.; Kansagra, K.V.; Stoll, B.; Burrin, D.G.; Sheffield, C.L.; Anderson, R.C.; Nisbet, D.J. Qualitative and quantitative comparison of gut bacterial colonization in enterally and parenterally fed neonatal pigs. *Curr. Issues Intest. Microbiol.* **2006**, *7*, 61–64. [PubMed]

52. Stephen, A.M.; Wiggins, H.S.; Cummings, J.H. Effect of changing transit time on colonic microbial metabolism in man. *Gut* **1987**, *28*, 601–609. [CrossRef] [PubMed]

53. Nagalingam, N.A.; Lynch, S.V. Role of the microbiota in inflammatory bowel diseases. *Inflamm. Bowel Dis.* **2012**, *18*, 968–984. [CrossRef] [PubMed]

54. Frank, D.N.; Robertson, C.E.; Hamm, C.M.; Kpadeh, Z.; Zhang, T.; Chen, H.; Zhu, W.; Sartor, R.B.; Boedeker, E.C.; Harpaz, N.; et al. Disease phenotype and genotype are associated with shifts in intestinal-associated microbiota in inflammatory bowel diseases. *Inflamm. Bowel Dis.* **2011**, *17*, 179–184. [CrossRef] [PubMed]

55. Sartor, R.B. The intestinal microbiota in inflammatory bowel diseases. *Nestle Nutr. Inst. Workshop Ser.* **2014**, *79*, 29–39. [CrossRef] [PubMed]

56. Cuív, P.Ó.; Begun, J.; Keely, S.; Lewindon, P.J.; Morrison, M. Towards an integrated understanding of the therapeutic utility of exclusive enteral nutrition in the treatment of Crohn's disease. *Food Funct.* **2016**, *7*, 1741–1751. [CrossRef] [PubMed]

57. Sasaki, M.; Sitaraman, S.V.; Babbin, B.A.; Gerner-Smidt, P.; Ribot, E.M.; Garrett, N.; Alpern, J.A.; Akyildiz, A.; Theiss, A.L.; Nusrat, A.; Klapproth, J.M. Invasive *Escherichia coli* are a feature of Crohn's disease. *Lab. Investig.* **2007**, *87*, 1042–1054. [CrossRef] [PubMed]

58. Rolhion, N.; Darfeuille-Michaud, A. Adherent-invasive *Escherichia coli* in inflammatory bowel disease. *Inflamm. Bowel Dis.* **2007**, *13*, 1277–1283. [CrossRef] [PubMed]

59. Martinez-Medina, M.; Aldeguer, X.; Lopez-Siles, M.; González-Huix, F.; López-Oliu, C.; Dahbi, G.; Blanco, J.E.; Blanco, J.; Garcia-Gil, L.J.; Darfeuille-Michaud, A. Molecular diversity of *Escherichia coli* in the human gut: New ecological evidence supporting the role of adherent-invasive *E. coli* (AIEC) in Crohn's disease. *Inflamm. Bowel Dis.* **2009**, *15*, 872–878. [CrossRef] [PubMed]

nutrients

MDPI

Review

Evidence of the Anti-Inflammatory Effects of Probiotics and Synbiotics in Intestinal Chronic Diseases

Julio Plaza-Díaz [1,2,3], Francisco Javier Ruiz-Ojeda [1,2,3], Laura Maria Vilchez-Padial [2] and Angel Gil [1,2,3,4,*]

[1] Department of Biochemistry and Molecular Biology II, School of Pharmacy, University of Granada, Granada 18071, Spain; jrplaza@ugr.es (J.P.-D.); fruizojeda@ugr.es (F.J.R.-O.)
[2] Institute of Nutrition and Food Technology "José Mataix", Biomedical Research Center, University of Granada, Armilla, Granada 18016, Spain; lauramvilchez@gmail.com
[3] Instituto de Investigación Biosanitaria ibs., GRANADA, Complejo Hospitalario Universitario de Granada, Granada 18014, Spain
[4] CIBEROBN (Physiopathology of Obesity and Nutrition CB12/03/30038), Instituto de Salud Carlos III (ISCIII), Madrid 28029, Spain
* Correspondence: agil@ugr.es; Tel.: +34-958-241-000 (ext. 20307)

Received: 30 April 2017; Accepted: 24 May 2017; Published: 28 May 2017

Abstract: Probiotics and synbiotics are used to treat chronic diseases, principally due to their role in immune system modulation and the anti-inflammatory response. The present study reviewed the effects of probiotics and synbiotics on intestinal chronic diseases in in vitro, animal, and human studies, particularly in randomized clinical trials. The selected probiotics exhibit in vitro anti-inflammatory properties. Probiotic strains and cell-free supernatants reduced the expression of pro-inflammatory cytokines via action that is principally mediated by toll-like receptors. Probiotic administration improved the clinical symptoms, histological alterations, and mucus production in most of the evaluated animal studies, but some results suggest that caution should be taken when administering these agents in the relapse stages of IBD. In addition, no effects on chronic enteropathies were reported. Probiotic supplementation appears to be potentially well tolerated, effective, and safe in patients with IBD, in both CD and UC. Indeed, probiotics such as *Bifidobacterium longum* 536 improved the clinical symptoms in patients with mild to moderate active UC. Although it has been proposed that probiotics can provide benefits in certain conditions, the risks and benefits should be carefully assessed before initiating any therapy in patients with IBD. For this reason, further studies are required to understand the precise mechanism by which probiotics and synbiotics affect these diseases.

Keywords: probiotics; intestinal diseases; anti-inflammatory effects; inflammatory bowel diseases

1. Introduction

In 1907, the Russian Ilya Ilyich Mechnikov suggested that microbial ingestion improved host health. Indeed, he hypothesized that the consumption of lactic-acid-producing bacteria (LAB) strains found in yogurt might enhance longevity [1].

LAB is a heterogeneous group of microorganisms that are often present in a person's gut, introduced through the ingestion of fermented foods, as well as in the gastrointestinal and urogenital tract of animals. Some of these strains have probiotic effects [2]. In particular, strains belonging to *Bifidobacterium*, *Enterococcus*, and *Lactobacillus* are the most widely used probiotic bacteria [3–5].

Werner Kollath was probably the first person to use the word "probiotic" in 1953 [6]. In current use, the term refers to microorganisms that confer a health benefit to the host when administered in adequate amounts [7,8]. In addition, dead bacteria and bacterial molecular components may exhibit probiotic properties [9]. In 2014, the International Scientific Association for Probiotics and Prebiotics stated that the development of metabolic by-products, dead microorganisms, or other microbial-based, nonviable products has potential; however, these do not fall under the probiotic construct [10].

When ingested, probiotics produce microbial transformation in the intestinal microbiota and exert several health-promoting properties, including maintenance of the gut barrier function and modulation of the host immune system [11–16].

By contrast, a prebiotic is a non-viable food component that confers a health benefit on the host that is associated with the modulation of the intestinal microbiota. Prebiotics may be a fiber, but a fiber is not necessarily a prebiotic. Using prebiotics and probiotics in combination is often described as synbiotics, but only if the net health benefit is synergistic [17,18].

Hence, probiotics and synbiotics are consumed in numerous and diverse forms, such as yogurt and fermented milks, cheese, and other fermented foods. The use of probiotics and synbiotics in preventive medicine to maintain a healthy intestinal function is well documented. In addition, both probiotics and synbiotics have been proposed as therapeutic agents for gastrointestinal disorders and other pathologies [19,20].

Intestinal diseases, particularly infectious illnesses, were first recognized as a major health issue in developing countries. However, intestinal chronic diseases are more prevalent in developed regions, and their incidence has continued to increase over the past several decades in various regions worldwide [21]. The exact pathogenic mechanism of the onset of selected intestinal chronic diseases remains mostly unexplained [22]. The principal clinical manifestations of intestinal chronic diseases are inflammatory bowel diseases (IBDs), necrotizing enterocolitis (NEC), and malabsorption syndromes.

IBD is a term used to describe four pathologies: ulcerative colitis (UC), Crohn's disease (CD), pouchitis, and microscopic colitis. These conditions are systemic disorders that affect the gastrointestinal tract and have frequent extraintestinal manifestations [23,24], in which the epithelial barrier function is a critical factor for onset. In addition, native immunity and commensal enteric bacteria play a role [25,26].

The current treatment of IBD first involves the induction of remission, which is followed by maintaining remission. Patients with an active disease are treated with topical or systemic 5-aminosalicylic acids (5-ASA), corticosteroids, or immunomodulators, such as azathioprine and 6-mercaptopurine, in addition to anti-TNF monoclonal antibodies [27,28].

Data from clinical trials indicate that certain intestinal disease conditions, including NEC, pouchitis, UC, and irritable bowel syndrome (IBS), have yielded clinical benefits with some probiotic and synbiotic interventions [29], because probiotics largely act directly or indirectly on the intestinal microbiota [30].

Several experimental methods are available to assess the effect of probiotics or synbiotics on intestinal diseases, especially their anti-inflammatory properties. Both in vitro and in vivo studies have been conducted. In vitro studies principally involve intestinal porcine epithelial cells (IPEC) J2, CaCo-2 cells, human dendritic cells (DC) obtained from peripheral blood and umbilical cord blood, monocyte-derived DCs, peripheral blood mononuclear cells, and intestinal T cells [31–33]. In vivo studies involve animal models (mice, rats, and dogs) with chemical inflammation induction or human patients. Despite this variety of studies, however, the mechanisms underlying the beneficial effects of probiotics or synbiotics remain incompletely understood.

Therefore, the present review was conducted to investigate the anti-inflammatory effects of probiotics and synbiotics on chronic intestinal diseases in in vitro and in vivo studies, as well as current evidence from human randomized clinical trials (RCTs).

2. Materials and Methods

A comprehensive search of the relevant literature was performed using electronic databases, including MEDLINE (PubMed), EMBASE, and the Cochrane Library. MEDLINE through PubMed was searched for scientific articles that were published between 2010 and 2017 in English using the MeSH terms "probiotics" and "synbiotics", combined with "intestinal diseases," "Crohn's disease," and "ulcerative colitis". We evaluated the results obtained using the following equation search: ("intestinal diseases"(All Fields) OR "Crohn's disease"(All Fields) OR "colitis, Ulcerative"(All Fields)) and ("probiotics"(All Fields) OR "synbiotics"(All Fields)). Our search yielded 326 articles. A total of 38 articles were selected. Additionally, we searched the reference lists of the included articles for potential relevant literature.

3. Results

3.1. In Vitro Studies

Substantial evidence from in vitro studies suggests that known and potential probiotics exhibit strain-specific anti-inflammatory effects. A large inventory of animal and human cell lines is available as models of the gut [3], including DCs, porcine intestinal epith:liocyte (PIE) cells, intestinal epithelial cells (IEC)-6, HT-29, and IPEC-J2. In most of the in vitro experimental models, epithelial cells were cultivated as monolayers in which the establishment of a functional epithelial feature was not achieved.

DCs generate primary T-cell responses and mediate intestinal immune tolerance to prevent overt inflammation in response to gut microbiota. Indeed, DCs play a key role in UC pathogenesis [32,33]. *Lactobacillus casei* Shirota (LcS) was tested on human DCs from healthy controls and active UC patient samples. DCs from UC patients exhibit a reduced stimulatory capacity for the T-cell response and an enhanced expression of skin-homing markers, such as cutaneous lymphocyte-associated antigen (CLA) and C-C motif chemokine receptor 4 (CCR4) on stimulated T-cells. Those responses were characterized by increased interleukin (IL)-4 production and a loss of IL-22 and interferon (IFN)-γ secretion. LcS treatment restored the normal stimulatory capacity via a reduction in Toll-like receptor (TLR)-2 and TLR4 expression [32,33].

TLRs are transmembrane proteins expressed on various immune and non-immune cells, such as B-cells, natural killer cells, DCs, macrophages, fibroblast cells, epithelial cells, and endothelial cells. They are members of a family of evolutionarily conserved pattern recognition receptors that identify a wide range of microbial components [34].

Lactobacillus plantarum strain CGMCC1258 has a dual effect in an IPEC-J2 model that involves epithelial permeability, the expression of inflammatory cytokines, and an abundance of tight junction proteins. In this model, the damage was induced by enterotoxigenic *Escherichia coli* K88. The aforementioned probiotic strain decreased the transcript levels of IL-8, tumor necrosis factor (TNF-α), and negative regulators of TLRs, such as the single Ig Il-1-related receptor (SIGIRR), B-cell CLL/lymphoma 3 (Bcl3), and mitogen-activated protein kinase phosphatase-1 (MKP-1). Moreover, *L. plantarum* treatment reduced the gene and protein expression of occludin [35]. These results indicated that *L. plantarum* reduced the expression of pro-inflammatory cytokines induced by *E. coli* K88, possibly by modulating the TLR, nuclear factor kappa-B (NF-κB), and mitogen-activated protein kinase (MAPK) pathways [35].

The anti-inflammatory effects of *Lactobacillus delbrueckii* subsp. *delbrueckii* TUA4408L and its extracellular polysaccharide against *E. coli* 987P were evaluated in PIE cells. The activation of the MAPK and NF-κB pathways induced by *E. coli* 987P was downregulated via the upregulation of TLR negative regulators. In fact, TLR2 had a principal role in the immunomodulatory action of the probiotic strain [36].

Cell-free supernatants (CFS) from *E. coli* Nissle 1917 and *Lactobacillus rhamnosus* GG were evaluated for their capacity to prevent 5-fluorouracil-induced damage to IEC-6. Pre-treatment with the supernatants of those strains prevents or inhibits enterocyte apoptosis and the loss of the intestinal barrier function induced by 5-fluorouracil, potentially forming the basis of a preventative treatment modality for mucositis [37].

Finally, our investigation group assessed in vitro studies related to probiotics and their anti-inflammatory effects. Indeed, selected probiotics have been shown to modulate immune responses and inflammatory biomarkers in human DCs generated from CD34+ progenitor cells (hematopoietic stem cells) harvested from umbilical cord blood. The DCs exhibited surface antigens of dendritic Langerhans cells similar to the lamina propria DCs in the gut [38–40]. We co-incubated these intestinal-like human DCs with *Bifidobacterium breve* CNCM I-4035 or its CFS, *Salmonella typhi* CECT 725, or a combination of these treatments for 4 h. These treatments up-regulated TLR-9 gene transcription. In addition, CFS was a more powerful inducer of TLR-9 expression compared with probiotic bacteria in the presence of *S. typhi*. Both treatments induced Toll-interacting protein (TOLLIP) gene expression. Furthermore, CFS decreased the pro-inflammatory cytokines and chemokines in DCs that were challenged with *S. typhi*. By contrast, *B. breve* CNCM I-4035 was a potent inducer of the pro-inflammatory cytokines TNF-α, IL-8, and RANTES (regulated upon the activation of normal T cells, expressed, and presumably secreted), and anti-inflammatory cytokines, including IL-10. CFS restored the transforming growth factor (TGF)-β levels in the presence of *S. typhi*. These results indicate that *B. breve* CNCM I-4035 affects the intestinal immune response, whereas its supernatant exerts anti-inflammatory effects that are mediated by DCs. Similarly, *Lactobacillus paracasei* CNCM I-4034 and its CFS decreased pro-inflammatory cytokines and chemokines in human intestinal DCs that were challenged with *S. typhi* CECT 725. CFS was as effective as bacteria in reducing pro-inflammatory cytokine expression. These treatments strongly induced the transcription of the TLR-9 gene. In addition, an upregulation of the CASP8 and TOLLIP genes was observed. *L. paracasei* CNCM I-4034 was a potent inducer of TGF-β2 secretion, whereas the supernatant enhanced the innate immunity through the activation of TLR signaling. Moreover, *L. rhamnosus* CNCM I-4036 and its CFS were challenged with *E. coli* CECT 742, CECT 515, and CECT 729. *L. rhamnosus* treatment induced the production of TGF-β1 and TGF-β2, whereas the CFS increased TGF-β1 secretion. The two treatments induced the gene transcription of TLR-9. *L. rhamnosus* activated TLR-2 and TLR-4 gene expression, whereas CFS increased TLR-1 and TLR-5 gene expression [38–40].

Other selected probiotics exhibit in vitro anti-inflammatory properties. Both probiotic strains and CFS reduced the expression of pro-inflammatory cytokines via an action principally mediated by TLRs. Table 1 summarizes the principal investigations of probiotic effects in in vitro studies.

Table 1. Summary of probiotic anti-inflammatory effects in In Vitro studies.

Reference	Cell Type	Probiotic Strain	Type of Study	Main Outcome
Mann et al. 2014, 2013 [32,33]	human DC	*L. casei* Shirota	In vitro	DC from UC patients samples have an increase of IL-4 production and loss of IL-22 and IFN-γ secretion. *L. casei* Shirota treatment restored the normal stimulatory capacity through a reduction in the TLR-2 and TLR4 expression
Wu et al., 2016 [35]	IPEC-J2 model	*L. plantarum* strain CGMCC1258	In vitro	*L. plantarum* decreased transcript abundances of IL-8, TNF-α, and negative regulators of TLRs. Moreover, *L. plantarum* treatment decreased the gene and protein expression of occludin
Wachi et al., 2014 [36]	PIE cells	*L. delbrueckii* subsp. *delbrueckii* TUA4408L	In vitro	The activation of MAPK and NF-κB pathways induced by *E. coli* 987P were downregulated through upregulation of TLR negative regulators, principally by TLR2
Prisciandaro et al., 2012 [37]	IEC-6	*E. coli* Nissle 1917 and *L. rhamnosus* GG	In vitro	Pre-treatment with these probiotics could prevent or inhibit enterocyte apoptosis and loss of intestinal barrier function induced by 5-FU
Bermudez-Brito et al., 2012, 2013, 2014 [13,38,39]	DC	*L. paracasei* CNCM I-4034, *B. breve* CNCM I-4035, and *L. rhamnosus* CNCM I-4036	In vitro	Induction of TLR-9 expression and TGF-β2 secretion. CFS treatment decreased the pro-inflammatory cytokines and chemokines

CFS, cell-free supernatant; DC, dendritic cells; FU, fluorouracil; IEC, intestinal epithelial cells; IL, interleukin; IFN, interferon; IPEC, intestinal porcine epithelial cells, MAPK, mitogen-activated protein kinase; NF-κB, nuclear factor κ-B; TGF, transforming growth factor; TNF-α, tumor factor necrosis alpha; TLR, toll-like receptor; UC, ulcerative colitis.

3.2. In Vivo Studies

3.2.1. Animals

The anti-inflammatory effects of probiotics have been demonstrated in experimental models. Probiotic supplementation provides protective effects during spontaneous and chemically induced colitis by downregulating the production of inflammatory cytokines or by inducing regulatory mechanisms in a strain-specific manner [3].

Dextran Sulfate Sodium

The anti-inflammatory effects of *Lactobacillus acidophilus*, *L. plantarum*, *Bifidobacterium lactis*, *B. breve*, and inulin on UC colitis have been investigated. Acute UC was induced in Swiss mice using dextran sulfate sodium (DSS). The production of nitric oxide (NO) was evaluated in the supernatants of peritoneal macrophage cultures. The oral administration of probiotic strains and inulin reduced the severity of DSS-induced colitis. These treatments lead to a reduction in NO levels in peritoneal macrophage cultures [41]. The same mixture of strains was tested in another DSS-induced colitis animal model for seven days. Probiotic administration improved clinical symptoms and histological alterations observed in the colitis group, reduced NO production by peritoneal macrophages in DSS-treated mice, and enhanced mucus production in both DSS-treated and healthy mice [42].

Lactobacillus reuteri BR11 reduces the severity of experimental IBD, principally via a mechanism of thiol production [43]. Male Sprague–Dawley rats were administered 2% DSS to induce colitis. *L. reuteri* BR11 or *L. reuteri* BR11 mutants deficient in the cystine-uptake system were administered for 12 days. DSS administration resulted in significant colonic deterioration, including a reduced crypt area and increased damage severity. Probiotic administration partially alleviated the DSS effects, with a minor improvement in the crypt area. The administration of the mutant strain to colitic animals failed to produce significant differences when compared with the DSS control [43].

Lactobacillus fermentum CCTCC M206110, *Lactobacillus crispatus* CCTCC M206119, and *L. plantarum* NCIMB8826 were selected to assess the therapeutic effects on experimental colitis in BALB/c mice treated with DSS. *L. fermentum* CCTCC M206110 treatment resulted in reduced weight loss, colon length shortening, disease activity index scores, and histologic scores, whereas the *L. crispatus* CCTCC M206119 treatment group exhibited greater weight loss and colon length shortening, histologic scores, and more severe inflammatory infiltration. *L. plantarum* NCIMB8826 treatment improved the weight loss and colon length shortening, with no significant influence on the disease activity index and histologic damage in the colitis model [44]. The administration of an *L. crispatus* CCTCC M206119 supplement aggravated DSS-induced colitis, whereas *L. fermentum* CCTCC M206110 effectively attenuated DSS-induced colitis. The potential probiotic effect of *L. plantarum* NCIMB8826 on UC has not been assessed to date [44].

A previous study demonstrated that the intrarectal administration of mouse cathelin-related antimicrobial peptide (mCRAMP) alleviates DSS-induced colitis by preserving the mucus layer and reducing pro-inflammatory cytokines production [45]. A mutant of *Lactococcus lactis* NZ3900 that produces mCRAMP was tested in a murine model of DSS-induced colitis for seven days. Compared with the control group with colitis, cathelicidin-transformed *L. lactis* improved the clinical symptoms, maintained crypt integrity, and preserved the mucus content. The number of apoptotic cells, myeloperoxidase (MPO) activity, and malondialdehyde level were also significantly reduced. The increases in fecal microbiota in colitis animals were markedly prevented [45].

Hong et al., 2010 evaluated a mixture of *Lactobacillus brevis* HY7401, *Lactobacillus* sp. HY7801, and *Bifidobacterium longum* HY8004 in an acute DSS-induced colitis model for seven days [46]. Increased levels of acetate, butyrate, and glutamine, in addition to decreased levels of trimethylamine, were noted in the feces of the probiotic group compared with the DSS-alone-treated mice. The increased short chain fatty acid levels in the feces of mice fed the mixture indicates that probiotics have protective effects against DSS-induced colitis via modulation of the gut microbiota [46].

E. coli Nissle, 1917 was tested in a mouse model of reactivated colitis. Colitis was induced by adding DSS for five days. Two weeks later, colitis was reactivated by subsequent exposure to DSS. *E. coli* Nissle, 1917 administration exerted intestinal anti-inflammatory effects and attenuated colitis reactivation, as shown by reduced disease activity index values. Moreover, probiotic administration decreased the expression of pro-inflammatory cytokines and increased intestinal mucin-like and zona occludens-1 expression [47].

On the other hand, the effects of *L. rhamnosus* NutRes 1 and *B. breve* NutRes 204 on a DSS-induced chronic murine colitis model were assessed. Chronic colitis was induced by two DSS treatment cycles with a 10-day rest period. The probiotic supplementation was started after the first DSS treatment cycle and continued until the end of the experiment. *L. rhamnosus* NutRes 1, but not *B. breve* NutRes 204, rapidly and effectively improved the DSS-induced bloody diarrhea during the resolution phase. However, an increased expression of TLR2, TLR6, chemokine (C-C motif) ligand 2, IL-1β, TNF-α, and IL-6 was found in DSS-treated mice with *L. rhamnosus* supplementation. These results suggest caution in the use of probiotics in the relapse stages of IBD [48].

Capsules with bifidobacteria, lactobacilli, and *Streptococcus thermophilus* DSM24731 were administered to mice exposed to 5, 10, and 15 cycles of DSS. A probiotic mixture attenuated the disease activity index score and colon inflammation after 5, 10, and 15 cycles of DSS and reduced the histological alterations and the incidence of colonic dysplastic lesions in the three periods studied. In addition, the probiotic reduced the proliferating cell nuclear antigen labeling index and TNF-α, IL-1β, IL-6 production, and cyclooxygenase (COX)-2 expression, and increased IL-10 levels in colon tissue in the three periods assayed [49]. Additionally, in rats treated with DSS for seven days, the probiotic mixture exhibited anti-inflammatory properties, including reducing the disease activity index, MPO activity, iNOS, COX-2, NF-κB, TNF-α, IL-6, and p-Akt expression, and increasing IL-10 expression in colonic tissue. In addition, probiotic administration decreased TNF-α and IL-6, and increased IL-10 serum levels [50]. Moreover, probiotic administration was evaluated in acute intestinal ischemia/reperfusion injury in adult 129/SvEv mice. The mixture of strains reduced local tissue inflammation and injury. The reduction in local inflammation after a two-week course of the mixture was correlated with a significant reduction in active IL-1β levels and tissue levels of MPO. Active NF-κB levels were significantly higher in the control group, consistent with the tissue inflammation. Inflammation was attenuated by probiotic administration. Finally, the administration of bifidobacteria, lactobacilli, and *S. thermophilus* did not cause any systemic inflammation or lung injury [51].

Bacillus subtilis R179 also exhibits a protective effect in IBD. The effects of *B. subtilis* were analyzed using a mouse DSS model of colitis in which a higher dose ameliorated gut inflammation and dysbiosis [52].

2,4,6-Trinitrobenzenesulfonic Acid

The impact of *L. plantarum* 21 on inflammatory mediators in 2,4,6-trinitrobenzenesulfonic acid (TNBS)-induced colitis in rats has been evaluated. Treatment with *L. plantarum* 21 for 14 days after the induction of colitis decreased thiobarbituric acid reactive substances (TBARS) and NO, and increased glutathione concentrations. The IL-1β and TNF-α proteins, in addition to mRNA expression, were down-regulated, whereas IL-10 protein and mRNA expression was up-regulated in *L. plantarum* 21-treated rats. In addition, probiotic treatment attenuated macroscopic colonic damage and histopathological changes produced by TNBS [53].

The effects of lactobacilli and bifidobacteria administration on TNF-α and TLR4 expression in a rat colitis model induced by TNBS were also investigated. No significant differences were found in TLR4 and TNF-α expression between the two-week probiotics treatment group and the colitis group, whereas significant reductions were found in rats treated with probiotics for four weeks compared with the TNBS group [54].

The effects of *Butyricicoccus pullicaecorum* CCUG 55,265 in a rat colitis model with TNBS and a Caco-2 cell model were analyzed. *B. pullicaecorum* administration resulted in a significant protective effect based on macroscopic and histological criteria and decreased intestinal MPO, TNF-α, and IL-12 levels. *B. pullicaecorum* supernatant prevented the increase in IL-8 secretion induced by TNF-α and IFN-γ in a Caco-2 cell model [55].

Other Intestinal Inflammation Models

Chronic enteropathies (CE) are believed to be caused by an aberrant immune response towards the intestinal microbiome. The in vitro effects of the probiotic *Enterococcus faecium* NCIMB 10,415 E1707 were previously evaluated using canine cells (e.g., whole blood, intestinal biopsies), but data on in vivo efficacy are lacking. Dogs diagnosed with CE were prospectively recruited to receive a hydrolyzed elimination diet, in addition to either a synbiotic product containing *E. faecium* NCIMB 10,415 E1707 or a placebo for six weeks. Both veterinary staff and owners were blinded to the treatment. Of the 45 cases recruited, 12 completed the clinical trial. Seven dogs received the symbiotic and five received the placebo product. No difference was noted between groups or treatments regarding the clinical efficacy and histology scores [56]. Casp-1 and NLRP3 gene expression was reduced in the CE samples when compared with the controls. Ex vivo treatment with *E. faecium* NCIMB 10,415 E1707 reduced NLRP3 expression in the control samples [57].

The effects of *L. delbrueckii* subsp. *bulgaricus* in an intestinal malfunction mouse model induced by lincomycin hydrochloride were tested. Consequently, *L. delbrueckii* administration increased secretory immunoglobulin A and decreased intestinal pathological damage [58].

L. plantarum LS/07 CCM7766 alone or in combination with inulin was assessed in rats with chronic inflammation. *N,N*-dimethylhydrazine administration triggered the production of IL-2, IL-6, IL-17, and TNF-α, as well as the expression of NF-κB, COX-2, and iNOS, and caused the depletion of goblet cells. *L. plantarum* LS/07 CCM7766 alone and in combination with inulin abolished the inflammatory process in the jejunal mucosa by inhibiting the production of pro-inflammatory cytokines and stimulating IL-10 cytokine synthesis, whereas the TGF-β1 levels did not change significantly [59].

Ogita et al., 2015 tested the effects of *L. rhamnosus* OLL2838, *Bifidobacterium infantis* ATCC 15,697, and *S. thermophilus* Sfi 39 on the maturation of bone marrow-derived DCs from mice [60]. *L. rhamnosus* OLL2838 induced appreciable levels of IL-10 and NO production, whereas *S. thermophilus* Sfi 39 essentially elicited IL-12 and TNF-α [60]. In addition, *L. rhamnosus* OLL2838 was evaluated in an in vivo model of gluten-specific enteropathy characterized by villus blunting, crypt hyperplasia, high levels of intestinal IFN-γ, increased cell apoptosisin lamina propria, and reduced intestinal glutathione S-transferase activity. Probiotic administration enhanced the total glutathione and glutathione S-transferase activity, whereas caspase-3 activity was reduced. However, the probiotic strain failed to recover the normal histology and further increased intestinal IFN-γ [60].

Finally, Wu et al. identified a novel role of probiotics in activating vitamin D receptor (VDR), thus inhibiting inflammation, using cell models and VDR knockout mice [61]. The probiotics *L. rhamnosus* GG ATCC 53,103 and *L. plantarum* increased VDR protein expression in both mouse and human intestinal epithelial cells. Moreover, the role of probiotics in regulating VDR signaling was assessed in vivo using a *Salmonella typhimurium* ATCC 14,028-induced colitis model in VDR knockout mice. Probiotic treatment conferred physiological and histologic protection from colitis in mice, whereas probiotics did not affect the knockout mice. Probiotic treatment also enhanced the number of Paneth cells, which secrete AMPs for host defense [61].

Animal studies seem to be more extensively used than cell models in the evaluation of probiotic properties. Probiotic administration might improve clinical symptoms, histological alterations, and mucus production in the majority of the evaluated studies, but some results suggest that caution should be taken when administering these agents in the relapse stages of IBD. In addition, no effects on chronic enteropathies were noted. Table 2 shows the main investigations regarding the probiotic anti-inflammatory effects in animal studies.

Table 2. Summary of probiotic anti-inflammatory effects in animal studies.

Reference	Animal Species	Probiotic Strain/Treatment	Type of Study	Main Outcome	Adverse Event/Adverse Effects
Abdelouhab et al., 2012 [41]	Swiss mice	L. acidophilus, L. plantarum, B. lactis, B. breve, and inulin	In vivo, DSS-induced colitis	Oral administration of probiotic strains and inulin decreased severity colitis	-
Toumi et al., 2013 [42]	Swiss mice	L. acidophilus, L. plantarum, B. lactis, B. breve	In vivo, DSS-induced colitis	Probiotic administration improved clinical symptoms, histological alterations, and mucus production	-
Atkins et al., 2012 [43]	Male Sprague–Dawley rats	L. reuteri BR11	In vivo, DSS-induced colitis	Probiotic administration partially alleviated the DSS effects, with a minor improvement in crypt area	-
Cui et al., 2016 [44]	BALB/c mice	L. fermentum CCTCC M206110, L. crispatus CCTCC M206119, and L. plantarum NCIMB8826	In vivo, DSS-induced colitis	L. fermentum CCTCC M206110 proved to be effective at attenuating DSS-induced colitis. Administration of L. crispatus CCTCC M206119 aggravated DSS-induced colitis	Administration of L. crispatus CCTCC M206119 aggravated DSS-induced colitis
Wong et al., 2012 [45]	BALB/c mice	A mutant of L. lactis	In vivo, DSS-induced colitis	L. lactis could improve the clinical symptoms, maintain crypt integrity and preserve mucus content. The number of apoptotic cells, MPO activity and malondialdehyde level were also significantly reduced	-
Zhang et al., 2016 [52]	Male C57 mice	B. subtilis	In vivo, DSS-induced colitis	B. subtilis treatment ameliorated gut inflammation and dysbiosis	-
Hong et al., 2010 [46]	Male ICR mice	L. brevis HY7401, L. sp. HY7801 and B. longum HY8004	In vivo, DSS-induced colitis	Increased levels of acetate, butyrate, and glutamine and decreased levels of trimethylamine	-
Garrido-Mesa et al., 2011 [47]	C57BL/6J mice	E. coli Nissle 1917	In vivo, DSS-induced colitis	E. coli Nissle 1917 administration exerted intestinal anti-inflammatory effect and attenuated the reactivation of the colitis	-
Zheng et al., 2016 [48]	Female C57BL/6 mice	L. rhamnosus NutRes 1 and B. breve NutRes 204	In vivo, DSS-induced colitis	An increased expression of inflammation markers were found in DSS-treated mice with L. rhamnosus supplementation	-
Talero et al., 2015 [49]	Female C57BL/6 mice	Capsules with bifidobacteria, lactobacilli, and S. thermophilus	In vivo, DSS-induced colitis	Probiotic mixture attenuated the disease activity index score and colon inflammation and also inflammation markers	-

Table 2. *Cont.*

Reference	Animal Species	Probiotic Strain/Treatment	Type of Study	Main Outcome	Adverse Event/Adverse Effects
Dai et al., 2013 [50]	Male Wistar rats	Capsules with bifidobacteria, lactobacilli, and *S. thermophilus*	In vivo, DSS-induced colitis	The probiotic mixture have anti-inflammatory properties reducing the disease activity index, MPO activity, inflammation biomarkers, and also increasing of IL-10 expression	-
Salim et al., 2013 [51]	Adult male 129/SvEv mice	Capsules with bifidobacteria, lactobacilli, and *S. thermophilus*	In vivo, acute intestinal ischemia/reperfusion injury	Levels of active NF-κB were significantly higher in the control group, corroborating with the inflammation of the tissue, which was attenuated by probiotic administration	-
Satish Kumar et al., 2015 [53]	Wistar female rats	*L. plantarum* 21	In vivo, TNBS-induced colitis	Treatment with *L. plantarum* 21 for 14 days after induction of colitis decreased TBARS, NO, IL-1β and TNF-α and increased glutathione concentration and IL-10 expression	-
Yang et al., 2013 [54]	Sprague-Dawley Rats	Lactobacilli and bifidobacteria	In vivo, TNBS-induced colitis	TLR4 and TNF-α expression were reduced with probiotics	-
Eeckhaut et al., 2013 [55]	Male Wistar rats	*B. pullicaecorum*	In vivo, TNBS-induced colitis	*B. pullicaecorum* administration resulted in a decreased intestinal MPO, TNF-α and IL-12 levels	-
Schmitz et al., 2015 [56,57]	Dogs	*E. faecium* NCIMB 10415 E1707	Chronic enteropathies	There was no difference between groups or treatments regarding clinical efficacy, histology scores	-
Sun et al., 2015 [58]	BALB/c mice	*L. delbrueckii*	Intestinal malfunction induced by Lincomycin hydrochloride	*L. delbrueckii* administration increased secretory immunoglobulin A and decreased the intestine pathological damage	-
Štofilová et al., 2015 [59]	Female Sprague Dawley rats	*L. plantarum* LS/07 CCM7766	In vivo, N,N-dimethylhydrazine-induced colitis	*L. plantarum* LS/07 CCM7766 and its combination with inulin abolished inflammatory process in the jejunal mucosa	-
Ogita et al., 2015 [60]	DQ8 transgenic mice	*L. rhamnosus* OLL2838, *B. infantis* ATCC 15697, and *S. thermophilus* Sfi 39	In vivo, model of gluten-specific enteropathy	Probiotic administration enhanced total glutathione and glutathione S-transferase activity, whereas caspase-3 activity was reduced	-
Wu et al., 2015 [61]	Female C57J3L/6 mice	*L. rhamnosus* GG and *L. plantarum*	In vivo, vitamin D receptor knockout mice	Probiotic treatment conferred physiological and histologic protection from colitis	-

AE, adverse event; DSS, dextran sulfate sodium; IL, interleukin; MPO, myeloperoxidase; NF-κB, nuclear factor kappa-B; NO nitric oxide; TBARS, thiobarbituric acid reactive substances; TNF-α, tumor factor necrosis alpha; TNBS, 2,4,6 trinitrobenzenesulfonic acid; TLR, toll-like receptor.

3.2.2. Humans

Ulcerative Colitis

UC is a chronic IBD of unknown etiology that is characterized by acute exacerbations of intestinal complications, followed by remissions. Additionally, one of the main hypotheses is that UC is caused by an excessive immune response to endogenous bacteria in genetically predisposed individuals. Therefore, the manipulation of the mucosal microbiota to reduce the inflammatory potential of colonizing bacteria is an attractive therapy for UC. Recently, probiotic therapy has been demonstrated to be potentially effective and safe in patients with UC. Tamaki et al., 2016 investigated the efficacy and safety of probiotic treatment with *B. longum* 536, which is a probiotic isolated in 1969 from the feces of a breast-fed infant, in Japanese patients with active UC using an RCT [62]. The probiotic improved clinical symptoms, such as the UC disease activity index and Rachmilewitz endoscopic index, in patients with mild to moderately active UC, but further studies are needed to clarify the efficacy and safety of *B. longum* 536 for UC [62].

A single-center, randomized, double-blind, placebo-controlled study was conducted to examine whether 12 months of probiotic therapy, including a mixture of the strains *Streptococcus faecalis* T-110, *Clostridium butyricum* TO-A, and *Bacillus mesentericus* TO-A, was useful for preventing the relapse of UC in patients who were already in remission. The relapse rates in the probiotic therapy group increased significantly at three and nine months. However, no differences were noted at 12 months. Moreover, in a cluster analysis of fecal microbiota, which is a molecular technique that compares the diversity and colony structure of microbial complexes, seven patients belonged to cluster I; 32 to cluster II, which is the "appropriate intestinal microbiota"; and seven to cluster III. Therefore, probiotics may be effective for UC, especially in cluster I patients [63].

With regards to UC, a RCT was conducted to assess the clinical efficacy of profermin, a food with fermented oats containing *L. plantarum* 299v and other ingredients such as barley malt and lecithin, in relapsing UC. The patients with a mild-to-moderate flare-up of UC, which was defined as a Simple Clinical Colitis Activity Index (SCCAI) score ≥5 and ≤1, showed a significant decrease in the SCCAI score after probiotic supplementation. Thus, *L. plantarum* 299v administration was safe, well tolerated, and palatable, and induced a clinically significant reduction in the SCCAI score compared with a placebo in patients with mild-to-moderate flare-up of UC [64].

A recent case report of bacteremia was caused by *L. rhamnosus* GG in an adult patient affected by severe active UC under treatment with corticosteroids and mesalazine. *Lactobacillus* species are ubiquitous Gram-positive commensals of the normal human microbiota, but their role as opportunistic pathogens is emerging. The case of bacteremia was apparently associated with the translocation of bacteria and fungi from the intestinal lumen to the blood. Thus, *Candida* infection was likely promoted by previous extensive antibiotic use, whereas *L. rhamnosus* GG infection was most likely associated with the probiotic strain administered to the patient and possibly favored by the use of vancomycin, to which the strain was resistant. Notably, this observation based on pending conclusive evidence suggests that the use of probiotics should be considered with caution in cases of active severe IBD with mucosal disruption [65].

UC is also associated with fecal dysbiosis, and different human and animal studies suggest that the gastrointestinal microbiome may trigger the intestinal immune response. Fecal microbial transplantation (FMT) may be a therapeutic option. However, a clinical report published by Suskind et al. (2015) described that single-dose FMT via a nasogastric tube was well tolerated in four patients with UC, but no clinical benefit was demonstrated [66]. Recently, Paramsothy et al., 2017 have reported that intensive-dosing, multidonor FMT induces clinical remission and endoscopic improvement in active UC and is associated with distinct microbial changes that relate to the outcome [67].

Conversely, a 2014 case report by Brace et al. indicated that FMT may be a promising therapy for *Clostridium difficile* infection (CDI) in patients with IBD, because colonization with toxigenic *C. difficile* is significantly higher in IBD patients compared with the general population [68]. Thus, an IBD patient with two CDIs was treated 18 months apart, and each infection was successfully treated with FMT with no IBD flares or complications. Only this study described sequential FMT from a single donor for an IBD patient with CDI recurrences. Microbiota composition analysis indicated that the patient's pre-transplant samples exhibited reduced diversity, with deficiencies in the usually dominant populations of *Firmicutes* and *Bacteroidetes*. This microbiota pattern is consistent with microbial analyses of non-IBD patients during CDI, thus supporting the theory that predominating *Firmicutes* and *Bacteroidetes* groups may confer colonization resistance against *C. difficile*. The authors described only one patient-donor set, which is the main limitation of this study; therefore, further studies are needed to approach the vulnerability profile of patients at risk of relapse [68].

A case report described by Vahabnezhad et al., 2013 reported that probiotic strains of *Lactobacillus*, *L. rhamnosus* GG, caused bacteremia in a 17-year-old boy with UC managed with systemic corticosteroids and infliximab, which is a tumor necrosis factor-α antagonist [69]. *L. rhamnosus* GG was assessed via blood culture on day 2, but the subsequent blood cultures on day 3 and 5 were negative. The patient was treated with antibiotics for five days and defervesced by day eight of his illness. The authors hypothesize that the immunosuppressive effects from systemic corticosteroids and infliximab may have also predisposed the patient to a higher risk of infection.

The risk of infection due to lactobacilli and bifidobacteria is extremely rare and represents 0.05–0.4% of cases of infective endocarditis and bacteremia. Meanwhile, historically, *Lactobacillus* spp. found in food has been considered to be insignificant and they are often regarded as contaminants when isolated from patient samples. Therefore, although probiotics can offer potential benefits in certain healthy conditions, their risks and benefits should still be carefully assessed in patients with some complications, especially when the patients might be immunocompromised [69,70].

Crohn's Disease

CD is a systemic disorder in which the development of host genetic susceptibility represents an important etiological factor. Multiple studies have observed differences in the microbiotas of individuals with CD, with a reduction in anti-inflammatory bacteria and an increase in pro-inflammatory bacteria compared with the microbiotas of healthy subjects. In this sense, different studies have focused on reporting the effect of some strains in CD in recent years. Petersen et al., 2014 investigated the effects of *E. coli*, which is a member of the phylogenetic group B2, and its association with both CD and UC [71]. Interestingly, the probiotic *E. coli* Nissle, 1917 has an equivalent effect to mesalazine in preventing disease flares in UC patients. Moreover, antibiotics seem to have some effect in the treatment of IBD patients. Thus, these authors designed a study to investigate whether ciprofloxacin for one week, followed by therapy with *E. coli* Nissle, 1917 for seven weeks, or either of these treatments alone, influence the remission rate among UC patients. However, no benefit in the use of *E. coli* Nissle 1917 as an add-on treatment to conventional therapies for active UC was noted [71].

Fedorak et al., 2015 investigated the effects of capsules with bifidobacteria, lactobacilli, and *S. thermophilus* DSM24731 in preventing the recurrence of CD after surgery [72]. The recurrence after intestinal resection in CD is quite common. Thus, in this study, within 30 days of ileocolonic resection and re-anastomosis, patients with CD were randomly assigned to groups administered capsules versus a placebo. Although there were no differences in the endoscopic recurrence rates at day 90 between patients who received the probiotics strains, the mucosal levels of inflammatory cytokines, such as IL-8 and IL-1β, were lower among patients who received the probiotics. However, the authors concluded that additional studies are necessary to confirm the effect of the probiotic mixture in the prevention of postoperative recurrence [72].

Hevia et al., 2014 explored the levels of antibodies (IgG and IgA) raised against extracellular proteins produced by LAB and its association with IBD [73]. The presence of serum antibodies, such as IgG and IgA produced by food bacteria from the genera *Bifidobacterium* and *Lactobacillus*, which are used as serum biomarkers of CD or UC, were determined by western blot and ELISA in sera collections from healthy individuals, CD patients, and UC patients. The levels of IgA antibodies against a cell-wall hydrolase from *L. casei* subsp. *rhamnosus* GG (CWH) were significantly higher in the IBD group and appeared to have different immune responses to food bacteria. Specifically, IgA antibodies developed against an extracellular protein of *L. casei* are associated with IBD. Therefore, these results suggest that anti-CWH IgA levels have a potential use for the early detection of CD and UC. However, studies with larger sample sizes are necessary. In addition, the identification of other extracellular protein targets present in food and probiotic bacteria is needed [73].

Ahmed et al., 2013 performed a pilot study in patients with colitis, randomized to either receive a synbiotics for a month and then "crossed over" to receive a placebo or alternatively to receive the placebo first followed by the symbiotic [74]. The main results showed that there were no differences in colonic microbiota between patients with CD or UC, and the spectrum bacteria were not altered by synbiotic administration [74].

Pouchitis is a nonspecific inflammation of the pouch that is present in UC patients and remains the most common post-operative long-term complication. Inflammation of the pouch is characterized by increased stool frequency, rectal bleeding, abdominal cramping, urgency, and fever, and the pathogenesis of pouchitis is still poorly understood. However, there is evidence that implicates the gut microbiota, suggesting that probiotics could reduce the risk of the recurrence of pouchitis, but the mechanisms are not fully understood. Persborn et al., 2013 reported that treatment with probiotic mixture for eight weeks after antibiotics restored the increased permeation to *E. coli* K12 in 16 patients with chronic pouchitis [75]. Thirteen individuals served as a control. This finding could be an important factor for the prevention of recurrence during maintenance treatment with probiotics for this inflammatory status [75].

The bacteria *B. infantis* 35,624 exerts beneficial immunoregulatory effects by mimicking commensal-immune interactions. In an RCT, *B. infantis* 35,624 was used to assess the impact of oral administration for six to eight weeks on inflammatory biomarker and plasma cytokine levels in patients with UC, chronic fatigue syndrome, and psoriasis. *B. infantis* 35,624 reduced plasma CRP levels in all three inflammatory disorders compared with the placebo. Interestingly, plasma IL-6 was reduced in UC patients and chronic fatigue syndrome. Furthermore, in healthy subjects, LPS-stimulated TNF-α and IL-6 secretion by peripheral blood mononuclear cells was significantly reduced in the *B. infantis* 35,624-treated groups compared with the placebo following eight weeks of feeding. These findings demonstrate the reduction of systemic pro-inflammatory biomarkers by *B. infantis* 35,624 and the immunomodulatory effects of the microbiota in humans [76].

In humans, an RCT was described by Bourreille et al., 2013, which was a prospective study with 165 patients with CD who achieved remission after treatment with steroids or salicylates [77]. The patients were randomly assigned to groups that received *S. boulardii* or placebo for 52 weeks. No differences in the median time to relapse were noted between the groups. Thus, although the probiotic yeast *S. boulardii* effects were positive in animals, it does not appear to have any beneficial effects for patients with CD in remission after steroid or salicylate therapies [77].

In summary, the supplementation of selected probiotics appears to be potentially well tolerated, effective, and safe in patients with IBD, both CD and UC. Indeed, probiotics such as *B. longum* 536 improved clinical symptoms in patients with mild to moderately active UC. Although it has been proposed that probiotics can provide benefits in certain conditions, such as healthy individuals or individuals with obesity or metabolic syndrome, the risks and benefits should be carefully assessed before initiating any therapy in patients with IBD. Table 3 summarizes the studies of probiotics and synbiotics in humans.

Table 3. Summary of probiotic effects on IBD in human studies.

Reference	Subjects	Probiotic Strains/Treatment	Time	Main Outcome	Adverse Event/Adverse Effects
Tamaki et al., 2016 [62]	56 with mild to moderate UC	*B. longum* 536	8 weeks	Probiotics administration improved clinical symptoms in the patients with mild to moderately active UC	-
Yoshimatsu et al., 2015 [63]	60 outpatients with UC in remission	*S. faecalis, C. butyricum* and *B. mesentericus*	12 months	Probiotic may be effective for maintaining clinical remission in patients with UC	-
Krag et al., 2013 [64]	74 patients with a mild-to-moderate UC	*L. plantarum* 299v	8 weeks	Probiotic supplementation was safe, well tolerated, palatable, and able to reduce disease index scores in patients with mild-to-moderate UC	-
Petersen et al., 2014 [71]	100 patients with UC	*E. coli* Nissle 1917	7 weeks	There is no benefit in the use of *E. coli* Nissle as an add-on treatment to conventional therapies for active UC	-
Fedorak et al., 2015 [72]	119 patients with CD (within 30 days of ileocolonic resection and re-anastomosis	Capsules with bifidobacteria, lactobacilli, and *S. thermophilus*	90 days	There were no differences in endoscopic recurrence, but mucosal levels of inflammatory cytokines such as IL-8, IL-1β were lower among patients who received the probiotic	-
Hevia et al., 2014 [73]	50 healthy individuals, 37 CD patients and 15 UC patients	*L. casei* subsp. *rhamnosus* GG	90 days	Levels of IgA antibodies developed against a cell-wall hydrolase from *L. casei* subsp. *rhamnosus* GG were significantly higher in the IBD group	-
Ahmed et al., 2013 [74]	8 patients with CD and 8 patients with UC	*L. acidophilus* LA-5, *L. delbrueckii* subsp. *bulgaricus* LBY-27, *B. animalis* subsp. *lactis* BB-12, *S. thermophilus* STY-31 and 15 g oligofructose	1 month	There were no differences in colonic microbiota between patients with CD or UC and the spectrum a bacterium was not altered by synbiotics administration	-
Persborn et al., 2013 [75]	16 patients with chronic pouchitis and 13 individuals as a control	*L. acidophilus* Ecologic 825: *B. bifidum* (W23), *B. lactis* (W51), *B. lactis* (W52), *L. acidophilus* (W22), *L. casei* (W56), *L. paracasei* (W20), *L. plantarum* (W62), *L. salivarius* (W24) and *L. lactis* (W19)	8 weeks	Probiotics restored the mucosal barrier to *E. coli* in patients with pouchitis	-
Groeger et al., 2013 [76]	22 UC patients, 48 patients with chronic fatigue syndrome and 26 psoriasis patients	*B. infantis* 35,624	6–8 weeks	Probiotics administration reduced the systemic pro-inflammatory biomarkers in both gastrointestinal and non-gastrointestinal conditions	-
Bourreille et al., 2013 [77]	165 patients with CD	*S. boulardii*	52 weeks	Probiotics were well tolerated but it did not show any effect. Twenty-one AEs occurred during the treatment, these affected 17 patients, 9 in the *S. boulardii* group and 8 in placebo group	Twenty-one AEs occurred during the treatment, these affected 17 patients, 9 in the *S. boulardii* group and 8 in placebo group

AE, adverse event; CD, Crohn's disease; UC, ulcerative colitis.

4. Further Research and Directions

For a selected number of probiotics and synbiotics, the intestinal anti-inflammatory effects have been documented using in vitro and experimental approaches, as well as human studies. However, the appropriate doses and time of treatment have not been defined, and for a majority of them, the molecular mechanism of action has not been ascertained. Further research should evaluate the most appropriate doses and time of treatment for each probiotic and synbiotic in terms of efficacy in the control of specific chronic inflammatory diseases, and how they contribute to ameliorate the symptoms of the disease as related to the decrease of intestinal and systemic biomarkers of inflammation. In addition, the interactions of particular probiotics with cell receptors and how cell signaling cascades are affected, as well as how the expression of intestinal host genes involved in the immune and inflammatory responses are modulated, are key aspects to understand the action of probiotics. Moreover, the next generation sequence systems should contribute to knowledge on the individual intestinal ecology of patients affected with inflammatory intestinal diseases and the commensal and probiotics strains that actually have a key role in the control of biodiversity in the intestine. This should include the sequencing not only of the bacterial metagenomes, but also the intestinal viromes, thus envisaging the potentially efficient and safe fecal transplantation of healthy subjects to intestinal chronic inflamed patients.

Finally, Figure 1 represents the summary of anti-inflammatory effects of probiotics and synbiotics in intestinal chronic diseases.

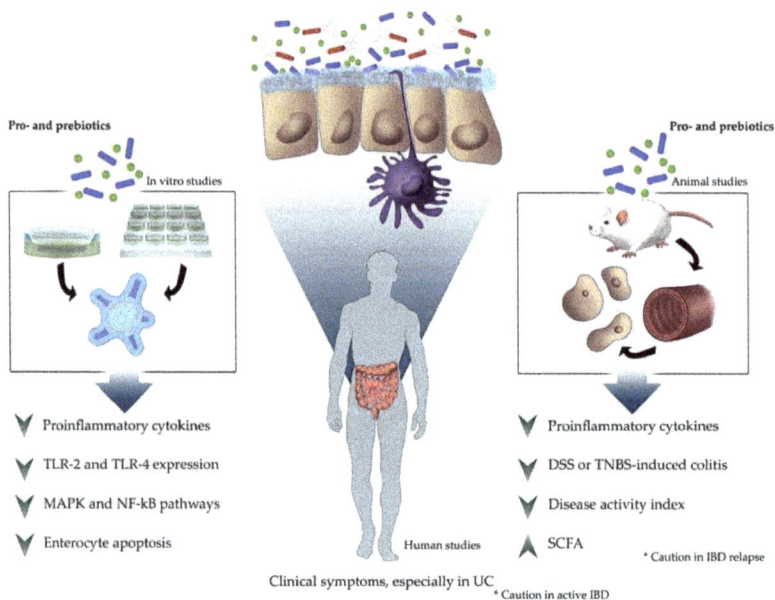

Figure 1. Summary of probiotic anti-inflammatory effects in intestinal chronic diseases in different scientific approaches. DSS, dextran sulfate sodium; IBD, inflammatory bowel disease; MAPK, mitogen-activated protein kinase; NF-κB, nuclear factor kappa-B; SCFA, short-chain fatty acids; TNBS, 2,4,6 trinitrobenzenesulfonic acid; TLR, toll-like receptor; UC, ulcerative colitis.

Author Contributions: Julio Plaza-Diaz, Francisco Javier Ruiz-Ojeda, Laura Maria Vilchez Padial, and Angel Gil contributed to the planning of the literature search, designed the analysis and presentation of the results, created the tool for assessing the quality of the articles, and were involved in the analyses of the articles. Julio Plaza-Diaz, Francisco Javier Ruiz-Ojeda, Laura Maria Vilchez Padial, and Angel Gil wrote the draft. All authors discussed and revised all drafts and approved the final manuscript.

Conflicts of Interest: The authors declare no conflict of interest.

References

1. Metchnikoff, E. *The Prolongation of Life: Optimistic Studies*, 1st ed.; Mitchell, P.C., Ed.; G.P. Putnam's Sons: New York, NY, USA, 1908.
2. De Moreno de LeBlanc, A.; LeBlanc, J.G. Effect of probiotic administration on the intestinal microbiota, current knowledge and potential applications. *World J. Gastroenterol.* **2014**, *20*, 16518–16528. [CrossRef] [PubMed]
3. Fontana, L.; Bermudez-Brito, M.; Plaza-Diaz, J.; Muñoz-Quezada, S.; Gil, A. Sources, isolation, characterisation and evaluation of probiotics. *Br. J. Nutr.* **2013**, *109*, 35–50. [CrossRef] [PubMed]
4. Gibson, G.R.; Roberfroid, M.B. Dietary modulation of the human colonic microbiota: Introducing the concept of prebiotics. *J. Nutr.* **1995**, *125*, 1401–1412. [PubMed]
5. Ouwehand, A.C.; Salminen, S.; Isolauri, E. Probiotics: An overview of beneficial effects. *Antonie Van Leeuwenhoek* **2002**, *82*, 279–289. [CrossRef] [PubMed]
6. Hamilton-Miller, J.M.; Gibson, G.R.; Bruck, W. Some insights into the derivation and early uses of the word 'probiotic'. *Br. J. Nutr.* **2003**, *90*, 845. [CrossRef] [PubMed]
7. Food and Agriculture Organization/World Health Organization. *Health and Nutritional Properties of Probiotics in Food Including Powder Milk with Live Lactic Acid Bacteria*; American Cordoba Park Hotel: Cordoba, Argentina, 2001; pp. 1–2.
8. Roberfroid, M.B. Prebiotics and probiotics: are they functional foods? *Am. J. Clin. Nutr.* **2000**, *71*, 1682–1687.
9. Giahi, L.; Aumueller, E.; Elmadfa, I.; Haslberger, A.G. Regulation of TLR4, p38 MAPkinase, IκB and miRNAs by inactivated strains of lactobacilli in human dendritic cells. *Benef. Microbes* **2012**, *3*, 91–98. [CrossRef] [PubMed]
10. Hill, C.; Guarner, F.; Reid, G.; Gibson, G.R.; Merenstein, D.J.; Pot, B.; Morelli, L.; Canani, R.B.; Flint, H.J.; Salminen, S.; et al. Expert consensus document. The International Scientific Association for Probiotics and Prebiotics consensus statement on the scope and appropriate use of the term probiotic. *Nat. Rev. Gastroenterol. Hepatol.* **2014**, *11*, 506–514. [CrossRef] [PubMed]
11. Collado, M.C.; Isolauri, E.; Salminen, S.; Sanz, Y. The impact of probiotic on gut health. *Curr. Drug. Metab.* **2009**, *10*, 68–78. [CrossRef] [PubMed]
12. Plaza-Díaz, J.; Robles-Sánchez, C.; Abadía-Molina, F.; Morón-Calvente, V.; Sáez-Lara, M.J.; Ruiz-Bravo, A.; Jiménez-Valera, M.; Gil, A.; Gómez-Llorente, C.; Fontana, L. Adamdec1, Ednrb and Ptgs1/Cox1, inflammation genes upregulated in the intestinal mucosa of obese rats, are downregulated by three probiotic strains. *Sci. Rep.* **2017**, *7*, 1939. [CrossRef] [PubMed]
13. Bermudez-Brito, M.; Plaza-Díaz, J.; Muñoz-Quezada, S.; Gómez-Llorente, C.; Gil, A. Probiotic mechanisms of action. *Ann. Nutr. Metab.* **2012**, *61*, 160–174. [CrossRef] [PubMed]
14. Plaza-Díaz, J.; Fernandez-Caballero, J.Á.; Chueca, N.; Garcia, F.; Gómez-Llorente, C.; Sáez-Lara, M.J.; Fontana, L.; Gil, A. Pyrosequencing analysis reveals changes in intestinal microbiota of healthy adults who received a daily dose of immunomodulatory probiotic strains. *Nutrients* **2015**, *7*, 3999–4015. [CrossRef] [PubMed]
15. Plaza-Diaz, J.; Gomez-Llorente, C.; Abadia-Molina, F.; Saez-Lara, M.J.; Campaña-Martin, L.; Muñoz-Quezada, S.; Romero, F.; Gil, A.; Fontana, L. Effects of *Lactobacillus paracasei* CNCM I-4034, *Bifidobacterium breve* CNCM I-4035 and *Lactobacillus rhamnosus* CNCM I-4036 on hepatic steatosis in Zucker rats. *PLoS ONE* **2014**, *9*, e98401. [CrossRef] [PubMed]
16. Plaza-Diaz, J.; Gomez-Llorente, C.; Campaña-Martin, L.; Matencio, E.; Ortuño, I.; Martínez-Silla, R.; Gomez-Gallego, C.; Periago, M.J.; Ros, G.; Chenoll, E.; et al. Safety and immunomodulatory effects of three probiotic strains isolated from the feces of breast-fed infants in healthy adults: SETOPROB study. *PLoS ONE* **2013**, *8*, e78111. [CrossRef] [PubMed]
17. Pineiro, M.; Asp, N.G.; Reid, G.; Macfarlane, S.; Morelli, L.; Brunser, O.; Tuohy, K. FAO Technical meeting on prebiotics. *J. Clin. Gastroenterol.* **2008**, *42*, S156–S159. [CrossRef] [PubMed]
18. Sáez-Lara, M.J.; Robles-Sanchez, C.; Ruiz-Ojeda, F.J.; Plaza-Diaz, J.; Gil, A. Effects of Probiotics and Synbiotics on Obesity, Insulin Resistance Syndrome, Type 2 Diabetes and Non-Alcoholic Fatty Liver Disease: A Review of Human Clinical Trials. *Int. J. Mol. Sci.* **2016**, *17*, 928. [CrossRef] [PubMed]

19. Tojo, R.; Suárez, A.; Clemente, M.G.; De los Reyes-Gavilán, C.G.; Margolles, A.; Gueimonde, M.; Ruas-Madiedo, P. Intestinal microbiota in health and disease: Role of bifidobacteria in gut homeostasis. *World J. Gastroenterol.* **2014**, *20*, 15163–15176. [CrossRef] [PubMed]

20. Upadhyay, N.; Moudgal, V. Probiotics: A Review. *J. Clin. Outcomes Manag.* **2012**, *19*, 76–84.

21. Molodecky, N.A.; Soon, I.S.; Rabi, D.M.; Ghali, W.A.; Ferris, M.; Chernoff, G.; Benchimol, E.I.; Panaccione, R.; Ghosh, S.; Barkema, H.W.; et al. Increasing incidence and prevalence of the inflammatory bowel diseases with time, based on systematic review. *Gastroenterology* **2012**, *142*, 46–54. [CrossRef] [PubMed]

22. Wedlake, L.; Slack, N.; Andreyev, H.J.; Whelan, K. Fiber in the treatment and maintenance of inflammatory bowel disease: A systematic review of randomized controlled trials. *Inflamm. Bowel Dis.* **2014**, *20*, 576–586. [CrossRef] [PubMed]

23. Podolsky, D.K. Inflammatory bowel disease. *N. Engl. J. Med.* **2002**, *347*, 417–429. [CrossRef] [PubMed]

24. Münch, A.; Aust, D.; Bohr, J.; Bonderup, O.; Fernández Bañares, F.; Hjortswang, H.; Madisch, A.; Munck, L.K.; Ström, M.; Tysk, C.; et al. Microscopic colitis: Current status, present and future challenges: Statements of the European Microscopic Colitis Group. *J. Crohns Colitis.* **2012**, *6*, 932–945. [CrossRef] [PubMed]

25. Kaser, A.; Zeissig, S.; Blumberg, R.S. Inflammatory bowel disease. *Annu. Rev. Immunol.* **2010**, *28*, 573–621. [CrossRef] [PubMed]

26. Abraham, C.; Cho, J.H. Inflammatory bowel disease. *N. Engl. J. Med.* **2009**, *361*, 2066–2078. [CrossRef] [PubMed]

27. Derikx, L.A.; Dieleman, L.A.; Hoentjen, F. Probiotics and prebiotics in ulcerative colitis. *Best Pract. Res. Clin. Gastroenterol.* **2016**, *30*, 55–71. [CrossRef] [PubMed]

28. Magro, F.; Gionchetti, P.; Eliakim, R.; Ardizzone, S.; Armuzzi, A.; Barreiro-de Acosta, M.; Burisch, J.; Gecse, K.B.; Hart, A.L.; Hindryckx, P.; et al. Third European Evidence-Based Consensus on Diagnosis and Management of Ulcerative Colitis. Part 1: Definitions, diagnosis, extra-intestinal manifestations, pregnancy, cancer surveillance, surgery, and ileo-anal pouch disorders. *J. Crohns Colitis.* **2017**, *11*, 1–39. [CrossRef] [PubMed]

29. Ringel, Y.; Carroll, I.M. Alterations in the intestinal microbiota and functional bowel symptoms. *Gastrointest. Endosc. Clin. N. Am.* **2009**, *19*, 141–150. [CrossRef] [PubMed]

30. O'Toole, P.W.; Cooney, J.C. Probiotic bacteria influence the composition and function of the intestinal microbiota. *Interdiscip. Perspect. Infect. Dis.* **2008**, *2008*. [CrossRef] [PubMed]

31. Butler, M.; Ng, C.Y.; Van Heel, D.A.; Lombardi, G.; Lechler, R.; Playford, R.J.; Ghosh, S. Modulation of dendritic cell phenotype and function in an in vitro model of the intestinal epithelium. *Eur. J. Immunol.* **2006**, *36*, 864–874. [CrossRef] [PubMed]

32. Mann, E.R.; Bernardo, D.; Ng, S.C.; Rigby, R.J.; Al-Hassi, H.O.; Landy, J.; Peake, S.T.C.; Spranger, H.; English, N.R.; Thomas, L.V.; et al. Human gut dendritic cells drive aberrant gut-specific T-cell responses in ulcerative colitis, characterized by increased IL-4 production and loss of IL-22 and IFN-γ. *Inflamm. Bowel Dis.* **2014**, *20*, 2299–2307. [CrossRef] [PubMed]

33. Mann, E.R.; You, J.; Horneffer-van der Sluis, V.; Bernardo, D.; Omar Al-Hassi, H.; Landy, J.; Peake, S.T.; Thomas, L.V.; Tee, C.T.; Lee, G.H.; et al. Dysregulated circulating dendritic cell function in ulcerative colitis is partially restored by probiotic strain *Lactobacillus casei* Shirota. *Mediators Inflamm.* **2013**, *2013*, 573–576. [CrossRef] [PubMed]

34. Gómez-Llorente, C.; Muñoz, S.; Gil, A. Role of toll-like receptors in the development of immunotolerance mediated by probiotics. *Proc. Nutr. Soc.* **2010**, *69*, 381–389. [CrossRef] [PubMed]

35. Wu, Y.; Zhu, C.; Chen, Z.; Chen, Z.; Zhang, W.; Ma, X.; Wang, L.; Yang, X.; Jiang, Z. Protective effects of *Lactobacillus plantarum* on epithelial barrier disruption caused by enterotoxigenic *Escherichia coli* in intestinal porcine epithelial cells. *Vet. Immunol. Immunopathol.* **2016**, *172*, 55–63. [CrossRef] [PubMed]

36. Wachi, S.; Kanmani, P.; Tomosada, Y.; Kobayashi, H.; Yuri, T.; Egusa, S.; Shimazu, T.; Suda, Y.; Aso, H.; Sugawara, M.; et al. *Lactobacillus delbrueckii* TUA4408L and its extracellular polysaccharides attenuate enterotoxigenic *Escherichia coli*-induced inflammatory response in porcine intestinal epitheliocytes via Toll-like receptor-2 and 4. *Mol. Nutr. Food Res.* **2014**, *58*, 2080–2093. [CrossRef] [PubMed]

37. Prisciandaro, L.D.; Geier, M.S.; Chua, A.E.; Butler, R.N.; Cummins, A.G.; Sander, G.R.; Howarth, G.S. Probiotic factors partially prevent changes to caspases 3 and 7 activation and transepithelial electrical resistance in a model of 5-fluorouracil-induced epithelial cell damage. *Support Care Cancer* **2012**, *20*, 3205–3210. [CrossRef] [PubMed]

38. Bermudez-Brito, M.; Muñoz-Quezada, S.; Gomez-Llorente, C.; Romero, F.; Gil, A. *Lactobacillus rhamnosus* and its cell-free culture supernatant differentially modulate inflammatory biomarkers in *Escherichia coli*-challenged human dendritic cells. *Br. J. Nutr.* **2014**, *111*, 1727–1737. [CrossRef] [PubMed]

39. Bermudez-Brito, M.; Muñoz-Quezada, S.; Gomez-Llorente, C.; Matencio, E.; Bernal, M.J.; Romero, F.; Gil, A. Cell-free culture supernatant of *Bifidobacterium breve* CNCM I-4035 decreases pro-inflammatory cytokines in human dendritic cells challenged with *Salmonella typhi* through TLR activation. *PLoS ONE* **2013**, *8*, e59370. [CrossRef] [PubMed]

40. Bermudez-Brito, M.; Muñoz-Quezada, S.; Gomez-Llorente, C.; Matencio, E.; Bernal, M.J.; Romero, F.; Gil, A. Human intestinal dendritic cells decrease cytokine release against *Salmonella* infection in the presence of *Lactobacillus paracasei* upon TLR activation. *PLoS ONE* **2012**, *7*, e43197. [CrossRef] [PubMed]

41. Abdelouhab, K.; Rafa, H.; Toumi, R.; Bouaziz, S.; Medjeber, O.; Touil-Boukoffa, C. Mucosal intestinal alteration in experimental colitis correlates with nitric oxide production by peritoneal macrophages: Effect of probiotics and prebiotics. *Immunopharmacol. Immunotoxicol.* **2012**, *34*, 590–597. [CrossRef] [PubMed]

42. Toumi, R.; Abdelouhab, K.; Rafa, H.; Soufli, I.; Raissi-Kerboua, D.; Djeraba, Z.; Touil-Boukoffa, C. Beneficial role of the probiotic mixture Ultrabiotique on maintaining the integrity of intestinal mucosal barrier in DSS-induced experimental colitis. *Immunopharmacol. Immunotoxicol.* **2013**, *35*, 403–409. [CrossRef] [PubMed]

43. Atkins, H.L.; Geier, M.S.; Prisciandaro, L.D.; Pattanaik, A.K.; Forder, R.E.; Turner, M.S.; Howarth, G.S. Effects of a *Lactobacillus reuteri* BR11 mutant deficient in the cystine-transport system in a rat model of inflammatory bowel disease. *Dig. Dis. Sci.* **2012**, *57*, 713–719. [CrossRef] [PubMed]

44. Cui, Y.; Wei, H.; Lu, F.; Liu, X.; Liu, D.; Gu, L.; Ouyang, C. Different effects of three selected *Lactobacillus* strains in dextran sulfate sodium-induced colitis in BALB/c mice. *PLoS ONE* **2016**, *11*, e0148241. [CrossRef] [PubMed]

45. Wong, C.C.; Zhang, L.; Li, Z.J.; Wu, W.K.; Ren, S.X.; Chen, Y.C.; Ng, T.B.; Cho, C.H. Protective effects of cathelicidin-encoding *Lactococcus lactis* in murine ulcerative colitis. *J. Gastroenterol. Hepatol.* **2012**, *27*, 1205–1212. [CrossRef] [PubMed]

46. Hong, Y.S.; Ahn, Y.T.; Park, J.C.; Lee, J.H.; Lee, H.; Huh, C.S.; Kim, D.H.; Ryu, D.H.; Hwang, G.S. ^{1}H NMR-based metabonomic assessment of probiotic effects in a colitis mouse model. *Arch. Pharm. Res.* **2010**, *33*, 1091–1101. [CrossRef] [PubMed]

47. Garrido-Mesa, N.; Utrilla, P.; Comalada, M.; Zorrilla, P.; Garrido-Mesa, J.; Zarzuelo, A.; Rodríguez-Cabezas, M.E.; Gálvez, J. The association of minocycline and the probiotic *Escherichia coli* Nissle 1917 results in an additive beneficial effect in a DSS model of reactivated colitis in mice. *Biochem. Pharmacol.* **2011**, *82*, 1891–1900. [CrossRef] [PubMed]

48. Zheng, B.; van Bergenhenegouwen, J.; van de Kant, H.J.; Folkerts, G.; Garssen, J.; Vos, A.P.; Morgan, M.E.; Kraneveld, A.D. Specific probiotic dietary supplementation leads to different effects during remission and relapse in murine chronic colitis. *Benef. Microbes* **2016**, *7*, 205–213. [CrossRef] [PubMed]

49. Talero, E.; Bolivar, S.; Ávila-Román, J.; Alcaide, A.; Fiorucci, S.; Motilva, V. Inhibition of chronic ulcerative colitis-associated adenocarcinoma development in mice by VSL#3. *Inflamm. Bowel. Dis.* **2015**, *21*, 1027–1037. [PubMed]

50. Dai, C.; Zheng, C.Q.; Meng, F.J.; Zhou, Z.; Sang, L.X.; Jiang, M. VSL#3 probiotics exerts the anti-inflammatory activity via PI3k/Akt and NF-κB pathway in rat model of DSS-induced colitis. *Mol. Cell Biochem.* **2013**, *374*, 1–11. [PubMed]

51. Salim, S.Y.; Young, P.Y.; Lukowski, C.M.; Madsen, K.L.; Sis, B.; Churchill, T.A.; Khadaroo, R.G. VSL#3 probiotics provide protection against acute intestinal ischaemia/reperfusion injury. *Benef. Microbes* **2013**, *4*, 357–365. [PubMed]

52. Zhang, H.L.; Li, W.S.; Xu, D.N.; Zheng, W.W.; Liu, Y.; Chen, J.; Qiu, Z.B.; Dorfman, R.G.; Zhang, J.; Liu, J. Mucosa-reparing and microbiota-balancing therapeutic effect of *Bacillus subtilis* alleviates dextrate sulfate sodium-induced ulcerative colitis in mice. *Exp. Ther. Med.* **2016**, *12*, 2554–2562. [CrossRef] [PubMed]

53. Satish Kumar, C.; Kondal Reddy, K.; Reddy, A.G.; Vinoth, A.; Ch, S.R.; Boobalan, G.; Rao, G.S. Protective effect of *Lactobacillus plantarum* 21, a probiotic on trinitrobenzenesulfonic acid-induced ulcerative colitis in rats. *Int. Immunopharmacol.* **2015**, *25*, 504–510. [CrossRef] [PubMed]

54. Yang, X.; Fu, Y.; Liu, J.; Ren, H.Y. Impact of probiotics on toll-like receptor 4 expression in an experimental model of ulcerative colitis. *J. Huazhong Univ. Sci. Technol. Med. Sci.* **2013**, *33*, 661–665. [CrossRef] [PubMed]

55. Eeckhaut, V.; Machiels, K.; Perrier, C.; Romero, C.; Maes, S.; Flahou, B.; Steppe, M.; Haesebrouck, F.; Sas, B.; Ducatelle, R.; et al. *Butyricicoccus pullicaecorum* in inflammatory bowel disease. *Gut* **2013**, *62*, 1745–1752. [CrossRef] [PubMed]

56. Schmitz, S.; Werling, D.; Allenspach, K. Effects of ex-vivo and in-vivo treatment with probiotics on the inflammasome in dogs with chronic enteropathy. *PLoS ONE* **2015**, *10*, e0120779. [CrossRef] [PubMed]

57. Schmitz, S.; Glanemann, B.; Garden, O.A.; Brooks, H.; Chang, Y.M.; Werling, D.; Allenspach, K. A prospective, randomized, blinded, placebo-controlled pilot study on the effect of *Enterococcus faecium* on clinical activity and intestinal gene expression in canine food-responsive chronic enteropathy. *J. Vet. Intern. Med.* **2015**, *29*, 533–543. [CrossRef] [PubMed]

58. Sun, Q.; Shi, Y.; Wang, F.; Han, D.; Lei, H.; Zhao, Y.; Sun, Q. Study on the effects of microencapsulated *Lactobacillus delbrueckii* on the mouse intestinal flora. *J. Microencapsul.* **2015**, *32*, 669–676. [CrossRef] [PubMed]

59. Štofilová, J.; Szabadosová, V.; Hrčková, G.; Salaj, R.; Bertková, I.; Hijová, E.; Strojný, L.; Bomba, A. Co-administration of a probiotic strain *Lactobacillus plantarum* LS/07 CCM7766 with prebiotic inulin alleviates the intestinal inflammation in rats exposed to N,N-dimethylhydrazine. *Int. Immunopharmacol.* **2015**, *24*, 361–368. [CrossRef] [PubMed]

60. Ogita, T.; Bergamo, P.; Maurano, F.; D'Arienzo, R.; Mazzarella, G.; Bozzella, G.; Luongo, D.; Sashihara, T.; Suzuki, T.; Tanabe, S.; et al. Modulatory activity of *Lactobacillus rhamnosus* OLL2838 in a mouse model of intestinal immunopathology. *Immunobiology* **2015**, *220*, 701–710. [CrossRef] [PubMed]

61. Wu, S.; Yoon, S.; Zhang, Y.G.; Lu, R.; Xia, Y.; Wan, J.; Petrof, E.O.; Claud, E.C.; Chen, D.; Sun, J. Vitamin D receptor pathway is required for probiotic protection in colitis. *Am. J. Physiol. Gastrointest. Liver Physiol.* **2015**, *309*, 341–349. [CrossRef] [PubMed]

62. Tamaki, H.; Nakase, H.; Inoue, S.; Kawanami, C.; Itani, T.; Ohana, M.; Kusaka, T.; Uose, S.; Hisatsune, H.; Tojo, M.; et al. Efficacy of probiotic treatment with *Bifidobacterium longum* 536 for induction of remission in active ulcerative colitis: A randomized, double-blinded, placebo-controlled multicenter trial. *Dig. Endosc.* **2016**, *28*, 67–74. [CrossRef] [PubMed]

63. Yoshimatsu, Y.; Yamada, A.; Furukawa, R.; Sono, K.; Osamura, A.; Nakamura, K.; Aoki, H.; Tsuda, Y.; Hosoe, N.; Takada, N.; et al. Effectiveness of probiotic therapy for the prevention of relapse in patients with inactive ulcerative colitis. *World J. Gastroenterol.* **2015**, *21*, 5985–5994. [PubMed]

64. Krag, A.; Munkholm, P.; Israelsen, H.; von Ryberg, B.; Andersen, K.K.; Bendtsen, F. Profermin is efficacious in patients with active ulcerative colitis-a randomized controlled trial. *Inflamm. Bowel Dis.* **2013**, *19*, 2584–2592. [CrossRef] [PubMed]

65. Meini, S.; Laureano, R.; Fani, L.; Tascini, C.; Galano, A.; Antonelli, A.; Rossolini, G.M. Breakthrough *Lactobacillus rhamnosus* GG bacteremia associated with probiotic use in an adult patient with severe active ulcerative colitis: case report and review of the literature. *Infection* **2015**, *43*, 777–781. [CrossRef] [PubMed]

66. Suskind, D.L.; Namita Singh, N.; Nielson, H.; Wahbeh, G. Fecal microbial transplant via nasogastric tube for active pediatric ulcerative colitis. *J. Pediatr. Gastroenterol. Nutr.* **2015**, *60*, 27–29. [CrossRef] [PubMed]

67. Paramsothy, S.; Kamm, M.A.; Kaakoush, N.O.; Walsh, A.J.; van den Bogaerde, J.; Samuel, D.; Leong, R.W.; Connor, S.; Ng, W.; Paramsothy, R.; et al. Multidonor intensive faecal microbiota transplantation for active ulcerative colitis: A randomised placebo-controlled trial. *Lancet* **2017**, *389*, 1218–1228. [CrossRef]

68. Brace, C.; Gloor, G.B.; Ropeleski, M.; Allen-Verco, E.; Petrof, E.O. Microbial composition analysis of *Clostridium difficile* infections in an ulcerative colitis patient treated with multiple fecal microbiota transplantations. *J. Crohns Colitis* **2014**, *8*, 1133–1137. [CrossRef] [PubMed]

69. Vahabnezhad, E.; Mochon, A.B.; Wozniak, L.Y.; Ziring, D.A. *Lactobacillus* bacteremia associated with probiotic use in a pediatric patient with ulcerative colitis. *Clin. Gastroenterol.* **2013**, *47*, 437–439. [CrossRef] [PubMed]

70. Gouriet, F.; Million, M.; Henri, M.; Fournier, P.E.; Raoult, D. *Lactobacillus rhamnosus* bacteremia: An emerging clinical entity. *Eur. J. Clin. Microbiol. Infect. Dis.* **2012**, *31*, 2469–2480. [CrossRef] [PubMed]

71. Petersen, A.M.; Mirsepasi, H.; Halkjær, S.I.; Mortensen, E.M.; Nordgaard-Lassen, I.; Krogfelt, K.A. Ciprofloxacin and probiotic *Escherichia coli* Nissle add-on treatment in active ulcerative colitis: A double-blind randomized placebo controlled clinical trial. *J. Crohns Colitis* **2014**, *8*, 1498–1505. [CrossRef] [PubMed]

72. Fedorak, R.N.; Feagan, B.G.; Hotte, N.; Leddin, D.; Dieleman, L.A.; Petrunia, D.M.; Enns, R.; Bitton, A.; Chiba, N.; Paré, P. The probiotic VSL#3 has anti-inflammatory effects and could reduce endoscopic recurrence after surgery for crohn's disease. *Clin. Gastroenterol. Hepatol.* **2015**, *13*, 928–935. [PubMed]

73. Hevia, A.; López, P.; Suárez, A.; Jacquot, C.; Urdaci, M.C.; Margolles, A.; Sánchez, B. Association of levels of antibodies from patients with inflammatory bowel disease with extracellular proteins of food and probiotic bacteria. *Biomed. Res. Int.* **2014**, *2014*, 351204. [CrossRef] [PubMed]

74. Ahmed, J.; Reddy, B.S.; Mølbak, L.; Leser, T.D.; MacFie, J. Impact of probiotics on colonic microflora in patients with colitis: A prospective double blind randomised crossover study. *Int. J. Surg.* **2013**, *11*, 1131–1136. [CrossRef] [PubMed]

75. Persborn, M.; Gerritsen, J.; Wallon, C.; Carlsson, A.; Akkermans, L.M.A.; Söderholm, J.D. The effects of probiotics on barrier function and mucosal pouch microbiota during maintenance treatment for severe pouchitis in patients with ulcerative colitis. *Aliment Pharmacol. Ther.* **2013**, *38*, 772–783. [CrossRef] [PubMed]

76. Groeger, D.; O'Mahony, L.; Murphy, E.F.; Bourke, J.F.; Dinan, T.G.; Kiely, B.; Shanahan, F.; Quigley, E.M.M. *Bifidobacterium infantis* 35624 modulates host inflammatory processes beyond the gut. *Gut Microbes* **2013**, *4*, 325–339. [CrossRef] [PubMed]

77. Bourreille, A.; Cadiot, G.; Le Dreau, G.; Laharie, D.; Beaugerie, L.; Dupas, J.L.; Marteau, P.; Rampal, P.; Moyse, D.; Saleh, A. *Saccharomyces boulardii* does not prevent relapse of Crohn's disease. *Clin. Gastroenterol. Hepatol.* **2013**, *11*, 982–987. [CrossRef] [PubMed]

Article

Pre-Pregnancy Body Mass Index Is Associated with Dietary Inflammatory Index and C-Reactive Protein Concentrations during Pregnancy

Dayeon Shin [1], Junguk Hur [2], Eun-Hee Cho [3], Hae-Kyung Chung [4], Nitin Shivappa [5,6], Michael D. Wirth [5,6], James R. Hébert [5,6] and Kyung Won Lee [7,†,*]

[1] Department of Nutrition & Dietetics, College of Nursing & Professional Disciplines, University of North Dakota, Grand Forks, ND 58202, USA; dayeon.shin@und.edu
[2] Department of Biomedical Sciences, University of North Dakota School of Medicine and Health Sciences, Grand Forks, ND 58202, USA; junguk.hur@med.und.edu
[3] Division of Endocrinology and Metabolism, Department of Internal Medicine, School of Medicine, Kangwon National University, Chuncheon 24289, Korea; ehcho@kangwon.ac.kr
[4] Department of Food and Nutrition, Hoseo University, Asan 31499, Korea; hkchung@hoseo.edu
[5] Department of Epidemiology and Biostatistics, and the Cancer Prevention and Control Program, Arnold School of Public Health, University of South Carolina, Columbia, SC 29208, USA; shivappa@mailbox.sc.edu (N.S.); wirthm@mailbox.sc.edu (M.D.W.); jhebert@mailbox.sc.edu (J.R.H.)
[6] Connecting Health Innovations, LLC, Columbia, SC 29201, USA
[7] Department of Food Science and Human Nutrition, Michigan State University, East Lansing, MI 48824, USA
*   Correspondence: kyungwlee@korea.kr; Tel.: +82-43-719-6741
†   Division of Epidemiology and Health Index, Center for Genome Science, Korea National Institute of Health, Korea Centers for Disease Control and Prevention, Chungcheongbuk-do 28160, Korea (Current affiliation).

Received: 4 December 2016; Accepted: 29 March 2017; Published: 1 April 2017

Abstract: There have been a limited number of studies examining the association between pre-pregnancy body mass index (BMI) and dietary inflammation during pregnancy. Our aim is to examine the association between pre-pregnancy BMI and the Dietary Inflammatory Index (DII)™ and C-reactive protein (CRP) concentrations during pregnancy. The study included 631 pregnant American women from the National Health and Nutrition Examination Survey (NHANES) cross-sectional examinations from 2003 to 2012. Pre-pregnancy BMI was calculated based on self-reported pre-pregnancy weight and measured height. The cut-offs of <18.5 (underweight), 18.5–24.9 (normal), 25.0–29.9 (overweight), and ≥ 30 kg/m^2 (obese) were used to categorize the weight status of pregnant women prior to pregnancy. The DII, a literature-based dietary index to assess the inflammatory properties of diet, was estimated based on a one-day 24-h recall. Multivariable linear and logistic regressions were performed to estimate beta coefficients and the adjusted odds ratios (AORs) and 95% confidence intervals (95% CIs) on the association of pre-pregnancy BMI categories with the DII and CRP concentrations during pregnancy. After controlling for variables including: race/ethnicity, family poverty income ratio, education, marital status, month in pregnancy, and smoking status during pregnancy; women who were obese before pregnancy ($n = 136$) had increased odds for being in the highest tertile of the DII and CRP concentrations compared to women with normal weight (AORs 2.40, 95% CIs 1.01–5.71; AORs 24.84, 95% CIs 6.19–99.67, respectively). These findings suggest that women with pre-pregnancy obesity had greater odds of reporting higher DII and having elevated CRP. In conclusion, high pre-pregnancy BMI was associated with increased odds of pro-inflammatory diet and elevated CRP levels during pregnancy in the USA.

Keywords: dietary inflammatory index; C-reactive protein; pregnancy body mass index; NHANES; reproductive health

1. Introduction

Pre-pregnancy body mass index (BMI) has been associated with increased risks for adverse pregnancy outcomes such as gestational hypertension, gestational diabetes mellitus (GDM), pre-term birth, and small- and large-for-gestational-age infants [1]. Regardless of how much weight pregnant women gained during pregnancy, women with obese pre-pregnancy BMIs had increased odds for GDM compared with women with normal pre-pregnancy BMIs (adjusted odds ratios (AORs) 2.78; 95% confidence intervals (95% CIs) 2.60–2.96) [1]. Elevated inflammation during pregnancy has also been found as a risk factor for pregnancy complications such as GDM [2,3], pre-term delivery [4], and pre-eclampsia [5]. C-reactive protein (CRP), a sensitive marker of inflammation, is associated with various adverse birth outcomes during the entire pregnancy where pregnant women with elevated CRP during the first trimester had an increased risk of developing GDM [2]. In addition, elevated mid-pregnancy CRP was found in women who delivered pre-term compared to those without pre-term births [4].

Elevated inflammation during pregnancy may mediate the relationship between weight status before pregnancy and adverse pregnancy outcomes. It has been found that the relationship between pre-pregnancy BMI and the risk of pre-eclampsia was partially mediated by inflammation during pregnancy [6]. Elevated levels of CRP at ≤20 weeks of gestation accounted for 31% of the effect of pre-pregnancy BMI on the risk of pre-eclampsia [6]; however, the underlying mechanisms between pre-pregnancy BMI and inflammation during pregnancy are unclear. Interestingly, pre-pregnancy BMI and inflammation are both associated with dietary factors. Women with obese pre-pregnancy BMIs had significantly lower quality diets during pregnancy compared with women with normal pre-pregnancy BMIs [7–9]. Although there are multiple determinants for inflammation, one of the major modifiable determinants for inflammation is diet. Fruit and vegetable intake [10] and dietary fiber [11] were negatively associated with plasma CRP concentrations, which may be due to antioxidants present in fruit and vegetables [12] and decreased lipid oxidation levels [13].

The Dietary Inflammatory Index (DII)™ was developed to estimate the overall inflammatory effects of diet, and was based on an extensive literature review (around 2000 peer-reviewed articles) [14,15]. Higher DII scores were associated with increased levels of inflammation [16,17] and health outcomes such as mortality among adults in the USA [18,19], cardiovascular disease in men and women aged 55–80 years of the multicenter, randomized, nutritional intervention trial [20], lung cancer in the Melbourne Collaborative cohort study [21], laryngeal cancer [22], and lower cognitive functioning in a French population [23].

Dietary patterns considering the overall combinations of food groups and nutrients [24] among pregnant women have been studied using factor analysis or principal component analysis [25–28], reduced rank regression [29–31], cluster analysis [32,33], index analysis [7–9,34–36], or latent class analysis [37,38] in relation to small-for-gestational-age infants or pre-term births. To the best of our knowledge, there has been no research investigating the overall inflammatory potential of diet during pregnancy in relation to pre-pregnancy BMI. Furthermore, the association of pre-pregnancy BMI with the DII and CRP, the subjective and objective measures of inflammation, respectively, has not been explored.

Considering that both pre-pregnancy BMI and inflammation are closely associated with dietary factors; it is important to understand how pre-pregnancy BMI is associated with diet-induced inflammation during pregnancy. If the association between pre-pregnancy BMI to inflammation is made clear, it may be possible to reduce inflammation during pregnancy through pre-pregnancy weight status. This will eventually decrease the inflammation and the risk of pregnancy complications as is the main goal under the Healthy People 2020 Maternal and Child Health initiative [39]. It is of great significance to cross-examine the relationship between pre-pregnancy BMI and inflammation during pregnancy to reduce the risk of pregnancy complications and short- and long-term adverse birth outcomes. Therefore, our aim was to examine the relationship of pre-pregnancy BMI with dietary inflammation measured by the DII and concentrations of CRP during pregnancy, and our hypothesis was that pre-pregnancy BMI is associated with the DII and CRP concentrations during pregnancy.

2. Material and Methods

2.1. Study Population

Our study utilized public domain data from the continuous National Health and Nutrition Examination Survey (NHANES) 2003–2004, 2005–2006, 2007–2008, 2009–2010, and 2011–2012. The NHANES is a program of biennial data collection designed to assess the health and nutritional status of the civilian, non-institutionalized population of the USA, conducted by the National Center for Health Statistics (NCHS) and the Centers for Disease Control and Prevention (CDC). The NHANES uses a stratified multi-stage probability sample based on the selection of counties, blocks, households, and finally persons within households. The NHANES survey is unique in that it combines in-home interviews and physical examinations conducted at Mobile Examination Centers (MECs). Written informed consent was obtained from each survey participant for both the interview and examination. The participants were interviewed for information regarding age, race/ethnicity, education level, marital status, physical activity, and family poverty income ratio, which is a poverty measure in the poverty guidelines developed by the Department of Health Human Services (HHS) [40]. Reproductive health interviews obtained information during the gestation period at the time of the survey and pregnancy status was based on a positive urine pregnancy test. A complete description of the data-collection procedures and analytic guidelines have been provided elsewhere [41].

The 2003–2012 NHANES dataset included 856 pregnant women. Subjects were excluded if they reported incomplete or unreliable dietary data or did not meet the minimum criteria without records in individual foods files, as defined by the NCHS [42] ($n = 61$), extreme energy intake values (≤ 2092 kJ (500 kcal) per day and ≥ 20920 kJ (5000 kcal) per day, $n = 9$); or missing information on one or more of the following: self-reported weight before pregnancy ($n = 23$), self-reported height ($n = 24$), marital status ($n = 1$), or month in pregnancy ($n = 107$). The final analytic sample size was 561 pregnant women (Figure 1). The study has been reviewed and approved by the Institutional Review Board at the University of North Dakota (IRB-201610-100).

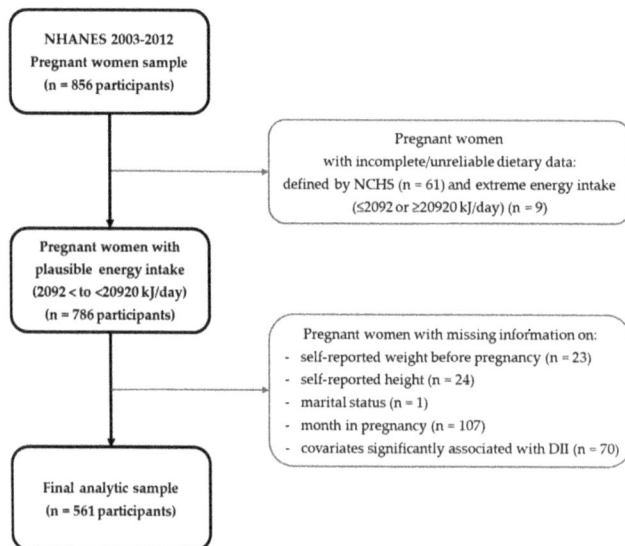

Figure 1. Flow chart describing National Health and Nutrition Examination Survey (NHANES) 2003–2012 pregnant women sample selection.

2.2. Exposure Variable

Pre-pregnancy BMI was calculated based on self-reported weight before pregnancy and self-reported height. Pre-pregnancy BMI from self-reported height and weight from pregnant women has been validated using the gold reference from multiple imputations [43]. The agreement in pre-pregnancy BMI classification from height and weight data between self-reported versus those measured, was high in pregnant women with the imputed weight status ($\kappa = 0.78$) and to the measured weight in the first trimester ($\kappa = 0.76$). A potential recall bias and BMI misclassification level can exist since Cohen's kappa values were not extremely high. Self-reported pre-pregnancy BMI status was stratified into four categories based on the WHO criteria [44]: <18.5 (underweight), 18.5–24.9 (normal), 25.0–29.9 (overweight), or ≥ 30 kg/m^2 (obese).

2.3. Outcome Variables

2.3.1. Dietary Inflammatory Index (DII)

The DII was developed by researchers at the University of South Carolina. Development and validation of the DII has been published previously [15,16]. In brief, the literature (approximately 2000 articles) between 1950 and 2010 was reviewed in terms of the relationship between various micronutrients, macronutrients, and whole food items (termed food parameters) and inflammation for the purposes of deriving inflammatory effect scores of the food parameters. At the same time, DII scores were standardized to a world database, which contains the means and standard deviations of intake for these 45 food parameters from 11 populations around the world [15]. It is rare that all 45 food parameters available in any given dataset: the NHANES dietary data includes 27 DII food parameters including vitamins A, B1, B2, B3 (niacin), B6, B9 (folic acid), B12, C, D, E; iron; magnesium; zinc; selenium; carbohydrates; protein; fat; saturated, monounsaturated, and polyunsaturated fatty acids; omega-3 and omega-6 polyunsaturated fatty acids; alcohol; fiber; cholesterol; beta carotene; and caffeine. The world mean value for that food parameter was subtracted from the actual intake value for each food parameter and then divided by the world standard deviation to create a z-score. The next step converted the z-scores to percentiles using the probnorm function in SAS, which were then centered by doubling the value and subtracting one. This value was then multiplied by the inflammatory effect score for each food parameter. These were then summarized across all food parameters to derive the overall DII score. The more positive scores are more pro-inflammatory, and the negative scores are more anti-inflammatory [15]. Lastly, it should be noted that DII scores were calculated per 1000 calories consumed and a single 24-h recall was used to derive the dietary information.

2.3.2. C-Reactive Protein (CRP)

CRP is an acute-phase protein produced by the liver in response to inflammation. In the NHANES 2003–2010, CRP (mg/dL) was measured by latex-enhanced nephelometry by a Behring Nephelometer for quantitative CRP determination [45–48].

2.4. Covariates

Multivariable models were adjusted for variables that were found to be significantly associated with the DII or pre-pregnancy BMI. Maternal age, family poverty income ratio and month in pregnancy were included in the multivariable models as continuous variables. Maternal education was grouped by the number of completed years of school: Less than high school graduate and more than college level. Race/ethnicity was divided into four groups: Hispanic, non-Hispanic white, non-Hispanic black and other (including multi-racial). Smoking status during pregnancy was defined by serum cotinine concentrations (non-smoker: ≤ 10 mg/L; smoker: >10 mg/L).

2.5. Statistical Analyses

Survey design procedures which considered the complex sampling design of NHANES were used for all analyses (SAS, version 9.4, Cary, NC, USA). Given that four cycles of 2-year data were used, 10-year sampling weights were developed by multiplying the 2-year weights by 0.2. Descriptive statistics for maternal age, family poverty income ratio, month in pregnancy, race/ethnicity, education, marital status, smoking status during pregnancy, physical activity and parity across categories of pregnancy BMI and DII tertiles were conducted with chi-square tests for categorical variables and an ANOVA test for continuous variables. Beta estimates (95% CIs) for association of pre-pregnancy BMI with the DII and CRP was calculated in both unadjusted and adjusted models. ORs and beta estimates (95% CIs) were calculated for the association between the DII and CRP in both unadjusted and adjusted models. Log-transformed CRP was used as an outcome for linear regression analyses. CRP was dichotomized at >3 mg/L and ≤3 mg/L for logistic regression analyses.

ORs and 95% CIs were estimated using logistic regression for the association of pre-pregnancy BMI categories with the DII tertiles and the CRP tertiles, respectively, using unadjusted and adjusted models. In the multivariable model, maternal age (continuous), family poverty income ratio (continuous), month in pregnancy (continuous), race/ethnicity (Hispanic, non-Hispanic white, non-Hispanic black, and other (including multi-racial groups)), education (less than high school graduate, more than college level), and smoking status during pregnancy (yes/no) were controlled. Due to missing variables in family poverty income ratio and physical activity, there were 561 pregnant women available for multivariable analyses for the association between pre-pregnancy BMI and the DII tertiles. To investigate the association between pre-pregnancy BMI and CRP tertiles, 551 women were available with CRP values, and 528 pregnant women were available for multivariable models due to missing variables in family poverty income ratio. A test for linear trend was performed using the median approach while calculating the median for each tertile of the DII and CRP, respectively, and as a continuous variable in analyses. A two-sided *p* value < 0.05 was declared as statistically significant.

3. Results

The distribution of study subjects by pre-pregnancy BMI categories and tertiles of the DII is presented in Table 1. Distributions of family poverty income ratio, month in pregnancy and race/ethnicity significantly differ in the categories of pre-pregnancy BMI. Women with a level of obese pre-pregnancy BMI had low family poverty income ratios and were relatively early in their pregnancies, and were more likely be non-Hispanic black (*p* < 0.05). There were significant differences in mean DII scores in sociodemographic and lifestyle factors. Pregnant women with greater odds of having a higher DII score (i.e., more pro-inflammatory) were those with an obese pre-pregnancy BMI. Women with advanced maternal age, Hispanics and non-smokers were less likely to have a higher DII score (*p* < 0.05).

Table 1. Distributions of socio-demographics and lifestyle factors by pre-pregnancy body mass index (BMI) and Dietary Inflammatory Index (DII) categories.

Socio-Demographics and Lifestyle Factors	Underweight (n = 31)		Normal (n = 311)		Overweight (n = 153)		Obese (n = 136)		p Value	Tertile 1 (n = 210)		Tertile 2 (n = 211)		Tertile 3 (n = 210)		p Value
	Mean	SEM	Mean	SEM	Mean	SEM	Mean	SEM		Mean	SEM	Mean	SEM	Mean	SEM	
Maternal Age (years)	27.5	1.4	28.1	0.6	28.5	0.9	28.1	0.8	0.8520	30.2	0.8	27.4	0.6	26.5	0.6	<0.0001
Family Poverty Income Ratio (n = 601)	2.5	0.5	3.1	0.2	2.7	0.3	2.3	0.2	0.0013	3.1	0.2	2.7	0.2	2.6	0.2	0.0775
Month in Pregnancy	5.2	0.7	5.9	0.3	5.6	0.3	4.7	0.3	0.0037	5.8	0.3	5.2	0.2	5.5	0.3	0.3511
	n	(Wt'd %[2])	n	(Wt'd %)	n	(Wt'd %)	n	(Wt'd %)		n	(Wt'd %)	n	(Wt'd %)	n	(Wt'd %)	
Race/Ethnicity									<0.0001							0.0002
Hispanic	4	2.0	93	47.4	52	26.7	48	23.9		80	43.4	75	35.0	42	21.6	
Non-Hispanic white	17	3.3	152	55.8	67	18.9	48	22.0		91	36.3	85	28.3	108	35.4	
Non-Hispanic black	4	5.0	39	25.8	25	19.4	37	49.8		17	15.4	39	38.9	49	45.7	
Other (including multi-racial)	6	16.6	27	63.2	9	15.8	3	4.4		22	56.7	12	31.4	11	11.9	
Education									0.2622							0.2086
≤High school graduate	12	4.4	140	42.8	85	24.5	77	28.4		91	29.5	108	34.0	115	36.5	
≥College	19	4.5	171	54.6	68	17.9	59	23.0		119	40.4	103	30.2	95	29.4	
Marital Status									0.3537							0.2410
Married/living with partner	26	4.6	244	51.4	122	21.2	94	22.8		181	38.6	168	32.0	137	29.4	
Widowed/divorced/separated/single	5	3.8	67	45.8	31	16.8	42	33.5		29	27.7	43	29.8	73	42.5	
Smoking Status during Pregnancy[3] (n = 585)									0.4225							0.0016
No	26	4.0	255	53.5	127	18.2	109	24.3		185	38.7	173	29.5	159	31.8	
Yes	3	8.5	38	42.9	12	28.0	15	20.5		8	10.1	22	35.0	38	54.9	
Physical Activity (n = 354)									0.8083							0.9646
Light (0–500 MET[4]-min/week)	10	3.8	88	53.8	48	20.4	41	22.0		72	39.6	47	24.3	68	36.1	
Moderate (500–1000 MET-min/week)	3	2.9	49	54.2	12	28.6	12	14.3		30	44.2	28	26.5	18	29.3	
Active (≥1000 MET-min/week)	3	2.4	53	64.0	21	18.2	14	15.4		36	45.7	30	20.5	25	33.8	
Parity (n = 297)									n/a							0.2143
None	0	.	11	77.2	2	8.1	3	14.7		3	11.6	5	39.5	8	48.9	
1	7	3.0	73	49.1	37	23.1	31	24.8		53	28.7	55	41.0	40	30.4	
2	3	4.3	35	56.1	25	15.7	21	23.9		33	49.1	26	24.1	25	26.8	
≥3	0	.	26	61.5	15	29.2	8	9.3		16	43.0	14	22.3	19	34.7	

[1] Pre-pregnancy BMI was stratified into four categories based on the WHO criteria: <18.5 kg/m² (underweight), 18.5–24.9 kg/m² (normal), 25.0–29.9 kg/m² (overweight), and ≥30 kg/m² (obese); [2] Wt'd % = Weighed percentage. Sample weights were created in NHANES to account for the complex survey design (including oversampling of some subgroups), survey non-responses, and post-stratification. When a sample was weighted in NHANES, it was representative of the US civilian non-institutionalized census population; [3] Smoking status during pregnancy was defined by serum cotinine concentrations (non-smoker: ≤10 mg/L; smoker >10 mg/L); [4] MET (Metabolic Equivalent of Task): Total MET-min/week from self-reported leisure-time physical activities. Tertile 1 was the highest anti-inflammatory group, and Tertile 3 was the most pro-inflammatory group. DII ranges for Tertile 1, Tertile 2 and Tertile 3 were −4.98–0.07, 0.08–1.67 and 1.68–4.14, respectively. p value: ANOVA test for continuous variables, and Chi-square test for categorical variables. n/a: Not available.

Beta estimates for the association of pre-pregnancy BMI with DII and CRP are presented in Table 2. Each kg/m^2 increase in pre-pregnancy BMI was associated with a 0.02 (95% CIs −0.03–4.05) increase in DII during pregnancy and a 0.07 (95% CIs 0.05–0.08) increase in CRP in the covariate-adjusted model.

Table 2. Beta estimates for association of pre-pregnancy BMI with DII and C-reactive protein (CRP).

	DII (*n* = 631 [1]; 561 [2])	CRP [3] (*n* = 551 [1]; 528 [2])
	Beta (95% CIs)	Beta (95% CIs)
Pre-pregnancy BMI [4]		
Unadjusted	0.03 (−0.35–0.90)	0.06 (0.05–0.07)
Multivariable [5]	0.02 (−0.03–4.05)	0.07 (0.05–0.08)

[1] Unadjusted model; [2] Multivariable model; [3] CRP was log-transformed; [4] Pre-pregnancy BMI (kg/m^2) was calculated as a continuous variable; [5] Adjusted for age (continuous), family poverty income ratio (continuous), month in pregnancy (continuous), race/ethnicity (Hispanic, non-Hispanic white, non-Hispanic black, other (including multi-racial)), education (≤high school graduate, ≥college) and smoking status during pregnancy (yes/no).

ORs and beta estimates for associations between the DII and CRP were calculated (Table 3), and there were no significant associations between the DII and CRP using logistic regression models. Each increase in the DII was associated with a 0.01 (95% CIs −0.04–0.06) increase in CRP in the unadjusted model and a 0.01 (95% CIs −0.03–0.06) increase in CRP in a covariate-adjusted model.

Table 3. Odds ratio and beta estimates for associations between DII and CRP.

	CRP (*n* = 551 [1]; 528 [2])	
	OR (95% CIs) [3]	Beta (95% CIs) [4]
DII continuous		
Unadjusted	0.97 (0.77–1.22)	0.01 (−0.04–0.06)
Multivariable [5]	0.94 (0.75–1.19)	0.01 (−0.03–0.06)

[1] Unadjusted model; [2] Multivariable model; [3] CRP was dichotomized at >0.3 mg/dL vs. ≤0.3 mg/dL for logistic regression analyses; [4] CRP was log-transformed; [5] Adjusted for age (continuous), family poverty income ratio (continuous), month in pregnancy (continuous), race/ethnicity (Hispanic, non-Hispanic white, non-Hispanic black, other (including multi-racial)), education (≤high school graduate, ≥college) and smoking status during pregnancy (yes/no).

Unadjusted and adjusted ORs for the highest tertile of the DII by pre-pregnancy BMI are presented in Table 4. No significant associations were found between pre-pregnancy BMI and the DII in the unadjusted model (*p*-trend = 0.4112); however, women with obese pre-pregnancy BMIs had increased odds of being in the highest tertile of the DII (pro-inflammatory) compared to women with normal pre-pregnancy BMIs (AORs 2.40, 95% CIs 1.01–5.71, *p*-trend = 0.009), when adjusted for age, family poverty income ratio, month in pregnancy, race/ethnicity, education, and smoking status during pregnancy. Unadjusted and adjusted ORs for the highest tertile of CRP by pre-pregnancy BMIs are presented in Table 5. In the unadjusted model, women with overweight and obese pre-pregnancy BMIs had increased odds of being in the highest tertile (AORs 3.69, 95% CIs 1.21–11.24; AORs 14.67, 95% CIs 4.80–44.83, respectively, *p*-trend < 0.0001). Women with overweight and obese pre-pregnancy BMIs had increased odds for being in the highest tertile of CRP in the multivariable model (AORs 3.95, 95% CIs 1.49–10.45; AORs 24.84, 95% CIs 6.19–99.67, respectively, *p*-trend < 0.0001).

Table 4. Unadjusted and adjusted odds ratios (ORs) and 95% CIs g in the highest DII tertile by pre-pregnancy BMI categories.

	Unadjusted (n = 631) Tertile 3 vs. Tertile 1 (Reference)			Adjusted [1] (n = 561) Tertile 3 vs. Tertile 1 (Reference)		
	ORs	95% CIs		AORs	95% CIs	
Pre-pregnancy BMI						
Underweight	2.26	0.58	8.85	3.11	0.85	11.45
Normal	1.00			1.00		
Overweight	1.31	0.56	3.11	1.44	0.56	3.73
Obese	2.15	0.96	4.83	2.40	1.01	5.71
p trend [2]	0.4112			0.009		
p trend [3]	0.116			0.037		
Age (continuous)				0.89	0.83	0.96
Family Poverty Income Ratio (continuous) (n = 601)				1.11	0.81	1.53
Month in Pregnancy (continuous)				0.98	0.81	1.18
Race/Ethnicity						
Hispanic				0.45	0.15	1.32
Non-Hispanic white				1.00		
Non-Hispanic black				2.30	0.63	8.45
Other (including multi-racial groups)				0.37	0.11	1.18
Education						
≤High school graduate				1.68	0.60	4.71
≥College				1.00		
Smoking Status during Pregnancy [4] (n = 585)						
Yes				4.25	1.25	14.51
No				1.00		

Mean ± SE for Tertile 1 (reference), Tertile 2 and Tertile 3 was −1.6 ± 0.1, 1.0 ± 0.1 and 2.4 ± 0.1, respectively. [1] Adjusted for age (continuous), month in pregnancy (continuous), race/ethnicity (Hispanic, non-Hispanic white (reference), non-Hispanic black, other including multi-racial), family poverty income ratio (continuous), education (≤high school graduate, ≥college (reference)) and smoking status during pregnancy (yes/no). p trend was obtained by [2] using the median approach, calculating median for each tertile of the DII as a continuous variable in analyses and [3] treating each DII as a continuous variable in the linear regression model. [4] Smoking status during pregnancy was defined by serum cotinine concentrations (non-smoker: ≤10 mg/L (reference); smoker >10 mg/L). Self-reported pre-pregnancy BMI was stratified into four categories based on the WHO criteria: <18.5 (underweight, reference), 18.5–24.9 (normal), 25.0–29.9 (overweight), and ≥30 kg/m² (obese).

Table 5. Unadjusted and adjusted odds ratios (ORs) and 95% CIs for being in the highest CRP tertile by pre-pregnancy BMI categories.

	Unadjusted (n = 551)			Adjusted[1] (n = 528)		
	Tertile 3 vs. Tertile 1 (Reference)			Tertile 3 vs. Tertile 1 (Reference)		
	ORs	95% CIs		AORs	95% CIs	
Pre-pregnancy BMI						
Underweight	0.25	0.06	1.08	0.36	0.08	1.56
Normal	1.00			1.00		
Overweight	3.69	1.21	11.24	3.95	1.49	10.45
Obese	14.67	4.80	44.83	24.84	6.19	99.67
p-trend[2]	<0.0001			<0.0001		
p-trend[3]	<0.0001			<0.0001		
Age (continuous)				1.02	0.95	1.10
Family Poverty Income Ratio (continuous) (n = 528)				1.41	1.07	1.85
Month in Pregnancy (continuous)				1.13	0.98	1.30
Race/Ethnicity						
Hispanic				2.62	1.14	6.03
Non-Hispanic white				1.00		
Non-Hispanic black				2.04	0.76	5.47
Other (including multi-racial groups)				1.88	0.22	16.12
Education						
≤High school graduate				3.45	1.17	10.20
≥College				1.00		
Smoking Status during Pregnancy[4]						
Yes				1.91	0.61	6.00
No				1.00		

[1] Adjusted for age (continuous), family poverty income ratio (continuous), month in pregnancy (continuous), race/ethnicity (Hispanic, non-Hispanic white (reference), non-Hispanic black, other (including multi-racial)), education (≤high school graduate, ≥college (reference)) and smoking status during pregnancy (yes/no). p-trend was obtained by[2] using the median approach, calculating median for each tertile of CRP values as a continuous variable in analyses and[3] treating each DII as a continuous variable in the linear regression model. [4] Smoking status during pregnancy was defined by serum cotinine concentrations (non-smoker: ≤10 mg/L (reference); smoker >10 mg/L). Self-reported pre-pregnancy BMI was stratified into four categories based on the WHO criteria: <18.5 (underweight), 18.5–24.9 (normal, reference), 25.0–29.9 (overweight), and ≥30 kg/m² (obese).

4. Discussion

The present study found that pre-pregnancy BMI was associated with the DII during pregnancy in pregnant women in the USA. This was consistent with the previous finding [18] that adults who were in the highest tertile of the DII (with pro-inflammatory diets) had significantly higher BMI compared to the lowest tertile of the DII (with anti-inflammatory diets) in the NHANES III. In another study conducted by Panagos et al. [49], women with obese pre-pregnancy BMI ($\geq$30 kg/m^2) (n = 21) demonstrated a greater pro-inflammatory diet, as indicated by a higher DII score compared to women with a lean pre-pregnancy BMI (18–25 kg/m^2) (n = 21) (−0.13 ± 0.82 vs. −0.68 ± 1.01, p = 0.06) in a follow-up study from late pregnancy (between 34 and 40 weeks gestational age) to 4–10 weeks postpartum.

As mentioned earlier, one of the most important issues relating to pre-pregnancy BMI and the DII is the quality of diet. Previous studies have examined the relationship between overall the quality of diet and pre-pregnancy BMI [7–9]. From the cross-sectional study of the NHANES 2003–2012, the Healthy Eating Index (HEI)-2010 was inversely associated with pre-pregnancy BMI [7]. In parallel with this finding, the cohort of Pregnancy, Infection, and Nutrition (PIN) study, the Diet Quality Index for Pregnancy (DQI-P) was inversely associated with pre-pregnancy BMI [8]. Whole grains, fruits [50], dietary fiber [51], and polyunsaturated fatty acids [52], which are major components in representative measures of the quality of diet, have shown inverse associations with inflammatory biomarkers, such as high-sensitive CRP (hs-CRP) and interleukin-6 (IL-6), in the general adult population. However, studies investigating the relationship of the quality of diet with the DII among pregnant women are limited. It would be important to cross-examine the relationship between the quality of diet and the DII during pregnancy in relation to pre-pregnancy BMI in future studies.

A strong and positive association between BMI and CRP among various populations such as non-pregnant adults and children has been well established [53–55]. In pregnant women, a few small-scale studies reported a positive correlation CRP with either pre-pregnancy BMI [56], or BMI at the first trimester [57]. Our present study was based on a larger study cohort (n = 551) and confirmed the positive significant association between pre-pregnancy BMI and CRP concentrations during pregnancy. Even after adjusting for DII, the beta estimates and standard error for the association between pre-pregnancy BMI and CRP was 0.046 and 0.007 (p < 0.0001), for which the underlying mechanisms are not clearly understood.

The distribution of the DII scores significantly differed according to maternal age, race/ethnicity and smoking status during pregnancy. By using multivariable logistic regressions, the DII was not significantly associated with CRP in our study, which was in contrast with previous findings where significantly increasing values of CRP were found with increasing DII scores, which in turn was associated with all-cause, cardiovascular and cancer mortality [18,58]. A pro-inflammatory diet is associated with increases in CRP concentrations and shortened telomere length in American adults [58], possibly due to elevated level of oxygen stress [59] and through the elevated level of CRP. In adults, higher DII scores significantly predicted higher plasma concentrations of IL-6 [17], tumor necrosis factor (TNF)-alpha [17], and CRP [16,17,23]. Contradicting results may be due to different target study populations such as pregnant women vs general adults. Specifically, a lack of statistically significant difference in CRP in relation to the DII among pregnant women may be partially due to elevated inflammatory responses triggered by the progression of pregnancy [60].

The present study; however, has several strengths. To the best of our knowledge, this is the first study that assessed the inflammatory status based on the diets of pregnant women with respect to pre-pregnancy BMI status. Second, the present study also included CRP as well as the DII to comprehensively assess inflammatory biomarkers and the properties of diet among pregnant women. Finally, various confounding factors (including maternal sociodemographic factors and smoking status during pregnancy) were controlled when examining the relationship of pre-pregnancy BMI with the DII during pregnancy.

There were also a few notable limitations in this study. First, only one 24-h recall was used to assess dietary information in the analysis. Using only one day of dietary information may not have accounted for the day-to-day variability in diet, leading to imprecise estimates [61]; Second, the relatively small sample size, particularly the number of underweight pregnant women, may be one of the limitations of this cross-sectional study. It may have been difficult to detect the significant difference on the DII and inflammatory biomarker, CRP across all the pre-pregnancy BMI categories. Third, potential under-reporting of dietary intakes by obese women might have occurred. Additionally, diets and metabolic profiles may have evolved during pregnancy. Fourth, although sociodemographic factors and pre-pregnancy BMI did not significantly differ between two groups of pregnant women with or without CRP values (631 vs. 551), the DII was significantly lower in women without a CRP value compared to those with CRP values (-0.2 ± 0.3 vs. 0.7 ± 0.2). Finally, due to the cross-sectional study design, the cause-effect relationship between BMI before pregnancy and the DII during pregnancy could not be drawn from this study.

5. Conclusions

In conclusion, high pre-pregnancy BMI was associated with increased risks of pro-inflammatory diet and elevated CRP levels in pregnant women. Future research is warranted to explore whether increased inflammation in obese pre-pregnancy women may adversely affect offspring health in the long-term.

Acknowledgments: The authors would like to thank the University of North Dakota College of Nursing & Professional Disciplines (CNPD) for providing the grant support through the CNPD Office of Research Seed Award in 2017.

Author Contributions: D.S. conceptualized the study question, conducted data analysis, interpreted the results, and wrote the first draft of the manuscript. J.H. critically reviewed and revised the manuscript and provided scientific advice. E.-H.C. provided scientific advice and contributed to the manuscript editing. H.-K.C. contributed to the data interpretation and critical revision of the manuscript. N.S., M.D.W. and J.R.H. contributed to the data analysis, calculated the DII, and critically revised the manuscript. K.W.L. reviewed and substantially edited the manuscript and guided the manuscript development. All authors read and approved the final manuscript.

Conflicts of Interest: J.R.H. owns controlling interest in Connecting Health Innovations LLC (CHI), a company to license the rights to his invention of the DII from the University of South Carolina to develop computer and smart phone applications for patient counseling and dietary intervention in clinical settings. N.S. and M.D.W. are employees of CHI.

References

1. Shin, D.; Song, W.O. Prepregnancy body mass index is an independent risk factor for gestational hypertension, gestational diabetes, preterm labor, and small- and large-for-gestational-age infants. *J. Matern. Fetal Neonatal Med.* **2015**, *28*, 1679–1686. [CrossRef] [PubMed]
2. Wolf, M.; Sandler, L.; Hsu, K.; Vossen-Smirnakis, K.; Ecker, J.L.; Thadhani, R. First-trimester C-reactive protein and subsequent gestational diabetes. *Diabetes Care* **2003**, *26*, 819–824. [CrossRef] [PubMed]
3. Qiu, C.; Sorensen, T.K.; Luthy, D.A.; Williams, M.A. A prospective study of maternal serum C-reactive protein (CRP) concentrations and risk of gestational diabetes mellitus. *Paediatr. Perinat. Epidemiol.* **2004**, *18*, 377–384. [CrossRef] [PubMed]
4. Bullen, B.L.; Jones, N.M.; Holzman, C.B.; Tian, Y.; Senagore, P.K.; Thorsen, P.; Skogstrand, K.; Hougaard, D.M.; Sikorskii, A. C-reactive protein and preterm delivery: Clues from placental findings and maternal weight. *Reprod. Sci.* **2013**, *20*, 715–722. [CrossRef] [PubMed]
5. Teran, E.; Escudero, C.; Moya, W.; Flores, M.; Vallance, P.; Lopez-Jaramillo, P. Elevated C-reactive protein and pro-inflammatory cytokines in Andean women with pre-eclampsia. *Int. J. Gynaecol. Obstet.* **2001**, *75*, 243–249. [CrossRef]
6. Bodnar, L.M.; Ness, R.B.; Harger, G.F.; Roberts, J.M. Inflammation and triglycerides partially mediate the effect of prepregnancy body mass index on the risk of preeclampsia. *Am. J. Epidemiol.* **2005**, *162*, 1198–1206. [CrossRef] [PubMed]
7. Shin, D.; Lee, K.W.; Song, W.O. Pre-pregnancy weight status is associated with diet quality and nutritional biomarkers during pregnancy. *Nutrients* **2016**, *8*, 162. [CrossRef] [PubMed]

8. Laraia, B.A.; Bodnar, L.M.; Siega-Riz, A.M. Pregravid body mass index is negatively associated with diet quality during pregnancy. *Public Health Nutr.* **2007**, *10*, 920–926. [CrossRef] [PubMed]
9. Tsigga, M.; Filis, V.; Hatzopoulou, K.; Kotzamanidis, C.; Grammatikopoulou, M.G. Healthy Eating Index during pregnancy according to pre-gravid and gravid weight status. *Public Health Nutr.* **2011**, *14*, 290–296. [CrossRef] [PubMed]
10. Esmaillzadeh, A.; Kimiagar, M.; Mehrabi, Y.; Azadbakht, L.; Hu, F.B.; Willett, W.C. Fruit and vegetable intakes, C-reactive protein, and the metabolic syndrome. *Am. J. Clin. Nutr.* **2006**, *84*, 1489–1497. [PubMed]
11. Ma, Y.; Griffith, J.A.; Chasan-Taber, L.; Olendzki, B.C.; Jackson, E.; Stanek, E.J.; Li, W.; Pagoto, S.L.; Hafner, A.R.; Ockene, I.S. Association between dietary fiber and serum C-reactive protein. *Am. J. Clin. Nutr.* **2006**, *83*, 760–766. [PubMed]
12. Liu, R.H. Health-promoting components of fruits and vegetables in the diet. *Adv. Nutr.* **2013**, *4*, 384S–392S. [CrossRef] [PubMed]
13. Hermsdorff, H.H.M.; Barbosa, K.B.; Volp, A.C.P.; Puchau, B.; Bressan, J.; Zulet, M.A.; Martínez, J.A. Vitamin C and fibre consumption from fruits and vegetables improves oxidative stress markers in healthy young adults. *Br. J. Nutr.* **2012**, *107*, 1119–1127. [CrossRef] [PubMed]
14. Cavicchia, P.P.; Steck, S.E.; Hurley, T.G.; Hussey, J.R.; Ma, Y.; Ockene, I.S.; Hébert, J.R. A new dietary inflammatory index predicts interval changes in serum high-sensitivity C-reactive protein. *J. Nutr.* **2009**, *139*, 2365–2372. [CrossRef] [PubMed]
15. Shivappa, N.; Steck, S.E.; Hurley, T.G.; Hussey, J.R.; Hebert, J.R. Designing and developing a literature-derived, population-based dietary inflammatory index. *Public Health Nutr.* **2014**, *17*, 1689–1696. [CrossRef] [PubMed]
16. Shivappa, N.; Steck, S.E.; Hurley, T.G.; Hussey, J.R.; Ma, Y.; Ockene, I.S.; Tabung, F.; Hebert, J.R. A population-based dietary inflammatory index predicts levels of C-reactive protein in the Seasonal Variation of Blood Cholesterol Study (SEASONS). *Public Health Nutr.* **2014**, *17*, 1825–1833. [CrossRef] [PubMed]
17. Tabung, F.K.; Steck, S.E.; Zhang, J.; Ma, Y.; Liese, A.D.; Agalliu, I.; Hingle, M.; Hou, L.; Hurley, T.G.; Jiao, L.; et al. Construct validation of the dietary inflammatory index among postmenopausal women. *Ann. Epidemiol.* **2015**, *25*, 398–405. [CrossRef] [PubMed]
18. Shivappa, N.; Steck, S.E.; Hussey, J.R.; Ma, Y.; Hebert, J.R. Inflammatory potential of diet and all-cause, cardiovascular, and cancer mortality in National Health and Nutrition Examination Survey III Study. *Eur. J. Nutr.* **2015**, *56*, 683. [CrossRef] [PubMed]
19. Deng, F.E.; Shivappa, N.; Tang, Y.; Mann, J.R.; Hebert, J.R. Association between diet-related inflammation, all-cause, all-cancer, and cardiovascular disease mortality, with special focus on prediabetics: Findings from NHANES III. *Eur. J. Nutr.* **2016**, *56*, 1085. [CrossRef] [PubMed]
20. Garcia-Arellano, A.; Ramallal, R.; Ruiz-Canela, M.; Salas-Salvado, J.; Corella, D.; Shivappa, N.; Schroder, H.; Hebert, J.R.; Ros, E.; Gomez-Garcia, E.; et al. Dietary Inflammatory Index and Incidence of Cardiovascular Disease in the PREDIMED Study. *Nutrients* **2015**, *7*, 4124–4138. [CrossRef] [PubMed]
21. Hodge, A.M.; Bassett, J.K.; Shivappa, N.; Hebert, J.R.; English, D.R.; Giles, G.G.; Severi, G. Dietary inflammatory index, Mediterranean diet score, and lung cancer: A prospective study. *Cancer Causes Control* **2016**, *27*, 907–917. [CrossRef] [PubMed]
22. Shivappa, N.; Hebert, J.R.; Rosato, V.; Serraino, D.; La Vecchia, C. Inflammatory potential of diet and risk of laryngeal cancer in a case-control study from Italy *Cancer Causes Control* **2016**, *27*, 1027–1034. [CrossRef] [PubMed]
23. Kesse-Guyot, E.; Assmann, K.E.; Andreeva, V.A.; Touvier, M.; Neufcourt, L.; Shivappa, N.; Hebert, J.R.; Wirth, M.D.; Hercberg, S.; Galan, P.; et al. Long-term association between the dietary inflammatory index and cognitive functioning: Findings from the SU.VI.MAX study. *Eur. J. Nutr.* **2016**. [CrossRef] [PubMed]
24. Hu, F.B. Dietary pattern analysis: A new direction in nutritional epidemiology. *Curr. Opin. Lipidol.* **2002**, *13*, 3–9. [CrossRef] [PubMed]
25. Shin, D.; Lee, K.W.; Song, W.O. Dietary patterns during pregnancy are associated with gestational weight gain. *Matern. Child Health J.* **2016**, *20*, 2527–2538. [CrossRef] [PubMed]
26. Grieger, J.A.; Grzeskowiak, L.E.; Clifton, V.L. Preconception dietary patterns in human pregnancies are associated with preterm delivery. *J. Nutr.* **2014**, *144*, 1075–1080. [CrossRef] [PubMed]

27. Thompson, J.M.; Wall, C.; Becroft, D.M.; Robinson, E.; Wild, C.J.; Mitchell, E.A. Maternal dietary patterns in pregnancy and the association with small-for-gestational-age infants. *Br. J. Nutr.* **2010**, *103*, 1665–1673. [CrossRef] [PubMed]

28. Rasmussen, M.A.; Maslova, E.; Halldorsson, T.I.; Olsen, S.F. Characterization of dietary patterns in the danish national birth cohort in relation to preterm birth. *PLoS ONE* **2014**, *9*, e93644. [CrossRef] [PubMed]

29. Shin, D.; Lee, K.W.; Song, W.O. Dietary patterns during pregnancy are associated with risk of gestational diabetes mellitus. *Nutrients* **2015**, *7*, 9369–9382. [CrossRef] [PubMed]

30. Vujkovic, M.; Steegers, E.A.; Looman, C.W.; Ocke, M.C.; van der Spek, P.J.; Steegers-Theunissen, R.P. The maternal Mediterranean dietary pattern is associated with a reduced risk of spina bifida in the offspring. *BJOG Int. J. Obstet. Gynaecol.* **2009**, *116*, 408–415. [CrossRef] [PubMed]

31. Obermann-Borst, S.A.; Vujkovic, M.; de Vries, J.H.; Wildhagen, M.F.; Looman, C.W.; de Jonge, R.; Steegers, E.A.; Steegers-Theunissen, R.P. A maternal dietary pattern characterised by fish and seafood in association with the risk of congenital heart defects in the offspring. *BJOG Int. J. Obstet. Gynaecol.* **2011**, *118*, 1205–1215. [CrossRef] [PubMed]

32. Okubo, H.; Miyake, Y.; Sasaki, S.; Tanaka, K.; Murakami, K.; Hirota, Y.; Kanzaki, H.; Kitada, M.; Horikoshi, Y.; Ishiko, O.; et al. Maternal dietary patterns in pregnancy and fetal growth in Japan: The Osaka Maternal and Child Health Study. *Br. J. Nutr.* **2012**, *107*, 1526–1533. [CrossRef] [PubMed]

33. McGowan, C.A.; McAuliffe, F.M. Maternal dietary patterns and associated nutrient intakes during each trimester of pregnancy. *Public Health Nutr.* **2013**, *16*, 97–107. [CrossRef] [PubMed]

34. Pick, M.E.; Edwards, M.; Moreau, D.; Ryan, E.A. Assessment of diet quality in pregnant women using the Healthy Eating Index. *J. Am. Diet. Assoc.* **2005**, *105*, 240–246. [CrossRef] [PubMed]

35. Rifas-Shiman, S.L.; Rich-Edwards, J.W.; Kleinman, K.P.; Oken, E.; Gillman, M.W. Dietary quality during pregnancy varies by maternal characteristics in Project Viva: A US cohort. *J. Am. Diet. Assoc.* **2009**, *109*, 1004–1011. [CrossRef] [PubMed]

36. Shin, D.; Bianchi, L.; Chung, H.; Weatherspoon, L.; Song, W.O. Is gestational weight gain associated with diet quality during pregnancy? *Matern. Child Health J.* **2014**, *18*, 1433–1443. [CrossRef] [PubMed]

37. Sotres-Alvarez, D.; Herring, A.H.; Siega-Riz, A.M. Latent class analysis is useful to classify pregnant women into dietary patterns. *J. Nutr.* **2010**, *140*, 2253–2259. [CrossRef] [PubMed]

38. Sotres-Alvarez, D.; Siega-Riz, A.M.; Herring, A.H.; Carmichael, S.L.; Feldkamp, M.L.; Hobbs, C.A.; Olshan, A.F. Maternal dietary patterns are associated with risk of neural tube and congenital heart defects. *Am. J. Epidemiol.* **2013**, *177*, 1279–1288. [CrossRef] [PubMed]

39. U.S. Department of Health and Human Services, Office of Disease Prevention and Health Promotion. Healthy People 2020. Available online: https://www.healthypeople.gov/2020/topics-objectives/topic/maternal-infant-and-child-health/objectives (accessed on 10 November 2016).

40. Centers for Disease Control and Prevention, National Center for Health Statistics. National Health and Nutrition Examination Survey 2003–2004 Data Documentation, Codebook, and Frequency: Demographic Variables and Sample Weights (DEMO_C). Available online: https://wwwn.cdc.gov/Nchs/Nhanes/2003-2004/DEMO_C.htm#Component_Description (accessed on 27 February 2016).

41. Johnson, C.L.; Paulose-Ram, R.; Ogden, C.E.; Carroll, M.D.; Kruszon-Moran, D.; Dohrmann, S.M.; Curtin, L.R. *National Health and Nutrition Examination Survey: Analytic guidelines, 1999–2010*; National Center for Health Statistics, Vital and Health Statistics: Washington, DC, USA, 2013; Volume 2.

42. Centers for Disease Control and Prevention, National Center for Health Statistic. Dietary Interview—Total nutrient Intakes, First Day. Available online: http://wwwn.cdc.gov/Nchs/Nhanes/2011--2012/DR1TOT_G.htm#DR1DRSTZ (accessed on 29 September 2016).

43. Shin, D.; Chung, H.; Weatherspoon, L.; Song, W.O. Validity of prepregnancy weight status estimated from self-reported height and weight. *Matern. Child Health J.* **2014**, *18*, 1667–1674. [CrossRef] [PubMed]

44. World Health Organization. Physical status: The use and interpretation of anthropometry. Report of a WHO Expert Committee. World Health Organization Technical Report Series; World Health Organization: Geneva, Switzerland, 1995; Volume 854, pp. 1–452.

45. Centers for Disease Control and Prevention; National Center for Health Statistics. National Health and Nutrition Examination Survey 2003–2004 Data Documentation, Codebook, and Frequency: C-Reactive Protein (CRP), Bone Alkaline Phosphatase (BAP) & Parathyroid Hormone (PTH) (L11_C). Available online: http://wwwn.cdc.gov/Nchs/Nhanes/2003--2004/L11_C.htm (accessed on 10 November 2016).

46. Centers for Disease Control and Prevention; National Center for Health Statistic. National Health and Nutiriton Examination Survey 2005–2006 Data Documentation, Codebook, and Frequencies: C-Reactive Protein (CRP) (CRP_D). Available online: http://wwwn.cdc.gov/Nchs/Nhanes/2005--2006/CRP_D.htm# LBXCRP (accessed on 10 November 2016).

47. Centers for Disease Control and Prevention; National Center for Health Statistics. National Health and Nutrition Examination Survey 2007–2008 Data Documentation, Codebook, and Frequencies: C-Reactive Protein (CRP) (CRP_E). Available online: http://wwwn.cdc.gov/Nchs/Nhanes/2007-2008/CRP_E.htm (accessed on 10 November 2016).

48. Centers for Disease Control and Prevention, National Center for Health Statistics. National Health and Nutrition Examination Survey 2009–2010 Data Documentation, Codebook, and Frequencies: C-Reactive Protein (CRP) (CRP_F). Available online: http://wwwn.cdc.gov/Nchs/Nhanes/2009-2010/CRP_F.htm (accessed on 10 November 2016).

49. Panagos, P.G.; Vishwanathan, R.; Penfield-Cyr, A.; Matthan, N.R.; Shivappa, N.; Wirth, M.D.; Hebert, J.R.; Sen, S. Breastmilk from obese mothers has pro-inflammatory properties and decreased neuroprotective factors. *J. Perinatol.* **2016**, *36*, 284–290. [CrossRef] [PubMed]

50. Nettleton, J.A.; Steffen, L.M.; Mayer-Davis, E.J.; Jenny, N.S.; Jiang, R.; Herrington, D.M.; Jacobs, D.R., Jr. Dietary patterns are associated with biochemical markers of inflammation and endothelial activation in the Multi-Ethnic Study of Atherosclerosis (MESA). *Am. J. Clin. Nutr.* **2006**, *83*, 1369–1379. [PubMed]

51. North, C.J.; Venter, C.S.; Jerling, J.C. The effects of dietary fibre on C-reactive protein, an inflammation marker predicting cardiovascular disease. *Eur. J. Clin. Nutr.* **2009**, *63*, 921–933. [CrossRef] [PubMed]

52. Clarke, R.; Shipley, M.; Armitage, J.; Collins, R.; Harris, W. Plasma phospholipid fatty acids and CHD in older men: Whitehall study of London civil servants. *Br. J. Nutr.* **2009**, *102*, 279–284. [CrossRef] [PubMed]

53. Timpson, N.J.; Nordestgaard, B.G.; Harbord, R.M.; Zacho, J.; Frayling, T.M.; Tybjaerg-Hansen, A.; Smith, G.D. C-reactive protein levels and body mass index: Elucidating direction of causation through reciprocal Mendelian randomization. *Int. J. Obes.* **2011**, *35*, 300–308. [CrossRef] [PubMed]

54. Rawson, E.S.; Freedson, P.S.; Osganian, S.K.; Matthews, C.E.; Reed, G.; Ockene, I.S. Body mass index, but not physical activity, is associated with C-reactive protein. *Med. Sci. Sports Exerc.* **2003**, *35*, 1160–1166. [CrossRef] [PubMed]

55. Ford, E.S.; Galuska, D.A.; Gillespie, C.; Will, J.C.; Giles, W.H.; Dietz, W.H. C-reactive protein and body mass index in children: Findings from the Third National Health and Nutrition Examination Survey, 1988–1994. *J. Pediatr.* **2001**, *138*, 486–492. [CrossRef] [PubMed]

56. Challier, J.C.; Basu, S.; Bintein, T.; Minium, J.; Hotmire, K.; Catalano, P.M.; Hauguel-de Mouzon, S. Obesity in pregnancy stimulates macrophage accumulation and inflammation in the placenta. *Placenta* **2008**, *29*, 274–281. [CrossRef] [PubMed]

57. Ramsay, J.E.; Ferrell, W.R.; Crawford, L.; Wallace, A.M.; Greer, I.A.; Sattar, N. Maternal obesity is associated with dysregulation of metabolic, vascular, and inflammatory pathways. *J. Clin. Endocrinol. Metab.* **2002**, *87*, 4231–4237. [CrossRef] [PubMed]

58. Shivappa, N.; Wirth, M.D.; Hurley, T.G.; Hebert, J.R. Association between the dietary inflammatory index (DII) and telomere length and C-reactive protein from the National Health and Nutrition Examination Survey-1999–2002. *Mol. Nutr. Food Res.* **2017**. [CrossRef] [PubMed]

59. Von Zglinicki, T. Oxidative stress shortens telomeres. *Trends Biochem. Sci.* **2002**, *27*, 339–344. [CrossRef]

60. Borzychowski, A.; Sargent, I.; Redman, C. Inflammation and Pre-Eclampsia. *Semin. Fetal Neonatal Med.* **2006**, *11*, 309–316. [CrossRef] [PubMed]

61. Basiotis, P.P.; Welsh, S.O.; Cronin, F.J.; Kelsay, J.L.; Mertz, W. Number of days of food intake records required to estimate individual and group nutrient intakes with defined confidence. *J. Nutr.* **1987**, *117*, 1638–1641. [PubMed]

![nutrients logo] *nutrients*

MDPI

Review

Zinc in Infection and Inflammation

Nour Zahi Gammoh and Lothar Rink *

Institute of Immunology, Faculty of Medicine, RWTH Aachen University, University Hospital, Pauwelstrasse 30, 52074 Aachen, Germany; nour.gammoh@rwth-aachen.de
* Correspondence: LRink@UKAachen.de; Tel.: +49-2418-080-208

Received: 29 April 2017; Accepted: 11 June 2017; Published: 17 June 2017

Abstract: Micronutrient homeostasis is a key factor in maintaining a healthy immune system. Zinc is an essential micronutrient that is involved in the regulation of the innate and adaptive immune responses. The main cause of zinc deficiency is malnutrition. Zinc deficiency leads to cell-mediated immune dysfunctions among other manifestations. Consequently, such dysfunctions lead to a worse outcome in the response towards bacterial infection and sepsis. For instance, zinc is an essential component of the pathogen-eliminating signal transduction pathways leading to neutrophil extracellular traps (NET) formation, as well as inducing cell-mediated immunity over humoral immunity by regulating specific factors of differentiation. Additionally, zinc deficiency plays a role in inflammation, mainly elevating inflammatory response as well as damage to host tissue. Zinc is involved in the modulation of the proinflammatory response by targeting Nuclear Factor Kappa B (NF-κB), a transcription factor that is the master regulator of proinflammatory responses. It is also involved in controlling oxidative stress and regulating inflammatory cytokines. Zinc plays an intricate function during an immune response and its homeostasis is critical for sustaining proper immune function. This review will summarize the latest findings concerning the role of this micronutrient during the course of infections and inflammatory response and how the immune system modulates zinc depending on different stimuli.

Keywords: zinc; infection; inflammation; homeostasis

---

1. Introduction

Zinc is a nutritionally fundamental trace element and is the second most abundant trace metal in the human body after iron. The influence of zinc on human health was first observed and described by Prasad et al. in the 1960s [1]. Zinc research has come a long way since then; its significance as a structural component in many proteins and its participation in numerous cellular functions is now well established. Such functions include, but are not limited to, cell proliferation and differentiation [2], RNA and DNA synthesis [3], as well as cell structures and cell membrane stabilization [4]. Its multifaceted role in the regulation of the immune system is particularly interesting and will be discussed in more detail in this review [5]. Consequently, zinc's implication in an array of functions demonstrates how a defect in nutritional absorption may lead to the manifestation of various diseases.

Zinc is involved in many metabolic and chronic diseases such as: diabetes, cancer (esophageal, oral small cell carcinoma, breast cancer), and neurodegenerative diseases. There is also strong evidence between zinc deficiency and several infectious diseases such as malaria, HIV, tuberculosis, measles, and pneumonia [6].

2. Zinc and Nutrition

The total zinc content in the human body amounts to 2–4 g, with a plasma concentration of 12–16 μM [2]. Albeit it is a small plasma pool, it is rapidly exchangeable and mobile. Sufficient daily intake of zinc is necessary to maintain a steady state because, unlike iron, the body has no specialized zinc storage system. The highest concentrations of zinc are found in the muscles, bones, skin, and liver [7,8].

Recommended daily intake of zinc depends on several factors such as age, sex, weight, and phytate content of diet. Those recommended values differ in each country. The US Food and Nutrition Board recommended intake of 11 mg/day and 8 mg/day for adult males and females, respectively [9]. While the German Society of Nutrition's recommendation comprised of 10 mg/day and 7 mg/day for adult males and females, respectively [10]. Both the World Health Organization (WHO) and the European Food Safety Authority (EFSA) consider the inhibitory effect of dietary phytate on zinc absorption when setting the recommended zinc intake values. WHO categorizes diets according to their potential absorption efficiency of zinc, per phytate-zinc molar ratio, into 3 groups; high (<5), moderate (5–15), and low (>15) zinc bioavailability [11]. Whereas EFSA provides different zinc reference recommendations for diets containing phytate intake levels of 300, 600, 900, and 1200 mg/day [12].

As mentioned, zinc bioavailability depends on the composition of the diet. Non-digestible plant ligands such as phytate, some dietary fibers, and lignin chelate zinc and inhibit its absorption. Other factors that influence the absorption of zinc are calcium and iron. Zinc is present in many food groups and its concentration and bioavailability varies considerably. Foods with the highest zinc concentration include red meat, some shellfish, legumes, fortified cereals, and whole grains [9]. Zinc from animal sources has higher bioavailability compared to zinc sourced from plant products. People who abstain from eating red meats, vegetarians, vegans, and people living in developing country who rely mainly on plant-based foods are at higher risk of developing zinc deficiency due to inadequate zinc intake [13]. Oral zinc supplements are readily available but not all offer the same zinc bioavailability. Zinc bound to amino acids such as aspartate, cysteine, and histidine shows the highest absorption concentration, followed by zinc chloride, sulfate, and acetate, whereas zinc oxide show the lowest bioavailability [14].

According to the WHO, zinc deficiency is currently the fifth leading cause of mortality and morbidity in developing countries. It is estimated that it affects about one-third of the world's population. Worldwide, zinc deficiency accounts for approximately 16% of lower respiratory tract infections, 18% of malaria, and 10% of diarrheal diseases. While severe zinc deficiency is rare, mild to moderate deficiency is more common worldwide [15,16].

3. Zinc Homeostasis

3.1. Zinc Transporters

There are two major protein families that mammalian zinc transporters belong to. The first group of transporter are ZIP (Zrt/Irt-like proteins), which are responsible for transporting zinc into the cytosol from either extracellular space or from intracellular compartments. There are 14 ZIP transporters, designated as Solute Carrier family SLC39A1-A14. The second group of 10 transporters are ZnT (Zinc transporters), which are designated as SLC30A1-A10. They generally transport zinc out of the cytosol into extracellular space or intracellular organelles such as zincosomes. Zincosomes are vesicles that can sequester high levels of zinc [17–19].

Each of the ZIP and ZnT transporters show tissue specificity and developmental and stimulus responsive expression patterns. On a cellular and subcellular level, they are also localized in specific compartments. Both transporter families respond to various stimuli such as zinc deficiency and excess by displaying specific changes in cellular localization and protein stability [20].

There are many zinc transporter mutations that are reported to be involved in inherited diseases. Most notably, a mutation in ZIP4 (SLC39A4), which is an intestinal zinc transporter, is responsible for the rare lethal autosomal-recessive inherited zinc deficiency disease, acrodermatitis enteropathica (AE). AE is characterized by severe dermatological manifestations, gastrointestinal disturbances, weight loss, growth retardation, male hypogonadism, and high susceptibility to infections among other clinical features. Complete recovery occurred after high-dose zinc supplementation [21]. In addition to AE, there is another genetic disease associated to mutations in ZIP transporters. Mutations in SLC39A13, encoding ZIP13, cause a novel subtype of Ehlers-Danlos Syndrome (EDS). EDS is a spectrum of connective tissue disorders caused by mutations affecting collagen synthesis and modification [22].

There are also some genetic diseases affecting the ZnT family. One example is a mutation influencing ZnT2. A substitution of a histidine with an arginine at amino acid 54 (H54R) in the human SLC30A2 gene causes a defect in zinc secretion. The H54R mutation of ZnT2 is autosomal dominant as women who are heterozygous for the allele present the phenotype of reduced zinc production in the breast milk during lactation. Infants that are exclusively breast-fed suffer from zinc deficiency due to low zinc levels in the breast milk, which in turn predisposes them to multiple infections. Zinc deficiency symptoms can be alleviated by oral zinc supplementation to the nursing babies [23].

Expression modulation of ZIPs and ZnTs during inflammatory processes has been documented. Intracellular zinc requirements are altered during an inflammatory event or when an invading pathogen gains access to the cell. For instance, in an allergic inflammation mouse model, expression of zinc transporters was altered, including increases in ZIP1 and ZIP14 and decreases in ZIP4 and ZnT4 [24]. Furthermore, zinc deprivation of phagocytosed *Histoplasma capsulatum* by sequestration into the Golgi apparatus is controlled by ZnT4 and ZnT7 and uptake of extracellular zinc is controlled by ZIP2 [25]. The role of zinc transporters during inflammation and infection is under active investigation and more will be revealed of how those transporters influence zinc homeostasis in different conditions.

3.2. Metallothioneins and Other Zinc Binding Proteins

Metallothioneins (MTs) are cysteine-rich 6–7 kDa proteins that bind metal ions such as zinc [26]. Up to 20% of intracellular zinc is bound to MTs, and can be rapidly released. They can form a complex with up to 7 zinc ions. There are four different MT classes; MT-1 and MT-2 are ubiquitous throughout the body, their main function is to maintain cellular zinc homeostasis and chelate heavy metals to reduce cytotoxicity and lower their intracellular concentrations, and due to their reactive oxygen species (ROS) scavenging properties, they help protect against several types of environmental stress. MT-3 and MT-4 expression is restricted to a cell type-specific pattern, with MT-3 predominantly found in the brain and MT-4 is primarily located in stratified epithelial tissues [20,27].

In addition to MTs, there are other zinc binding proteins that act as a storage system and control the release of zinc. Albumin binds around 80% of all plasma zinc and is thought to act as a major zinc transporter. Albumin modulates zinc uptake into certain cell types, such as endothelial cells. These cells co-transport albumin-bound zinc via a specific receptor-mediated pathway. Albumin has several metal-binding sites that are specific for different metal ions, for instance, site A binds specifically to zinc at high affinity. However, the fatty acid content of albumin influences metal binding, particularly during conditions of increased free fatty acid mobilization such as after exercise. Zinc-binding capacity may be reduced in such conditions when about four fatty acids may be bound to albumin. Even though zinc albumin complexes do not dissociate as easily as other zinc–protein complexes, they are still considered to have rapid exchange kinetics and contribute to the modulation of free zinc in the plasma [28].

Furthermore, the S100 protein family, which consists of more than 20 members, is composed of EF-hand calcium regulated proteins, and they are distributed in a cell-specific, tissue-specific, and cell cycle-specific manner in humans and other vertebrates. S100 proteins have diverse functions ranging from calcium buffering, intracellular functions such as modulating enzyme activities and secretions, nuclear functions such as apoptosis and transcription, and extracellular activities related to secretion and chemotaxis, among other functions [29]. S100 proteins are also regulated by zinc and several members show higher affinity towards zinc compared to calcium such as the case with S100A3. The zinc-binding site in S100B, S100A6, S100A7, S100A8/A9, S100A12, and S100A15 consist of either three histidine residues and one aspartate or four histidine residues and is distinct from the calcium binding sites in the EF-hands, while S100A2, S100A3, and S100A4 have cysteine-containing zinc-binding sites [30]. A brief description of the role of some S100 protein in nutritional immunity will be mentioned later.

α2-macroglobulin (A2M) is another zinc-binding protein which is an inhibitor of matrix metalloproteases (MMPs) and it is required to remove proteolytic potential, when MMPs increase, forming A2M-proteinase complexes. It has a very high affinity to zinc, where zinc is required for the activation of A2M and also for the binding of A2M with cytokines [31].

The cytokine interleukin 6 (IL6) induces the expression of MT and A2M and consequently reduces zinc availability. IL-6 is released during the acute phase of an inflammatory response. This mechanism is beneficial to the acute immune response, however, a long-term decrease in zinc availability may contribute to pathological processes in conditions of chronic inflammation (e.g., diabetes and dementia). IL6, MT, and A2M expression increase in old age and impaired zinc availability contributes to immunosenescence. A mutation in the IL6 promoter up-regulates its expression leading to increased MT, low plasma zinc, impaired innate immunity and increased risk of Alzheimer disease [32].

4. Zinc and Immunity

Many organs are affected by zinc deficiency, especially the immune system which is markedly susceptible to changes of zinc levels. It seems that every immunological event is influenced by zinc somehow. The immune response involves two mechanisms; innate and adaptive immunity. The first cells to encounter invading pathogens and eliminate them are the cells of the innate immunity. Polymorphonuclear cells (PMNs), macrophages (Mφ), and natural killer cells (NK) are some of those first responders. PMN chemotaxis and phagocytosis are reduced during zinc deficiency, whereas zinc supplementation has the opposite effect. After phagocytosis, pathogens are destroyed by the activity of nicotinamide adenine dinucleotide phosphate (NADPH) oxidases which has been shown to be inhibited by both zinc deficiency and excess [33,34]. Alternatively, PMN can kill pathogens by releasing neutrophil extracellular traps (NETs). These matrices of DNA, chromatin, and granule proteins can capture extracellular bacteria [35]. Chelating free zinc abolishes NET formation was observed in vitro. Moreover, before macrophages can mature into tissue resident cells, circulating monocytes must be attracted to the target tissues and adhere to endothelial cells. In vitro, this process of adhesion is augmented by zinc. Zinc deficiency increases the production of proinflammatory cytokines, such as interleukins IL-1β, IL-6, and tumor necrosis factor (TNF)-α. Recognition of major histocompatibility complex (MHC) class I by NK cells and the lytic activity of NK cells is influenced by zinc depletion. In terms of the adaptive immune response, zinc deficiency causes thymic atrophy and subsequent T-cell lymphopenia as well as reduction of premature and immature B cells, and consequently antibody production is also reduced [36]. The following sections will delve deeper into how zinc influences immune cells and their mediators during inflammation and infection.

It is worth mentioning that most zinc deficiency studies have been conducted on animal models and various cell culture types. Since zinc does not have major storage depot in the body, zinc deficiency is easily and rapidly produced. As mentioned previously, severe zinc deficiency is rare but mild to moderate deficiency is more prevalent. It remains a challenging task to attribute certain clinical manifestations to mild to moderate zinc deficiency, particularly because depletion of zinc from tissue is usually accompanied by deficiencies in other nutrients. This can be further rationalized by understanding the influence of zinc deficiency on food intake; this effect was observed in rats that were either starved or fed zinc deficient diets resulting in reduced weight and metabolic activity among other indicators. Thus common zinc deficiency symptoms can be attributed to either low zinc or to reduced overall nutrient intake [37]. However, that does not undermine the vital role zinc plays in the immune system.

5. Zinc in Inflammation

Inflammation is a natural process required to protect the host from tissue damage and infections, which leads to the resolution of the inflammatory response and the restoration of homeostasis.

However, sometimes the inflammation does not resolve and later becomes chronic, leading to loss of tissue function [38]. This section will review how the body mounts an inflammatory response and how zinc influences those processes.

5.1. NF-κB and Other Signalling Pathways

Proper modulation of inflammatory pathways is required to achieve adequate response to various stimuli such as stress, free radicals, cytokines, or bacterial and viral antigens. One of the main inflammatory pathways is the nuclear factor kappa-light-chain-enhancer of activated B cells (NF-κB) signaling pathway. It regulates the genes controlling apoptosis, cell adhesion, proliferation, tissue remodeling, the innate and adaptive immune responses, inflammatory processes, and cellular-stress responses. Subsequently, it influences the expression of proinflammatory cytokines such as TNF-α, IL-1β, IL-6, IL-8, and MCP (monocyte chemoattractant protein)-1. NF-κB is one of the most versatile regulators of gene expression [39].

The NF-κB protein family in mammalian cells consists of five members, RelA (p65), RelB, c-Rel, p50/p105 (NF-κB1), and p52/p100 (NF-κB2). Different NF-κB complexes are also formed from homo- and heterodimers. Non-active NF-κB complexes are typically found in the cytoplasm, where they are bound to and silenced by a family of inhibitory proteins known as inhibitors of NF-κB (IκBs). There are several IκBs; IκBα, IκBβ, IκBγ, IκBε, Bcl-3 along with p100 and p105 which function as IκB-like proteins that inhibit their NF-κB-subunit dimeric partners, p50 and p52, respectively. Activation of NF-κB requires the phosphorylation of IκBs by the IκB kinase (IKK) complex, which degrades the IκB and releases NF-κB and allows it to translocate freely to the nucleus so it can induce targeted gene expression [39,40].

There are several conflicting studies regarding the effect of zinc on this process. In vitro studies using different cell types and zinc concentrations as well as the impact of chelating agents have made it obvious that the influence of zinc cannot be interpreted using a unilateral approach [41]. For instance, a study by Haase et al. has revealed that zinc is necessary for the activation of lipopolysaccharide (LPS)-induced NF-κB signaling pathway, whereas chelating zinc with membrane permeable zinc specific chelator TPEN (*N,N,N′,N′*-tetrakis-(2-pyridyl-methyl) ethylenediamine) completely blocked this pathway [42]. On the other hand, there is a growing body of literature that supports the role of zinc as a negative regulator of NF-κB signaling pathways. There are several inhibitory mechanisms that have been suggested. One of the major inhibitory mechanisms relies on how zinc affects the expression of protein A20. A20 is a zinc-finger protein that is recognized as an anti-inflammatory protein which also negatively regulates tumor necrosis factor receptor (TNFR)- and toll like receptor (TLR)-initiated NF-κB pathways (Figure 1). During TNFR signaling, A20 is able to deubiquitinate receptor interacting protein 1 (RIP1), which prevents its interaction with NF-κB essential modulator IKKγ. It also inhibits TLR signaling by removing polyubiquitin chains from TNF receptor associated factor 6 (TRAF6). Although the deubiquitinase activity of A20 remains unchanged by zinc chelator [19], Prasad et al. demonstrated that the induction of A20 mRNA and generation of the protein in pre-monocytic, endothelial, and cancer cell is zinc-dependent [43]. Additionally, zinc supplementation was able to downregulate inflammatory cytokines by decreasing gene expression of IL-1β and TNF-α through upregulation of mRNA and DNA-specific binding for A20, subsequently inhibiting NF-κB activation [44].

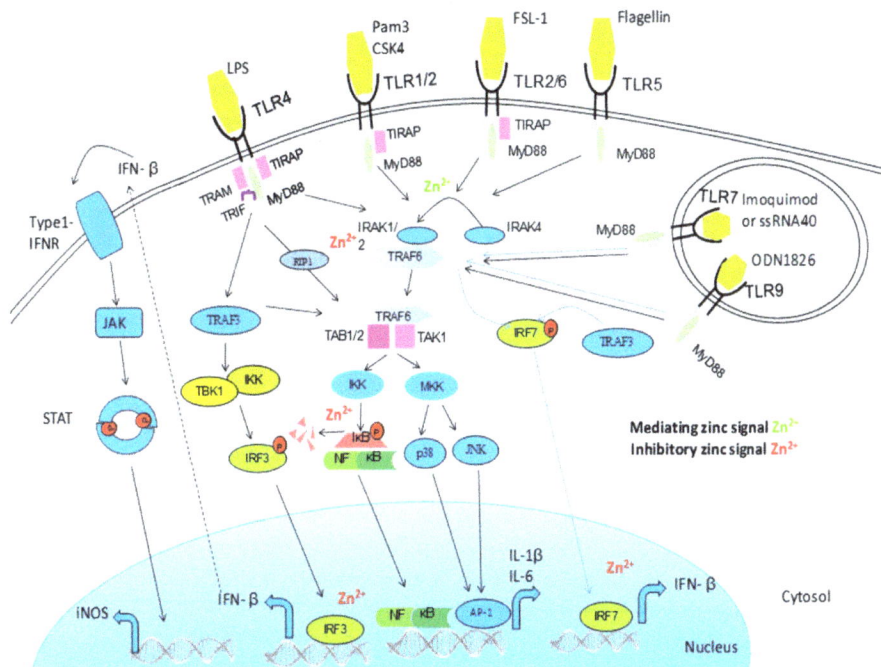

Figure 1. Activation of Toll-like receptor (TLR) signaling pathways is mediated by a complex array of proteins. When TLRs bind ligands, dimerization of the ectodomain of TLRs is induced, bringing their cytoplasmic Toll/IL-1R domains together, resulting in the recruitment of intracellular adapter proteins and initiation of downstream signaling events. Following this, the Toll–IL-1-resistence (TIR) domains of TLRs engage TIR domain-containing adaptor proteins (either myeloid differentiation primary-response protein 88 (MYD88) or TIR domain-containing adaptor protein inducing IFNβ (TRIF) and TRIF-related adaptor molecule (TRAM)). This in turn stimulates downstream signaling pathways that involve interactions between IL-1R-associated kinases (IRAKs) and the adaptor molecules TNF receptor-associated factors (TRAFs), and that lead to the activation of the mitogen-activated protein kinases (MAPKs) JUN *N*-terminal kinase (JNK) and p38, and to the activation of transcription factors. Two groups of transcription factors are activated downstream of TLR signaling; nuclear factor-κB (NF-κB) and interferon-regulatory factors (IRFs). This TLR signaling pathway leads to the induction of proinflammatory cytokines. Zinc influences this pathway on multiple levels. It augments MyD88-dependent signaling whereas inhibiting TRIF-mediated activation of IRF3/7. It has been shown that zinc inhibits IRAK thereby inhibiting further signal transductions. In vitro, zinc also directly impaired LPS-induced IKK activity [45].

There are several studies that have examined the influence of zinc on A20-mediated NF-κB inhibition [46–48]. Moreover, zinc inhibits NF-κB activation in the DNA nuclear binding levels by increasing the expression of peroxisome proliferator-activated receptor α (PPAR-α), which is a mediator for lipoprotein metabolism, inflammation, and glucose homeostasis. PPAR-α increase leads to the down-regulation of inflammatory cytokines and adhesion molecule [49]. Additionally, zinc acts as an inhibitor of cyclic nucleotide phosphodiesterase (PDE). When PDE is inhibited, cyclic nucleotide cGMP (Cyclic guanosine monophosphate) is elevated leading to the activation of PKA (protein kinase A) and subsequent inhibition of NF-κB [50]. Similarly, zinc can bind to a zinc finger-like motif found on protein kinase C (PKC) and inhibit PMA-mediated PKC translocation to the membrane. When this occurs in mast cells, NF-κB activity is indirectly inhibited [51]. Those mentioned studies demonstrates how zinc influences inflammation via NF-κB signaling pathways through several mechanisms and at many levels. Figures 2 and 3 provide more details on how zinc influences several pro- and anti-inflammatory pathways.

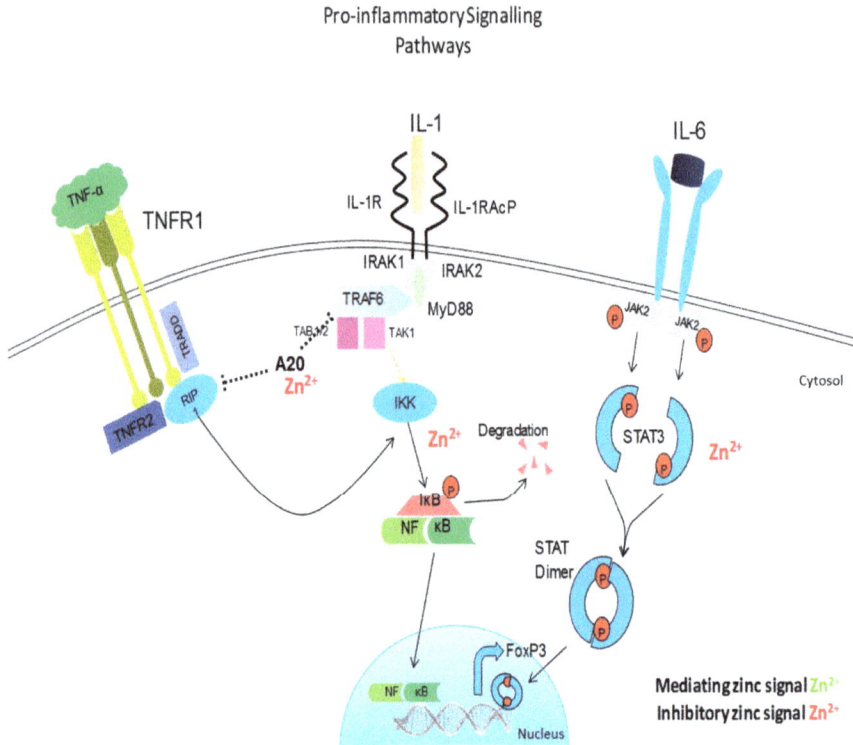

Figure 2. Pro-inflammatory signaling pathway influences by zinc. Similar to TLR signaling, IL-1, and TNF-R signaling pathways converge on a common IκB kinase complex that phosphorylates the NF-κB inhibitory protein, resulting in the release of NF-κB and its translocation to the nucleus. Zinc prevents the dissociation of NF-κB from its corresponding inhibitory protein, thus preventing the nuclear translocation of NF-κB and inhibiting subsequent inflammation. Zinc also inhibits IL-6-mediated activation of STAT3. Zinc acts as anti-inflammatory element influencing major pro-inflammatory signaling pathways.

Zinc can simultaneously modulate inflammation through TLR signaling at different levels and pathways [45]. TLRs are single, membrane-spanning, noncatalytic receptors which are expressed on many cells such as macrophages and dendritic cells and can distinguish and be activated by structurally conserved molecules derived from microbes. In monocytes, it has been observed that TLR4 activation initiates zinc-mediated signaling in a MyD88 (Myeloid differentiation primary response gene 88) and TRIF independent manner. Zinc has been shown to influence cellular signal transduction by inhibition of several dephosphorylating enzymes like protein tyrosine phosphatase (PTPs), cyclic nucleotide phosphodiesterases, and dual specificity phosphatases. Therefore, an alternative mechanism was purposed in which zinc acts as a permissive signal. In this mechanism, the level of intracellular free zinc regulates the rate of dephosphorylation, and thereby the signal intensity of phosphorylation-dependent signaling [19].

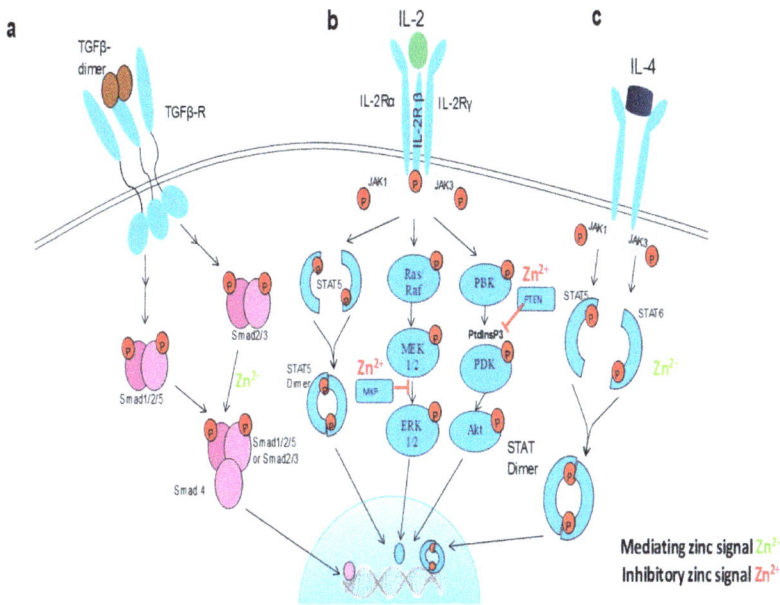

Figure 3. Anti-inflammatory signaling pathways influenced by free zinc. (**a**) TGFβ signaling is dependent on a dynamic on and off switch in Smad activity. Free zinc is a cofactor in Smad proteins and promote Smad 2/3 nuclear translocation and transcriptional activity. (**b**) zinc regulates IL-2 signaling pathway via blocking MAP kinase phosphatase (MKP) in extracellular signal-regulated kinases (ERK) 1/2 pathways and Phosphatase and tensin homologue (PTEN) which opposes phosphoinositide 3-kinase (PI3K) function in PI3k/Akt pathway. (**c**) free zinc phosphorylates STAT6 and promotes translocation of STAT dimers into the nucleus, hence promote the anti-inflammatory effects of Il-4.

Zinc also plays a role in modulating the apoptotic and inflammatory processes of caspases, which are a family of endoproteases. Apoptosis is an intricate process where intracellular components are dismantled in a controlled fashion whilst avoiding inflammation and damage to the surrounding cells. Dysregulation in apoptosis influences the pathogenesis of many diseases such as AIDS, diabetes mellitus, autoimmune diseases, malignancies, and most prominently, neurodegenerative disorders. Thus, understanding the regulatory factors involved in this process is crucial for developing therapeutic strategies [52]. There are several conflicting results showing how zinc can be anti- or pro-apoptotic depending on the zinc concentration used. Moreover, it has been observed that both zinc deficiency and supplementation induced apoptosis in the same cellular model [53]. There are two anti-apoptotic mechanisms of action exerted by zinc on caspase-dependent processes. Firstly, zinc suppresses some of the signaling pathways that lead to caspase activation and apoptosis by limiting the damage caused by free oxygen radicals and other toxins. Secondly, zinc directly affects apoptotic caspase enzymes. Influence of zinc on limiting free radicals will be discussed in the following section. With regard to the direct effect zinc has on caspases, many studies established the inhibitory role of zinc on apoptotic caspases 3, 6, 7, 8, and 9 [54–56]. Caspase-3 is particularly interesting due to its dominant role in the apoptotic pathway. Zinc is a potent inhibitor of caspase-3 with an IC_{50} below 10 nM [57]. Zinc containing compounds such as ziram (a zinc-containing dithiocarbamate) caused degradation of pro-caspase-1 which is a precursor of the inflammatory caspase-1 [58]. Active caspase-1 cleaves pro-IL-1β and pro-IL-18, which facilitates the secretion of these proinflammatory cytokines [59]. This indicated that zinc can also modulate caspase-controlled inflammatory processes.

Recent findings have established that zinc metabolism and its role in immune function are directly linked to zinc transporter function. This is demonstrated by ZIP8 which functions as a critical negative feedback regulator. When inflammation starts, NF-κB directly activates the expression of ZIP8, which then localizes to the plasma membrane, thereby mediating Zn uptake. When zinc enters the cytosol via ZIP8 it goes on to inhibit IKKβ kinase activity, which leads to the attenuation of the pro-inflammatory response [60].

5.2. Oxidative Stress

Under non-pathological conditions, cells produce ROS during cellular respiration. However, excessive production of ROS and the decreased rate of its neutralization and removal by antioxidant defense mechanisms lead to an imbalance between oxidants and antioxidants which results in oxidative stress [61]. Accumulation of those free radicals leads to cell and tissue damage. Oxidative stress is responsible for the development of many chronic diseases such as cancer, cardiovascular disease, atherosclerosis, hypertension, ischemia/reperfusion injury, diabetes mellitus, neurodegenerative diseases (Alzheimer's disease and Parkinson's disease), rheumatoid arthritis, and ageing [62].

Within the cell, there are several sites and processes responsible for ROS production, these include; mitochondrial electron transport chain, peroxisomal long-chain fatty acid oxidation, and respiratory burst primarily through activation of NADPHs, which are plasma membrane-associated enzymes. Other enzymes generate ROS during enzymatic reaction cycles, such enzymes are cytochrome P450 monooxygenase, nitric oxide synthase (NOS), xanthine oxidase, cyclooxygenase (COX), and lipoxygenase (LOX). Apart from those endogenous sources of free radicals, there are also exogenous or environmental sources of ROS. Air pollutants, tobacco smoke, ionizing and nonionizing radiations, foods and drugs, as well as xenobiotics can all contribute to oxidative stress [63]. Heavy metals such as lead, arsenic, mercury, chromium, and cadmium; organic solvents; and pesticides are common exogenous sources of ROS [64].

In order to neutralize and remove excess ROS, cells have endogenous non-enzymatic and enzymatic antioxidant defense mechanisms. Non-enzymatic antioxidants include glutathione (GSH), thioredoxin (Trx), and melatonin. While Antioxidant enzymatic mechanisms involve enzymes, such as superoxide dismutase (SOD), glutathione peroxidase (GPx), glutathione reductase (GR), catalase (CAT), and heme oxygenase (HO). Exogenous antioxidants include vitamin C, vitamin E, carotenoids including vitamin A, polyphenols including flavonoids and minerals such as zinc, copper, manganese, iron, and selenium [63,65].

Zinc has several antioxidant effects. It is a cofactor of the Cu/Zn-SOD enzyme, which catalyzes the dismutation of superoxide radical ($O_2^{\bullet-}$) into the less harmful O_2 and H_2O_2, which is then detoxified by CAT and GPx. It also inhibits NADPH oxidases, causing reduced generation of ROS. Furthermore, zinc induces the production of MTS, which are excellent ROS scavengers due to the high cysteine content [66]. Stabilization of protein sulfhydryls against oxidation by zinc is another mechanism by which zinc hinders the oxidative processes. Zinc binds directly to the thiol group, creates steric hindrance by binding close to the sulfhydryl group in the protein, and binds to other sites leading to conformational changes in the protein. This all results in the reduced activity of sulfhydryl [67]. Moreover, zinc antagonizes redox-active transition metals such as copper and iron that catalyze formation of free radicals, primarily through Fenton reactions. Those transition metals form complexes with cellular components, such as nucleotides, glucose, and citrates for iron and carbohydrates, DNA and enzymes for copper. When the metal is complexed, it becomes trapped and reacts with H_2O_2 and forms reactive hydroxyl radical ($HO^{\bullet}$). This leads to lipid peroxidation, DNA and protein damage, and subsequent severe tissue damage. Zinc is able to replace copper and iron and reduce the localized oxidative injury [68].

Zinc supplementation of healthy human subjects reduced the oxidative stress-related by-products malondialdehyde (MDA), 4-hydroxyalkenals (HAE), and 8-hydroxydeoxyguanine in the plasma [44]. This effect was also seen with elderly subjects who usually have lower plasma zinc concentration compared to younger people. Supplementation with zinc has reduced oxidative stress markers and lowered inflammatory cytokines and infection incidence [69]. Involvement of oxidative stress in other processes is discussed in more detail elsewhere [70].

5.3. Zinc Status and Inflammatory Cytokines

Zinc can influence the production and signaling of numerous inflammatory cytokines in a variety of cell types. Plasma zinc concentrations rapidly decline during acute phase response to different stimuli such as stress, infection, and trauma. Consequently, zinc is shuttled into cellular compartments, where it is utilized for protein synthesis, neutralization of free radicals, and to prevent microbial invasion. This redistribution of zinc during inflammatory events seems to be mediated by cytokines. Several studies have demonstrated how patients with acute illnesses present with hypozincemia along with elevated cytokine production [71,72].

Chronic inflammation is characterized by increased levels of inflammatory cytokine production. Some conditions are associated with chronic inflammation such as obesity, where patients with lower zinc dietary intake present with lower plasma and intracellular zinc concentrations along with upregulated gene expression of IL-1α, IL-1β, and IL-6 compared to patients with higher zinc intake [73].

The effect of zinc deficiency and supplementation is being reported regularly in in vitro and in vivo models. Cell specific responses and varying zinc dosage plays an important role in determining the outcome when subjecting cells to zinc depletion or supplementation. Various study models have been used. There are some noteworthy examples, for instance, the study mentioned above involving elderly subjects revealed that the elderly have significantly lower plasma zinc and higher ex vivo generation of inflammatory cytokines. Zinc supplementation partially reversed the levels of many inflammatory cytokines [69]. Yet zinc supplementation is not always beneficial, adding zinc at very high concentrations (>100 μM) increases cytokine production in some cell types, this was seen in human peripheral blood mononuclear cells (PBMCs) harvested from healthy adults [74] and in human promonocytic HL-CZ cells [75]. As mentioned previously, zinc deficiency induces production of IL-1β. This was seen in LPS-stimulated PBMCs from zinc-deficient adults compared to their zinc-sufficient counterparts [76]. This observation was confirmed through studying the epigenetic mechanisms that are implicated during zinc deficiency and it was reported that long term zinc deprivation promoted changes of the chromatin structures of IL-1β and TNFα promoters enabling the expression of both genes [77]. Additionally, phytohemagglutinin (PHA)-induced production of IL-2 was reduced in PBMCs of zinc-deficient patients with head and neck cancer and in zinc-deficient healthy volunteers [78]. IL-2 and IL-2Rα mRNA were also reduced in zinc deficient elderly people [79]. Similarly, inducing deficiency in healthy men via a zinc restricted diet (4.6 mg/day for 10 weeks) led to a reduction of PHA-stimulated IL-2R production [80]. Furthermore, IL-6 production in zinc-deficient elderly subjects was increased [81]. These observations were later confirmed via DNA methylation profile characterization, which showed that zinc deficiency induced a progressive demethylation of the IL6 promoter in THP1 cells that correlated with increased IL6 expression [82].

Cytokines are produced by different types of cells, primarily by T lymphocytes and macrophages. The influence of zinc on these cells may explain the observed effect zinc has on cytokine production. Zinc deficiency affects immune cells in several ways. It reduces thymulin activity, which is required for the maturation of T-helper cells (Th), it causes an imbalance between Th1 and Th2 cells which decrease recruitment of T naïve cells and percentage of cytotoxic T cells [83,84]. It also lowers the cell lytic activity of NK cells [85]. During zinc deficiency, there is also reduced production of IFN-γ and IL-2, which are products of Th1 cells, while the production of IL-4, IL-6, and IL-10 (products of Th2 cells) remain unchanged [86]. On the other hand, zinc excess (50–100 μM) inhibits T cell activity which is demonstrated by the suppression of IL-1β-stimulated IFN-γ expression [87]. This shows how both zinc deficiency and increased zinc levels influence T-cell function. Additionally, regulatory T cells (Treg) are induced and stabilized in zinc-supplemented mixed lymphocyte culture (MLC), in lymphocytes from patients allergic to pollen [88] and in experimental autoimmune encephalomyelitis (EAE) [89]. Furthermore, zinc enhances the capacity of TGFβ1 to induce Treg cells [90,91]. However, the capacity of zinc to induce Tregs is dependent on the activation status of the cells. Activated zinc-supplemented T cells are driven to differentiate to Treg, whereas resting T cells are pushed to Th1 response [89]. Hence, the influence of zinc is expansive yet specific to each cell type.

6. Zinc in Infection

6.1. Nutritional Immunity

Nutritional immunity is a process by which the host organism sequesters trace minerals during an infection so it can be of limited availability to pathogens. It was first described for iron but has now been extended to other trace elements. Zinc is an essential trace element for both host and pathogens. Pathogens require zinc for survival, propagation and disease establishment. This prompts a competitive process between the host and the invading pathogens. There are three mechanisms by which the host is able to compete for zinc and achieve a zinc limited environment for the pathogen, however, some pathogens have developed tactics to overcome some of those mechanisms [5,92].

On a systematic level, zinc distribution in the body is altered. This mainly involves the small pool of free zinc that is unbound within plasma as 99.9% of zinc remains inside cells and cannot be directly accessed by pathogens. During an infection, plasma zinc levels are markedly reduced. This is achieved by several approaches, but primarily involves the secretion of inflammatory cytokines such as IL-6 that upregulates expression of ZIP14 in hepatocytes which leads to the accumulation of zinc bound to metallothionein in the liver [93]. Furthermore, zinc concentrations can be altered on an extracellular level through the release of some antimicrobial peptides from the S100 family. Several cell types secrete different peptides, keratinocytes secrete S100A7 that can kill *Escherichia coli* by sequestering zinc [94]. Additionally, neutrophils also secrete calprotectin (a heterodimer of S100A8 and S100A9). It can inhibit the growth of *Staphylococcus aureus* by sequestering zinc as well [95]. As mentioned previously, neutrophils also release NETs, in which they cast out DNA, chromatins, and granular proteins to capture and kill microorganisms [35]. Calprotectin is also released in very high concentrations during NETosis, where it is either incorporated into NET or in the surrounding fluids. On an intracellular level, macrophages have evolved two opposing strategies to kill phagocytosed pathogens. Macrophages can deprive *Histoplasma capsulatum* of zinc by reducing the phagosome zinc content [25]. On the other hand, they kill *Mycobacterium tuberculosis* by intoxicating it with excess amounts of zinc and copper [96].

Some pathogens have developed defense strategies against several of those mechanisms. For example, *Neisseria meningitidis* uses a high affinity zinc uptake receptor ZnuD which allows it to escape NET-mediated nutritional immunity [97]. It also responds to zinc deprivation by expressing CbpA, which is an outer membrane protein that acts as a receptor for calprotectin and enables *N. meningitides* to acquire zinc bound to calprotectin. *N. meningitidis* effectively undermines an important defense mechanism used by the host and utilizes it in its favor [98]. *Yersinia pestis* utilizes the siderophore yersiniabactin (Ybt) as a zincophore, and ZnuABC, which is the high-affinity zinc transporter found in bacteria and fungi, to acquire zinc and develop a lethal infection in a septicemic plague mouse model [99]. *Salmonella typhimurium* can also overcome calprotectin-mediated zinc chelation by expressing ZnuABC, thereby allowing it to compete with commensal bacteria and thrive in the inflamed gut [100].

Although low zinc is associated with an impaired immune system and poor prognosis in conditions such as sepsis, supplementation during sepsis was not beneficial, whereas prophylactic zinc administration showed some positive results [101]. In contrast, administering zinc via lozenges at concentrations of ≥ 75 mg/day reduced the duration of common cold symptoms in healthy individuals [102]. It is worth mentioning that the decision to administer zinc during an infection must take into account the risks of creating a zinc microenvironment that is more favorable for pathogen growth, meanwhile downgrading the efforts carried out by the innate system to reduce free zinc. This only demonstrates the needs for more research to investigate and understand the therapeutic potentials of zinc for infectious diseases.

Despite zinc being recognized as a crucial factor for the proper functioning of the immune system, increased susceptibility to infections and other immune dysfunctions could be the consequence of one or more nutrient deficiencies. The same dietary factors leading to deficiency of one micronutrient often cause deficiency of other micronutrients. Additionally, supplementation of one micronutrient in high doses may exacerbate the status of other micronutrients, which can be seen in the case of iron supplementation impairing zinc uptake. Furthermore, excess supplementation of either zinc or copper influences the tightly controlled ratio of those two metals. Supplementation with high doses of zinc for prolonged periods lead to copper deficiency [53]. Also, altered zinc and copper homeostasis is implicated in several conditions, for example, high copper/zinc ratio was observed in the elderly, particularly those with neurodegenerative disorders [103,104]. The copper/zinc ratio maybe considered a biomarker for mortality in the elderly population [105], which further emphasizes the importance of understanding the complex interplay between different nutrients and their collective influence on health. Maintaining the homeostasis of micronutrients and vitamins is fundamental for a healthy immune response and overall wellness of the body. However, the interaction between micronutrients must be explored further in order to design supplementation interventions that target multiple deficiencies.

Tables 1–3 provide an accumulative record of selected publications on zinc supplementation trails. The influence of zinc on viral (Table 1), bacterial (Table 2), and parasitic (Table 3) infections depends on many factors that include, but not limited to; zinc status at baseline measurement, zinc supplementation concentration and frequency, zinc species and age. Results must be interpreted cautiously and analytically. A continuous joint effort between members of the scientific community who conduct such clinical trials will only shed more light on the influence zinc has on infectious diseases and how it can be utilized further.

Table 1. Zinc supplementation and viral diseases (updated from [106]).

Disease	Zinc Species	Zinc Dosage	Period	Participation	Effect of Zinc Supplementation	Reference
Common cold	more than 12 different studies, analyzing the therapeutic effects of zinc				variable results, reduced duration of symptoms if administered within 24 h of onset	[107]
	zinc sulfate	15 mg daily	7 months	100 (Z) 100 (P)	lower mean number of colds demonstrating the prophylactic effect of zinc	[108]
	Not specified	12 mg for women and 15 mg for men/day	18 months	115 (Z) 116 (P)	no effect on viral load. four-fold reduction in the likelihood of immunological failure. Reduced the rate of diarrhea by more than half. No significant difference in mortality	[109]
	Chelated zinc	15 mg daily	12 months	Low: 5 (Z)/7 (P) Normal: 8 (Z)/10 (P)	CD4+ cell count significantly increased	[110]
	Zinc sulfate	20 mg daily	24 weeks	26 (Z) 26 (P)	no effect on the increase in CD4%, decrease in viral load, anthropometric indices, and morbidity profile in HIV-infected children started on ART	[111]
	Zinc sulfate	45.5 mg daily	1 month	29 (Z) 28 (P)	increase or stabilization in body weight; increase in plasma zinc levels, CD4+ T cells and plasma active zinc-bound thymulin; reduced or delayed frequency of opportunistic infections due to Pneumocystis jiroveci and Candida, not to Cytomegalovirus and Toxoplasma	[112]
HIV/AIDS	Zinc gluconate	45 mg three time daily	15 days	5 (Z) 5 (C)	increased zinc concentrations in red blood, HLA-DR + cells, stimulation of lymphocyte transformation and phagocytosis of opsonized zymosan by neutrophils	[113]
	Zinc sulfate	10 mg daily	6 months	44 (Z) 41 (P)	no effect on HIV viral load; decreased morbidity from diarrhea	[114]
	Zinc sulfate	50 mg daily	1 month	31 (Z) 34 (P)	no improvements in immune responses to tuberculosis, CD4/CD8 ratio, lymphocyte subsets, and viral load	[115]
	Zinc sulfate	25 mg daily	6 months	200 (Z) 200 (P)	when supplemented to pregnant HIV-positive women, no effect on birth outcomes or T-lymphocyte counts, and negative effects on hematological indicators	[116]
	Zinc sulfate	25 mg daily	6 months	200 (Z) 200 (P)	increased risk of wasting	[117]
				50 (Z) 50 (P)	no effect on viral load	
	Zinc gluconate	50 mg daily	6 days	44 (Z) 45 (P)	no improvements in antibody responses to a pneumococcal conjugate vaccine	[118]

Table 1. *Cont.*

Disease	Zinc Species	Zinc Dosage	Period	Participation	Effect of Zinc Supplementation	Reference
	Not specified	10 mg	60 days	26 (Z + 6400 mg/day Branched-chain amino acids) 27 (P)	BCAA-to-tyrosine ratio (BTR) and zinc levels were significantly increased compared with the placebo group. supplementation reduced the serum α-fetoprotein AFP levels in patients who had elevated serum AFP levels at baseline	[119]
hepatitis C virus	Polapre-zinc	150 mg	48 weeks	11 (Z) 12 (C)	serum alanine aminotransferase (ALT) level is lower in zinc group compared to control group. HCV RNA disappeared in all patients in the zinc group and in 80% control patients at 48 week. Polaprezinc supplementation decreased plasma thiobarbituric acid reactive substances and prevented the decrease of polyunsaturated fatty acids of erythrocyte membrane phospholipids	[120]
	Polapre-zinc	17 mg twice a day	24 weeks	40 (Z) 35 (C)	zinc supplementation increases serum zinc levels and improves the response to IFN-α therapy	[121]
	Zinc gluconate	50 mg daily	6 months	18 (Z) 35 (P) 20 (C)	increased serum zinc levels; decreased incidences of gastrointestinal disturbances, body weight loss, and mild anemia	[122]

Z—zinc, P—placebo, C—control.

Table 2. Zinc supplementation and bacterial infectious diseases (updated from [106]).

Disease	Zinc Species	Zinc Dosage	Period	Participation	Effect of Zinc Supplementation	Reference
Diarrhea		multiple different studies			decreased duration, severity and occurrence of diarrhea	[123]
	Not specified	20 mg daily	14 days	41 (Z) 39 (micronutrient combination * + Vit A) 44 (Z+ Vit A) 43(P)	supplementation with a combination of micronutrients and vitamins was not superior to zinc alone, confirming clinical benefit of zinc in children with diarrhea	[124]

Table 2. Cont.

Disease	Zinc Species	Zinc Dosage	Period	Participation	Effect of Zinc Supplementation	Reference
Respiratory tract infections	Zinc sulfate	20 mg daily	5 months	134 (Z) 124 (P)	reduced acute lower respiratory tract infection morbidity	[125]
	zinc gluconate	10 mg daily	60 days	48 (Z) 48 (P)	reduced episodes of acute lower respiratory infections and severe acute lower respiratory infections. Increased infection free days	[126]
	Zinc oxide	5 mg daily	12 months	162 (Z) 167 (C)	decreased incidence of upper respiratory tract infections and diarrhoeal disease episodes	[127]
	zinc gluconate	10 mg daily	6 months	298 (Z) 311 (P)	increased plasma zinc levels; decreased episodes of infection	[128]
	Zinc acetate	10 mg twice a day	5 days	76 (Z) 74 (P)	increased serum zinc levels and recovery rates from illness and fever in boys	[129]
	Zinc sulfate	15 mg daily	6 months	40 (Z) 40 (P)	increased plasma retinol concentrations; earlier sputum conversion and resolution of X-ray lesion area	[130]
Tuberculosis	zinc sulfate	220 mg daily	18 months	8 (Z)	reduced dose of clofazimine; withdrawal of steroids; toleration of dapsone; reduced incidence and severity of erythema nodosum leprosum; gradual decrease in the size of granuloma; gradual increase in the number of lymphocytes	[131]
	zinc sulfate	220 mg daily	18 months	15 (Z) 10 (P)	decreased erythema, edema, and infiltration; regrowth of eyebrows; reduced bacterial index of granuloma; increased serum zinc levels, neovascularization, and endothelial cell proliferation	[132]
Lepromatous leprosy	Zinc acetate	200 mg twice a day	13 weeks	17 (Z) 10 (P) 10 (C)	increased serum zinc levels and delayed hypersensitivity reactions; decreased size of skin nodules; disappearance of erythema; regrowth of eyebrows	[133]
	zinc sulfate	220 mg daily	4 months	40 (Z)	improvements on frequency, duration, and severity of erytheme nodosum leprosum reactions; reduction in steroid requirement	[134]
	zinc acetate	1.3 mg/kg three times a day	1 month	16 (Z) 16 (P)	increased intestinal mucosal permeability and better nitrogen absorption; increased serum zinc and alkaline phosphatase activity	[135]
	zinc acetate	20 mg daily	2 weeks	28 (Z) 28 (P)	increased serum zinc levels, lymphocyte proliferation in response to phytohemagglutinin and plasma invasion plasmid-encoded antigen-specific IgG titers	[136]
Shigellosis	zinc acetate	20 mg daily	2 weeks	28 (Z) 28 (P)	increased serum zinc levels, serum shigellacidal antibody titers, CD20+ cells, and CD20+CD38+ cells	[137]
	Not specified	20 mg daily	2 weeks	14 (Z) 16 (C)	faster recovery from acute illness. Increased mean body weight. Fewer episodes of diarrhoea	[138]
Helicobacter pylori infection	polapre zinc	150 mg twice a day	7 days	33 (Z) 28 (C)	administration of zinc together with antimicrobial therapy increased cure rate of Helicobacter pylori infection compared with antibiotic treatment alone	[139]

Z–zinc, P–placebo, C–control; Diarrhea and respiratory infections can be caused by nonbacterial pathogens. The mentioned studies do not specific the causative agent; * micronutrient combination: zinc, 20 mg; iron, 10 mg; copper, 2 mg; selenium, 40 mg; vitamin B12, 1.4 mg; folate, 100 mg.

Table 3. Zinc supplementation and parasites (updated from [103]).

Disease	Zinc Species	Zinc Dosage	Period	Participation	Effect of Zinc Supplementation	Reference
Malaria	Not specified	10 mg 6 times/week	6 months	74 (Z + 1 single dose of 200,000 IU Vit A) 74 (P)	significant decrease in the prevalence malaria. Lower malaria episodes. Time to first malaria episode was longer. 22% fewer fever episodes than the placebo group	[140]
	Zinc gluconate	10 mg 6 times/week	46 weeks	136 (Z) 138 (P)	reduction in *Plasmodium falciparum*-mediated febrile episodes	[141]
	Zinc acetate/zinc gluconate	70 mg twice a week	15 months	55 (Z) 54 (P)	not statistically significant trend towards fewer malaria episodes; no effect on plasma and hair zinc, diarrhea, and respiratory illness	[142]
	Zinc sulfate	12.5 mg 6 times/week	6 months	336 (Z) 344 (P)	increased serum zinc levels; reduced prevalence of diarrhea	[143]
	Zinc sulfate	20 or 40 mg daily	4 days	473 (Z) 483 (P)	increased plasma zinc, no effect on fever, parasitemia, or hemoglobin concentration	[144]
	Zinc sulfate	20 mg daily	7 months	191 (Z) 189 (P)	no significant effect on P. vivax incidence; significantly reduced diarrhea morbidity	[145]

Z–zinc, P–placebo.

6.2. Zinc as a Critical Component of the Membrane Barrier

Zinc may contribute to the host defense by maintaining the membrane barrier structure and function. This is very crucial in sites like the lung and intestine that become continuously exposed to a myriad of pathogens and noxious agents. Zinc deficiency may contribute to the severity of infectious diseases and mortality in malnourished children. In case of diarrhea, which is the leading cause of death globally among children under five years of age, zinc supplementation has been shown to reduce the severity and duration of symptoms [146]. WHO recommends zinc supplementation of 20 mg per day for 10–14 days for the management of diarrhea [147]. Several studies have examined the influence of zinc depletion and supplementation on the permeability of the endothelial cell barrier. The intestinal epithelium barrier consists of intercellular junctional complexes between neighboring cells that provide a continuous seal around the apical region of the cells. These complexes are composed of several units, including the tight junctions (TJ) and adherens junctions (AJ), that form circumferential zones of contact between adjacent cells. E-cadherin is the main transmembrane adhesion molecule localized at the AJ and its binding to β-catenin is fundamental for appropriate AJ organization. Zinc depletion disrupted the TJ and AJ through several mechanisms. One way zinc affected structural proteins was by enhancing the degradation of E-cadherin and β-catenin [148]. This was also seen in zinc-deprived airway epithelial cells, where there was accelerated proteolysis of E-cadherin and β-catenin leading to increased leakage across the monolayer of upper and alveolar lung epithelial cultures [149]. Hypozincemia may induce uncontrolled neutrophil migration through the disrupted junctional complexes by inducing chemokine production. Exacerbated inflammation may develop and lead to mucosal damage which further contributes to intestinal and lung disease. On the other hand, zinc supplementation has been seen to preserve and restore membrane function and structure [148].

Zinc is also an integral part of the epidermal and dermal tissues, where it acts as a stabilizer of cell membranes and as an essential cofactor in numerous transcription factors and enzymes. This includes the zinc-dependent matrix metalloproteinases that enhance autodebridement and keratinocyte migration during wound repair. Moreover, zinc confers resistance to epithelial apoptosis through cytoprotection against reactive oxygen species and bacterial toxins possibly through antioxidant activity of the cysteine-rich metallothioneins [150]. Hence, zinc deficiency may result in delayed wound healing, particularly in the elderly with impaired nutritional status. Delayed wound healing in the elderly constitutes a major clinical and economic challenge, especially as the aging population grows [151]. Topical zinc therapy has shown promising results in enhancing wound healing due to zinc's role in reducing superinfections and necrotic material by augmenting local defense systems and collagenolytic activity, as well as promoting epithelialization of wounds [150]. Thus, utilizing topical zinc treatments to support wound healing provides a therapeutic advantage and enhances quality of life.

6.3. Peptidoglycan Regulation Proteins (PGLYRPs)

Another beneficial effect that zinc has on secretory molecules is how it is involved in the bactericidal activity of peptidoglycan recognition proteins (PGRPs or PGLYRPs). Those innate immunity pattern recognition molecules have effector functions and are expressed in either PMN molecules or in the skin, eyes, salivary glands, throat, tongue, esophagus, stomach, and intestine. Along with other antimicrobial peptides they protect the body from pathogen at the first line of exposure [152]. The activity of PGLYRPs against Gram-positive and Gram-negative bacteria is dependent on free zinc [153].

7. Conclusions

Zinc is a multipurpose metal that is vital for the growth and function of all cells. The immune system is especially affected by the modification of zinc homeostasis. Achieving an optimal immune response to different stimuli and avoiding damage of tissues and organs is a delicate balance that relies, amongst other factors, on the regulation of zinc in extracellular and intracellular compartments.

It would be highly beneficial to attain a standard by which zinc supplementation can be administered in different conditions, however, creating such standard might be challenging due to many conflicting findings on various cell types, experimental models, and zinc concentrations, and due to the lack of a good biomarker for zinc status in the body. Nevertheless, resolving zinc deficiency and the adverse manifestations associated with it can be achieved through adequate diet and supplementation.

Acknowledgments: L.R. is a member of Zinc-Net. N.Z.G. is supported by a scholarship from the German Academic Exchange Service (DAAD).

Author Contributions: N.Z.G. wrote the manuscript. L.R. made critical revisions.

Conflicts of Interest: The authors declare no conflict of interest.

References

1. Prasad, A.S.; Miale, A., Jr.; Farid, Z.; Sandstead, H.H.; Schulert, A.R. Zinc metabolism in patients with the syndrome of iron deficiency anemia, hepatosplenomegaly, dwarfism, and hypogonadism. *J. Lab. Clin. Med.* **1963**, *61*, 537–549. [PubMed]
2. Rink, L.; Gabriel, P. Zinc and the immune system. *Proc. Nutr. Soc.* **2000**, *59*, 541–552. [CrossRef] [PubMed]
3. Vallee, B.L.; Falchuk, K.H. The biochemical basis of zinc physiology. *Physiol. Rev.* **1993**, *73*, 79–118. [PubMed]
4. Maret, W.; Li, Y. Coordination dynamics of zinc in proteins. *Chem. Rev.* **2009**, *109*, 4682–4707. [CrossRef] [PubMed]
5. Haase, H.; Rink, L. Multiple impacts of zinc on immune function. *Metall. Integr. Biomet. Sci.* **2014**, *6*, 1175–1180. [CrossRef] [PubMed]
6. Maret, W. Zinc and Human Disease. In *Interrelations between Essential Metal Ions and Human Diseases*; Sigel, A., Sigel, H., Sigel, R.K., Eds.; Springer: Dordrecht, The Netherlands, 2013; pp. 389–414.
7. Mills, C.F. (Ed.) Zinc in Human Biology. In *Physiology of Zinc: General Aspects*; Springer: London, UK, 1989. [CrossRef]
8. Favier, A.; Favier, M. Effects of zinc deficiency in pregnancy on the mother and the newborn infant. *Rev. Fr. Gynecol. Obstet.* **1990**, *85*, 13–27. [PubMed]
9. Otten, J.J.; Hellwig, J.P.; Meyers, L.D. (Eds.) *Dietary Reference Intakes: The Essential Guide to Nutrient Requirements*; The National Academies Press: Washington, DC, USA, 2006. Available online: https://doi.org/10.17226/11537 (accessed on 7 July 2017).
10. Deutsche Gesellschaft für Ernährung. *Österreichische Gesellschaft für Ernährung*; Schweizerische Gesellschaft für Ernährungsforschung; Schweizerische Vereinigung für Ernährung; Referenzwerte für die Nährstoffzufuhr: Bonn, Germany, 2016. (In German)
11. World Health Organization (WHO). *Trace Elements in Human Nutrition and Health*; World Health Organization: Geneva, Switzerland, 1996.
12. EFSA Panel on Dietetic Products Nutrition and Allergies. Scientific Opinion on Dietary Reference Values for zinc. *EFSA J.* **2014**, *12*. [CrossRef]
13. King, J.C.; Brown, K.H.; Gibson, R.S.; Krebs, N.F.; Lowe, N.M.; Siekmann, J.H.; Raiten, D.J. Biomarkers of Nutrition for Development (BOND)—Zinc Review. *J. Nutr.* **2016**, *146*, 858S–885S. [CrossRef] [PubMed]
14. Brieger, A.; Rink, L. Zink und Immunfunktionen. *Ernährung Medizin* **2010**, *25*, 156–160. [CrossRef]
15. Sandstead, H.H. Zinc Deficiency. *Am. J. Dis. Child.* **1991**, *145*, 853–859. [CrossRef] [PubMed]
16. World Health Organisation. *The World Health Report*; World Health Organization: Geneva, Switzerland, 2002.
17. Wellenreuther, G.; Cianci, M.; Tucoulou, R.; Meyer-Klaucke, W.; Haase, H. The ligand environment of zinc stored in vesicles. *Biochem. Biophys. Res. Commun.* **2009**, *380*, 198–203. [CrossRef] [PubMed]
18. Kambe, T.; Hashimoto, A.; Fujimoto, S. Current understanding of ZIP and ZnT zinc transporters in human health and diseases. *Cell. Mol. Life Sci.* **2014**, *71*, 3281–3295. [CrossRef] [PubMed]
19. Haase, H.; Rink, L. Functional significance of zinc-related signaling pathways in immune cells. *Annu. Rev. Nutr.* **2009**, *29*, 133–152. [CrossRef] [PubMed]

20. Kimura, T.; Kambe, T. The Functions of Metallothionein and ZIP and ZnT Transporters: An Overview and Perspective. *Int. J. Mol. Sci.* **2016**, *17*. [CrossRef] [PubMed]
21. Prasad, A.S. Discovery of human zinc deficiency: Its impact on human health and disease. *Adv. Nutr.* **2013**, *4*, 176–190. [CrossRef] [PubMed]
22. Jeong, J.; Eide, D.J. The SLC39 family of zinc transporters. *Mol. Asp. Med.* **2013**, *34*, 612–619. [CrossRef] [PubMed]
23. Huang, L.; Tepaamorndech, S. The SLC30 family of zinc transporters—A review of current understanding of their biological and pathophysiological roles. *Mol. Asp. Med.* **2013**, *34*, 548–560. [CrossRef] [PubMed]
24. Lang, C.; Murgia, C.; Leong, M.; Tan, L.-W.; Perozzi, G.; Knight, D.; Ruffin, R.; Zalewski, P. Anti-inflammatory effects of zinc and alterations in zinc transporter mRNA in mouse models of allergic inflammation. *Am. J. Physiol. Lung Cell. Mol. Physiol.* **2007**, *292*, L577–L584. [CrossRef] [PubMed]
25. Subramanian Vignesh, K.; Landero Figueroa, J.A.; Porollo, A.; Caruso, J.A.; Deepe, G.S., Jr. Granulocyte macrophage-colony stimulating factor induced Zn sequestration enhances macrophage superoxide and limits intracellular pathogen survival. *Immunity* **2013**, *39*, 697–710. [CrossRef] [PubMed]
26. Maret, W. The function of zinc metallothionein: A link between cellular zinc and redox state. *J. Nutr.* **2000**, *130*, 1455S–1458S. [PubMed]
27. King, J.C. Zinc: An essential but elusive nutrient. *Am. J. Clin. Nutr.* **2011**, *94*, 679–684. [CrossRef] [PubMed]
28. Lu, J.; Stewart, A.J.; Sadler, P.J.; Pinheiro, T.J.; Blindauer, C.A. Albumin as a zinc carrier: Properties of its high-affinity zinc-binding site. *Biochem. Soc. Trans.* **2008**, *36*, 1317–1321. [CrossRef] [PubMed]
29. Heizmann, C.W.; Cox, J.A. New perspectives on S100 proteins: A multi-functional Ca(2+)-, Zn(2+)- and Cu(2+)-binding protein family. *Biometals* **1998**, *11*, 383–397. [CrossRef] [PubMed]
30. Gilston, B.A.; Skaar, E.P.; Chazin, W.J. Binding of transition metals to S100 proteins. *Sci. China Life Sci.* **2016**, *59*, 792–801. [CrossRef] [PubMed]
31. Mocchegiani, E.; Costarelli, L.; Giacconi, R.; Cipriano, C.; Muti, E.; Malavolta, M. Zinc-binding proteins (metallothionein and alpha-2 macroglobulin) and immunosenescence. *Exp. Gerontol.* **2006**, *41*, 1094–1107. [CrossRef] [PubMed]
32. Nuttall, J.R.; Oteiza, P.I. Zinc and the aging brain. *Genes Nutr.* **2014**, *9*. [CrossRef] [PubMed]
33. DeCoursey, T.E.; Morgan, D.; Cherny, V.V. The voltage dependence of NADPH oxidase reveals why phagocytes need proton channels. *Nature* **2003**, *422*, 531–534. [CrossRef] [PubMed]
34. Hasegawa, H.; Suzuki, K.; Nakaji, S.; Sugawara, K. Effects of zinc on the reactive oxygen species generating capacity of human neutrophils and on the serum opsonic activity in vitro. *Luminescence* **2000**, *15*, 321–327. [CrossRef]
35. Brinkmann, V.; Reichard, U.; Goosmann, C.; Fauler, B.; Uhlemann, Y.; Weiss, D.S.; Weinrauch, Y.; Zychlinsky, A. Neutrophil Extracellular Traps Kill Bacteria. *Science* **2004**, *303*, 1532–1535. [CrossRef] [PubMed]
36. Haase, H.; Rink, L. Zinc signals and immune function. *BioFactors* **2014**, *40*, 27–40. [CrossRef] [PubMed]
37. Gaetke, L.M.; Frederich, R.C.; Oz, H.S.; McClain, C.J. Decreased food intake rather than zinc deficiency is associated with changes in plasma leptin, metabolic rate, and activity levels in zinc deficient rats. *J. Nutr. Biochem.* **2002**, *13*, 237–244. [CrossRef]
38. Nathan, C.; Ding, A. Nonresolving Inflammation. *Cell* **2010**, *140*, 871–882. [CrossRef] [PubMed]
39. Jarosz, M.; Olbert, M.; Wyszogrodzka, G.; Młyniec, K.; Librowski, T. Antioxidant and anti-inflammatory effects of zinc. Zinc-dependent NF-κB signaling. *Inflammopharmacology* **2017**, *25*, 11–24. [CrossRef] [PubMed]
40. Perkins, N.D. Integrating cell-signalling pathways with NF-κB and IKK function. *Nat. Rev. Mol. Cell Biol.* **2007**, *8*, 49–62. [CrossRef] [PubMed]
41. Foster, M.; Samman, S. Zinc and Regulation of Inflammatory Cytokines: Implications for Cardiometabolic Disease. *Nutrients* **2012**, *4*, 676–694. [CrossRef] [PubMed]
42. Haase, H.; Ober-Blöbaum, J.L.; Engelhardt, G.; Hebel, S.; Heit, A.; Heine, H.; Rink, L. Zinc Signals Are Essential for Lipopolysaccharide-Induced Signal Transduction in Monocytes. *J. Immunol.* **2008**, *181*, 6491–6502. [CrossRef] [PubMed]
43. Prasad, A.S.; Bao, B.; Beck, F.W.; Sarkar, F.H. Zinc-suppressed inflammatory cytokines by induction of A20-mediated inhibition of nuclear factor-κB. *Nutrition* **2011**, *27*, 816–823. [CrossRef] [PubMed]
44. Prasad, A.S.; Bao, B.; Beck, F.W.; Kucuk, O.; Sarkar, F.H. Antioxidant effect of zinc in humans. *Free Radic. Biol. Med.* **2004**, *37*, 1182–1190. [CrossRef] [PubMed]

45. Brieger, A.; Rink, L.; Haase, H. Differential Regulation of TLR-Dependent MyD88 and TRIF Signaling Pathways by Free Zinc Ions. *J. Immunol.* **2013**, *191*, 1808–1817. [CrossRef] [PubMed]
46. Morgan, C.I.; Ledford, J.R.; Zhou, P.; Page, K. Zinc supplementation alters airway inflammation and airway hyperresponsiveness to a common allergen. *J. Inflamm.* **2011**, *8*, 36. [CrossRef]
47. Yan, Y.-W.; Fan, J.; Bai, S.-L.; Hou, W.-J.; Li, X.; Tong, H. Zinc Prevents Abdominal Aortic Aneurysm Formation by Induction of A20-Mediated Suppression of NF-κB Pathway. *PLoS ONE* **2016**, *11*, e0148536. [CrossRef] [PubMed]
48. Li, C.; Guo, S.; Gao, J.; Guo, Y.; Du, E.; Lv, Z.; Zhang, B. Maternal high-zinc diet attenuates intestinal inflammation by reducing DNA methylation and elevating H3K9 acetylation in the A20 promoter of offspring chicks. *J. Nutr. Biochem.* **2015**, *26*, 173–183. [CrossRef] [PubMed]
49. Bao, B.; Prasad, A.S.; Beck, F.W.J.; Fitzgerald, J.T.; Snell, D.; Bao, G.W.; Singh, T.; Cardozo, L.J. Zinc decreases C-reactive protein, lipid peroxidation, and inflammatory cytokines in elderly subjects: A potential implication of zinc as an atheroprotective agent. *Am. J. Clin. Nutr.* **2010**, *91*, 1634–1641. [CrossRef] [PubMed]
50. Von Bulow, V.; Dubben, S.; Engelhardt, G.; Hebel, S.; Plumakers, B.; Heine, H.; Rink, L.; Haase, H. Zinc-dependent suppression of TNF-alpha production is mediated by protein kinase A-induced inhibition of Raf-1, I kappa B kinase beta, and NF-kappa B. *J. Immunol.* **2007**, *179*, 4180–4186. [CrossRef] [PubMed]
51. Nishida, K.; Hasegawa, A.; Nakae, S.; Oboki, K.; Saito, H.; Yamasaki, S.; Hirano, T. Zinc transporter Znt5/Slc30a5 is required for the mast cell-mediated delayed-type allergic reaction but not the immediate-type reaction. *J. Exp. Med.* **2009**, *206*, 1351–1364. [CrossRef] [PubMed]
52. Truong-Tran, A.Q.; Carter, J.; Ruffin, R.E.; Zalewski, P.D. The role of zinc in caspase activation and apoptotic cell death. *Biometals* **2001**, *14*, 315–330. [CrossRef] [PubMed]
53. Plum, L.M.; Rink, L.; Haase, H. The essential toxin: Impact of zinc on human health. *Int. J. Environ. Res. Public Health* **2010**, *7*, 1342–1365. [CrossRef] [PubMed]
54. Stennicke, H.R.; Salvesen, G.S. Biochemical Characteristics of Caspases-3, -6, -7, and -8. *J. Biol. Chem.* **1997**, *272*, 25719–25723. [CrossRef] [PubMed]
55. Huber, K.L.; Hardy, J.A. Mechanism of zinc-mediated inhibition of caspase-9 *Protein Sci.* **2012**, *21*, 1056–1065. [CrossRef] [PubMed]
56. Velazquez-Delgado, E.M.; Hardy, J.A. Zinc-mediated allosteric inhibition of caspase-6. *J. Biol. Chem.* **2012**, *287*, 36000–36011. [CrossRef] [PubMed]
57. Maret, W.; Jacob, C.; Vallee, B.L.; Fischer, E.H. Inhibitory sites in enzymes: Zinc removal and reactivation by thionein. *Proc. Natl. Acad. Sci. USA* **1999**, *96*, 1936–1940. [CrossRef] [PubMed]
58. Muroi, M.; Tanamoto, K.-I. Zinc- and oxidative property-dependent degradation of pro-caspase-1 and NLRP3 by ziram in mouse macrophages. *Toxicol. Lett.* **2015**, *235*, 199–205. [CrossRef] [PubMed]
59. Ting, J.P.-Y.; Willingham, S.B.; Bergstralh, D.T. NLRs at the intersection of cell death and immunity. *Nat. Rev. Immunol.* **2008**, *8*, 372–379. [CrossRef] [PubMed]
60. Liu, M.-J.; Bao, S.; Galvez-Peralta, M.; Pyle, C.J.; Rudawsky, A.C.; Pavlovicz, R.E.; Killilea, D.W.; Li, C.; Nebert, D.W.; Wewers, M.D.; et al. ZIP8 regulates host defense through zinc-mediated inhibition of NF-kappaB. *Cell Rep.* **2013**, *3*, 386–400. [CrossRef] [PubMed]
61. Marreiro, D.D.N.; Cruz, K.J.C.; Morais, J.B.S.; Beserra, J.B.; Severo, J.S.; de Oliveira, A.R.S. Zinc and Oxidative Stress: Current Mechanisms. *Antioxidants* **2017**, *6*, 24. [CrossRef] [PubMed]
62. Valko, M.; Leibfritz, D.; Moncol, J.; Cronin, M.T.; Mazur, M.; Telser, J. Free radicals and antioxidants in normal physiological functions and human disease. *Int. J. Biochem. Cell Biol.* **2007**, *39*, 44–84. [CrossRef] [PubMed]
63. Bhattacharyya, A.; Chattopadhyay, R.; Mitra, S.; Crowe, S.E. Oxidative Stress: An Essential Factor in the Pathogenesis of Gastrointestinal Mucosal Diseases. *Physiol. Rev.* **2014**, *94*, 329–354. [CrossRef] [PubMed]
64. Sharma, B.; Singh, S.; Siddiqi, N.J. Biomedical Implications of Heavy Metals Induced Imbalances in Redox Systems. *BioMed Res. Int.* **2014**, *2014*. [CrossRef] [PubMed]
65. Rahman, K. Studies on free radicals, antioxidants, and co-factors. *Clin. Interv. Aging* **2007**, *2*, 219–236. [PubMed]
66. Prasad, A.S. Zinc: An antioxidant and anti-inflammatory agent: Role of zinc in degenerative disorders of aging. *J. Trace Elem. Med. Biol.* **2014**, *28*, 364–371. [CrossRef] [PubMed]
67. Gibbs, P.N.; Gore, M.G.; Jordan, P.M. Investigation of the effect of metal ions on the reactivity of thiol groups in human 5-aminolaevulinate dehydratase. *Biochem. J.* **1985**, *225*, 573–580. [CrossRef] [PubMed]

68. Powell, S.R. The antioxidant properties of zinc. *J. Nutr.* **2000**, *130*, 1447S–1454S. [PubMed]
69. Prasad, A.S.; Beck, F.W.J.; Bao, B.; Fitzgerald, J.T.; Snell, D.C.; Steinberg, J.D.; Cardozo, L.J. Zinc supplementation decreases incidence of infections in the elderly: Effect of zinc on generation of cytokines and oxidative stress. *Am. J. Clin. Nutr.* **2007**, *85*, 837–844. [PubMed]
70. Kloubert, V.; Rink, L. Zinc as a micronutrient and its preventive role of oxidative damage in cells. *Food Funct.* **2015**, *6*, 3195–3204. [CrossRef] [PubMed]
71. Young, B.; Ott, L.; Kasarskis, E.; Rapp, R.; Moles, K.; Dempsey, R.J.; Tibbs, P.A.; Kryscio, R.; McClain, C. Zinc supplementation is associated with improved neurologic recovery rate and visceral protein levels of patients with severe closed head injury. *J. Neurotrauma* **1996**, *13*, 25–34. [CrossRef] [PubMed]
72. Besecker, B.Y.; Exline, M.C.; Hollyfield, J.; Phillips, G.; DiSilvestro, R.A.; Wewers, M.D.; Knoell, D.L. A comparison of zinc metabolism, inflammation, and disease severity in critically ill infected and noninfected adults early after intensive care unit admission. *Am. J. Clin. Nutr.* **2011**, *93*, 1356–1364. [CrossRef] [PubMed]
73. Costarelli, L.; Muti, E.; Malavolta, M.; Cipriano, C.; Giacconi, R.; Tesei, S.; Piacenza, F.; Pierpaoli, S.; Gasparini, N.; Faloia, E.; et al. Distinctive modulation of inflammatory and metabolic parameters in relation to zinc nutritional status in adult overweight/obese subjects. *J. Nutr. Biochem.* **2010**, *21*, 432–437. [CrossRef] [PubMed]
74. Chang, K.-L.; Hung, T.-C.; Hsieh, B.-S.; Chen, Y.-H.; Chen, T.-F.; Cheng, H.-L. Zinc at pharmacologic concentrations affects cytokine expression and induces apoptosis of human peripheral blood mononuclear cells. *Nutrition* **2006**, *22*, 465–474. [CrossRef] [PubMed]
75. Tsou, T.-C.; Chao, H.-R.; Yeh, S.-C.; Tsai, F.-Y.; Lin, H.-J. Zinc induces chemokine and inflammatory cytokine release from human promonocytes. *J. Hazard. Mater.* **2011**, *196*, 335–341. [CrossRef] [PubMed]
76. Beck, F.W.J.; Li, Y.; Bao, B.; Prasad, A.S.; Sarkar, F.H. Evidence for reprogramming global gene expression during zinc deficiency in the HUT-78 cell line. *Nutrition* **2006**, *22*, 1045–1056. [CrossRef] [PubMed]
77. Wessels, I.; Haase, H.; Engelhardt, G.; Rink, L.; Uciechowski, P. Zinc deficiency induces production of the proinflammatory cytokines IL-1beta and TNFalpha in promyeloid cells via epigenetic and redox-dependent mechanisms. *J. Nutr. Biochem.* **2013**, *24*, 289–297. [CrossRef] [PubMed]
78. Prasad, A.S.; Beck, F.W.; Grabowski, S.M.; Kaplan, J.; Mathog, R.H. Zinc deficiency: Changes in cytokine production and T-cell subpopulations in patients with head and neck cancer and in noncancer subjects. *Proc. Assoc. Am. Physicians* **1997**, *109*, 68–77. [PubMed]
79. Prasad, A.S.; Bao, B.; Beck, F.W.J.; Sarkar, F.H. Correction of interleukin-2 gene expression by in vitro zinc addition to mononuclear cells from zinc-deficient human subjects: A specific test for zinc deficiency in humans. *Transl. Res. J. Lab. Clin. Med.* **2006**, *148*, 325–333. [CrossRef] [PubMed]
80. Pinna, K.; Kelley, D.S.; Taylor, P.C.; King, J.C. Immune functions are maintained in healthy men with low zinc intake. *J. Nutr.* **2002**, *132*, 2033–2036. [PubMed]
81. Kahmann, L.; Uciechowski, P.; Warmuth, S.; Plumakers, B.; Gressner, A.M.; Malavolta, M.; Mocchegiani, E.; Rink, L. Zinc supplementation in the elderly reduces spontaneous inflammatory cytokine release and restores T cell functions. *Rejuvenation Res.* **2008**, *11*, 227–237. [CrossRef] [PubMed]
82. Wong, C.P.; Rinaldi, N.A.; Ho, E. Zinc deficiency enhanced inflammatory response by increasing immune cell activation and inducing IL6 promoter demethylation. *Mol. Nutr. Food Res.* **2015**, *59*, 991–999. [CrossRef] [PubMed]
83. Prasad, A.S. Zinc: Mechanisms of Host Defense. *J. Nutr.* **2007**, *137*, 1345–1349. [PubMed]
84. Beck, F.W.; Prasad, A.S.; Kaplan, J.; Fitzgerald, J.T.; Brewer, G.J. Changes in cytokine production and T cell subpopulations in experimentally induced zinc-deficient humans. *Am. J. Physiol.* **1997**, *272*, E1002–E1007. [PubMed]
85. Tapazoglou, E.; Prasad, A.S.; Hill, G.; Brewer, G.J.; Kaplan, J. Decreased natural killer cell activity in patients with zinc deficiency with sickle cell disease. *J. Lab. Clin. Med.* **1985**, *105*, 19–22. [PubMed]
86. Honscheid, A.; Rink, L.; Haase, H. T-lymphocytes: A target for stimulatory and inhibitory effects of zinc ions. *Endocr. Metab. Immune Disord. Drug Targets* **2009**, *9*, 132–144. [CrossRef] [PubMed]
87. Wellinghausen, N.; Martin, M.; Rink, L. Zinc inhibits interleukin-1-dependent T cell stimulation. *Eur. J. Immunol.* **1997**, *27*, 2529–2535. [CrossRef] [PubMed]
88. Rosenkranz, E.; Hilgers, R.-D.; Uciechowski, P.; Petersen, A.; Plumakers, B.; Rink, L. Zinc enhances the number of regulatory T cells in allergen-stimulated cells from atopic subjects. *Eur. J. Nutr.* **2017**, *56*, 557–567. [CrossRef] [PubMed]

89. Rosenkranz, E.; Maywald, M.; Hilgers, R.-D.; Brieger, A.; Clarner, T.; Kipp, M.; Plümäkers, B.; Meyer, S.; Schwerdtle, T.; Rink, L. Induction of regulatory T cells in Th1-/Th17-driven experimental autoimmune encephalomyelitis by zinc administration. *J. Nutr. Biochem.* **2016**, *29*, 116–123. [CrossRef] [PubMed]

90. Maywald, M.; Meurer, S.K.; Weiskirchen, R.; Rink, L. Zinc supplementation augments TGF-β1-dependent regulatory T cell induction. *Mol. Nutr. Food Res.* **2017**, *61*. [CrossRef] [PubMed]

91. Rosenkranz, E.; Metz, C.H.; Maywald, M.; Hilgers, R.; Weßels, I.; Senff, T.; Haase, H.; Jäger, M.; Ott, M.; Aspinall, R.; et al. Zinc supplementation induces regulatory T cells by inhibition of Sirt-1 deacetylase in mixed lymphocyte cultures. *Mol. Nutr. Food Res.* **2016**, *60*. [CrossRef] [PubMed]

92. Hennigar, S.R.; McClung, J.P. Nutritional Immunity. *Am. J. Lifestyle Med.* **2016**, *10*, 170–173. [CrossRef]

93. Aydemir, T.B.; Chang, S.-M.; Guthrie, G.J.; Maki, A.B.; Ryu, M.-S.; Karabiyik, A.; Cousins, R.J. Zinc transporter ZIP14 functions in hepatic zinc, iron and glucose homeostasis during the innate immune response (endotoxemia). *PLoS ONE* **2012**, *7*, e48679. [CrossRef]

94. Glaser, R.; Harder, J.; Lange, H.; Bartels, J.; Christophers, E.; Schroder, J.-M. Antimicrobial psoriasin (S100A7) protects human skin from *Escherichia coli* infection. *Nat. Immunol.* **2005**, *6*, 57–64. [CrossRef] [PubMed]

95. Corbin, B.D.; Seeley, E.H.; Raab, A.; Feldmann, J.; Miller, M.R.; Torres, V.J.; Anderson, K.L.; Dattilo, B.M.; Dunman, P.M.; Gerads, R.; et al. Metal chelation and inhibition of bacterial growth in tissue abscesses. *Science* **2008**, *319*, 962–965. [CrossRef] [PubMed]

96. Botella, H.; Stadthagen, G.; Lugo-Villarino, G.; Chastellier, C.; de Neyrolles, O. Metallobiology of host-pathogen interactions: An intoxicating new insight. *Trends Microbiol.* **2012**, *20*, 106–112. [CrossRef] [PubMed]

97. Lappann, M.; Danhof, S.; Guenther, F.; Olivares-Florez, S.; Mordhorst, I.L.; Vogel, U. In vitro resistance mechanisms of Neisseria meningitidis against neutrophil extracellular traps. *Mol. Microbiol.* **2013**, *89*, 433–449. [CrossRef] [PubMed]

98. Stork, M.; Grijpstra, J.; Bos, M.P.; Manas Torres, C.; Devos, N.; Poolman, J.T.; Chazin, W.J.; Tommassen, J. Zinc piracy as a mechanism of Neisseria meningitidis for evasion of nutritional immunity. *PLoS Pathog.* **2013**, *9*, e1003733. [CrossRef] [PubMed]

99. Bobrov, A.G.; Kirillina, O.; Fetherston, J.D.; Miller, M.C.; Burlison, J.A.; Perry, R.D. The Yersinia pestis siderophore, yersiniabactin, and the ZnuABC system both contribute to zinc acquisition and the development of lethal septicaemic plague in mice. *Mol. Microbiol.* **2014**, *93*, 759–775. [CrossRef] [PubMed]

100. Liu, J.Z.; Jellbauer, S.; Poe, A.J.; Ton, V.; Pesciaroli, M.; Kehl-Fie, T.E.; Restrepo, N.A.; Hosking, M.P.; Edwards, R.A.; Battistoni, A.; et al. Zinc sequestration by the neutrophil protein calprotectin enhances Salmonella growth in the inflamed gut. *Cell Host Microbe* **2012**, *11*, 227–239. [CrossRef] [PubMed]

101. Nowak, J.E.; Harmon, K.; Caldwell, C.C.; Wong, H.R. Prophylactic zinc supplementation reduces bacterial load and improves survival in a murine model of sepsis. *Pediatr. Crit. Care Med.* **2012**, *13*, e323–e329. [CrossRef] [PubMed]

102. Singh, M.; Das, R.R. Zinc for the common cold. *Cochrane Database Syst. Rev.* **2013**, CD001364. [CrossRef]

103. Mezzetti, A.; Pierdomenico, S.D.; Costantini, F.; Romano, F.; de Cesare, D.; Cuccurullo, F.; Imbastaro, T.; Riario-Sforza, G.; Di Giacomo, F.; Zuliani, G.; et al. Copper/zinc ratio and systemic oxidant load: Effect of aging and aging-related degenerative diseases. *Free Radic. Biol. Med.* **1998**, *25*, 676–681. [CrossRef]

104. Kozlowski, H.; Luczkowski, M.; Remelli, M.; Valensin, D. Copper, zinc and iron in neurodegenerative diseases (Alzheimer's, Parkinson's and prion diseases). *Coord. Chem. Rev.* **2012**, *256*, 2129–2141. [CrossRef]

105. Malavolta, M.; Giacconi, R.; Piacenza, F.; Santarelli, L.; Cipriano, C.; Costarelli, L.; Tesel, S.; Pierpaoli, S.; Basso, A.; Galeazzi, R.; et al. Plasma copper/zinc ratio: An inflammatory/nutritional biomarker as predictor of all-cause mortality in elderly population. *Biogerontology* **2010**, *11*, 309–319. [CrossRef] [PubMed]

106. Overbeck, S.; Rink, L.; Haase, H. Modulating the immune response by oral zinc supplementation: A single approach for multiple diseases. *Arch. Immunol. Ther. Exp.* **2008**, *56*, 15–30. [CrossRef] [PubMed]

107. Hulisz, D. Efficacy of zinc against common cold viruses: An overview. *J. Am. Pharm. Assoc.* **2004**, *44*, 594–603. [CrossRef]

108. Kurugol, Z.; Akilli, M.; Bayram, N.; Koturoglu, G. The prophylactic and therapeutic effectiveness of zinc sulphate on common cold in children. *Acta Paediatr.* **2006**, *95*, 1175–1181. [CrossRef] [PubMed]

109. Baum, M.K.; Lai, S.; Sales, S.; Page, J.B.; Campa, A. Randomized Controlled Clinical Trial of Zinc Supplementation to Prevent Immunological Failure in HIV-Positive Adults1,2. *Clin. Infect. Dis.* **2010**, *50*, 1653–1660. [CrossRef] [PubMed]

110. Asdamongkol, N.; Phanachet, P.; Sungkanuparph, S. Low Plasma Zinc Levels and Immunological Responses to Zinc Supplementation in HIV-Infected Patients with Immunological Discordance after Antiretroviral Therapy. *Jpn. J. Infect. Dis.* 2013, *66*, 469–474. [CrossRef] [PubMed]

111. Lodha, R.; Shah, N.; Mohari, N.; Mukherjee, A.; Vajpayee, M.; Singh, R.; Singla, M.; Saini, S.; Bhatnagar, S.; Kabra, S.K. Immunologic effect of zinc supplementation in HIV-infected children receiving highly active antiretroviral therapy: A randomized, double-blind, placebo-controlled trial. *J. Acquir. Immune Defic. Syndr.* 2014, *66*, 386–392. [CrossRef] [PubMed]

112. Mocchegiani, E.; Veccia, S.; Ancarani, F.; Scalise, G.; Fabris, N. Benefit of oral zinc supplementation as an adjunct to zidovudine (AZT) therapy against opportunistic infections in AIDS. *Int. J. Immunopharmacol.* 1995, *17*, 719–727. [CrossRef]

113. Zazzo, J.F.; Rouveix, B.; Rajagopalon, P.; Levacher, M.; Girard, P.M. Effect of zinc on the immune status of zinc-depleted AIDS related complex patients. *Clin. Nutr.* 1989, *8*, 259–261. [CrossRef]

114. Bobat, R.; Coovadia, H.; Stephen, C.; Naidoo, K.L.; McKerrow, N.; Black, R.E.; Moss, W.J. Safety and efficacy of zinc supplementation for children with HIV-1 infection in South Africa: A randomised double-blind placebo-controlled trial. *Lancet* 2005, *366*, 1862–1867. [CrossRef]

115. Green, J.A.; Lewin, S.R.; Wightman, F.; Lee, M.; Ravindran, T.S.; Paton, N.I. A randomised controlled trial of oral zinc on the immune response to tuberculosis in HIV-infected patients. *Int. J. Tuberc. Lung Dis.* 2005, *9*, 1378–1384. [PubMed]

116. Fawzi, W.W.; Villamor, E.; Msamanga, G.I.; Antelman, G.; Aboud, S.; Urassa, W.; Hunter, D. Trial of zinc supplements in relation to pregnancy outcomes, hematologic indicators, and T cell counts among HIV-1-infected women in Tanzania. *Am. J. Clin. Nutr.* 2005, *81*, 161–167. [PubMed]

117. Villamor, E.; Aboud, S.; Koulinska, I.N.; Kupka, R.; Urassa, W.; Chaplin, B.; Msamanga, G.; Fawzi, W.W. Zinc supplementation to HIV-1-infected pregnant women: Effects on maternal anthropometry, viral load, and early mother-to-child transmission. *Eur. J. Clin. Nutr.* 2006, *60*, 862–869. [CrossRef] [PubMed]

118. Deloria-Knoll, M.; Steinhoff, M.; Semba, R.D.; Nelson, K.; Vlahov, D.; Meinert, C.L. Effect of zinc and vitamin A supplementation on antibody responses to a pneumococcal conjugate vaccine in HIV-positive injection drug users: A randomized trial. *Vaccine* 2006, *24*, 1670–1679. [CrossRef] [PubMed]

119. Kawaguchi, T.; Nagao, Y.; Abe, K.; Imazeki, F.; Honda, K.; Yamasaki, K.; Miyanishi, K.; Taniguchi, E.; Kakuma, T.; Kato, J.; et al. Effects of branched-chain amino acids and zinc-enriched nutrients on prognosticators in HCV-infected patients: A multicenter randomized controlled trial. *Mol. Med. Rep.* 2015, *11*, 2159–2166. [CrossRef] [PubMed]

120. Murakami, Y.; Koyabu, T.; Kawashima, A.; Kakibuchi, N.; Kawakami, T.; Takaguchi, K.; Kita, K.; Okita, M. Zinc supplementation prevents the increase of transaminase in chronic hepatitis C patients during combination therapy with pegylated interferon alpha-2b and ribavirin. *J. Nutr. Sci. Vitaminol.* 2007, *53*, 213–218. [CrossRef] [PubMed]

121. Takagi, H.; Nagamine, T.; Abe, T.; Takayama, H.; Sato, K.; Otsuka, T.; Kakizaki, S.; Hashimoto, Y.; Matsumoto, T.; Kojima, A.; et al. Zinc supplementation enhances the response to interferon therapy in patients with chronic hepatitis C. *J. Viral Hepat.* 2001, *8*, 367–371. [CrossRef] [PubMed]

122. Ko, W.-S.; Guo, C.-H.; Hsu, G.-S.W.; Chiou, Y.-L.; Yeh, M.-S.; Yaun, S.-R. The effect of zinc supplementation on the treatment of chronic hepatitis C patients with interferon and ribavirin. *Clin. Biochem.* 2005, *38*, 614–620. [CrossRef] [PubMed]

123. Hoque, K.M.; Binder, H.J. Zinc in the treatment of acute diarrhea: Current status and assessment. *Gastroenterology* 2006, *130*, 2201–2205. [CrossRef] [PubMed]

124. Dutta, P.; Mitra, U.; Dutta, S.; Naik, T.N.; Rajendran, K.; Chatterjee, M.K. Zinc, vitamin A, and micronutrient supplementation in children with diarrhea: A randomized controlled clinical trial of combination therapy versus monotherapy. *J. Pediatr.* 2011, *159*, 633–637. [CrossRef] [PubMed]

125. Malik, A.; Taneja, D.K.; Devasenapathy, N.; Rajeshwari, K. Zinc supplementation for prevention of acute respiratory infections in infants: A randomized controlled trial. *Indian Pediatr.* 2014, *51*, 780–784. [CrossRef] [PubMed]

126. Shah, U.H.; Abu-Shaheen, A.K.; Malik, M.A.; Alam, S.; Riaz, M.; AL-Tannir, M.A. The efficacy of zinc supplementation in young children with acute lower respiratory infections: A randomized double-blind controlled trial. *Clin. Nutr.* 2013, *32*, 193–199. [CrossRef] [PubMed]

127. Martinez-Estevez, N.S.; Alvarez-Guevara, A.N.; Rodriguez-Martinez, C.E. Effects of zinc supplementation in the prevention of respiratory tract infections and diarrheal disease in Colombian children: A 12-month randomised controlled trial. *Allergol. Immunopathol.* **2016**, *44*, 368–375. [CrossRef] [PubMed]

128. Sazawal, S.; Black, R.E.; Jalla, S.; Mazumdar, S.; Sinha, A.; Bhan, M.K. Zinc supplementation reduces the incidence of acute lower respiratory infections in infants and preschool children: A double-blind, controlled trial. *Pediatrics* **1998**, *102*, 1–5. [CrossRef] [PubMed]

129. Mahalanabis, D.; Lahiri, M.; Paul, D.; Gupta, S.; Gupta, A.; Wahed, M.A.; Khaled, M.A. Randomized, double-blind, placebo-controlled clinical trial of the efficacy of treatment with zinc or vitamin A in infants and young children with severe acute lower respiratory infection. *Am. J. Clin. Nutr.* **2004**, *79*, 430–436. [PubMed]

130. Karyadi, E.; West, C.E.; Schultink, W.; Nelwan, R.H.H.; Gross, R.; Amin, Z.; Dolmans, W.M.V.; Schlebusch, H.; van der Meer, J.W.M. A double-blind, placebo-controlled study of vitamin A and zinc supplementation in persons with tuberculosis in Indonesia: Effects on clinical response and nutritional status. *Am. J. Clin. Nutr.* **2002**, *75*, 720–727. [PubMed]

131. Mathur, N.K.; Bumb, R.A.; Mangal, H.N. Oral zinc in recurrent Erythema Nodosum Leprosum reaction. *Lepr. India* **1983**, *55*, 547–552. [PubMed]

132. Mathur, N.K.; Bumb, R.A.; Mangal, H.N.; Sharma, M.L. Oral zinc as an adjunct to dapsone in lepromatous leprosy. *Int. J. Lepr. Other Mycobact. Dis. Off. Org. Int. Lepr. Assoc.* **1984**, *52*, 331–338.

133. El-Shafei, M.M.; Kamal, A.A.; Soliman, H.; el-Shayeb, F.; Abdel Baqui, M.S.; Faragalla, S.; Sabry, M.K. Effect of oral zinc supplementation on the cell mediated immunity in lepromatous leprosy. *J. Egypt. Public Health Assoc.* **1988**, *63*, 311–336. [PubMed]

134. Mahajan, P.M.; Jadhav, V.H.; Patki, A.H.; Jogaikar, D.G.; Mehta, J.M. Oral zinc therapy in recurrent erythema nodosum leprosum: A clinical study. *Indian J. Lepr.* **1994**, *66*, 51–57. [PubMed]

135. Alam, A.N.; Sarker, S.A.; Wahed, M.A.; Khatun, M.; Rahaman, M.M. Enteric protein loss and intestinal permeability changes in children during acute shigellosis and after recovery: Effect of zinc supplementation. *Gut* **1994**, *35*, 1707–1711. [CrossRef] [PubMed]

136. Raqib, R.; Roy, S.K.; Rahman, M.J.; Azim, T.; Ameer, S.S.; Chisti, J.; Andersson, J. Effect of zinc supplementation on immune and inflammatory responses in pediatric patients with shigellosis. *Am. J. Clin. Nutr.* **2004**, *79*, 444–450. [PubMed]

137. Rahman, M.J.; Sarker, P.; Roy, S.K.; Ahmad, S.M.; Chisti, J.; Azim, T.; Mathan, M.; Sack, D.; Andersson, J.; Raqib, R. Effects of zinc supplementation as adjunct therapy on the systemic immune responses in shigellosis. *Am. J. Clin. Nutr.* **2005**, *81*, 495–502. [PubMed]

138. Roy, S.K.; Raqib, R.; Khatun, W.; Azim, T.; Chowdhury, R.; Fuchs, G.J.; Sack, D.A. Zinc supplementation in the management of shigellosis in malnourished children in Bangladesh. *Eur. J. Clin. Nutr.* **2008**, *62*, 849–855. [CrossRef] [PubMed]

139. Kashimura, H.; Suzuki, K.; Hassan, M.; Ikezawa, K.; Sawahata, T.; Watanabe, T.; Nakahara, A.; Mutoh, H.; Tanaka, N. Polaprezinc, a mucosal protective agent, in combination with lansoprazole, amoxycillin and clarithromycin increases the cure rate of Helicobacter pylori infection. *Aliment. Pharmacol. Ther.* **1999**, *13*, 483–487. [CrossRef] [PubMed]

140. Zeba, A.N.; Sorgho, H.; Rouamba, N.; Zongo, I.; Rouamba, J.; Guiguemdé, R.T.; Hamer, D.H.; Mokhtar, N.; Ouedraogo, J.-B. Major reduction of malaria morbidity with combined vitamin A and zinc supplementation in young children in Burkina Faso: A randomized double blind trial. *Nutr. J.* **2008**, *7*, 7. [CrossRef] [PubMed]

141. Shankar, A.H.; Genton, B.; Baisor, M.; Paino, J.; Tamja, S.; Adiguma, T.; Wu, L.; Rare, L.; Bannon, D.; Tielsch, J.M.; et al. The influence of zinc supplementation on morbidity due to *Plasmodium falciparum*: A randomized trial in preschool children in Papua New Guinea. *Am. J. Trop. Med. Hyg.* **2000**, *62*, 663–669. [CrossRef] [PubMed]

142. Bates, C.J.; Evans, P.H.; Dardenne, M.; Prentice, A.; Lunn, P.G.; Northrop-Clewes, C.A.; Hoare, S.; Cole, T.J.; Horan, S.J.; Longman, S.C. A trial of zinc supplementation in young rural Gambian children. *Br. J. Nutr.* **1993**, *69*, 243–255. [CrossRef] [PubMed]

143. Muller, O.; Becher, H.; van Zweeden, A.B.; Ye, Y.; Diallo, D.A.; Konate, A.T.; Gbangou, A.; Kouyate, B.; Garenne, M. Effect of zinc supplementation on malaria and other causes of morbidity in west African children: Randomised double blind placebo controlled trial. *BMJ Clin. Res.* **2001**, *322*, 1567. [CrossRef]

144. Zinc Against Plasmodium Study Group. Effect of zinc on the treatment of *Plasmodium falciparum* malaria in children: A randomized controlled trial. *Am. J. Clin. Nutr.* **2002**, *76*, 805–812.
145. Richard, S.A.; Zavaleta, N.; Caulfield, L.E.; Black, R.E.; Witzig, R.S.; Shankar, A.H. Zinc and iron supplementation and malaria, diarrhea, and respiratory infections in children in the Peruvian Amazon. *Am. J. Trop. Med. Hyg.* **2006**, *75*, 126–132. [PubMed]
146. Lazzerini, M.; Wanzira, H. Oral zinc for treating diarrhoea in children. *Cochrane Database Syst. Rev.* **2016**. [CrossRef]
147. World Health Organisation. *Zinc Supplementation in the Management of Diarrhoea*; World Health Organisation: Geneva, Switzerland, 2017.
148. Finamore, A.; Massimi, M.; Conti Devirgiliis, L.; Mengheri, E. Zinc deficiency induces membrane barrier damage and increases neutrophil transmigration in Caco-2 cells. *J. Nutr.* **2008**, *138*, 1664–1670. [PubMed]
149. Bao, S.; Knoell, D.L. Zinc modulates cytokine-induced lung epithelial cell barrier permeability. *Am. J. Physiol. Lung Cell. Mol. Physiol.* **2006**, *291*, L1132-41. [CrossRef] [PubMed]
150. Lansdown, A.B.G.; Mirastschijski, U.; Stubbs, N.; Scanlon, E.; Agren, M.S. Zinc in wound healing: Theoretical, experimental, and clinical aspects. *Wound Repair Regen.* **2007**, *15*, 2–16. [CrossRef] [PubMed]
151. Gosain, A.; DiPietro, L.A. Aging and Wound Healing. *World J. Surg.* **2004**, *28*, 321–326. [CrossRef] [PubMed]
152. Lu, X.; Wang, M.; Qi, J.; Wang, H.; Li, X.; Gupta, D.; Dziarski, R. Peptidoglycan Recognition Proteins Are a New Class of Human Bactericidal Proteins. *J. Biol. Chem.* **2006**, *281*, 5895–5907. [CrossRef] [PubMed]
153. Wang, M.; Liu, L.-H.; Wang, S.; Li, X.; Lu, X.; Gupta, D.; Dziarski, R. Human Peptidoglycan Recognition Proteins Require Zinc to Kill Both Gram-Positive and Gram-Negative Bacteria and Are Synergistic with Antibacterial Peptides. *J. Immunol.* **2007**, *178*, 3116–3125. [CrossRef] [PubMed]

nutrients

MDPI

Article

Copper to Zinc Ratio as Disease Biomarker in Neonates with Early-Onset Congenital Infections

Monika Wisniewska [1,†], Malte Cremer [2,†], Lennart Wiehe [1,†], Niels-Peter Becker [1], Eddy Rijntjes [1], Janine Martitz [1], Kostja Renko [1], Christoph Bührer [2] and Lutz Schomburg [1,*]

[1]   Institute for Experimental Endocrinology, Charité-Universitätsmedizin Berlin, Augustenburger Platz 1, CVK, D-13353 Berlin, Germany; Monika.Wisniewska@charite.de (M.W.); Lennart.Wiehe@charite.de (L.W.); Niels-Peter.Becker@charite.de (N.-P.B.); Eddy.Rijntjes@charite.de (E.R.); Janine.Martitz@charite.de (J.M.); Kostja.Renko@charite.de (K.R.)
[2]   Department of Neonatology, Charité-Universitätsmedizin Berlin, Augustenburger Platz 1, CVK, D-13353 Berlin, Germany; Malte.Cremer@charite.de (M.C.); christoph.buehrer@charite.de (C.B.)
*   Correspondence: lutz.schomburg@charite.de; Tel.: +49-30-450-524289
†   These authors contributed equally.

Received: 24 January 2017; Accepted: 28 March 2017; Published: 30 March 2017

Abstract: Copper (Cu) and zinc (Zn) are essential trace elements for regular development. Acute infections alter their metabolism, while deficiencies increase infection risks. A prospective observational case-control study was conducted with infected ($n = 21$) and control ($n = 23$) term and preterm newborns. We analyzed trace element concentrations by X-ray fluorescence, and ceruloplasmin (CP) by Western blot. Median concentration of Cu at birth (day 1) was 522.8 [387.1–679.7] µg/L, and Zn was 1642.4 ± 438.1 µg/L. Cu and Zn correlated positively with gestational age in control newborns. Cu increased in infected newborns from day 1 to day 3. CP correlated positively to Cu levels at birth in both groups and on day 3 in the group of infected neonates. The Cu/Zn ratio was relatively high in infected newborns. Interleukin (IL)-6 concentrations on day 1 were unrelated to Cu, Zn, or the Cu/Zn ratio, whereas C-reactive protein (CRP) levels on day 3 correlated positively to the Cu/Zn -ratio at both day 1 and day 3. We conclude that infections affect the trace element homeostasis in newborns: serum Zn is reduced, while Cu and CP are increased. The Cu/Zn ratio combines both alterations, independent of gestational age. It may, thus, constitute a meaningful diagnostic biomarker for early-onset infections.

Keywords: ceruloplasmin; preterm; C reactive protein; interleukin-6; micronutrient

1. Introduction

Copper (Cu) and zinc (Zn) are trace elements essential for life and constitute components of numerous enzymes with high importance for survival and function of eukaryotic cells [1–4]. Cu and Zn are necessary for cellular metabolism and the antioxidative defense systems [1–3,5]. The regular fetal development and growth critically depend on both Cu and Zn, especially during the maturation of the nervous system [2,4,5]. Similarly, the maturing immune system relies on these trace elements [2,3,5,6], especially for antibody production (Cu, Zn), function of neutrophils and monocytes (Cu) [6], the viability, proliferation, and differentiation of cells of both the innate and adaptive immune system (Zn), as well as for the maintenance of the skin and mucosal barriers (Zn) [3,6].

Both a deficiency and an excess of Cu or Zn can cause harm, so the homeostasis of both elements is strictly regulated [1–3]. The main source of both micronutrients is dietary intake [1,3]. The absorption from the gastrointestinal tract depends on the nutritional form and the micronutrient status of the subject [4,7]. The cellular homeostasis is controlled by import and export proteins, cytosolic metallochaperones, glutathione, and metallothioneins [1,2,5]. The latter also acts as a dynamic Cu and

Zn pool [3,5]. Around 95% of Cu in blood is bound to liver-derived ceruloplasmin (CP), which is used along with serum Cu concentrations as a biomarker of Cu status [4]. Serum Cu and CP concentrations constitute acute phase reactants [5]. Cu is mainly stored in liver, secreted as Cu-CP complex into blood, and an excess can be eliminated by biliary excretion [1,4]. In contrast to Cu, the protein-mediated transport, storage, and regulated excretion of Zn are more complex and less well understood [3,7].

Cu and Zn deficiencies constitute prevalent and under-diagnosed health risks [1,3]. Neonates and especially preterm infants have a notable risk of Cu and Zn deficiency due to their rapid growth and the concomitant increasing requirement for both micronutrients [4]. On the one hand, Cu or Zn deficiencies impair the immune defense and confer a high susceptibility to infectious diseases [3–6,8]. On the other hand, acute infections cause an increase in serum Cu in the context of an acute phase response [4,9] and a decrease in serum Zn due to a redistribution into liver and other tissues [3,7,10–12]. These two responses to inflammation may feed a vicious cycle of impaired immune defense and higher infection risk, which is of particular importance especially in very vulnerable patients, such as preterm and term neonates with an immature immune system.

Congenital infections and especially neonatal sepsis are a frequent cause for morbidity and mortality of newborns [13–16]. Vertical bacteria transmission from mother to child results in early-onset infections, defined by an onset within 72 h postnatum [16]. The symptoms of a systemic inflammation by newborns are unspecific [13], and the diagnosis is challenging. Isolation of bacteria from blood cultures of neonates is difficult due to the relatively high blood volumes and long incubation times required, and by the initially low bacteria counts in the majority of affected neonates [13,14]. The cytokine interleukin 6 (IL-6) and the acute-phase reactant C-reactive protein (CRP) are important diagnostic markers for early (IL-6) or later (CRP) phases of inflammation [15]. Despite their usefullness, the current clinical algorithm does not provide a satisfying diagnostic specificity in neonates [14], implying that additional biomarkers are needed to facilitate the correct diagnosis and to avoid unnecessary drug administrations. No consented diagnostic and therapeutic guidelines for newborn sepsis have been established. The current therapeutic strategy advocates an early application of an empiric antibiotic treatment if an infection is just suspected [17]. Some antibiotics are applied in terms of an off-label use due to a lack of intervention studies in newborns, raising concerns about their safety [17]. Despite the supportive and antibiotic treatments, congenital infections may still cause long-term complications such as brain damage and neurological sequelae [18].

From animal and clinical studies with adult patients, it is well established that inflammatory cytokines disturb the trace element homeostasis [9]. Furthermore, the Cu/Zn ratio is altered in certain diseases [19–23]. Therefore, we postulated that infections disrupt homeostasis of the trace elements Cu and Zn in neonates.

2. Materials and Methods

2.1. Study Design

The design of this prospective observational case-control study has been described before in the context of the effects of congenital infections on the selenium (Se) status [24]. Briefly, this explorative study was conducted on the neonatology wards of the Charité-Universitätsmedizin Berlin from February 2013 until April 2014, after clearance with the local ethics committee (approval no.: EA2/092/12). Informed written consent was obtained for each neonate enrolled in the study from the legal guardian(s) prior to analysis. Relevant clinical data were extracted from the electronic and traditional paper-based medical files. Trace elements were analyzed using residual plasma samples from routine laboratory evaluations ordered by the attending physician. Hereby, one sample was collected from each neonate (control or infection) at birth (day 1). The 2nd sample was obtained 48 h later (day 3) only from infected neonates.

2.2. Study Population

Neonates were screened for fulfilling the inclusion criteria of early-onset infection according to published recommendations and the diagnostic steps taken [25–27]. In brief, neonates qualifying for the infection group had to exhibit at least one of the following clinical signs: pneumonia, respiratory distress, tachycardia, bradycardia, fever (>38.5 °C), hypothermia (<36.0 °C), irritability, lethargy, hypotonia, poor feeding, increased frequency of apnea, and/or coagulation disorder in combination with laboratory evidence for an inflammation (IL-6 > 100 ng/L, or CRP > 10 mg/L). In neonates with suspected infection, blood cultures were performed prior to antibiotic treatment. Blood cultures were performed with the BacT/ALERT automated system (Organon Teknika, Eppelheim, Germany) in Pedi-BacT pediatric blood culture bottles capable of detecting anaerobic as well as aerobic bacteria. Neonates with suspected early-onset infection were immediately treated with ampicillin and gentamicin for at least 3 days, and a second blood drawing was conducted 48 h later to determine inflammation markers and gentamicin levels. Two residual plasma samples were, thus, available from each of the neonates with suspected infection (day 1 and day 3), and one from the control neonates at time of birth (day 1). The newborns in the control group showed no laboratory evidence for inflammation (IL-6 < 100 ng/L, and/or CRP < 10 mg/L), and were not receiving antibiotic treatment during the hospital stay. Several infants had to be excluded because of birth before 30 weeks of gestation, birth weight below 1000 g, a diagnosed genetic disease, severe congenital malformation, parenteral supplementation with trace elements, or a missing written consent. Details on the neonates enrolled into this study have been published earlier in relation to analyzing their Se status [24] and are provided below (Table 1). Neonates with suspected early-onset infection are summarized as the "infection group" and are denoted as "infected neonates" in this scientific report. However, it needs to be pointed out that a suspected early-onset infection in neonates cannot as safely be diagnosed as in adults for a number of reasons including unspecific symptoms, a higher variability of symptoms, and a very limited amount of blood to be analyzed by laboratory tests and blood culture.

Table 1. Characteristics of the study population, as previously described [24].

	Control Group	Infection Group	*p*-Value (Two-Sided)
Participants	*n* = 23	*n* = 21	-
Gestation age [weeks]	34.9 [33.9–37.1]	38.4 [35.1–39.8]	0.003
Term infants	*n* = 6 (26%)	*n* = 13 (62%)	
Preterm infants (<37 gestation weeks)	*n* = 17 (74%)	*n* = 8 (38%)	
Birth Weight [g]	2452.3 ± 694.1	3119.7 ± 733.7	0.003
Vaginal birth	*n* = 7 (30%)	*n* = 10 (48%)	-
Twins	*n* = 7 (30%)	*n* = 1 (5%)	-
Apgar 1 min *	9.0 [6.0–9.0]	8.0 [4.5–9.0]	0.505
Apgar 5 min *	9.0 [8.0–10.0]	9.0 [7.5–10.0]	0.951
Cord arterial pH	7.25 [7.2–7.29]	7.23 [7.19–7.3]	0.655
pH on admission #	7.33 [7.27–7.38]	7.33 [7.26–7.37]	0.673
Base excess on admission ##	−2.15 [(−3.75)–(−0.90)]	−2.8 [(−5.32)–(−1.28)]	0.283
IL-6 [pg/mL], day 1	5.8 [4.2–17.0]	498.5 [203.6–2523.0]	<0.001

* The Apgar score is used to characterize the condition of neonates after birth [28]. The Apgar score evaluates the condition of the neonate 1, 5, and 10 min after birth and guides subsequent interventions. Furthermore, the 5 min Apgar score has been used to predict neurological long-term-outcome. Each of five easily identifiable characteristics–respiration, heart rate, skin color, reflex irritability, and muscle tone—is assessed and assigned a value of 0 to 2. # The blood pH was measured in each neonate immediately after admission to the neonatal ward. ## The base excess refers to an excess or deficit in the amount of base in the blood. A negative base excess reflects an acid/base disturbance caused by the accumulation of lactate acid because of anaerobic glycolysis. IL stands for interleukin.

2.3. Trace Elements

Total plasma Cu and Zn concentrations were determined by total reflection X-ray fluorescence (TXRF) as described in [24]. The method was chosen because of the low sample volume requirements, which is necessary particularly for infants, and prospectively because of the short time needed for analysis, which is a prerequisite for considering the technique in routine clinical decision-making. Briefly, 10 μL of plasma samples were diluted with 10 μL of a gallium standard solution (f.c. 550 μg/L, Sigma-Aldrich, Steinheim, Germany) and mixed thoroughly. Duplicates of 8 μL each were applied to ultra clean quartz glass carriers, dried at 37 °C and measured using a TXRF spectrometer (S2 PICOFOX, Bruker nano GmbH, Berlin, Germany) as described in [29]. The inter-assay coefficient of variation (CV) was less than 10% for both Cu and Zn.

2.4. Ceruloplasmin

Western blot analysis was performed for assessment of plasma CP levels. Three Western blots were prepared containing plasma samples from neonates of the control group along with samples from infected neonates. An additional Western blot was prepared with samples from the group of infected neonates that were pre-selected with respect to the CRP levels. Plasma was diluted in ultrapure H_2O (Biochrom AG, Berlin, Germany) and 4x sample buffer (200 mM Tris-HCl, pH 7.5, 50% glycerin, 4% SDS, 0.04% bromophenol blue, and 125 mM DTT). Plasma proteins were size-fractionated by sodium dodecyl (lauryl) sulfate-polyacrylamide gel electrophoresis (SDS-PAGE) and subsequently transferred onto nitrocellulose membranes by semi-dry blotting (Optitran, Schleicher & Schuell, Dassel, Germany). Antibodies against CP (1:2000 dilution, ab19171, Abcam, Cambridge, UK) were used and bands visualized by chemiluminescence (Western-Bright Substrate Sirius, Biozym Scientific, Oldendorf, Germany) using the Fluor Chem FC2 detection system (Biozym Scientific). Quantification of Western blot signals was achieved by using the Java-based image processing program ImageJ (NIH, Bethesda, MD, USA).

2.5. Interleukin 6 and C-reactive Protein

The IL-6 and CRP values were measured by Labor Berlin, Charité Vivantes GmbH, Germany, by standard laboratory analyses as described earlier [24]. Briefly, IL-6 was measured by an electro-chemical luminescence immunoassay and CRP was determined by a turbidometric assay (COBAS 8000 or COBAS 6000; Roche Diagnostics, Mannheim, Germany).

2.6. Statistical Analysis

Statistical analysis was performed with the Statistical Package for the Social Sciences (SPSS Statistics 21®, IBM, Chicago, IL, USA) and GraphPad Prism (GraphPad Software Inc., San Diego, CA, USA). Normal distribution of values was assessed by the Shapiro-Wilk test. A two-tailed T-test for unpaired or paired variables and the bivariate Pearson correlation test were used for normally distributed values. For not-normally distributed variables, the Mann-Whitney-U-test, the Wilcoxon-test, and the Spearman´s correlation test were used. The quantified Western blot signals were analyzed using nonparametric tests (Mann-Whitney-U-Test, the Wilcoxon-Test, and the Spearman's correlation test). Linear regression analysis was conducted to specify associations of variables. Multiple logistic regression was performed to evaluate the results in consideration of relevant confounders. Odds ratios where calculated to examine the quality of the Cu/Zn ratio as a biomarker. The results were considered as statistically significant when the p-value was less than 0.05, and differences are marked as follows: $p < 0.05$ (*), $p < 0.01$ (**), and $p < 0.001$ (***). Parametric data are represented as means ± standard deviation (SD) or medians and interquartile range (IQR); median [IQR].

3. Results

3.1. Cu and Zn Status at Day of Birth in Control and Infected Newborns

Out of 108 newborns, a total of 72 neonates qualified for analysis. Written informed consent was provided for 44 of the samples, of which 23 fulfilled criteria for the control and 21 for the infection group (Table 1). Newborns in the control group had on average a lower gestational age, and lower birth weight. There were more preterm infants (<37 weeks of gestation) in the non-infection control than in the infection group, and more Caesarian sections in the control group (preterm: 13 in control and 5 in infection group vs. term: 3 in control vs. 6 in infection group). There was a variety of clinical symptoms displayed by the infected newborns (Table 2), whereas the newborns in the control group were not displaying any of these signs in combination with laboratory evidence of an inflammation. None of the neonates in the infection group had a positive blood culture. None of the infected or control children had severe thrombocytopenia or leukocytopenia, or displayed any specific signs for an infection with toxoplasma gondii, rubella virus, cytomegalovirus, or herpes simplex virus 1 or 2 (TORCH). No mother had a history of TORCH infection during pregnancy.

Table 2. Clinical characteristics of the neonates with suspected early-onset infection ($n = 21$). IL stands for interleukin; IRQ stands for interquartile range; CRP stands for C-reactive protein; SD stands for standard deviation.

Clinical Characteristics	
Pneumonia/Respiratory Distress	$n = 19$
Tachycardia/bradycardia	$n = 6$
Fever/hypothermia	$n = 2$
Irritability/lethargy	$n = 3$
Coagulation disorder	$n = 0$
IL-6 [ng/L], day 1 (median [IQR])	498.5 [203.6–2523.0]
CRP [mg/L], day 3, (mean $\pm$ SD)	12.1 $\pm$ 7.8
Antibiotic treatment [days] (median [IQR])	5.0 [3.0–5.0]

The median concentration of plasma Cu of all neonates on day 1 was 522.8 [387.1–679.7] µg/L. There was no significant difference between the control and the infection group. The distribution of plasma Cu concentrations was more heterogeneous in the infection group as compared to that of the control group (Figure 1A). Zn concentrations were normally distributed in both groups of newborns, and were higher in the control group as compared to that of the infection group (1804.6 $\pm$ 377.0 vs. 1464.8 $\pm$ 439.3 µg/L, $p = 0.009$) (Figure 1B). When separating the newborns according to infection and gestational age, Zn levels were significantly lower in the infected term infants as compared to that of the control term infants (1430.3 $\pm$ 374.4 vs. 2021.0 $\pm$ 488.7 µg/L; $p = 0.01$), but there was no difference in Zn concentration between the infected vs. control preterm neonates.

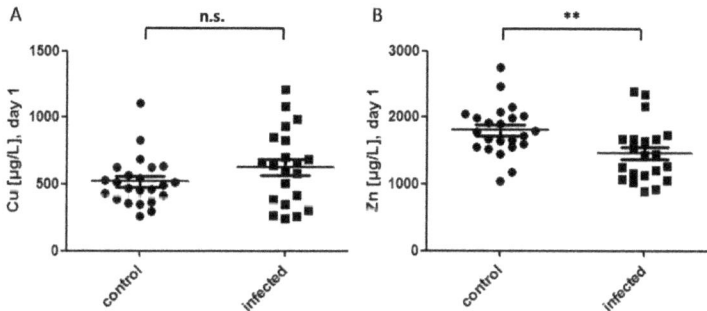

Figure 1. Plasma Cu and Zn in control and infected neonates. (**A**) The Cu concentrations in plasma are not significantly different between the groups of control ($n = 23$) and infected ($n = 21$) neonates on the day of birth (day 1); Mann-Whitney U Test: U = 184; Z = -1.351; $p = 0.177$; (**B**) Infected neonates exhibit significantly lower Zn concentrations as compared to that of control neonates; T-Test (two-sided, unpaired), **: $p < 0.01$.

3.2. Cu and CP in Relation to Gestational Age and Infection

Plasma Cu concentrations increased with gestational age in the neonates (Figure 2A). When comparing all samples, plasma Cu concentrations were higher in term neonates as compared to that of preterm neonates (715.8 ± 254.0 vs. 461.4 ± 160.5 µg/L; $p = 0.001$) (Figure 2B). There was a significant positive correlation of gestational age and plasma Cu concentrations in the group of control neonates (Figure 2C), but not in the group of infected neonates (Figure 2D). There were no significant differences of the plasma Cu between female and male neonates in the control (491.6 ± 144.1 µg/L (males) vs. 572.0 ± 247.5 µg/L (females); $p = 0.332$) or in the infection group (494.8 ± 195.1 µg/L (males) vs. 681.1 ± 299.7 µg/L (females); $p = 0.178$), when corrected for gestational age and sex.

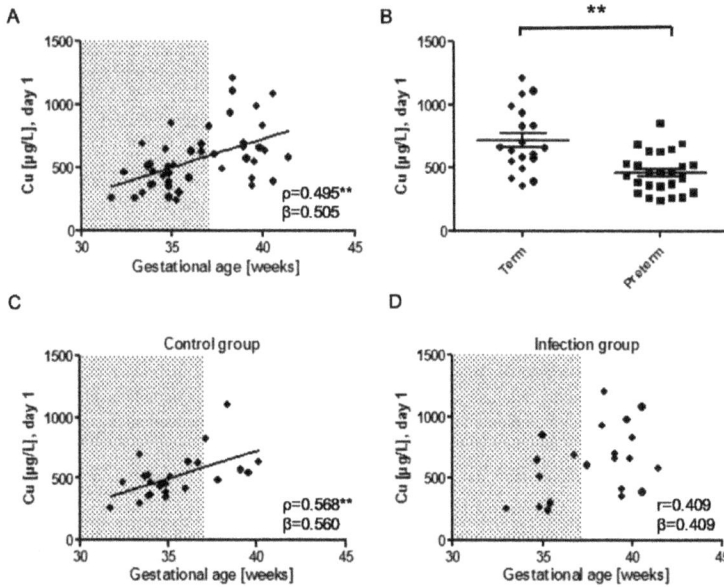

Figure 2. Association of plasma Cu at birth with gestational age. (**A**) In the combined (infection plus control) group, plasma Cu at birth (day 1) correlates positively with gestational age; (**B**) In the combined (infection plus control) group, mean Cu concentration at day 1 is higher in term as compared to that in preterm (<37 gestational weeks) neonates; $n = 44$, T-Test (two-sided, unpaired), $p = 0.001$; (**C**) In the control group, plasma Cu increases with gestational age; (**D**) In infected neonates, there is no significant association of plasma Cu with gestational age. The shaded area (▒) indicates pregnancy length until term, and includes the preterm neonates (<37 weeks of gestation); ρ: Spearman's rank correlation coefficient; r: Pearson correlation coefficient; β: standardised regression coefficient; **: $p < 0.01$.

Ceruloplasmin (CP) in plasma is an established transport protein for Cu. In order to test whether an early-onset infection affects plasma CP and Cu levels alike, both biomarkers were determined from plasma of infected and control neonates. Western blot analysis detected a single band for CP at the expected size of 130 kDa (Figure 3A). Relative CP units were determined by quantification of the detected signals and showed a significant correlation to the Cu concentrations both at birth (day 1) (Figure 3B) and at day 3 (Figure 3C).

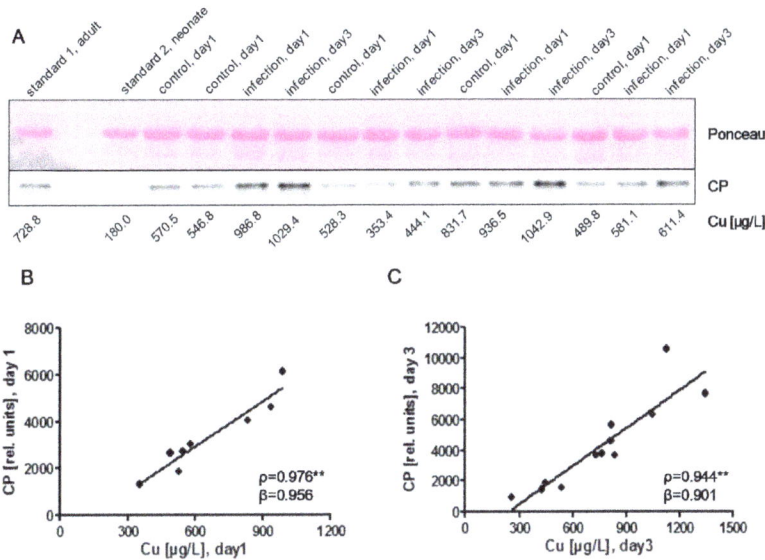

Figure 3. Biomarkers of Cu status in neonates at birth (day 1) and at day 3 after birth. (**A**) Western blot analysis of plasma ceruloplasmin (CP) indicates a single immunoreactive band of 130 kDa. Relatively strong CP signals are observed in plasma samples of infected neonates, especially at day 3. Plasma albumin staining by Ponceau illustrates equal loading of the lanes. The Cu concentrations of the samples are indicated below the Western blot; (**B**) Positive correlation of the CP signal intensities shown in (**A**) to the plasma Cu concentrations at day of birth (day 1), with some notable exceptions. This association was confirmed by a separate analysis of two further blots (2nd blot; $\rho = 0.741$ *, $\beta = 0.903$; 3rd blot; $\rho = 0.964$ **, $\beta = 0.918$), implying that both biomarkers are suitable to reflect the Cu status; (**C**) Positive correlation of CP signal intensities in infected neonates to plasma Cu concentrations, determined from a 4th Western blot analysis of plasma samples collected at day 3. ρ: Spearman's rank correlation coefficient; r: Pearson correlation coefficient; β: standardised regression coefficient; **: $p < 0.01$.

The pro-inflammatory cytokine IL-6 is an early diagnostic marker of congenital infection. In the group of infected newborns, IL-6 was not associated with plasma Cu concentrations at day of birth (day 1) (Figure 4A). Plasma Cu concentrations increased in infected newborns from day 1 to day 3 (Figure 4B). The acute phase protein CRP is determined as late biomarker of neonatal infection. CRP on day 3 was positively associated with plasma Cu concentrations both on day 1 (r = 0.720, $p < 0.001$, $\beta = 0.720$) and on day 3 (Figure 4C). A similar positive correlation was detected between plasma CP and CRP concentrations (Figure 4D). The CP detected in plasma of infected newborns ran as a single protein of the expected size, and it was relatively abundant in samples with elevated CRP levels (Figure 4E).

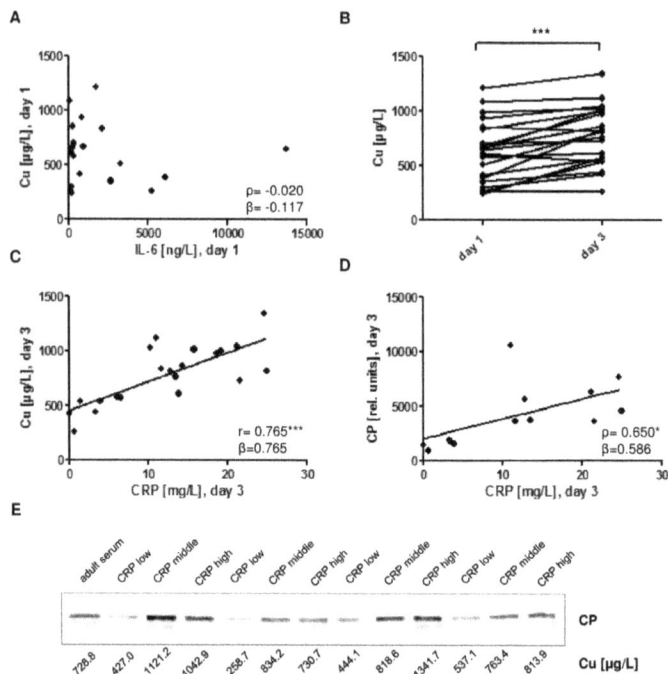

Figure 4. Associations of plasma Cu with established markers of inflammation. (**A**) Plasma Cu concentrations are not associated with the early inflammation marker IL-6 on the day of birth (day 1) in infected neonates; (**B**) Plasma Cu concentrations increase from day 1 to day 3 in infected neonates; $n = 21$, T-Test (two-sided, paired), 627.9 ± 282.5 µg/L vs. 777.1 ± 270.2 µg/L, $p < 0.001$; (**C**) Plasma Cu concentrations correlate positively to the late inflammation marker CRP on day 3 in infected neonates; (**D**) Similarly, CP levels show a positive correlation to CRP in plasma of infected neonates on day 3; (**E**) Newborns were categorized according to severity of infection based on CRP levels (top row; [CRP]: low, < 5 mg/L; moderate, 10–15 mg/L; high, > 20 mg/L). Western blot analysis detected a single immunoreactive CP band and indicated a positive association of CRP levels with CP concentrations in the infected neonates on day 3. Equal protein loading was evaluated by Ponceau staining prior to Western blot analysis. ρ: Spearman's rank correlation coefficient; r: Pearson correlation coefficient; β: standardised regression coefficient; *: $p < 0.05$; ***: $p < 0.001$.

3.3. Associations of Zn with Gestational Age, Birth Weight, and Infection

Plasma Zn concentrations correlated positively to gestational age in control neonates (Figure 5A). In comparison, no significant correlation was observed in the group of infected neonates (Figure 5B). Similarly, plasma Zn concentrations were positively correlated with birth weight in the group of controls (Figure 5C) but not in infected neonates (Figure 5D). There were no significant differences of the plasma Zn concentrations between female and male neonates either in the control group (1739.4 ± 316.2 µg/L (males) vs. 1926.8 ± 469.7 µg/L (females); $p = 0.266$) or in the infection group (1308.9 ± 234.3 µg/L (males) vs. 1527.2 ± 491.4 µg/L (females); $p = 0.316$). The diagnostic markers of inflammation determined in the group of infected neonates, i.e., IL-6 on day 1 and CRP on day 3, were not associated with the plasma Zn concentrations at day 1 (Figure 5E) or at day 3 (Figure 5F).

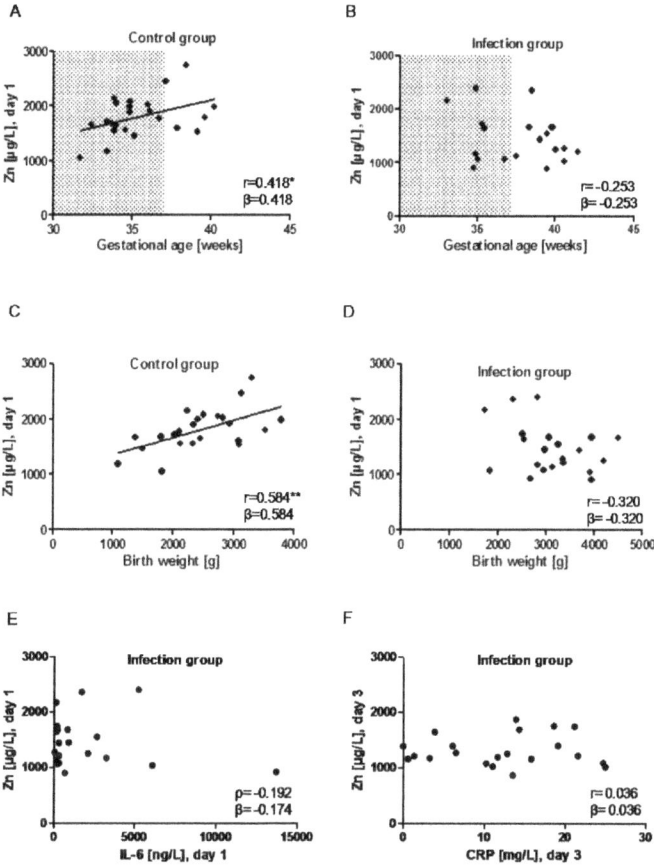

Figure 5. Associations of plasma Zn with gestational age, birth weight, and infection markers. (**A**) There is a positive correlation between plasma Zn and gestational age in the control group; (**B**) In infected neonates, plasma Zn and gestational age are not related; (**C**) Similarly, plasma Zn is positively associated with birth weight in the control group; (**D**) but not in the group of infected neonates; (**E**) There is neither an association of plasma Zn concentrations with the early marker of infection, i.e., IL-6; (**F**) nor with the late marker CRP. The shaded area ( ) indicates the preterm neonates (<37 weeks of gestation); ρ: Spearman's rank correlation coefficient; r: Pearson correlation coefficient; β: standardised regression coefficient; *: $p < 0.05$; **: $p < 0.01$.

3.4. Cu/Zn Ratio

In our quest for a more robust parameter, which may complement the current diagnostic tools for suspected early-onset infections, the Cu/Zn ratios were calculated. This ratio was significantly higher in the infection group as compared to that of the control group of neonates (Figure 6A). In infected neonates, there was no association of the Cu/Zn ratio with the IL-6 levels on day 1 (Figure 6B). However, the CRP levels determined at day 3 in infected neonates showed a positive correlation to both the Cu/Zn ratios at the day of birth (day 1) (Figure 6C) and at day 3 (Figure 6D). Furthermore, the plasma Cu/Zn ratio was correlated with gestational age in the control group ($\rho = 0.417^*$, $p = 0.048$) but not in the infection group (r = 0.332, $p = 0.141$).

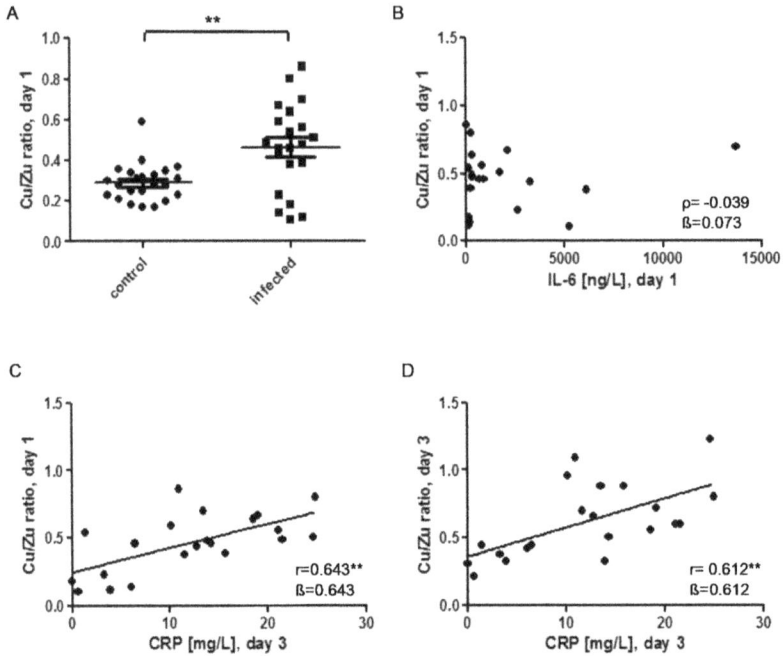

Figure 6. The Cu/Zn ratio as a diagnostic marker for early-onset congenital infections. (**A**) The Cu/Zn ratio is significantly elevated in the infection group (0.28 [0.23–0.34] vs. 0.48 [0.30–0.61]; U = 118; Z = −2.902; p = 0.004); (**B**) There is no correlation between the Cu/Zn ratio and IL-6 on the day of birth (day 1) in infected infants; (**C**) The Cu/Zn ratio on day 1 correlates positively with the CRP-values on day 3 after birth in the infection group; (**D**) Similarly, the Cu/Zn ratio on day 3 correlates positively with the CRP levels on day 3 in the infected neonates. ρ: Spearman's rank correlation coefficient; r: Pearson correlation coefficient; β: standardised regression coefficient; **: $p < 0.01$.

To verify the reliability of the Cu/Zn ratio as a biomarker for early-onset infection considering the different distribution of gestational age between the groups, we conducted a multinomial logistic regression with gestational age as a covariate. The Odds Ratio for the Cu/Zn ratio (day 1) was 9.067 (95% confidence interval 2.306–35.650), which indicates a high Cu/Zn ratio (>50th-percentile) has higher odds for infection than those in the <50th-percentile-group (Table 3).

Table 3. Association of Cu/Zn ratio and infection adjusted for gestational age by multiple logistic regression analysis. CI: confidence interval.

	Significance	Odds Ratio	95% CI
Gestational age	0.079	1.310	0.969–1.770
Cu/Zn ratio	0.034	164.224	1.457–18,506.693

4. Discussion

In this study, we evaluate the infection-related differences in plasma Cu and Zn concentrations in preterm and term neonates. Our results indicate that both trace elements increase in plasma with gestational age in control newborns, and that this correlation is lost in infection. Infected neonates show relatively low plasma Zn concentrations at birth, and develop elevated plasma Cu concentrations during infection. As the plasma concentrations of these two trace elements are regulated in opposite directions by infection, we speculated that the Cu/Zn ratio may provide a more robust marker of early-onset infection than either value alone. Our study indicates that the Cu/Zn ratio correlates to the CRP levels determined at day 3 as the established biomarker of infection. Interestingly, this correlation is already shown at day 1. It remains to be tested if it could be of value for the diagnosis of early-onset infections. The Cu/Zn ratio at day 1 may already reflect the severity and predict the potential course of the infection, which then becomes detected later by the elevated CRP levels determined at day 3. It may, thus, constitute a helpful early prognostic biomarker of early-onset infection in term and preterm neonates.

There is evidence that preterm neonates are at especially high risk for Cu and Zn deficiency [4,30]. This notion is supported by previous studies reporting that the maternal Cu levels rise with the length of gestation [31]. Consistent with former findings in neonates [32], the Cu concentrations of the infants in our study were lower, and Zn was higher than in adults (adult reference intervals according to [11]; Cu; 10–22 μM, i.e., 635.5–1398.0 μg/L, and Zn; 12–18 μM, i.e., 784.6–1176.8 μg/L). Our data are in line with previous studies in young children, especially with respect to an increase in plasma Cu levels upon infection [33]. Notably, also in preterm and term infants, a tight correlation of plasma Cu concentrations and plasma CP is reported [34]. These data along with our findings support the concept that plasma CP may serve as a surrogate marker of plasma Cu concentrations in children. This relation offers the option for bed-side testing of the Cu status by immunological assay procedures, as recently demonstrated by using quantum dots for fast CP quantification [35]. Unfortunately, no reliable protein biomarker of plasma Zn status is yet at hand, which would enable a fast multiplex bed-side quantification of the Cu/Zn ratio via these surrogate protein biomarkers that can be detected by point-of-care technologies.

Importantly, the inverse regulation of plasma Cu and Zn is a well-established characteristic of infections, and the Cu/Zn ratio has been proven of diagnostic value in a number of human disorders, including pediatric infectious diseases, such as giardiasis or amebiasis [20] and tuberculosis [23,36]. Furthermore, the diagnostic value of the Cu/Zn ratio as a disease marker was also shown in autism, attention-deficit hyperactivity disorder, hypertension, inflammatory, as well as neoplastic diseases [33,37–39]. The quotient was also described as a potential biomarker of inflammation and nutritional status as well as a mortality predictor in elderly people [40], and as a variable correlating with inflammation, disrupted immune system, and an increased oxidative stress in peritoneal dialysis [41]. Our data indicate that the Cu/Zn ratio may also be of diagnostic value in neonates with suspected infection, as it was associated with severity of inflammation at an early time point while being independent of gestational age.

4.1. Early-Onset Congenital Infections as Disruptors of the Trace Element Homeostasis

We found significantly lower plasma Zn concentrations in infected neonates compared with that of the control group, which is congruent with former findings in children, adults, and animals suffering from an acute inflammation and/or critical illness [3,7,10,11,42]. However, Zn levels were not associated with the early inflammation marker IL-6 or with the late acute phase reactant CRP. This lack of stringent interrelation is in line with studies in adults, where there are only marginal correlations of serum Zn with markers of inflammation [43]. This may indicate that low plasma Zn may not reflect the severity of the inflammation, i.e., it may not be directly regulated by the cytokines released in response to infection. However, in other studies, a respective correlation of plasma Zn with inflammation markers in critically ill children and adults has been reported [11,44]. The lower Zn levels in our infected neonates may, thus, not necessarily be the consequence of the inflammation, but potentially a risk factor for infection [8]. However, this hypothesis needs to be tested in other clinical trials, as our analysis is an observational study and not designed to identify causal relationships.

The Cu levels increased with gestational age. In general, the fetal hepatic tissue does not efficiently support an incorporation of Cu into the CP apoenzyme [32]. Thus, the increasing Cu levels with age may reflect the functional maturation of the liver. Due to ethical reasons, only residual plasma samples were available from infected neonates at day 3 but not from control neonates. Such an analysis would shed light on the relative importance of age and infection for the rising Cu concentrations observed in the study. Moreover, the diagnosis of infection in the newborns was based on clinical symptoms in combination with laboratory evidence for an inflammation, and not on positive blood cultures or additional laboratory analysis, which constitutes a general shortcoming of our study.

4.2. The Cu/Zn Ratio as a Potential Biomarker of Early-Onset Congenital Infections

The concentrations of plasma Cu or Zn considered separately did not qualify as useful biomarkers of early-onset infection. The correlation of Cu on day 1 with CRP on day 3 seems to impart a predictive value to the trace element, and the association on day 3 implicates its potential as a clinical marker of disease course. However, Cu levels were not significantly different between control and infected subjects at birth. The strong positive association of CP and CRP on day 3 suggests that CP is related to the severity of infection. CP is described as a positive acute phase reactant in adults [4,9]. However, it decreases with time during increasing severity of the inflammation in adults [11,45], suggesting that it does not steadily correlate to infection severity. Zn levels were significantly lower in infected neonates compared with that of the control group at birth. However, Zn neither correlated with IL-6 on day 1, nor with CRP on day 3. Considering these interactions, Zn alone does not seem to quality as an appropriate infection marker in neonates.

Nevertheless, the Cu/Zn ratio appears to provide additional information on the possible infection of newborns. This notion needs to be tested in prospective trials with a sufficient number of neonates, as both the infection and control groups were relatively small in our pilot study.

5. Conclusions

Infection as well as inflammation can affect the trace element homeostasis in newborns. Infected neonates may develop increased plasma Cu concentrations and display a relative plasma Zn deficit. The Cu/Zn ratio may thus constitute a useful biomarker of early-onset infection in neonates with some relation to the clinical course.

Acknowledgments: We express our gratitude to all parents and children who took part in this study and to our clinical colleagues in the Department of Neonatology, Charité-Universitätsmedizin Berlin, Germany. This work was supported by DFG Grants RE3038/1–1 (to Kostja Renko) and Scho849/4–1 (to Lutz Schomburg) and an Elsa-Neumann stipend from the City of Berlin (to Janine Martitz).

Author Contributions: Monika Wisniewska, Malte Cremer, Lennart Wiehe, Niels-Peter Becker, Christoph Bührer, and Lutz Schomburg formulated the research question and designed the study. Monika Wisniewska, Lennart Wiehe, Niels-Peter Becker, Eddy Rijntjes, Janine Martitz, and Kostja Renko conducted the experimental analyses. All authors contributed to the statistical analysis and interpretation of the data. All authors contributed to writing and editing the manuscript.

Conflicts of Interest: The authors declare no conflict of interest.

References

1. Stern, B.R. Essentiality and toxicity in copper health risk assessment: Overview, update and regulatory considerations. *J. Toxicol. Environ. Health Part A* **2010**, *73*, 114–127. [CrossRef] [PubMed]
2. Chasapis, C.T.; Loutsidou, A.C.; Spiliopoulou, C.A.; Stefanidou, M.E. Zinc and human health: An update. *Arch. Toxicol.* **2012**, *86*, 521–534. [CrossRef] [PubMed]
3. Bonaventura, P.; Benedetti, G.; Albarede, F.; Miossec, P. Zinc and its role in immunity and inflammation. *Autoimmun. Rev.* **2015**, *14*, 277–285. [CrossRef] [PubMed]
4. De Romana, D.L.; Olivares, M.; Uauy, R.; Araya, M. Risks and benefits of copper in light of new insights of copper homeostasis. *J. Trace Elem. Med. Biol.* **2011**, *25*, 3–13. [CrossRef] [PubMed]
5. Uriu-Adams, J.Y.; Keen, C.L. Copper, oxidative stress, and human health. *Mol. Asp. Med.* **2005**, *26*, 268–298. [CrossRef] [PubMed]

6. Maggini, S.; Wintergerst, E.S.; Beveridge, S.; Hornig, D.H. Selected vitamins and trace elements support immune function by strengthening epithelial barriers and cellular and humoral immune responses. *Br. J. Nutr.* **2007**, *98* (Suppl. S1), 29–35. [CrossRef] [PubMed]
7. Rink, L.; Gabriel, P. Zinc and the immune system. *Proc. Nutr. Soc.* **2000**, *59*, 541–552. [CrossRef] [PubMed]
8. Krebs, N.F.; Miller, L.V.; Hambidge, K.M. Zinc deficiency in infants and children: A review of its complex and synergistic interactions. *Paediatr. Int. Child Health* **2014**, *34*, 279–288. [CrossRef] [PubMed]
9. Milanino, R.; Marrella, M.; Gasperini, R.; Pasqualicchio, M.; Velo, G. Copper and zinc body levels in inflammation: An overview of the data obtained from animal and human studies. *Agents Actions* **1993**, *39*, 195–209. [CrossRef] [PubMed]
10. Besecker, B.Y.; Exline, M.C.; Hollyfield, J.; Phillips, G.; Disilvestro, R.A.; Wewers, M.D.; Knoell, D.L. A comparison of zinc metabolism, inflammation, and disease severity in critically ill infected and noninfected adults early after intensive care unit admission. *Am. J. Clin. Nutr.* **2011**, *93*, 1356–1364. [CrossRef] [PubMed]
11. Stefanowicz, F.; Gashut, R.A.; Talwar, D.; Duncan, A.; Beulshausen, J.F.; McMillan, D.C.; Kinsella, J. Assessment of plasma and red cell trace element concentrations, disease severity, and outcome in patients with critical illness. *J. Crit. Care* **2014**, *29*, 214–218. [CrossRef] [PubMed]
12. Rech, M.; To, L.; Tovbin, A.; Smoot, T.; Mlynarek, M. Heavy metal in the intensive care unit: A review of current literature on trace element supplementation in critically ill patients. *Nutr. Clin. Pract.* **2014**, *29*, 78–89. [CrossRef] [PubMed]
13. Shah, B.A.; Padbury, J.F. Neonatal sepsis: An old problem with new insights. *Virulence* **2014**, *5*, 170–178. [CrossRef] [PubMed]
14. Molyneux, E.; Gest, A. Neonatal sepsis: An old issue needing new answers. *Lancet Infect. Dis.* **2015**, *15*, 503–505. [CrossRef]
15. Hedegaard, S.S.; Wisborg, K.; Hvas, A.M. Diagnostic utility of biomarkers for neonatal sepsis-a systematic review. *Infect. Dis.* **2015**, *47*, 117–124. [CrossRef] [PubMed]
16. Raimondi, F.; Ferrara, T.; Maffucci, R.; Milite, P.; Del Buono, D.; Santoro, P.; Grimaldi, L.C. Neonatal sepsis: A difficult diagnostic challenge. *Clin. Biochem.* **2011**, *44*, 463–464. [CrossRef] [PubMed]
17. Bajcetic, M.; Spasic, S.; Spasojevic, I. Redox therapy in neonatal sepsis: Reasons, targets, strategy, and agents. *Shock* **2014**, *42*, 179–184. [CrossRef] [PubMed]
18. Alshaikh, B.; Yusuf, K.; Sauve, R. Neurodevelopmental outcomes of very low birth weight infants with neonatal sepsis: Systematic review and meta-analysis. *J. Perinatol.* **2013**, *33*, 558–564. [CrossRef] [PubMed]
19. Karahan, S.C.; Deger, O.; Orem, A.; Ucar, F.; Erem, C.; Alver, A.; Onder, E. The effects of impaired trace element status on polymorphonuclear leukocyte activation in the development of vascular complications in type 2 diabetes mellitus. *Clin. Chem. Lab. Med.* **2001**, *39*, 109–115. [CrossRef] [PubMed]
20. Karakas, Z.; Demirel, N.; Tarakcioglu, M.; Mete, N. Serum zinc and copper levels in southeastern turkish children with giardiasis or amebiasis. *Biol. Trace Elem. Res.* **2001**, *84*, 11–18. [CrossRef]
21. Oyama, T.; Matsuno, K.; Kawamoto, T.; Mitsudomi, T.; Shirakusa, T.; Kodama, Y. Efficiency of serum copper/zinc ratio for differential diagnosis of patients with and without lung cancer. *Biol. Trace Elem. Res.* **1994**, *42*, 115–127. [CrossRef] [PubMed]
22. Donma, M.M.; Donma, O.; Tas, M.A. Hair zinc and copper concentrations and zinc: Copper ratios in pediatric malignancies and healthy children from southeastern turkey. *Biol. Trace Elem. Res.* **1993**, *36*, 51–63. [CrossRef] [PubMed]
23. Luterotti, S.; Kordic, T.V.; Letoja, I.Z.; Dodig, S. Contribution to diagnostics/prognostics of tuberculosis in children. Ii. Indicative value of metal ions and biochemical parameters in serum. *Acta Pharm.* **2015**, *65*, 321–329. [CrossRef] [PubMed]
24. Wiehe, L.; Cremer, M.; Wisniewska, M.; Becker, N.P.; Rijntjes, E.; Martitz, J.; Hybsier, S.; Renko, K.; Buhrer, C.; Schomburg, L. Selenium status in neonates with connatal infection. *Br. J. Nutr.* **2016**, *116*, 504–513. [CrossRef] [PubMed]
25. Young Infants Clinical Signs Study Group. Clinical signs that predict severe illness in children under age 2 months: A multicentre study. *Lancet* **2008**, *371*, 135–142.
26. Goldstein, B.; Giroir, B.; Randolph, A.; International Consensus Conference on Pediatric Sepsis. International pediatric sepsis consensus conference: Definitions for sepsis and organ dysfunction in pediatrics. *Pediatr. Crit. Care Med.* **2005**, *6*, 2–8. [CrossRef] [PubMed]

27. Dollner, H.; Vatten, L.; Austgulen, R. Early diagnostic markers for neonatal sepsis: Comparing c-reactive protein, interleukin-6, soluble tumour necrosis factor receptors and soluble adhesion molecules. *J. Clin. Epidemiol.* **2001**, *54*, 1251–1257. [CrossRef]

28. Apgar, V. A proposal for a new method of evaluation of the newborn infant. *Curr. Res. Anesth. Analg.* **1953**, *32*, 260–267. [CrossRef] [PubMed]

29. Stosnach, H. Environmental trace-element analysis using a benchtop total reflection X-ray fluorescence spectrometer. *Anal. Sci.* **2005**, *21*, 873–876. [CrossRef] [PubMed]

30. Lombeck, I.; Fuchs, A. Zinc and copper in infants fed breast-milk or different formula. *Eur. J. Pediatr.* **1994**, *153*, 770–776. [CrossRef] [PubMed]

31. Zhang, Z.; Yuan, E.; Liu, J.; Lou, X.; Jia, L.; Li, X.; Zhang, L. Gestational age-specific reference intervals for blood copper, zinc, calcium, magnesium, iron, lead, and cadmium during normal pregnancy. *Clin. Biochem.* **2013**, *46*, 777–780. [CrossRef] [PubMed]

32. Salmenpera, L.; Perheentupa, J.; Pakarinen, P.; Siimes, M.A. Cu nutrition in infants during prolonged exclusive breast-feeding: Low intake but rising serum concentrations of cu and ceruloplasmin. *Am. J. Clin. Nutr.* **1986**, *43*, 251–257. [PubMed]

33. Shenkin, A. Trace elements and inflammatory response: Implications for nutritional support. *Nutrition* **1995**, *11*, 100–105. [PubMed]

34. Koo, W.W.K.; Succop, P.; Hambidge, K.M. Sequential concentrations of copper and ceruloplasmin in serum from preterm infants with rickets and fractures. *Clin. Chem.* **1991**, *37*, 556–559. [PubMed]

35. Li, Z.H.; Wang, Y.; Wang, J.; Tang, Z.W.; Pounds, J.G.; Lin, Y.H. Rapid and sensitive detection of protein biomarker using a portable fluorescence biosensor based on quantum dots and a lateral flow test strip. *Anal. Chem.* **2010**, *82*, 7008–7014. [CrossRef] [PubMed]

36. Mohan, G.; Kulshreshtha, S.; Dayal, R.; Singh, M.; Sharma, P. Effect of therapy on serum zinc and copper in primary complex of children. *Biol. Trace Elem. Res.* **2007**, *118*, 184–190. [CrossRef] [PubMed]

37. Canatan, H.; Bakan, I.; Akbulut, M.; Halifeoglu, I.; Cikim, G.; Baydas, G.; Kilic, N. Relationship among levels of leptin and zinc, copper, and zinc/copper ratio in plasma of patients with essential hypertension and healthy normotensive subjects. *Biol. Trace Elem. Res.* **2004**, *100*, 117–123. [CrossRef]

38. Faber, S.; Zinn, G.M.; Kern, J.C., 2nd; Kingston, H.M. The plasma zinc/serum copper ratio as a biomarker in children with autism spectrum disorders. *Biomarkers* **2009**, *14*, 171–180. [CrossRef] [PubMed]

39. Buntzel, J.; Glatzel, M.; Micke, O.; Mucke, R.; Schonekaes, K.; Bruns, F.; Frohlich, D. The copper-zinc-ratio as marker of tumor activity in head and neck cancer? *Strahlenther. Onkol.* **2005**, *181*, 122.

40. Malavolta, M.; Giacconi, R.; Piacenza, F.; Santarelli, L.; Cipriano, C.; Costarelli, L.; Tesei, S.; Pierpaoli, S.; Basso, A.; Galeazzi, R.; et al. Plasma copper/zinc ratio: An inflammatory/nutritional biomarker as predictor of all-cause mortality in elderly population. *Biogerontology* **2010**, *11*, 309–319. [CrossRef] [PubMed]

41. Guo, C.H.; Chen, P.C.; Yeh, M.S.; Hsiung, D.Y.; Wang, C.L. Cu/Zn ratios are associated with nutritional status, oxidative stress, inflammation, and immune abnormalities in patients on peritoneal dialysis. *Clin. Biochem.* **2011**, *44*, 275–280. [CrossRef] [PubMed]

42. Cvijanovich, N.Z.; King, J.C.; Flori, H.R.; Gildengorin, G.; Vinks, A.A.; Wong, H.R. A safety and dose escalation study of intravenous zinc supplementation in pediatric critical illness. *JPEN* **2016**, *40*, 860–868. [CrossRef] [PubMed]

43. Jung, S.; Kim, M.K.; Choi, B.Y. The relationship between zinc status and inflammatory marker levels in rural Korean adults aged 40 and older. *PLoS ONE* **2015**, *10*, e0130016. [CrossRef] [PubMed]

44. Cvijanovich, N.Z.; King, J.C.; Flori, H.R.; Gildengorin, G.; Wong, H.R. Zinc homeostasis in pediatric critical illness. *Pediatr. Crit. Care Med.* **2009**, *10*, 29–34. [CrossRef] [PubMed]

45. Duncan, A.; Talwar, D.; McMillan, D.C.; Stefanowicz, F.; O'Reilly, D.S. Quantitative data on the magnitude of the systemic inflammatory response and its effect on micronutrient status based on plasma measurements. *Am. J. Clin. Nutr.* **2012**, *95*, 64–71. [CrossRef] [PubMed]

nutrients

MDPI

Article

Impact of Maternal Selenium Status on Infant Outcome during the First 6 Months of Life

Kristin Varsi [1,*], Bjørn Bolann [1,2], Ingrid Torsvik [3], Tina Constanse Rosvold Eik [1], Paul Johan Høl [4] and Anne-Lise Bjørke-Monsen [1]

1   Laboratory of Clinical Biochemistry, Haukeland University Hospital, N-5021 Bergen, Norway; bjorn.bolann@uib.no (B.B.); tina.constanse.rosvold.eik@helse-bergen.no (T.C.R.E.); almo@helse-bergen.no (A.-L.B.-M.)
2   Department of Clinical Science, Faculty of Medicine and Dentistry, University of Bergen, N-5020 Bergen, Norway
3   Department of Pediatrics, Haukeland University Hospital, N-5021 Bergen, Norway; ingrid.kristin.torsvik@helse-bergen.no
4   Department of Clinical Medicine, Faculty of Medicine and Dentistry, University of Bergen, N-5020 Bergen, Norway; paul.hol@k1.uib.no
*   Correspondence: Kristin.varsi@helse-bergen.no; Tel.: +47-92-453-617

Received: 2 March 2017; Accepted: 5 May 2017; Published: 11 May 2017

Abstract: Pregnant women and infants are at risk for selenium deficiency, which is known to have negative effects on immune and brain function. We have investigated selenium levels in 158 healthy never-pregnant women and in 114 pregnant and lactating women and their infants at age 6 months and related this to clinical outcomes during the first 6 months of life. Neurodevelopment was assessed with the parental questionnaire Ages and Stages (ASQ) at 6 months. A maternal selenium level ≤ 0.90 µmol/L in pregnancy week 18 was negatively related to infant neurodevelopment at 6 months (B = −20, $p = 0.01$), whereas a selenium level ≤ 0.78 µmol/L in pregnancy week 36 was associated with an increased risk (odds ratio 4.8) of having an infant infection during the first 6 weeks of life. A low maternal selenium status in pregnancy was found to be associated with an increased risk of infant infection during the first 6 weeks of life and a lower psychomotor score at 6 months. We suggest a cutoff for maternal serum selenium deficiency of 0.90 µmol/L in pregnancy week 18 and 0.78 µmol/L in pregnancy week 36. This should be reevaluated in an intervention study.

Keywords: selenium; deficiency; infant; pregnancy; lactation; infection; neurodevelopment

1. Introduction

Selenium is a trace element essential for normal human metabolism, and even less overt selenium deficiency is reported to have negative health effects in humans [1]. Overt deficiency is typically associated with loss of immunocompetence, affecting both cell-mediated and humoral immune function [1,2], and selenium supplementation has been shown to improve immune function even in selenium replete individuals [3].

A low maternal selenium status during pregnancy has been associated with fetal malformations, like neural tube defects [4], and is disadvantageous for cognitive development in infants and toddlers [5,6]. However, due to extensive physiological changes during pregnancy [7], the definition of maternal selenium deficiency is difficult in this period. Serum selenium levels decrease during pregnancy with the lowest levels observed before delivery [6,8,9], and increase postpartum to levels observed in non-pregnant women [8]. Selenium levels in breastmilk depend on maternal selenium status, and are reported to decrease during the lactational period [10]. It has been estimated that recommended selenium intake is not achieved in approximately 30% of breastfed infants [10], and infants are considered to be particularly at risk for selenium deficiency [11]. Serum selenium decreases from birth to reach nadir levels between 2 to 4 months, increases thereafter and is higher in older children [11–13].

Seafood and meat are rich sources of selenium, whereas the selenium content in plant-based foods depends on where it is grown. As the selenium content in soil and food habits vary substantially in different world regions, this will affect the selenium status in different populations [1,14]. Northern Europe represents a low selenium area [1], and low selenium levels have been reported in inhabitants from the Nordic countries [14,15].

We have investigated selenium levels in Norwegian never-pregnant and pregnant women during pregnancy and postpartum, in breastmilk, and in their infants at age 6 months. Statistically based reference ranges may not be optimal for securing an adequate micronutrient level in pregnant women and their infants. Our purpose was to establish cut-off levels for maternal selenium deficiency during pregnancy based on clinical outcomes in the infants during the first 6 months of life.

2. Materials and Methods

2.1. Study Population and Design

Between June 2012 and March 2015, 140 healthy pregnant women with a singleton pregnancy were recruited at routine ultrasound examinations in pregnancy week (PW) 18 at the Obstetrical Department at Haukeland University Hospital, Bergen, Norway. The women were invited back at pregnancy weeks 28 and 36, 6 weeks and 4 and 6 months postpartum, and the final visit also included the infant. Women with pregnancy related or chronic disease were excluded, except those with well-regulated hypothyroidism ($n = 7$). Of the 140 pregnant women initially recruited, 114 met the inclusion criteria, and were included in the study.

During the same period, 158 healthy, never-pregnant women aged 18 to 40 years were recruited among students and employees at the University of Bergen and Haukeland University Hospital, Bergen, Norway.

Ethical approval of the protocol was granted by the Regional Committee on Medical Research Ethics, REK 2011/2447, and written informed consent was obtained from all women before enrollment.

2.2. Clinical Data

The subjects completed a questionnaire concerning body weight, nutrition, use of multiple micronutrient supplements (MMN), and health status at each visit. The postpartum visits included additional information about infant nutrition, growth parameters, and clinical symptoms concerning feeding difficulties, colic, constipation, dermatitis, and infections. Infections were defined as clinical symptoms compatible with infection with or without fever. Fever-episodes occurring after scheduled vaccinations were excluded. Use of MMN more than three days per week in two or more periods during pregnancy or postpartum period was defined as a regular supplement user.

Infant neurodevelopment was assessed by Ages and Stages Questionnaire: A Parent-Completed, Child-Monitoring System (ASQ), a screening tool that includes five developmental domains: communication, gross motor function, fine motor function, social functioning, and problem solving [16]. Each domain had six questions, and the parents assessed whether a milestone was achieved (yes, 10 points), partially achieved (sometimes, 5 points), or not achieved (no, 0 points). Partial scores of each domain and a total score (maximum 300 points) were calculated for each infant.

Nutrients **2017**, *9*, 486

2.3. Selenium Analysis of Blood and Breastmilk

Non-fasting blood samples from women and infants were obtained by antecubital venipuncture and collected into vacutainer tubes without additives and approved for trace metal analysis (Terumo, Tokyo, Japan). A complete set of blood samples was available for all, except nine women and 16 infants. Selenium in serum was measured by Inductively Coupled Plasma Mass Spectrometry (ICP-MS) on Elan® Dynamic Reaction Cell-e (Perkin Elmer MDS Sciex, Concord, ON, Canada) in standard mode [17]. The serum samples were diluted 1:15 with dilution solution containing 1% Triton® X-100 (Merck, Damstadt, Germany) 0.33% w/v pro analysis HNO_3 (Merck, Germany), and 5 ppm Gold (Perkin Elmer, Shelton, CT, USA). Seronorm Trace Elements serum Level 1 and 2 (Sero, Billingstad, Norway) were included in every analysis series. The analytical between run precision within the normal range was <4%. Twenty-two measurements of a low serum pool with a concentration of 0.39 µmol/L gave a coefficient of variation (CV) of 5.4 %, and this concentration was set as the limit of quantification (LOQ). No samples were below the LOQ. The average recovery for the serum samples was 99%.

At each postpartum visit, the mothers ($n = 61$) brought a sample of breastmilk taken the same day and stored in tubes without additives, and approved for trace metal analysis. Breastmilk samples were thawed and 1 mL was mixed with 2 mL ultrapure HNO_3, 65%, and 1 mL ultrapure H_2O_2, 30%, before microwave digestion (MLS 1200 Mega, Milestone, Sorisole, Italy). Thereafter, the samples were mixed with 0.25 mL suprapure HCl, 30%, before analysis by high resolution sector field ICP-MS on Element 2 (Thermo Finnigan, Bremen, Germany) [18]. For quality control, a Seronorm Trace Elements serum Level 2 (Sero, Billingstad, Norway) was used. One mL of serum followed the sample preparation as the breastmilk samples. In addition, a reference material of skimmed milk powder (ERM-BD-150, Geel, Belgium) was used. The analytical precision between runs was 4% for the serum control and 12% for the milk powder. For the breastmilk method, the standard deviation (SD) of the blank was 0.017 µmol/L. This gives a theoretical limit of detection (LOD) (defined as three times the SD of the blank) of 0.052 µmol/L. The average recovery for the serum and milk powder was 109% and 106%, respectively, indicating that the selenium concentration may have been slightly overestimated.

2.4. Statistical Analysis

Normally distributed data are presented as means and SD, and compared by Student's *t*-test, whereas non-normally distributed data are presented as medians and interquartile ranges (IQR) defined by 25 and 75 percentiles, or 2.5 and 97.5 percentiles, and compared by Mann-Whitney U test or Kruskal Wallis test. Categorical data are presented as percentages compared by Chi-square test. Spearman correlations were used to explore relationships between data. Selenium tertiles were used in linear or logistic regression models due to a small sample size and more interest in extreme quantiles than in the mean values. Multiple linear regression models were used to assess the relationship between ASQ scores at 6 months and maternal selenium status during pregnancy. The unstandardized coefficient (B) represents an estimate of the standardized β coefficient, and the models also included variables chosen due to their reported impact for neurodevelopment [19]. Logistic regressions models, including factors considered to be related to the risk of infection during the first weeks of life, were used to assess the risk of having an infant infection during the first 6 weeks of life in relation to maternal selenium levels during pregnancy and postpartum. The SPSS statistical package (version 23) was used, and two-sided *p*-values < 0.05 were considered statistically significant.

3. Results

3.1. Demographics and Nutrition

All of the mothers had an omnivore diet, and the majority used multiple micronutrient supplements (MMN) containing a low dose selenium (55–60 µg/tablet) at more than 2 periods during pregnancy (86/114, 75%) and postpartum (82/114, 72%).

Compared to the mothers, the never-pregnant controls (n = 158) were younger, 23% (39/158) were vegetarians, and a lower percentage were regular MMN users (Table 1).

All infants (n = 114) were healthy, 53% (60/114) were males, and all infants except one (born at gestational age 36 weeks) were born at term, with an appropriate for gestational age weight (mean 3573 + 418 g). All, except one infant, were breastfed, with a mean duration of exclusive breastfeeding of 3.8 + 1.5 months. At four months, 50% (70/114) of the infants had been introduced to solid food. Use of vitamin D supplements was reported in 60% of the infants at 6 weeks, increasing up to 90% at 6 months, and no infant was given MMNs or iron supplements at any time point.

3.2. Selenium Levels in Pregnant, Lactating, and Never-Pregnant Women

Serum selenium decreased significantly during pregnancy ($p < 0.001$), and the median level in pregnancy week 36 was reduced by 20% compared to never-pregnant women ($p < 0.001$) (Figure 1). The levels increased postpartum and remained unchanged from 6 weeks to 6 months where they were comparable to never-pregnant levels (Table 2, Figure 1). Spearman rank correlation coefficients (r) between maternal serum selenium levels during pregnancy and postpartum were all >0.5, $p < 0.001$, and maternal selenium levels were higher in regular users of MMN (Table 3).

No significant correlations were seen between serum selenium and body mass index (BMI), age, or the use of MMN in the never-pregnant controls.

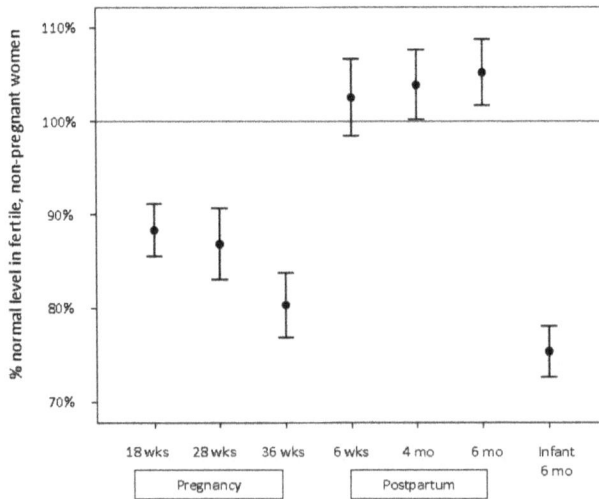

Figure 1. Selenium levels in pregnant and lactating women and their infants, expressed as the percentage of median selenium level in never-pregnant women (mean + 2 SD).

Table 1. Baseline characteristics of never-pregnant and pregnant women in pregnancy week 18. BMI = body mass index.

	Never-Pregnant Women n = 158	Pregnant Women n = 114	p Value
Age, year, mean + SD	25.3 ± 4.8	31.5 ± 4.3	<0.001 *
Prepregnancy BMI, kg/m², mean + SD	22.5 ± 3.0	22.8 ± 3.1	0.38 *
Higher education, n (%) **	93 (60)	67 (59)	0.96 ***
Para 0, n (%)	158 (100)	63 (55)	<0.001 ***
Smoking, n (%)	4 (3)	2 (2)	0.65 ***
Regular use of micronutrient supplements (≥3 days/week), n (%)	35 (22)	47 (41)	0.001 ***

* Comparison by Student's *t*-test; ** Higher education defined as ≥5 years at university or college. *** Comparison by Pearson Chi-square test.

Table 2. Serum selenium levels in never-pregnant women, women during pregnancy and postpartum, and infants at 6 months.

Serum Selenium, μmol/L	Never-Pregnant Women n = 158	Pregnant Women			Postpartum Women			Infants 6 Months n = 91
		Week 18 n = 108	Week 28 n = 114	Week 36 n = 114	6 Weeks n = 113	4 Months n = 113	6 Months n = 114	
Median (25, 75) *	1.07 (0.98, 1.16)	0.96 (0.86, 1.04)	0.92 (0.81, 1.03)	0.85 (0.75, 0.96)	1.07 (0.96, 1.20)	1.08 (0.98, 1.24)	1.11 (1.00, 1.25)	0.81 (0.74, 0.89)
(2.5, 97.5) *	(0.81, 1.59)	(0.71, 1.35)	(0.67, 1.37)	(0.63, 1.33)	(0.74, 1.58)	(0.86, 1.59)	(0.88, 1.51)	(0.56, 1.16)

* Percentiles.

Table 3. Maternal and infant serum selenium according to maternal use of micronutrients in pregnancy and postpartum.

Use of Micronutrient Supplements during Pregnancy	Maternal Serum Selenium, μmol/L, Median (25, 75) * (n = 114)			Infant Serum Selenium, μmol/L, At Age 6 Months, Median (25, 75) * (n = 91)	
	Pregnancy 18 Weeks	Pregnancy Week 28	Pregnancy Week 36		
Non-user, n = 28	0.92 (0.83, 1.00)	0.83 (0.77, 0.98)	0.77 (0.73, 0.86)	n = 25	0.76 (0.71, 0.87)
Regular user, n = 86	0.98 (0.89, 1.05)	0.98 (0.89, 1.04)	0.87 (0.77, 0.98)	n = 66	0.83 (0.76, 0.90)
p value **	0.06	0.02	0.008		0.05
Use of Micronutrient Supplements Postpartum	**6 Weeks Postpartum**	**4 Months Postpartum**	**6 Months Postpartum**	**Infants 6 Months (n = 91)**	
Non-user, n = 32	1.01 (0.95, 1.14)	1.04 (0.93, 1.14)	1.08 (0.92, 1.15)	n = 27	0.76 (0.69, 0.84)
Regular user, n = 82	1.10 (0.97, 1.27)	1.12 (0.99, 1.28)	1.12 (1.02, 1.29)	n = 64	0.83 (0.75, 0.90)
p value **	0.06	0.04	0.01		0.03

* Percentiles, ** Comparison with Mann-Whitney U test.

3.3. Selenium Levels in Breastmilk

Breastmilk selenium levels at 6 weeks and 4 and 6 months were significantly correlated to maternal selenium levels both during pregnancy and postpartum, with the highest correlations seen for maternal postpartum levels (r: 0.28–0.55, $p < 0.03$). No significant differences in breastmilk selenium levels according to maternal use of MMN were observed ($p > 0.22$). The median selenium level decreased by 23% between 6 weeks and 4 months ($p < 0.001$), with no change from 4 to 6 months ($p = 0.90$) (Table 4).

3.4. Selenium Levels in Infants at 6 Months

Infant selenium levels at age 6 months were lower than in never-pregnant, pregnant, and lactating mothers (Table 2), correlated to maternal levels during pregnancy and postpartum ($r = 0.28$–0.45, $p < 0.01$), and were higher in infants of mothers who were regular users of MMN during pregnancy ($p = 0.05$) and postpartum ($p = 0.03$) (Table 3).

Calculated median selenium intake from breastmilk decreased from 6 weeks to 6 months ($p = 0.02$) (Table 4), but the median selenium level in the infants at 6 months did not significantly differ according to months of exclusive breastfeeding, neither to gender or growth parameters during the first 6 months of life.

3.5. Selenium Levels and Infant Neurodevelopment

ASQ data were available for 98% (112/114) of the infants at 6 months. Median and IQR were 228 (210, 255) for total ASQ score, 50 (40, 55) for communication score, 50 (39, 55) for personal-social functioning score, 55 (46, 60) for problem solving score, 35 (30, 45) for gross motor score, and 45 (35, 55) for fine motor score.

In a multiple linear regression model, ASQ total (B = 11, $p = 0.007$), problem solving (B = 3, $p = 0.006$), and fine motor scores (B = 3, $p = 0.04$) increased significantly with maternal selenium tertiles in pregnancy week 18, whereas communication (B = 1, $p = 0.22$), personal-social functioning (B = 2, $p = 0.11$), and gross motor score (B = 1, $p = 0.56$) did not. The multiple linear regression model also included birthweight, weight at 6 months, gender, months of exclusive breastfeeding, maternal age, education, and parity.

Based on examination of scatterplots visualizing the relationship between maternal serum selenium in pregnancy week 18 and infant ASQ scores at 6 months, the lower tertile of maternal serum selenium (<0.90 µmol/L) was chosen as a cutoff level for defining maternal selenium deficiency in pregnancy week 18. Infants born to mothers with serum selenium <0.90 µmol/L in pregnancy week 18 (lower tertile) had significantly lower total, problem solving, personal-social functioning, and fine motor function ASQ scores compared to infants born to mothers with higher selenium levels (Table 5, Supplemental Figure S1).

No significant associations were observed between ASQ scores and maternal selenium status at other timepoints or for infant selenium status (data not shown).

3.6. Selenium Levels and Infant Infections

The prevalence of reported infant infections increased during the first 6 months of life from 18% (19/107) during the first 6 weeks to 35% (39/113) from 6 weeks to 4 months and to 44% (50/114) from 4 to 6 months. Of the infections in the first 6 weeks of life (sepsis ($n = 1$), lower airway infection ($n = 2$), upper airway infection ($n = 8$), lower urinary tract infection ($n = 1$), dermal or mucosal infections ($n = 8$)), 32% (6/19) were associated with fever and 26% (5/19) required antibiotics. Maternal serum selenium levels in pregnancy week 36 and 6 weeks postpartum were significantly lower in mothers of infants who had a reported infection during the first 6 weeks of life, compared to mothers of healthy infants (Table 6).

Table 4. Selenium in breastmilk and calculated infant intake at 6 weeks and 4 and 6 months postpartum.

Parameters	6 Weeks n = 59	4 Months n = 60	6 Months n = 61	p Value *
Breast-milk selenium, μmol/L				
Median	0.13	0.10	0.09	
(25, 75) **	(0.11–0.17)	(0.08–0.13)	(0.08–0.13)	<0.001
(2.5, 97.5) **	(0.06–0.27)	(0.04–0.36)	(0.03–0.29)	
Daily selenium intake, μmol ***				
Median	0.10	0.08	0.09	
(25, 75) **	(0.08, 0.12)	(0.06, 0.10)	(0.07, 0.12)	0.02
(2.5, 97.5) **	(0.05, 0.20)	(0.03, 0.32)	(0.03, 0.29)	

* Comparison by Kruskall-Wallis test; ** Percentiles; *** Daily selenium intake was calculated based on selenium levels in breastmilk and estimated milk intake (150 mL/kg/day at 6 weeks and 120 mL/kg/day at 4 to 6 months).

Table 5. Ages and Stages Questionnaire (ASQ) scores for the infants at 6 months in relation to maternal serum selenium levels in pregnancy week 18.

Maternal Serum Selenium in Week 18	Total	ASQ Scores				
		Communication	Problem Solving	Personal-Social Functioning	Gross Motor Function	Fine Motor Function
<0.90 μmol/L (Tertile 1) n = 35	213 (189, 236)	45 (40, 50)	50 (36, 59)	45 (30, 50)	35 (30, 40)	40 (35, 50)
>0.90 μmol/L (Tertile 2–3) n = 70	235 (215, 258)	50 (45, 55)	55 (50, 60)	50 (40, 55)	35 (30, 45)	50 (40, 55)
p value *	0.002	0.12	0.005	0.02	0.57	0.02

* Compared by Mann-Whitney U test.

Table 6. Maternal and infant serum selenium in relation to reported infant infection from birth to age 6 weeks.

Infant Infection from Birth to Age 6 Weeks	Maternal Serum Selenium, μmol/L, Median (25, 75) *, n = 107				Infant Serum Selenium at 6 Months, μmol/L, Median (25, 75) * n = 91
	Pregnancy Week 18	Pregnancy Week 28	Pregnancy Week 36	6 Weeks Postpartum	
Yes, n = 19	0.99 (0.86, 1.04)	0.87 (0.80, 0.96)	0.77 (0.71, 0.86)	0.98 (0.92, 1.10)	0.77 (0.74, 0.81)
No, n = 88	0.96 (0.87, 1.04)	0.93 (0.81, 1.02)	0.86 (0.77, 0.96)	1.09 (0.98, 1.26)	0.83 (0.74, 0.90)
p value **	0.94	0.52	0.03	0.03	0.10

* Percentiles. ** Comparison by Mann-Whitney U test.

Based on examination of scatterplots, box-plot and error bars visualizing the relation between maternal serum selenium in pregnancy week 36 and the occurrence of infant infection during the first 6 weeks of life, the lower tertile of maternal serum selenium (<0.78 µmol/L) was chosen as a cutoff level for defining maternal selenium deficiency in pregnancy week 36. The infants had an increased risk with an odds ratio (OR) of 4.8 (95% CI 1.2–20.5) for having an infection during the first 6 weeks of life if the maternal serum selenium level was in the lower tertile (≤0.78 µmol/L) in pregnancy week 36, and an OR of 2.5 (95% 0.7–9.5) if maternal selenium was in the lower tertile (<0.99 µmol/L) at 6 weeks postpartum. These results are from a logistic regression model which additionally included birthweight, weight at 6 weeks, gender, weeks of exclusive breastfeeding, parity, and incidence of maternal infections during the same period (Supplemental Figure S2).

No significant relationships were observed between maternal selenium status during pregnancy and lactation and the occurrence of maternal infections postpartum. We did not observe any significant correlations between maternal and infant selenium status to any other reported condition (feeding difficulties, colic, constipation, or dermatitis) in the infant during the first 6 months of life.

4. Discussion

Compared to never-pregnant women, median maternal selenium levels were reduced by 20% in later pregnancy, increased postpartum, and remained unchanged for the first 6 months. Maternal selenium levels in pregnancy and postpartum were highly correlated to breastmilk and infant selenium levels. A maternal selenium level in the lower tertile in pregnancy week 18 (≤0.90 µmol/L) was negatively related to infant neurodevelopment at 6 months, whereas a selenium level in the lower tertile in pregnancy week 36 (≤0.78 µmol/L) was associated with an increased risk (OR 4.8) of infant infection during the first 6 weeks of life.

4.1. Serum Selenium Levels in Never-Pregnant, Pregnant, and Lactating Women

In adults, a plateau in plasma selenoprotein P has been reached with serum selenium concentrations >1.57 µmol/L [20]; this is considered to indicate a selenium-replete status [21]. Median serum selenium levels are reported to be lower in females compared to men (1.18 vs. 1.27 µmol/L), lower in subjects younger than 40 years compared to older subjects (1.08 vs. 1.24 µmol/L), and lower in adult Norwegians compared to adult Americans (1.20 vs. 1.51 µmol/L) [14,21]. In our population of healthy never-pregnant women, the median level was 1.07 µmol/L and 98% (154/158) had a selenium level below 1.57 µmol/L, indicating a non-replete selenium status.

Median selenium levels in our pregnant population varied from 0.96 µmol/L in week 18 to 0.85 µmol/L in week 36, comparable to reported selenium levels in pregnancy week 15–22 ranging from 0.54 µmol/L in Polish women [6], to 1.25 µmol/L in Spanish women [9], with lower levels observed just before delivery [6,8,9]. The levels increased postpartum, and were higher in regular users of MMN throughout pregnancy and postpartum, as has been observed by others [8], however, almost all (97–99%) values were below 1.57 µmol/L at all time points.

4.2. Selenium Levels in Breastmilk

Milk selenium level is closely related to maternal selenium status and varies substantially between world regions and with time after birth [10]. Median selenium levels are reported to be 26 µg/L (0.33 µmol/L) in colostrum (0–5 days), decrease to 18 µg/L (0.23 µmol/L) in transitional milk (6–21 days), decrease further to nadir levels of 15 µg/L (0.19 µmol/L) in mature milk (1–3 months) and be slightly higher 17 µg/L (0.22 µmol/L) in late milk (>5 months) [10]. Compared to these data, selenium levels in breastmilk from mothers in Northern Europe is reported to be lower [10], as demonstrated by our data.

4.3. Selenium Intake and Serum Levels in Infants

The recommended daily selenium intake up to 4 months is set to 10 µg (0.13 µmol) and based on the assumption that selenium levels in breastmilk are optimal for this age-group [22]. For infants 4 to 12 months, the recommended daily intake is 15 µg (0.19 µmol), which is still based on selenium milk values and estimated increase in average body weight as no data is available on selenium intake from solid food in infants [22]. In our population, median selenium intake from breastmilk decreased during the lactational period and was 77% of the recommended level at 6 weeks, and 69% at 6 months, which is comparable to published data in breastfed Polish infants, in whom both plasma selenium and glutathione peroxidase-3 levels increased after maternal selenium supplementation, indicating a former selenium deficiency [23].

Infants are considered to be at risk for selenium deficiency, and the levels are reported to decrease during the first months of life [11]. In German children, median serum selenium decreased from 0.64 µmol/L in infants less than 1 month to 0.44 µmol/L at 4 months, increased to 0.62 µmol/L between 4 to 12 months, remained stable between 1 to 5 years (0.90 µmol/L), and increased slightly from 5 to 18 years (median 0.99 µmol/L) [11].

A pattern with nadir serum levels in infants between 3 to 6 months is observed also for other micronutrients, like cobalamin and iron, and is associated with a negative clinical outcome [24–26], merely reflecting the vulnerable nutrient status during the first months of life. The median selenium level in Norwegian infants at 6 months (0.81 µmol/L) was within the established reference range based on German children [11], however, reference ranges may not be an optimal method for evaluating micronutrient status, particularly when based on populations with a high prevalence of deficiency.

4.4. Selenium Levels and Infant Clinical Status

The obtained ASQ scores in our infant population resemble published data in healthy, Norwegian infants at 6 months [27]. A study from Poland reported a positive association between maternal selenium levels in the first trimester (mean 0.61 (SD 0.13) µmol/L) and motor development at 1 year of age, and language development at 2 years [6]. Similar associations between maternal selenium status and neurodevelopment have been reported in other human [5] and animal studies [28]. Our observed lower ASQ scores (total, problem solving, personal-social functioning, and fine motor function) in infants born to mothers with a selenium level ≤0.90 µmol/L in pregnancy week 18 are in line with these data.

Neurodevelopment is multifaceted, and factors like birthweight, weight increase after birth, gender, breastfeeding, maternal age, education, and parity are reported to have an impact on neurodevelopment [19]. However, ASQ total score, problem solving, personal-social functioning, and fine motor score still increased significantly with maternal selenium tertiles in pregnancy week 18, even when such factors were included in the multiple linear regression models. Neurodevelopment is also dependent on a range of nutritional factors, and we cannot exclude that nutrients other than selenium may have contributed to the observed differences in clinical outcome. Further, it is likely that women with an adequate selenium status also have an adequate status for other micronutrients due to a healthy diet or MMN supplements.

We also observed significantly lower maternal selenium levels in pregnancy week 36 and 6 weeks postpartum in mothers of infants who had an early infection. A logistic regression model, which additionally included factors considered to be related to the risk of infection, demonstrated a significantly increased risk (OR 4.8) of infant infection during the first 6 weeks of life if the maternal serum selenium level in pregnancy week 36 was ≤0.78 µmol/L. Studies have evaluated clinical infant outcomes related to selenium status in HIV-infected pregnant women [29], but apart from this, data on the association between maternal selenium status in pregnancy and infant immunocompetence has, to our knowledge, not been previously published.

While maternal selenium status in pregnancy week 18 was related to neurodevelopment, maternal status in the last trimester was related to risk of infection. Brain development begins in the third gestational week and continues after birth [30]. The preventive effect of folic acid supplementation on neural tube defects [31] has demonstrated the importance of an adequate micronutrient status in early pregnancy for normal neurodevelopment, and our findings may justify a role for selenium in early neurodevelopment as well.

Fetal micronutrient stores are formed during the last trimester, and a low maternal selenium status during this period will reduce fetal stores. A low selenium status may negatively influence both the humoral and cell-mediated immune function [32], and maternal antenatal and postpartum selenium supplementation have been associated with a reduced child mortality after 6 weeks of age [29].

4.5. Strength and Limitations

This was an observational study with a small sample size of mother and infant dyads, and the clinical data were reported by the mothers, factors known to be disadvantageous. The data were, however, collected prospectively throughout pregnancy and postpartum and the participation rate was high, which are strengths to the study. Evaluation of neurodevelopment in young infants is challenging [33], but ASQ is a validated screening tool with high sensitivity and specificity to detect children with developmental delay [16]. The maternal reported infant infections included diagnoses defined by quite distinct and well-known symptoms of infection, unlikely to be confused by other diagnoses, like allergy.

5. Conclusions

An adequate selenium status is important for fetal and infant development. It is, therefore, important to optimize maternal selenium status during pregnancy. As the interpretation of maternal selenium status is hampered in pregnancy, due to numerous physiological changes, we suggest to use clinical infant outcome in order to establish selenium cut off levels in pregnancy. In a Norwegian population of healthy pregnant women, a low maternal selenium status was associated with a lower psychomotor score at 6 months and an increased risk of infant infection during the first 6 weeks of life. Based on our observations, we suggest a cutoff for maternal serum selenium deficiency of 0.90 μmol/L in pregnancy week 18, and 0.78 μmol/L in pregnancy week 36. This should be evaluated in a randomized intervention study.

Supplementary Materials: The following are available online at www.mdpi.com/2072-6643/9/5/486/s1, Figure S1: Infant ASQ total score (mean ± 2 SD) at age 6 months in relation to maternal serum selenium levels in pregnancy week 36, Figure S2: Percentage of infants with infection (n = 19) and without infection (n = 88) during the first 6 weeks of life in relation to maternal serum selenium levels (tertiles) in pregnancy week 36 and 6 weeks postpartum.

Acknowledgments: We thank all mothers and infants for their willingness to participate in the study and the laboratory staff at the Laboratory of Clinical Biochemistry, Haukeland University Hospital, Norway for help with blood sampling.

Author Contributions: K. Varsi and A.-L. Bjørke-Monsen conceived, designed, and performed the study, analyzed the data, and wrote the paper; I.K. Torsvik and T.C. Rosvold Eik performed the experiments; B. Bolann and P.J. Høl wrote the paper.

Conflicts of Interest: The authors have no financial relationships to disclose or no conflicts of interest relevant to this article.

References

1. Rayman, M.P. The importance of selenium to human health. *Lancet* **2000**, *356*, 233–241. [CrossRef]
2. Spallholz, J.E.; Boylan, L.M.; Larsen, H.S. Advances in understanding selenium's role in the immune system. *Ann. N. Y. Acad. Sci.* **1990**, *587*, 123–139. [CrossRef] [PubMed]
3. Kiremidjian-Schumacher, L.; Roy, M.; Wishe, H.I.; Cohen, M.W.; Stotzky, G. Supplementation with selenium and human immune cell functions. II. Effect on cytotoxic lymphocytes and natural killer cells. *Biol. Trace Elem. Res.* **1994**, *41*, 115–127. [CrossRef] [PubMed]

4. Guvenc, H.; Karatas, F.; Guvenc, M.; Kunc, S.; Aygun, A.D.; Bektas, S. Low levels of selenium in mothers and their newborns in pregnancies with a neural tube defect. *Pediatrics* **1995**, *95*, 879–882. [PubMed]
5. Skroder, H.M.; Hamadani, J.D.; Tofail, F.; Persson, L.A.; Vahter, M.E.; Kippler, M.J. Selenium status in pregnancy influences children's cognitive function at 1.5 years of age. *Clin. Nutr.* **2015**, *34*, 923–930. [CrossRef] [PubMed]
6. Polanska, K.; Krol, A.; Sobala, W.; Gromadzinska, J.; Brodzka, R.; Calamandrei, G.; Chiarotti, F.; Wasowicz, W.; Hanke, W. Selenium status during pregnancy and child psychomotor development-Polish Mother and Child Cohort study. *Pediatr. Res.* **2016**, *79*, 863–869. [CrossRef] [PubMed]
7. Costantine, M.M. Physiologic and pharmacokinetic changes in pregnancy. *Front. Pharmacol.* **2014**, *5*, 65. [CrossRef] [PubMed]
8. Hansen, S.; Nieboer, E.; Sandanger, T.M.; Wilsgaard, T.; Thomassen, Y.; Veyhe, A.S.; Odland, J.Ø. Changes in maternal blood concentrations of selected essential and toxic elements during and after pregnancy. *J. Environ. Monit.* **2011**, *13*, 2143–2152. [CrossRef] [PubMed]
9. Izquierdo Alvarez, S.; Castanon, S.G.; Ruata, M.L.; Aragues, E.F.; Terraz, P.B.; Irazabal, Y.G.; González, E.G.; Rodríquez, B.G. Updating of normal levels of copper, zinc and selenium in serum of pregnant women. *J. Trace Elem. Med. Biol.* **2007**, *21*, 49–52. [CrossRef] [PubMed]
10. Dorea, J.G. Selenium and breast-feeding. *Br. J. Nutr.* **2002**, *88*, 443–461. [CrossRef] [PubMed]
11. Muntau, A.C.; Streiter, M.; Kappler, M.; Roschinger, W.; Schmid, I.; Rehnert, A.; Schramel, P.; Roscher, A.A. Age-related reference values for serum selenium concentrations in infants and children. *Clin. Chem.* **2002**, *48*, 555–560. [PubMed]
12. Jacobson, B.E.; Lockitch, G. Direct determination of selenium in serum by graphite-furnace atomic absorption spectrometry with deuterium background correction and a reduced palladium modifier: Age-specific reference ranges. *Clin. Chem.* **1988**, *34*, 709–714. [PubMed]
13. Rossipal, E.; Tiran, B. Selenium and glutathione peroxidase levels in healthy infants and children in Austria and the influence of nutrition regimens on these levels. *Nutrition* **1995**, *11*, 573–575. [PubMed]
14. Birgisdottir, B.E.; Knutsen, H.K.; Haugen, M.; Gjelstad, I.M.; Jenssen, M.T.; Ellingsen, D.G.; Thomassen, Y.; Alexander, J.; Meltzer, H.M.; Brantsaeter, A.L. Essential and toxic element concentrations in blood and urine and their associations with diet: results from a Norwegian population study including high-consumers of seafood and game. *Sci. Total Environ.* **2013**, *463–464*, 836–844. [CrossRef] [PubMed]
15. Gao, D.; He, Z.; Wu, J.; Ma, Q.; Song, H.; Mei, L.; Wu, Y. Long-term results of combined splenorenal shunt and porta-azygos devascularization in patients with portal hypertension. *Zhonghua Wai Ke Za Zhi* **1998**, *36*, 327–329. (In Chinese). [PubMed]
16. Schonhaut, L.; Armijo, I.; Schonstedt, M.; Alvarez, J.; Cordero, M. Validity of the ages and stages questionnaires in term and preterm infants. *Pediatrics* **2013**, *131*, 1468–1474. [CrossRef] [PubMed]
17. Bolann, B.J.; Distante, S.; Morkrid, L.; Ulvik, R.J. Bloodletting therapy in hemochromatosis: Does it affect trace element homeostasis? *J. Trace Elem. Med. Biol.* **2015**, *31*, 225–229. [CrossRef] [PubMed]
18. Matos, C.; Moutinho, C.; Almeida, C.; Guerra, A.; Balcao, V. Trace element compositional changes in human milk during the first four months of lactation. *Int. J. Food Sci. Nutr.* **2014**, *65*, 547–551. [CrossRef] [PubMed]
19. Lung, F.W.; Shu, B.C.; Chiang, T.L.; Lin, S.J. Twin-singleton influence on infant development: A national birth cohort study. *Child Care Health Dev.* **2009**, *35*, 409–418. [CrossRef] [PubMed]
20. Hurst, R.; Armah, C.N.; Dainty, J.R.; Hart, D.J.; Teucher, B.; Goldson, A.J.; Broadley, M.R.; Motley, A.K.; Fairweather-Tait, S.J. Establishing optimal selenium status: Results of a randomized, double-blind, placebo-controlled trial. *Am. J. Clin. Nutr.* **2010**, *91*, 923–931. [CrossRef] [PubMed]
21. Niskar, A.S.; Paschal, D.C.; Kieszak, S.M.; Flegal, K.M.; Bowman, B.; Gunter, E.W.; Pirkle, J.L.; Rubin, C.; Sampson, E.J.; McGeehin, M. Serum selenium levels in the US population: Third National Health and Nutrition Examination Survey, 1988–1994. *Biol. Trace Elem. Res.* **2003**, *91*, 1–10. [CrossRef]
22. Kipp, A.P.; Strohm, D.; Brigelius-Flohe, R.; Schomburg, L.; Bechthold, A.; Leschik-Bonnet, E.; Heseker, H. Revised reference values for selenium intake. *J. Trace Elem. Med. Biol.* **2015**, *32*, 195–199. [CrossRef] [PubMed]
23. Trafikowska, U.; Zachara, B.A.; Wiacek, M.; Sobkowiak, E.; Czerwionka-Szaflarska, M. Selenium supply and glutathione peroxidase activity in breastfed Polish infants. *Acta Paediatr.* **1996**, *85*, 1143–1145. [CrossRef] [PubMed]

24. Monsen, A.L.; Refsum, H.; Markestad, T.; Ueland, P.M. Cobalamin status and its biochemical markers methylmalonic acid and homocysteine in different age groups from 4 days to 19 years. *Clin Chem.* **2003**, *49*, 2067–2075. [CrossRef] [PubMed]
25. Greibe, E.; Lildballe, D.L.; Streym, S.; Vestergaard, P.; Rejnmark, L.; Mosekilde, L.; Nexo, E. Cobalamin and haptocorrin in human milk and cobalamin-related variables in mother and child: a 9-mo longitudinal study. *Am. J. Clin Nutr.* **2013**, *98*, 389–395. [CrossRef] [PubMed]
26. Lozoff, B.; Beard, J.; Connor, J.; Barbara, F.; Georgieff, M.; Schallert, T. Long-lasting neural and behavioral effects of iron deficiency in infancy. *Nutr. Rev.* **2006**, *64*, S34–S43. [CrossRef] [PubMed]
27. Alvik, A.; Groholt, B. Examination of the cut-off scores determined by the Ages and Stages Questionnaire in a population-based sample of 6 month-old Norwegian infants. *BMC Pediatr.* **2011**, *11*, 117. [CrossRef] [PubMed]
28. Watanabe, C.; Satoh, H. Brain selenium status and behavioral development in selenium-deficient preweanling mice. *Physiol. Behav.* **1994**, *56*, 927–932. [CrossRef]
29. Kupka, R.; Mugusi, F.; Aboud, S.; Msamanga, G.I.; Finkelstein, J.L.; Spiegelman, D.; Fawzi, W.W. Randomized, double-blind, placebo-controlled trial of selenium supplements among HIV-infected pregnant women in Tanzania: effects on maternal and child outcomes. *Am. J. Clin Nutr.* **2008**, *87*, 1802–1808. [PubMed]
30. Stiles, J.; Jernigan, T.L. The basics of brain development. *Neuropsychol. Rev.* **2010**, *20*, 327–348. [CrossRef] [PubMed]
31. MRC Vitamin Study Research Group. Prevention of neural tube defects: results of the Medical Research Council Vitamin Study. *Lancet* **1991**, *338*, 131–137.
32. Kiremidjian-Schumacher, L.; Roy, M.; Wishe, H.I.; Cohen, M.W.; Stotzky, G. Regulation of cellular immune responses by selenium. *Biol. Trace Elem. Res.* **1992**, *33*, 23–35. [CrossRef] [PubMed]
33. Heineman, K.R.; Hadders-Algra, M. Evaluation of neuromotor function in infancy—A systematic review of available methods. *J. Dev. Behav. Pediatr.* **2008**, *29*, 315–323. [CrossRef] [PubMed]

nutrients

MDPI

Article

Serum Insulin-Like Growth Factor Axis and the Risk of Pancreatic Cancer: Systematic Review and Meta-Analysis

Yuanfeng Gong [1,†], Bingyi Zhang [2,†], Yadi Liao [1], Yunqiang Tang [1,*], Cong Mai [1], Tiejun Chen [1] and Hui Tang [1]

1   Department of Hepatobiliary Surgery, The Affiliated Cancer Hospital& Institute of Guangzhou Medical University, Guangzhou 510095, China; medgongyf@126.com (Y.G.); medliaoyd@126.com (Y.L.); medmaic@126.com (C.M.); medchentj@126.com (T.C.); medtangh@126.com (H.T.)
2   Department of Ultrasound, the First People's Hospital of Yichang, China Three Gorges University, Yichang 443000, China; yczhangyb@126.com
*   Correspondence: medtangyq@126.com; Tel.: +86-206-667-3666
†   These authors contributed equally to this work.

Received: 29 December 2016; Accepted: 12 April 2017; Published: 18 April 2017

Abstract: Objective: To investigate the association between serum concentration of insulin-like growth factor (IGF) and the risk of pancreatic cancer (PaC). Methods: We identified eligible studies in Medline and EMBASE databases (no reference trials from 2014 to 2016) in addition to the reference lists of original studies and review articles on this topic. A summary of relative risks with 95% confidence intervals (CI) was calculated using a random-effects model. The heterogeneity between studies was assessed using Cochran Q and I^2 statistics. Results: Ten studies (seven nested case-control studies and three retrospective case-control studies) were selected as they met our inclusion criteria in this meta-analysis. All these studies were published between 1997 and 2013. The current data suggested that serum concentrations of IGF-I, IGF-II and insulin-like growth factor binding protein-3 (IGFBP-3)in addition to the IGF-I/IGFBP-3 ratio were not associated with an increased risk of PaC (Summary relative risks (SRRs) = 0.92, 95% CI: 0.67–1.16 for IGF-I; SRRs = 0.84, 95% CI: 0.54–1.15 for IGF-II; SRRs = 0.93, 95% CI: 0.69–1.17 for IGFBP-3; SRRs = 0.97, 95% CI: 0.71–1.23 for IGF-I/IGFBP-3 ratio). There was no publication bias in the present meta-analysis. Conclusion: Serum concentrations of IGF-I, IGF-II, IGFBP-1 and IGFBP-3 as well as the IGF-I/IGFBP-3 ratio were not associated with increased risk of PaC.

Keywords: insulin-like growth factor; insulin-like growth factor binding protein; pancreatic cancer; morbidity; meta-analysis

---

1. Introduction

Pancreatic cancer (PaC) ranks as the fourth most common cause of death from cancer in both men and women in the United States [1]. Despite decades of effort by clinicians and scientists, the five-year survival rate remains poor, as it has only reached a maximum of 5%. The incidence and mortality rates of PaC in the United States have remained stable over the past two decades [2]. However, the incidence rate of PaC in males from China rose in the period from 2000 to 2011 [3]. Radical resection is the only potentially curative therapy. As it is difficult to diagnose in its early stage, approximately 80% of patients cannot receive surgical resection. The five-year survival rate is about 20% after resection. Patients suffering from local, advanced, unresectable or metastatic disease can undergo chemotherapy or chemo-radiotherapy if these patients have a good performance status. A considerable number of epidemiological studies and meta-analyses have investigated possible risk factors of PaC. The association between cigarette smoking and PaC has been demonstrated by nearly all published studies [4–6]. Several meta-analyses have suggested that obesity is a risk factor for pancreatic cancer [7,8]. Pancreatitis, especially chronic pancreatitis, was associated with a significantly increased risk of PaC [9], with this risk appearing to be the highest in rare types of pancreatitis, such as hereditary pancreatitis and tropical pancreatitis [10]. New-onset diabetes was associated with a significantly increased risk, meaning it could be a potential clue for the early diagnosis of PaC [11]. Cholelithiasis [12], cholecystectomy [13] and gastrectomy [14] may also increase the risk of PaC.

The insulin-like growth factor (IGF) axis includes two growth factors, IGF-I and IGF-II, in addition to several IGF binding proteins (IGFBP-1 to IGFBP-6), which work together to regulate the amount of free IGF-I and IGF-II in serum. IGFs have long been known as nutritional biomarkers, which are dysregulated in states of under- and over-nutrition. Serum concentration of IGF-I falls rapidly in malnutrition and responds promptly to refeeding, so it may convey the messages of nutritional status, monitoring effect of nutritional support [15].

Recently, there has been growing interest in its role in health and disease, especially in cancers. IGF-I and the IGF-I receptor are highly expressed on the surface of pancreatic cancer cell lines, which initiate intracellular signaling transduction associated with proliferation, invasion and expression of mediators of angiogenesis. Meta-analyses of present studies confirmed previous reports regarding elevated serum levels of IGF-I and IGF-II to be associated with an increased risk of colorectal cancer [16,17], breast cancer [18] and prostate cancer [19]. A high level of IGFBP-3 was associated with a reduced risk of lung cancer [20]. However, the findings have been somewhat contrary to present studies of pancreatic cancer [21,22]. Therefore, we performed this first systematic review and meta-analysis of all available evidence of observational studies, following the meta-analysis of observational studies in epidemiological guidelines to clarify the association between the serum IGF axis and risk of PaC.

2. Materials and Methods

2.1. Data Sources and Searches

Two authors (Y.G. and Y.L.) independently performed a literature search using Medline and EMBASE databases for articles dated up to 1 May 2016. We searched the studies with the following text words and/or Medical Subject Heading (MeSH) terms: ("IGF" OR "insulin-like growth factor", "IGFBP" OR "insulin-like growth factor binding protein") AND ("pancreas", "cancer" or "adenocarcinoma" or "neoplasm" or "tumor").

2.2. Study Selection

We included studies that met all the following criteria: (1) published as an original article; (2) used a case-control, cross-sectional, nested case-control or cohort design; (3) explored the serum level of IGFs and IGFBPs; (4) studied outcome was incidence of mortality of pancreatic cancer; and (5) estimated odds ratio (OR) or relative risk (RR) with corresponding 95% confidence intervals (CIs) (or data to calculate them) for the highest versus non/lowest levels of insulin-like growth factors and insulin-like growth factor binding proteins were reported. Two authors (Y.G. and Y.L.) independently evaluated all the studies retrieved from the databases. If there were multiple publications from the same study, the most relevant was selected, with other publications used to clarify methodology or characteristics of the population. We did not contact the authors for detailed information of primary studies.

2.3. Data Extraction and Quality Assessment

Three authors (C.M., T.C. and H.T.) independently evaluated all the studies retrieved according to the prespecified selection criteria. Any discrepancies between reviewers were addressed by a joint reevaluation of the original article. The following information from each study was extracted using a standardized data collection form: the first author's last name, year of publication, geographic location, study design, sample size, quality of each study, exposure of interest, concentration levels, the effect estimates with 95% CIs and covariates adjusted in the statistical analysis.

The quality of each study was evaluated independently by three reviewers using the Newcastle-Ottawa Scale (NOS). The NOS consists of three parameters of quality: selection, comparability and outcome (cohort studies) or exposure (case-control studies). The NOS assigns a maximum of four points for selection, a maximum of two points for comparability and a maximum of three points for exposure or outcome. Any discrepancies between reviewers were addressed by a joint reevaluation of the original article.

2.4. Statistical Analysis

For simplicity, all measures were interpreted as relative risks (RR) with no distinction between the various estimates (i.e., OR, rate ratio, hazard ratio). As different studies might report different exposure categories (thirds or quarters), we used the study-specific relative risk for the highest versus the lowest category of IGFs and IGFBPs for the meta-analysis. We transformed the corresponding CIs into the log RRs, using the Greenland formula to calculate the corresponding variances. For studies that lacked estimates, we calculated crude estimates from tabular data [23–25]. We pooled these relative risks using a fixed effects model to get a summary relative risk for further meta-analysis. We used Woolf's formula to evaluate the standard error (SE) of the log RRs. Summary relative risks (SRR) with their corresponding 95% CIs were combined and weighted to produce pooled RRs using a fixed- or random-effects model, according to I^2 statistics.

To investigate the sources of heterogeneity across these studies, we carried out heterogeneity tests and sensitivity analyses. In heterogeneity tests, we used the Cochran Q and I^2 statistics [26], which were used to test the differences obtained between studies due to chance. For the Q statistic, a p-value of less than 0.10 was considered representative of statistically significant heterogeneity. The I^2 statistic is the proportion of total variation contributed by variation between studies. It has been suggested that I^2 values of 25%, 50% and 75% are assigned to low, moderate and high heterogeneity, respectively [27]. We conducted the sensitivity analyses to estimate the influence of each individual study on the summary results by repeating the random-effects meta-analysis after omitting one study at a time. We evaluated the role of several potential sources of heterogeneity by sub-group analyses according to adjustments for confounding variables: alcohol consumption, smoking and diabetes mellitus (DM).

Funnel plots and Egger's test were performed to test evidence of publication bias [28]. Meta-analyses were carried out using STATA12.0 (Stata Corp, College Station, TX, USA).

3. Results

3.1. Data Sources and Searches

The detailed steps of our literature search are presented in Figure 1. In brief, a total of 294 citations were obtained for review of the title and abstract. Of the 294 citations, 271 were not relevant. Full texts of the remaining 25 studies were retrieved for review. Thirteen studies were excluded because, when reviewed in detail, their data were not relevant. Two studies were excluded as they were review articles. Finally, 10 studies were included in the final meta-analysis (Figure 1).

3.2. Study Characteristics

Ten articles that met our inclusion criteria in this meta-analysis were published between 1997 and 2013 (no reference trials from 2014 to 2016). There were seven nested case-control studies [21,22,29–34] and three retrospective case-control studies [23–25]. Nine articles described the association between IGF-I concentration and PaC risk [21–25,30–33], three described the association between IGF-II concentration and PaC risk [22,30,31], six reported the association between IGFBP-3 concentration and PaC risk [21,22,30–33], while four reported the association between the IGF-I/IGFBP-3 ratio and PaC risk [21,22,30,32]. The average score for the quality assessment of included studies was 7.9 (Table 1).

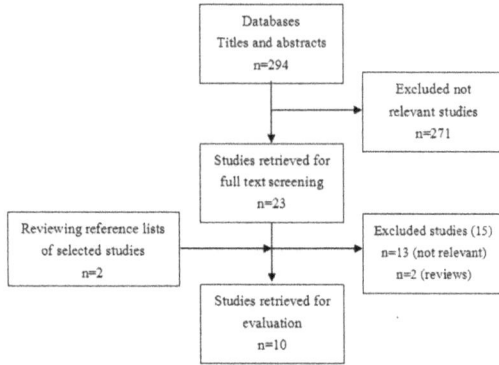

Figure 1. Flow chart of selection of studies included in the meta-analysis.

Meta-analysis of six nested case-control studies in a fixed-effects model found that the serum IGF-I concentration was not associated with the risk of PaC (SRRs = 0.92, 95% CI: 0.67–1.16; test for heterogeneity $p = 0.435$, $I^2 = 0.0\%$). A similar result was found in three retrospective case-control studies (Figure 2a,b). The serum IGF-II concentration was also not related with the risk of PaC (SRRs = 0.84, 95% CI: 0.54–1.15; test for heterogeneity $p = 0.574$, $I^2 = 0.0\%$) (Figure 3).

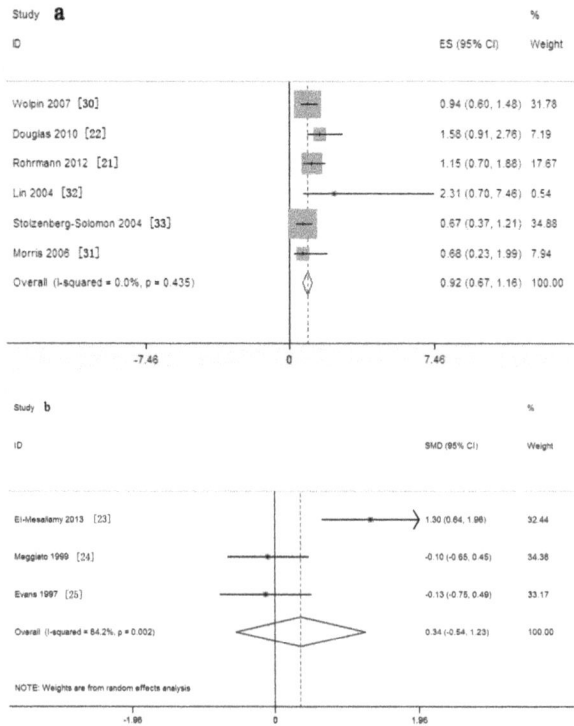

Figure 2. Forest plot of insulin-like growth factor (IGF)-I and pancreatic cancer (PaC) risk for: (**a**) nested case-control studies and (**b**) case-control studies.

Table 1. Characteristics of the ten included studies.

Author/References	Study Published/Location	Study Design	Cases	Controls	NOS	Exposure of Interest	Concentration Levels	Effect Estimate (95%CI)	Adjustments
Rohrmann [21]	2012/Europe	NCC	422	422	9	IGF-1 / IGFBP-3 / IGF-1/IGFBP-3	Q4 vs. Q1	1.15 (0.70, 1.88) / 1.06 (0.68, 1.65) / 1.29 (0.77, 2.16)	Age, sex, education, BMI, physical activity, alcohol consumption, smoking and DM
Douglas [22]	2010/USA	NCC	187	374	9	IGF-1 / IGF-2 / IGFBP-3 / IGF-1/IGFBP-3	Q4 vs. Q1	1.58 (0.91, 2.76) / M: 1.56 (0.78, 3.14) / F: 1.74 (0.67, 4.51) / 0.86 (0.49, 1.50) / M: 0.75 (0.36, 1.56) / F: 0.86 (0.49, 1.50) / 0.88 (0.51, 1.51) / M: 1.03 (0.51, 2.09) / F: 0.77 (0.32, 1.83) / 1.54 (0.89, 2.66) / M: 1.39 (0.69, 2.80) / F: 1.47 (0.58, 3.75)	Age, race, sex, smoking, education, BMI, physical activity DM, alcohol consumption and nutrients intake
Wolpin [30]	2007/USA	NCC	212	635	9	IGF-1 / IGF-2 / IGFBP-3 / IGF-1/IGFBP-3	Q4 vs. Q1	0.94 (0.60, 1.48) / 0.96 (0.61, 1.52) / 1.21 (0.75, 1.92) / 0.84 (0.54, 1.31)	Age, sex, smoking, BMI, physical activity, DM, vitamin use and energy intake
Wolpin [29]	2007/ USA	NCC	144	429	9	IGFBP-1	Q4 vs. Q1	0.56 (0.31,1.01)	Age, sex, smoking, BMI, physical activity, DM, vitamin use and energy intake
Morris [31]	2006/UK	NCC	38	114	8	IGF-1 / IGF-2 / IGFBP-3	T3 vs. T1	0.68 (0.23, 1.99) / 0.48 (0.14, 1.68) / 0.74 (0.26, 2.12)	Age, smoking
Lin [32]	2004/Japan	NCC	69	207	8	IGF-1 / IGFBP-3	Q4 vs. Q1	2.31 (0.70, 7.64) / 2.53 (0.93, 6.85)	Age, sex, BMI, smoking and DM
Stolzenberg-Solomon [33]	2004/Finland	NCC	93	400	8	IGF-1 / IGFBP-3 / IGF-1/IGFBP-3	T3 vs. T1	0.67 (0.37, 1.21) / 0.70 (0.38, 1.27) / 0.85 (0.50, 1.46)	Age, BMI, smoking, energy intake, alcohol consumption
El-Mesallamy [23]	2013/Egypt	CC	23	20	7	IGF-1 (ug/L) / IGFBP-3 (ug/L) / IGF-1/IGFBP-3	Continuous	1.30 (0.64, 1.96)[a] / −1.04 (−1.68, −0.40)[a] / 1.48 (0.80,2.16)[a]	Age, sex, race
Meggiato [24]	1999/Italy	CC	35	22	6	IGF-1 (ug/L)	Continuous	−0.10 (−0.65,0.45)[a]	Age, sex, race
Evans [25]	1997/UK	CC	20	20	6	IGF-1 (ug/L) / IGF-2 (U/mL) / IGFBP-3 (nmol/L)	Continuous	−0.13 (−0.75, 0.49)[a] / −0.50 (−1.13, 0.13)[a] / −0.20 (−0.82, 0.42)[a]	Age

Abbreviations: M = male; F = female; NCC = nested case-control study; CC = case-control study; Q = quartile; T = tertile. [a] Mean difference (95%CI). BMI = Body mass index; CI = Confidence interval; DM = Diabetes mellitus; IGF = Insulin-like growth factor; IGFBP = Insulin-like growth factor binding protein.

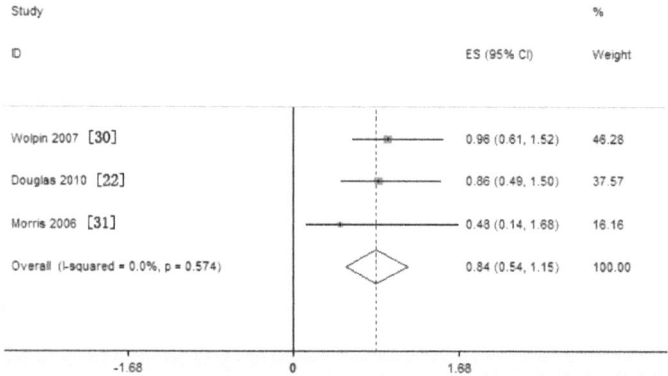

Figure 3. Forest plot of insulin-like growth factor (IGF-II) and pancreatic cancer (PaC) risk. CI = Confidence interval; ES = Effect size.

3.3. Meta-Analysis

Meta-analysis of six nested case-control studies in a fixed-effects model showed that serum IGFBP-3 concentration was also not associated with the risk of PaC (SRRs = 0.93, 95% CI: 0.69–1.17; test for heterogeneity $p = 0.623$, $I^2 = 0.0\%$) (Figure 4).

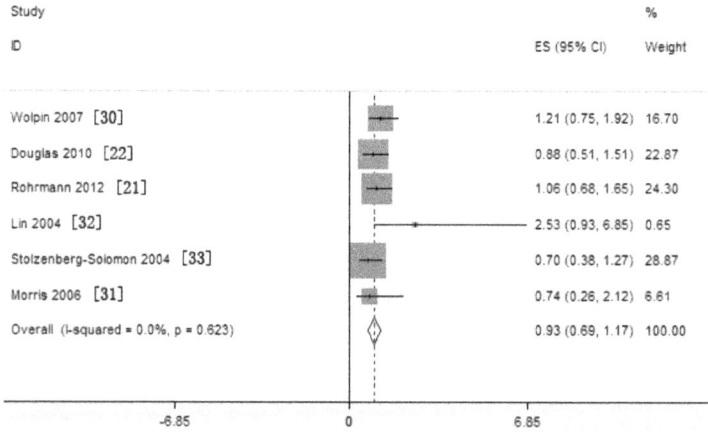

Figure 4. Forest plot of Insulin-like growth factor binding protein (IGFBP)-3 and pancreatic cancer (PaC) risk. CI = Confidence interval; ES = Effect size.

Similarly, meta-analysis of four nested case-control studies in a fixed-effects model showed that the serum IGF-I/IGFBP-3 ratio was not associated with the risk of PaC (SRRs = 0.97, 95% CI: 0.71–1.23; test for heterogeneity $p = 0.379$, $I^2 = 2.8\%$) (Figure 5).

In a sensitivity analysis, the overall homogeneity and effect size was calculated by removing one study at a time. The direction of effect did not change when any study was excluded, supporting the stability of the lack of correlation of IGF-I, IGF-II and IGFBP-3 concentration as well as the IGF-I/IGFBP-3 ratio with an increased risk of PaC.

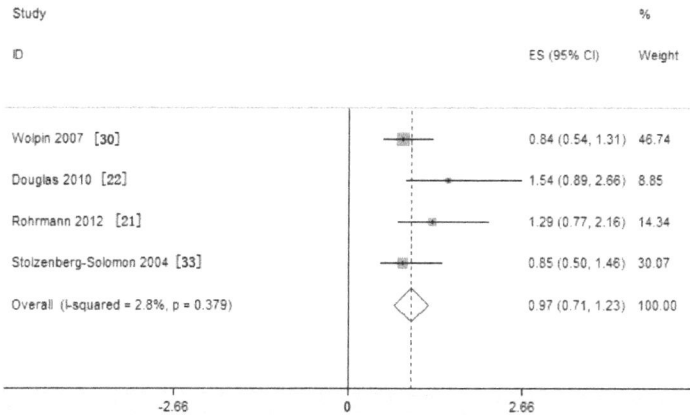

Figure 5. Forest plot of the insulin-like growth factor (IGF)-I/Insulin-like growth factor binding protein (IGFBP)-3 ratio and pancreatic cancer (PaC) risk. CI = Confidence interval; ES = Effect size.

We subsequently conducted a sub-group systematic review and meta-analysis according to adjustments for confounding variables: alcohol consumption, smoking and DM. Alcohol consumption, smoking and DM are important confounders for risk of PaC. When we limited the meta-analysis to studies that controlled for one of the above confounders or all of them, no positive association was found (Table 2).

Table 2. Sub-group analysis of relative risks for the association between IGF-I and IGFBP-3 with pancreatic cancer.

Subgroup	References	Relative Risk (95% CI)	Tests for Heterogeneity	
			I^2 (%)	*p*
IGF-I				
Adjustment for alcohol				
Yes	[21,22,33]	0.92 (0.60, 1.24)	48.7	0.142
No	[30–32]	0.91 (0.52, 1.30)	0	0.625
Adjustment for DM				
Yes	[21,22,30,32]	1.10 (0.77, 1.43)	0	0.560
No	[31,33]	0.67 (0.29, 1.05)	0	0.984
Adjustment for smoking, alcohol and DM				
Yes	[21,22]	1.27 (0.78, 1.77)	0	0.442
No	[30–33]	0.80 (0.51, 1.08)	0	0.661
IGFBP-3				
Adjustment for alcohol				
Yes	[21,22,33]	0.87 (0.59, 1.14)	0	0.562
No	[30–32]	1.12 (0.63, 1.60)	0	0.448
Adjustment for DM				
Yes	[21,22,30,32]	1.05 (0.75, 1.35)	0	0.638
No	[31,33]	0.71 (0.31, 1.11)	0	0.939
Adjustment for smoking, alcohol and DM				
Yes	[21]	0.97 (0.62, 1.32)	0	0.613
No	[30–33]	0.89 (0.56, 1.22)	4.1	0.372

CI = Confidence interval; I^2 = *I*-square; DM = Diabetes mellitus; IGF = Insulin-like growth factor.

One study reported the association between IGFBP-1 and risk of PaC. IGFBP-1 was not associated with risk of PaC (RR = 0.56, 95% CI: 0.31–1.01) (Table 1).

3.4. Publication Bias

The shape of the funnel plots for studies examining the association of IGF-I and IGFBP-3 concentration with PaC risk seemed symmetrical, indicating no publication bias (Figure 6).

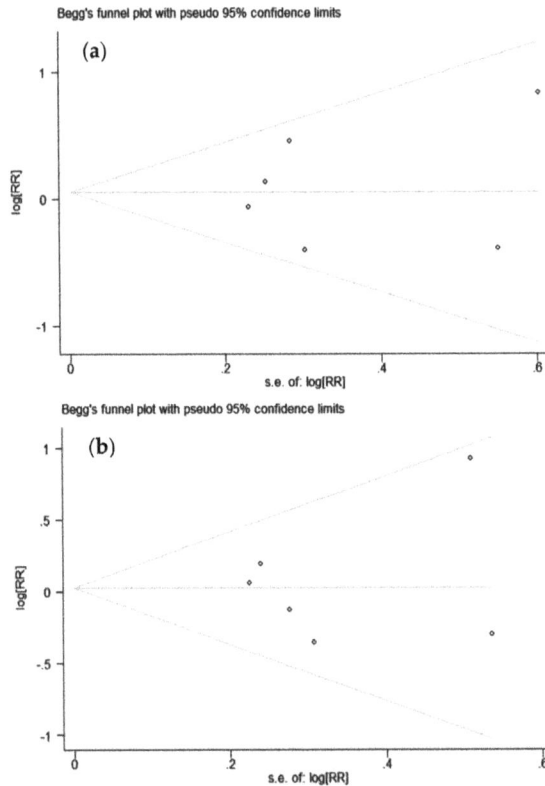

Figure 6. Funnel plot of studies evaluating the association of (a) insulin-like growth factor (IGF)-I with pancreatic cancer (PaC) risk (Begg's test (p = 1.00) [21,22,30–33], Egger's test (p = 0.794) [21,22,30–33]); and (b) IGFBP-3 with PaC risk (Begg's test (p = 0.707) [21,22,30–33], Egger's test (p = 0.785) [21,22,30–33]. RR = Relative risk.

4. Discussion

In this collaborative meta-analysis, the results showed for the first time that serum IGF-I, IGF-II, IGFBP-1 and IGFBP-3 concentrations as well as the IGF-I/IGFBP-3 ratio were not associated with risk of PaC. Sub-group analysis also did not show any significant associations. The conclusion was contrary to the results found in a meta-analysis of colorectal cancer[16], which showed that IGF-I and IGF-II significantly increased colorectal cancer risk (19 studies included, OR = 1.25, 95% CI: 1.08–1.45 for IGF-I; OR = 1.52, 95% CI: 1.16–2.01 for IGF-II), meta-analysis of breast cancer [17], which showed that IGF-I is positively associated with an increased risk of breast cancer (17 studies included, OR = 1.28, 95% CI: 1.14–1.44), and meta-analysis of lung cancer [35], which showed that IGFBP-3 was inversely associated with an increased risk of this cancer (six studies included, OR = 0.68, 95% CI: 0.48–0.88).

Interest in IGF-I utilized as a nutritional biomarker began in 1973 when its serum concentration was observed to fall in malnutrition [36]. A previous study showed that starvation, fasting and caloric restriction all resulted in a decrease in serum IGF-I concentration, the physiological function of which was to convert substrates to energy production. Low IGF-I concentration leads to protein catabolism in skeletal muscle, transferring amino acids for hepatic gluconeogenesis, which maintains the glucose level needed to keep the main organs functioning. The decrease in IGF-I concentration also results in enhanced growth hormone (GH) secretion, which enhances hepatic gluconeogenesis by antagonizing insulin's suppressive function and also by providing more amino acids from muscle [15]. Furthermore, the decrease in IGF-I concentration was more obvious in those with protein and energy malnutrition compared with protein malnutrition alone [37]. However, optimal intakes of both protein and energy are necessary for maintaining an appropriate IGF-I level [38]. The serum IGF-I level appears to be sensitive to both the amount and type of fat provided in nutritional support. Fish oil and low fat formula was significantly related to a faster recovery of the serum IGF-I concentration [39].

IGFs share structural homology and in vitro metabolic activity with insulin, both of which play an important role in proliferation and differentiation of normal and malignant cells. However, they have different receptors. The affinity of the IGF receptor for IGFs is 1000 times greater than that for insulin, while the insulin receptor shows 100 times greater affinity to insulin than that for IGFs [40]. The insulin-like growth factor axis is composed of two ligands (IGF-I and IGF-II), three cell-membrane receptors (insulin receptor (IR), IGF-I receptor (IGF-IR) and IGF-II receptor (IGF-IIR)) and six high-affinity IGF binding proteins (IGFBP-1 to IGFBP-6). Insulin is the main regulator of glucose metabolism, but the IGF axis also exerts insulin-like actions and increases insulin sensitivity. Recombinant human IGF-I could increase insulin sensitivity and improves glycemic control in type 2 diabetes mellitus (T2DM) [41]. The serum concentration of IGF-I was independently associated with insulin sensitivity in subjects with different degrees of glucose tolerance [42]. IGFBPs might influence the risk of DM. The decrease in IGF levels, controlled by the increase of IGFBP-1, served to protect against possible insulin-like activity of the IGFs during fasting [43]. An elevation in IGFBP-1 decreased free IGF-I in serum and muscle protein synthesis under stress conditions [44]. However, the association of the IGF axis with DM might not be causal or pathological. Basic research showed that early T2DM and impaired glucose tolerance are usually characterized by insulin resistance and hyperinsulinemia. Insulin could stimulate hepatic IGF-I synthesis, suppress hepatic IGFBP-1 synthesis in the liver, which could lead to an increase in the serum concentration of IGF-I. Thus, high serum IGF-I levels in patients with T2DM might be due to high insulin levels rather than the biological impact of the IGF axis on DM pathogenesis [45].

The association between IGF axis and risk of PaC is biologically plausible. About 99% of IGFs were combined with IGFBPs. Less than 1% of IGFs were free in serum. Free IGFs in circulation plays an important role in the regulation of cell behavior by binding to its receptor. Nevertheless, IGFBPs can inhibit the activities of IGFs by competitively binding to it and thereby reducing its bioavailability. Furthermore, in vitro experiments showed that exogenously adding insulin, IGF-I and IGF-II stimulated the growth of PaC cell lines via the PI3-kinase pathway, while the mitogenic effects were markedly blocked by providing anti-insulin receptor substrate-1 antibody or PI3-kinase inhibitor [46,47]. Small interfering RNA targeting IGF-IR [48] or anti-IGF-IR antibody [49,50] could be effective and efficient against the growth and metastasis of PaC cell lines.

Based on the compelling preclinical rationale, the IGF axis showed great promise in the diagnosis of PaC. The combination of the carbohydrate cancer antigen 19-9, IGF-I and albumin resulted in a combined area under the curve of 0.959 with 93.6% sensitivity and 95% specificity, much higher than CA 19-9 alone[51]. However, clinical trials targeting the IGF axis produced disappointment results. A phase II randomized, double-blind, placebo-controlled trial showed that ganitumab (monoclonal antibody inhibitor of IGF-IR, AMG 479) combined with gemcitabine had manageable toxicity but did not improve overall survival [52]. Similarly, adding another IGF-IR inhibitor, cixutumumab to erlotinib and gemcitabine also did not lead to longer progress-free survival or overall survival in metastatic PaC [53]. However, a randomized phase II study showed that higher levels of IGF-I, IGF-II and IGFBP-3 or lower levels of IGFBP-2 were associated with improved overall survival in metastatic PaC patients treated with ganitumab versus placebo [54]. Based on the results of the present study, scientists and clinicians should reconsider the role of the IGF axis in the development and progression of PaC in addition to the targeted therapies focused on the IGF axis.

There are three strengths of the present study. (1) To our knowledge, this study is the first to investigate the association between the IGF axis and risk of PaC; (2) Although limited studies were included, we have performed a comprehensive and systematic search of the literature by using an extensive search strategy; (3) The majority of the included studies were nested case-control studies, which could be effectively controlled for confounding factors.

This meta-analysis has limitations that affect interpretation of the true results. First, all studies in this meta-analysis used a nested case-control study or case-control design, which was more susceptible to recall and selection biases. Second, this investigation did not have sufficient information to perform sub-group analysis, which might affect the stability of the results due to heterogeneity across studies, and might miss some positive results in sub-groups. The carcinogenesis initiated or promoted by trace amounts of growth factors is complicated and naturally lasts for longer periods of time, so follow-up time is one of the most important factors affecting the result. For example, serum transforming growth factor-$\beta1$ was not associated with an increase in pancreatic cancer risk. However, this association differed significantly by follow-up time. Higher risk was observed during follow-up time of more than 10 years (OR = 2.13, 95% CI: 1.23–3.68) [55]. Finally, unmeasured or uncontrolled confounding inherited from original studies is a concern in this meta-analysis. Most estimate risks were derived from multivariable models, but individual studies did not adjust for potential confounding factors in a consistent way.

5. Conclusions

Our meta-analysis of observational studies provided evidence for the first time that serum IGF-I, IGF-II, IGFBP-1 and IGFBP-3 concentrations as well as the IGF-I/IGFBP-3 ratio were not associated with an increased risk of PaC. Given the small number of studies included in this meta-analysis, limited details and the non-randomized controlled study designs, further prospective cohort studies with a larger sample size and more accurate assessment of baseline characteristics in addition to being well-controlled for confounding factors are needed to affirm the effect of the IGF axis on PaC.

Acknowledgments: This work was supported by a grant from the Scientific Study Project Foundation of Guangzhou Medical University (No. 2015C38). We would like to acknowledge Zhiming Tan for her help in statistical analysis.

Author Contributions: Yuanfeng Gong and Yunqiang Tang conceived and designed the study. Yuanfeng Gong and Yadi Liao performed a literature search and identified eligible studies. Cong Mai, Tiejun Chen and Hui Tang extracted data from retrieved studies. Yuanfeng Gong carried out statistical analysis and interpreted results. The authors do not have any possible conflicts of interest. All drafts of the reports were written by Yuanfeng Gong and Bingyi Zhang. All authors read and approved the final paper.

Conflicts of Interest: The authors declare no conflict of interests.

Abbreviations

BMI	Body mass index
CI	Confidence interval
DM	Diabetes mellitus
IGF	Insulin-like growth factor
IGFBP	Insulin-like growth factor binding protein
OR	Odds ratio
PaC	Pancreatic cancer
RR	Relative risk
SRRs	Summary relative risks

References

1. Siegel, R.L.; Miller, K.D.; Jemal, A. Cancer statistics, 2016. *CA Cancer J. Clin.* **2016**, *66*, 7–30. [CrossRef] [PubMed]
2. Statbite, U.S. Pancreatic cancer rates. *J. Natl. Cancer Inst.* **2010**, *102*, 1822.
3. Chen, W.; Zheng, R.; Baade, P.D.; Zhang, S.; Zeng, H.; Bray, F.; Jemal, A.; Yu, X.Q.; He, J. Cancer statistics in china, 2015. *CA Cancer J. Clin.* **2016**, *66*, 115–132. [CrossRef] [PubMed]
4. Zou, L.; Zhong, R.; Shen, N.; Chen, W.; Zhu, B.; Ke, J.; Lu, X.; Zhang, T.; Lou, J.; Wang, Z.; et al. Non-linear dose-response relationship between cigarette smoking and pancreatic cancer risk: evidence from a meta-analysis of 42 observational studies. *Eur. J. Cancer* **2014**, *50*, 193–203. [CrossRef] [PubMed]
5. Bosetti, C.; Lucenteforte, E.; Silverman, D.T.; Petersen, G.; Bracci, P.M.; Ji, B.T.; Negri, E.; Li, D.; Risch, H.A.; Olson, S.H.; et al. Cigarette smoking and pancreatic cancer: An analysis from the international pancreatic cancer case-control consortium (PANC4). *Ann. Oncol.* **2012**, *23*, 1880–1888. [CrossRef] [PubMed]
6. Iodice, S.; Gandini, S.; Maisonneuve, P.; Lowenfels, A.B. Tobacco and the risk of pancreatic cancer: A review and meta-analysis. *Langenbecks Arch. Surg.* **2008**, *393*, 535–545. [CrossRef] [PubMed]
7. Aune, D.; Greenwood, D.C.; Chan, D.S.; Vieira, R.; Vieira, A.R.; Navarro, R.D.; Cade, J.E.; Burley, V.J.; Norat, T. Body mass index, abdominal fatness and pancreatic cancer risk: A systematic review and non-linear dose-response meta-analysis of prospective studies. *Ann. Oncol.* **2012**, *23*, 843–852. [CrossRef] [PubMed]
8. Arslan, A.A.; Helzlsouer, K.J.; Kooperberg, C.; Shu, X.O.; Steplowski, E.; Bueno-de-Mesquita, H.B.; Fuchs, C.S.; Gross, M.D.; Jacobs, E.J.; Lacroix, A.Z.; et al. Anthropometric measures, body mass index, and pancreatic cancer: A pooled analysis from the pancreatic cancer cohort consortium (PanScan). *Arch. Intern. Med.* **2010**, *170*, 791–802. [CrossRef] [PubMed]
9. Tong, G.X.; Geng, Q.Q.; Chai, J.; Cheng, J.; Chen, P.L.; Liang, H.; Shen, X.R.; Wang, D.B. Association between pancreatitis and subsequent risk of pancreatic cancer: A systematic review of epidemiological studies. *Asian Pac. J. Cancer Prev.* **2014**, *15*, 5029–5034. [CrossRef] [PubMed]
10. Raimondi, S.; Lowenfels, A.B.; Morselli-Labate, A.M.; Maisonneuve, P.; Pezzilli, R. Pancreatic cancer in chronic pancreatitis; Aetiology, incidence, and early detection. *Best Pract. Res. Clin. Gastroenterol.* **2010**, *24*, 349–358. [CrossRef] [PubMed]
11. Pannala, R.; Basu, A.; Petersen, G.M.; Chari, S.T. New-onset diabetes: a potential clue to the early diagnosis of pancreatic cancer. *Lancet Oncol.* **2009**, *10*, 88–95. [CrossRef]
12. Gong, Y.; Li, S.; Tang, Y.; Mai, C.; Ba, M.; Jiang, P.; Tang, H. Cholelithiasis and risk of pancreatic cancer: Systematic review and meta-analysis of 21 observational studies. *Cancer Causes Control* **2014**, *25*, 1543–1551. [CrossRef] [PubMed]
13. Lin, G.; Zeng, Z.; Wang, X.; Wu, Z.; Wang, J.; Wang, C.; Sun, Q.; Chen, Y.; Quan, H. Cholecystectomy and risk of pancreatic cancer: A meta-analysis of observational studies. *Cancer Causes Control* **2012**, *23*, 59–67. [CrossRef] [PubMed]
14. Gong, Y.; Zhou, Q.; Zhou, Y.; Lin, Q.; Zeng, B.; Chen, R.; Li, Z. Gastrectomy and risk of pancreatic cancer: Systematic review and meta-analysis of observational studies. *Cancer Causes Control* **2012**, *23*, 1279–1288. [CrossRef] [PubMed]
15. Livingstone, C. Insulin-like growth factor-I (IGF-I) and clinical nutrition. *Clin. Sci. (Lond.)* **2013**, *125*, 265–280. [CrossRef] [PubMed]
16. Chi, F.; Wu, R.; Zeng, Y.C.; Xing, R.; Liu, Y. Circulation insulin-like growth factor peptides and colorectal cancer risk: An updated systematic review and meta-analysis. *Mol. Biol. Rep.* **2013**, *40*, 3583–3590. [CrossRef] [PubMed]
17. Rinaldi, S.; Cleveland, R.; Norat, T.; Biessy, C.; Rohrmann, S.; Linseisen, J.; Boeing, H.; Pischon, T.; Panico, S.; Agnoli, C.; et al. Serum levels of IGF-I, IGFBP-3 and colorectal cancer risk: Results from the epic cohort, plus a meta-analysis of prospective studies. *Int. J. Cancer* **2010**, *126*, 1702–1715. [CrossRef] [PubMed]
18. Key, T.J.; Appleby, P.N.; Reeves, G.K.; Roddam, A.W. Insulin-like growth factor 1 (IGF-1), IGF binding protein 3 (IGFBP-3), and breast cancer risk: pooled individual data analysis of 17 prospective studies. *Lancet Oncol.* **2010**, *11*, 530–542. [PubMed]
19. Rowlands, M.A.; Gunnell, D.; Harris, R.; Vatten, L.J.; Holly, J.M.; Martin, R.M. Circulating insulin-like growth factor peptides and prostate cancer risk: a systematic review and meta-analysis. *Int. J. Cancer* **2009**, *124*, 2416–2429. [CrossRef] [PubMed]
20. Cao, H.; Wang, G.; Meng, L.; Shen, H.; Feng, Z.; Liu, Q.; Du, J. Association between circulating levels of IGF-1 and IGFBP-3 and lung cancer risk: A meta-analysis. *PLoS ONE* **2012**, *7*, e49884. [CrossRef] [PubMed]

21. Rohrmann, S.; Grote, V.A.; Becker, S.; Rinaldi, S.; Tjonneland, A.; Roswall, N.; Gronbaek, H.; Overvad, K.; Boutron-Ruault, M.C.; Clavel-Chapelon, F.; et al. Concentrations of IGF-I and IGFBP-3 and pancreatic cancer risk in the european prospective investigation into cancer and nutrition. *Br. J. Cancer* **2012**, *106*, 1004–1110. [CrossRef] [PubMed]

22. Douglas, J.B.; Silverman, D.T.; Pollak, M.N.; Tao, Y.; Soliman, A.S.; Stolzenberg-Solomon, R.Z. Serum IGF-I, IGF-II, IGFBP-3, and IGF-I/IGFBP-3 molar ratio and risk of pancreatic cancer in the prostate, lung, colorectal, and ovarian cancer screening trial. *Cancer Epidemiol. Biomark. Prev.* **2010**, *19*, 2298–2306. [CrossRef] [PubMed]

23. El-Mesallamy, H.O.; Hamdy, N.M.; Zaghloul, A.S.; Sallam, A.M. Clinical value of circulating lipocalins and insulin-like growth factor axis in pancreatic cancer diagnosis. *Pancreas* **2013**, *42*, 149–154. [CrossRef] [PubMed]

24. Meggiato, T.; Plebani, M.; Basso, D.; Panozzo, M.P.; Del, F.G. Serum growth factors in patients with pancreatic cancer. *Tumour Biol.* **1999**, *20*, 65–71. [CrossRef] [PubMed]

25. Evans, J.D.; Eggo, M.C.; Donovan, I.A.; Bramhall, S.R.; Neoptolemos, J.P. Serum levels of insulin-like growth factors (IGF-I and IGF-II) and their binding protein (IGFBP-3) are not elevated in pancreatic cancer. *Int. J. Pancreatol.* **1997**, *22*, 95–100. [PubMed]

26. Higgins, J.P.; Thompson, S.G. Quantifying heterogeneity in a meta-analysis. *Stat. Med.* **2002**, *21*, 1539–1558. [CrossRef] [PubMed]

27. Higgins, J.P.; Thompson, S.G.; Deeks, J.J.; Altman, D.G. Measuring inconsistency in meta-analyses. *BMJ* **2003**, *327*, 557–560. [CrossRef] [PubMed]

28. Sterne, J.A.; Egger, M. Funnel plots for detecting bias in meta-analysis: Guidelines on choice of axis. *J. Clin. Epidemiol.* **2001**, *54*, 1046–1055. [CrossRef]

29. Wolpin, B.M.; Michaud, D.S.; Giovannucci, E.L.; Schernhammer, E.S.; Stampfer, M.J.; Manson, J.E.; Cochrane, B.B.; Rohan, T.E.; Ma, J.; Pollak, M.N.; et al. Circulating insulin-like growth factor binding protein-1 and the risk of pancreatic cancer. *Cancer Res.* **2007**, *67*, 7923–7928. [CrossRef] [PubMed]

30. Wolpin, B.M.; Michaud, D.S.; Giovannucci, E.L.; Schernhammer, E.S.; Stampfer, M.J.; Manson, J.E.; Cochrane, B.B.; Rohan, T.E.; Ma, J.; Pollak, M.N.; et al. Circulating insulin-like growth factor axis and the risk of pancreatic cancer in four prospective cohorts. *Br. J. Cancer* **2007**, *97*, 98–104. [CrossRef] [PubMed]

31. Morris, J.K.; George, L.M.; Wu, T.; Wald, N.J. Insulin-like growth factors and cancer: No role in screening. Evidence from the BUPA study and meta-analysis of prospective epidemiological studies. *Br. J. Cancer* **2006**, *95*, 112–117. [CrossRef] [PubMed]

32. Lin, Y.; Tamakoshi, A.; Kikuchi, S.; Yagyu, K.; Obata, Y.; Ishibashi, T.; Kawamura, T.; Inaba, Y.; Kurosawa, M.; Motohashi, Y.; et al. Serum insulin-like growth factor-I, insulin-like growth factor binding protein-3, and the risk of pancreatic cancer death. *Int. J. Cancer* **2004**, *110*, 584–588. [CrossRef] [PubMed]

33. Stolzenberg-Solomon, R.Z.; Limburg, P.; Pollak, M.; Taylor, P.R.; Virtamo, J.; Albanes, D. Insulin-like growth factor (IGF)-1, IGF-binding protein-3, and pancreatic cancer in male smokers. *Cancer Epidemiol. Biomark. Prev.* **2004**, *13*, 438–444.

34. Karna, E.; Surazynski, A.; Orlowski, K.; Laszkiewicz, J.; Puchalski, Z.; Nawrat, P.; Palka, J. Serum and tissue level of insulin-like growth factor-I (IGF-I) and IGF-I binding proteins as an index of pancreatitis and pancreatic cancer. *Int. J. Exp. Pathol.* **2002**, *83*, 239–245. [CrossRef] [PubMed]

35. Chen, B.; Liu, S.; Xu, W.; Wang, X.; Zhao, W.; Wu, J. IGF-I and IGFBP-3 and the risk of lung cancer: A meta-analysis based on nested case-control studies. *J. Exp. Clin. Cancer Res.* **2009**, *28*, 89. [CrossRef] [PubMed]

36. Grant, D.B.; Hambley, J.; Becker, D.; Pimstone, B.L. Reduced sulphation factor in undernourished children. *Arch. Dis. Child.* **1973**, *48*, 596–600. [CrossRef] [PubMed]

37. Donahue, S.P.; Phillips, L.S. Response of IGF-1 to nutritional support in malnourished hospital patients: A possible indicator of short-term changes in nutritional status. *Am. J. Clin. Nutr.* **1989**, *50*, 962–969. [PubMed]

38. Thissen, J.P.; Ketelslegers, J.M.; Underwood, L.E. Nutritional regulation of the insulin-like growth factors. *Endocr. Rev.* **1994**, *15*, 80–101. [PubMed]

39. Abribat, T.; Nedelec, B.; Jobin, N.; Garrel, D.R. Decreased serum insulin-like growth factor-I in burn patients: Relationship with serum insulin-like growth factor binding protein-3 proteolysis and the influence of lipid composition in nutritional support. *Crit. Care Med.* **2000**, *28*, 2366–2372. [CrossRef] [PubMed]

40. Fujita-Yamaguchi, Y.; LeBon, T.R.; Tsubokawa, M.; Henzel, W.; Kathuria, S.; Koyal, D.; Ramachandran, J. Comparison of insulin-like growth factor I receptor and insulin receptor purified from human placental membranes. *J. Biol. Chem.* **1986**, *261*, 16727–16731. [PubMed]

41. Moses, A.C.; Young, S.C.; Morrow, L.A.; O'Brien, M.; Clemmons, D.R. Recombinant human insulin-like growth factor I increases insulin sensitivity and improves glycemic control in type II diabetes. *Diabetes* **1996**, *45*, 91–100. [CrossRef] [PubMed]

42. Sesti, G.; Sciacqua, A.; Cardellini, M.; Marini, M.A.; Maio, R.; Vatrano, M.; Succurro, E.; Lauro, R.; Federici, M.; Perticone, F. Plasma concentration of IGF-I is independently associated with insulin sensitivity in subjects with different degrees of glucose tolerance. *Diabetes Care* **2005**, *28*, 120–125. [CrossRef] [PubMed]

43. Katz, L.E.; DeLeon, D.D.; Zhao, H.; Jawad, A.F. Free and total insulin-like growth factor (IGF)-I levels decline during fasting: Relationships with insulin and IGF-binding protein-1. *J. Clin. Endocrinol. Metab.* **2002**, *87*, 2978–2983. [CrossRef] [PubMed]

44. Lang, C.H.; Vary, T.C.; Frost, R.A. Acute in Vivo elevation of insulin-like growth factor (IGF) binding protein-1 decreases plasma free IGF-I and muscle protein synthesis. *Endocrinology* **2003**, *144*, 3922–3933. [CrossRef] [PubMed]

45. Brismar, K.; Fernqvist-Forbes, E.; Wahren, J.; Hall, K. Effect of insulin on the hepatic production of insulin-like growth factor-binding protein-1 (IGFBP-1), IGFBP-3, and IGF-I in insulin-dependent diabetes. *J. Clin. Endocrinol. Metab.* **1994**, *79*, 872–878. [PubMed]

46. Kornmann, M.; Maruyama, H.; Bergmann, U.; Tangvoranuntakul, P.; Beger, H.G.; White, M.F.; Korc, M. Enhanced expression of the insulin receptor substrate-2 docking protein in human pancreatic cancer. *Cancer Res.* **1998**, *58*, 4250–4254. [PubMed]

47. Bergmann, U.; Funatomi, H.; Kornmann, M.; Beger, H.G.; Korc, M. Increased expression of insulin receptor substrate-1 in human pancreatic cancer. *Biochem. Biophys. Res. Commun.* **1996**, *220*, 886–890. [CrossRef] [PubMed]

48. Subramani, R.; Lopez-Valdez, R.; Arumugam, A.; Nandy, S.; Boopalan, T.; Lakshmanaswamy, R. Targeting insulin-like growth factor 1 receptor inhibits pancreatic cancer growth and metastasis. *PLoS ONE* **2014**, *9*, e97016. [CrossRef] [PubMed]

49. Awasthi, N.; Zhang, C.; Ruan, W.; Schwarz, M.A.; Schwarz, R.E. BMS-754807, a small-molecule inhibitor of insulin-like growth factor-1 receptor/insulin receptor, enhances gemcitabine response in pancreatic cancer. *Mol. Cancer Ther.* **2012**, *11*, 2644–2653. [CrossRef] [PubMed]

50. Moser, C.; Schachtschneider, P.; Lang, S.A.; Gaumann, A.; Mori, A.; Zimmermann, J.; Schlitt, H.J.; Geissler, E.K.; Stoeltzing, O. Inhibition of insulin-like growth factor-I receptor (IGF-IR) using NVP-AEW541, a small molecule kinase inhibitor, reduces orthotopic pancreatic cancer growth and angiogenesis. *Eur. J. Cancer* **2008**, *44*, 1577–1586. [CrossRef] [PubMed]

51. Ferri, M.J.; Saez, M.; Figueras, J.; Fort, E.; Sabat, M.; López-Ben, S.; de Llorens, R.; Aleixandre, R.N.; Peracaula, R. Improved pancreatic adenocarcinoma diagnosis in jaundiced and non-jaundiced pancreatic adenocarcinoma patients through the combination of routine clinical markers associated to pancreatic adenocarcinoma pathophysiology. *PLoS ONE* **2016**, *11*, e0147214. [CrossRef] [PubMed]

52. Kindler, H.L.; Richards, D.A.; Garbo, L.E.; Garon, E.B.; Stephenson, J.J.; Rocha-Lima, C.M.; Safran, H.; Chan, D.; Kocs, D.M.; Galimi, F.; et al. A randomized, placebo-controlled phase 2 study of ganitumab (AMG 479) or conatumumab (AMG 655) in combination with gemcitabine in patients with metastatic pancreatic cancer. *Ann. Oncol.* **2012**, *23*, 2834–2842. [CrossRef] [PubMed]

53. Philip, P.A.; Goldman, B.; Ramanathan, R.K.; Lenz, H.J.; Lowy, A.M.; Whitehead, R.P.; Wakatsuki, T.; Iqbal, S.; Gaur, R.; Benedetti, J.K.; et al. Dual blockade of epidermal growth factor receptor and insulin-like growth factor receptor-1 signaling in metastatic pancreatic cancer: Phase Ib and randomized phase II trial of gemcitabine, erlotinib and cixutumumab versus gemcitabine plus erlotinib (SWOG S0727). *Cancer* **2014**, *120*, 2980–2985. [PubMed]

54. McCaffery, I.; Tudor, Y.; Deng, H.; Tang, R.; Suzuki, S.; Badola, S.; Kindler, H.L.; Fuchs, C.S.; Loh, E.; Patterson, S.D.; et al. Putative predictive biomarkers of survival in patients with metastatic pancreatic adenocarcinoma treated with gemcitabine and ganitumab, an IGF1R inhibitor. *Clin. Cancer Res.* **2013**, *19*, 4282–4289. [CrossRef] [PubMed]

55. Jacobs, E.J.; Newton, C.C.; Silverman, D.T.; Nogueira, L.M.; Albanes, D.; Mannisto, S.; Pollak, M.; Stolzenberg-Solomon, R.Z. Serum transforming growth factor-beta1 and risk of pancreatic cancer in three prospective cohort studies. *Cancer Causes Control* **2014**, *25*, 1083–1091. [CrossRef] [PubMed]

nutrients

MDPI

Article

Chondroprotective Effects of Ginsenoside Rg1 in Human Osteoarthritis Chondrocytes and a Rat Model of Anterior Cruciate Ligament Transection

Wendan Cheng [1,2], **Juehua Jing** [1], **Zhen Wang** [3], **Dongying Wu** [4,*] **and Yumin Huang** [5,*]

1   Department of Orthopedics, The Second Hospital of Anhui Medical University, No. 678 Furong Road,
     Hefei 230601, China; sunyccc@126.com (W.C.); jjhhu@sina.com (J.J.)
2   Department of Orthopedics, Lu'an People's Hospital Affiliated to Anhui Medical University,
     Lu'an 237000, China
3   Department of Orthopedics, The Peoples Hospital of Luhe Affiliated to Yangzhou University Medical
     Academy, Nanjing 211500, China; dzwangzhen@126.com
4   Department of Orthopedics, The Affiliated Hospital of Xuzhou Medical University, No. 99 Huaihai West
     Road, Xuzhou 221000, China
5   Department of Orthopedics, The First Affiliated Hospital of Nanjing Medical University,
     No. 300 Guangzhou Road, Nanjing 210029, China
*   Correspondance: officialwdy@126.com (D.W.); officialzxz@126.com (Y.H.);
     Tel.: +86-135-1256-1916 (D.W.); +86-139-1303-8613 (Y.H.)

Received: 13 December 2016; Accepted: 3 March 2017; Published: 10 March 2017

Abstract: This study aimed to assess whether Ginsenoside Rg1 (Rg1) inhibits inflammatory responses in human chondrocytes and reduces articular cartilage damage in a rat model of osteoarthritis (OA). Gene expression and protein levels of type II collagen, aggrecan, matrix metalloproteinase (MMP)-13 and cyclooxygenase-2 (COX-2) were determined in vitro by quantitative real-time-polymerase chain reaction and Western blotting. Prostaglandin E2 (PGE_2) amounts in the culture medium were determined by enzyme-linked immunosorbent assay (ELISA). For in vivo assessment, a rat model of OA was generated by anterior cruciate ligament transection (ACLT). Four weeks after ACLT, Rg1 (30 or 60 mg/kg) or saline was administered by gavage once a day for eight consecutive weeks. Joint damage was analyzed by histology and immunohistochemistry. Ginsenoside Rg1 inhibited Interleukin (IL)-1β-induced chondrocyte gene and protein expressions of MMP-13, COX-2 and PGE_2, and prevented type II collagen and aggrecan degradation, in a dose-dependent manner. Administration of Ginsenoside Rg1 to OA rats attenuated cartilage degeneration, and reduced type II collagen loss and MMP-13 levels. These findings demonstrated that Ginsenoside Rg1 can inhibit inflammatory responses in human chondrocytes in vitro and reduce articular cartilage damage in vivo, confirming the potential therapeutic value of Ginsenoside Rg1 in OA.

Keywords: ginsenoside-Rg1; osteoarthritis; chondrocyte

1. Introduction

Osteoarthritis (OA) is a pathological process characterized by degenerative changes and inflammatory responses in chondrocytes. Interleukin (IL)-1β, a major pro-inflammatory cytokine produced by chondrocytes and synovial cells, contributes to increased chondrocyte apoptosis [1–3], along with the synthesis of other inflammatory mediators, including matrix metalloproteinase (MMP), cyclooxygenase-2 (COX-2) and prostaglandin E2 (PGE_2). These inflammatory mediators ultimately inhibit type II collagen and aggrecan synthesis, increase extracellular matrix (ECM) degradation, and cause articular cartilage damage [4–7]. According to the 2015 recommendations for knee OA, the primary pharmacotherapy for OA currently comprises non-steroidal anti-inflammatory drugs

(NSAIDs) and hormone-like drugs. However, these drugs are only temporarily effective and exhibit numerous side effects. Therefore, safe, effective and economical strategies to inhibit both apoptosis and inflammation in chondrocytes for the treatment of OA are urgently required.

Ginseng is broadly used as an herbal medicine because of its wide-range efficacy and low side-effect profile [8,9]. Among the 30 ginsenosides, ginsenoside Rg1 (Rg1) is one of the major active ingredients of ginseng [10]. Rg1 is a proven effective agent for neurodegenerative diseases such as Alzheimer's disease, and exerts remarkable neuroprotective effects against tert-Butylhydroperoxide induced oxidative stress [11,12]. In addition, our previous study demonstrated that Rg1 protects chondrocytes from IL-1β-induced apoptosis via the phosphatidylinositol 3-kinase/protein kinase B signaling pathway, by preventing caspase-3 release [13]. These findings suggest that Rg1 may serve as a novel pharmacotherapeutic drug for OA. However, in addition to chondrocyte apoptosis, inflammation in chondrocytes plays a critical role in the pathogenesis of OA. The anti- inflammatory properties in chondrocytes and protective effects on OA in vivo of Rg1 have not been reported.

The present study assessed the ability of Rg1 to inhibit inflammatory responses and reduce articular cartilage damage in OA by using an in vitro model of human chondrocytes and an in vivo model of rat OA, and confirmed the potential therapeutic value of Rg1 in OA.

2. Materials and Methods

2.1. Materials

Ginsenoside Rg1 (>98% purity) was purchased from the National Institutes for Food and Drug Control (Beijing, China). Recombinant human IL-1β was from R&D Systems (Minneapolis, MN, USA). Type II collagen and MMP-13 antibodies used in immunohistochemistry were obtained from Boster (Wuhan, Hubei, China) and Abcam (Cambridge, MA, USA), respectively. All other antibodies were purchased from Santa Cruz Biotechnology Inc. (Santa Cruz, CA, USA). Enzyme-linked immunosorbent assay (ELISA) kits were from Enzo life sciences (Farmingdale, NY, USA). 3-(4,5-dimethylthiazal-2-yl)-2,5-diphenyl-tetrazolium bromide (MTT) was obtained from Sigma (St. Louis, MO, USA).

2.2. Cell Culture

The study protocol was approved by the Ethics committee of The First Affiliated Hospital to Nanjing Medical University; all patients involved provided the necessary informed consent for participation and publication (permit number 20120413). Articular chondrocytes were harvested from 20 patients who underwent total knee replacement. NSAIDs were dissolved at least 1 week prior to surgery. Isolated chondrocytes were cultured in Dulbecco's modified Eagle medium (DMEM) supplemented with 10% fetal bovine serum (FBS) at 37 °C in a 5% CO_2 incubator. First-generation cells were cultured in tissue culture dishes at a density of 1×10^6 cells/mL.

2.3. Animals

Forty-eight male Sprague-Dawley (SD) rats (Animal Inc., Nanjing Medical University, Nanjing, Jiangsu, China) weighing 240–280 g were housed in a controlled-temperature room (21–22 °C) (six per cage), with free access to food and water. To confirm the therapeutic potential of Rg1 in an OA model, forty-eight rats were randomized into four groups of 12 rats each. All animal procedures were approved by the Animal Research Ethics Committee of Nanjing Medical University.

2.4. Anterior Cruciate Ligament Transaction (ACLT)

The ACLT model was established as previously described [14]. Briefly, rats were anesthetized with chloral hydrate. The right knee joint was exposed through a medial parapatellar approach. The anterior cruciate ligament (ACL) was then transected with micro-scissors. Complete transection was confirmed by a positive anterior drawer sign. The control group underwent arthrotomy without transection of the ACL. After surgery, all animals were allowed free exercise out of their cages for 20 min daily.

2.5. Experimental Design

For the in vitro study, first-generation human chondrocytes were cultured in serum-free DMEM supplemented with 2% serum-free bovine serum albumin (BSA) for 12 h. To assess the anti-inflammatory effects of different concentrations of Rg1, cells were treated with 10 ng/mL IL-1β, alone or in combination with Rg1 at 0.1, 1 and 10 μg/mL. A control group was left untreated except for medium change. Cells were harvested after 24 h of incubation.

For in vivo assessment, rats underwent anterior cruciate ligament transaction as described above. Four weeks after ACLT, Rg1 intervention groups (R1 group, 30 mg/kg/day; R2 group, 60 mg/kg/day) received Rg1 dissolved in sterile saline by gavage once a day for eight consecutive weeks. Meanwhile, rats in control and ACLT groups were administered an equivalent volume of sterile saline. All animals were sacrificed 12 weeks post-surgery, and right knees were dissected to observe articular cartilage changes and for histological and immunohistochemical analyses.

2.6. Quantitative Real-Time-Polymerase Chain Reaction (q-PCR)

Total RNA was extracted with RNAiso Plus (TaKaRa, Dalian, China)), and reverse transcribed into cDNA using the PrimeScript RT reagent kit (TaKaRa, Dalian, China). Quantitative real-time PCR analysis was performed with SYBR Premix Ex Taq II reagents from TaKaRa (Dalian, China), according to the manufacturer's instructions. PCR reactions were performed in a 20 μL mixtures containing 2 μL of cDNA. Aggrecan, type II collagen, MMP-13, COX-2, PGE$_2$ and β-actin cDNAs were amplified using specific primers (TaKaRa). Gene expression was normalized to glyceraldehyde-3-phosphatedehydrogenase (GADPH) and derived by the $2^{-\Delta\Delta CT}$ method.

2.7. Western Blotting

Total protein was prepared using RIPA buffer (Beyotime, Haimen, Jiangsu, China); nuclear protein isolation was carried out with Nuclear Protein Extraction Kit (Beyotime) according to the manufacturer's instructions. Protein mixtures were separated by sodium dodecyl sulfate-polyacrylamide gel electrophoresis and transferred onto nitrocellulose membranes. The membranes were incubated with antibodies raised against the proteins of interest and detected by enhanced chemiluminescence (Thermo Fisher Scientific, Rockford, IL, USA). The data were analyzed with the Image Lab software (Bio-Rad Laboratories, Hercules, CA, USA).

2.8. ELISA

An ELISA kit (Enzo Life Sciences, Farmingdale, NY, USA) was used to measure PGE$_2$ levels in the culture medium, according to the manufacturer's protocol.

2.9. Histology and Immunohistochemistry

The rats in each group were sacrificed by cervical dislocation under anesthesia at 12 weeks after the surgery. The patella was isolated to expose the knee joint cavity. The fixed specimens were decalcified using 10% ethylenediaminetetraacetic for 18 days. After decalcification, the specimens were paraffin embedded. Standardized 3 μm serial sections were cut in the sagittal plane and stained with hematoxylin and eosin (H&E), Safranin O/fast green, and toluidine blue.

To analyze type II collagen and MMP-13 distribution in the cartilage, the sections were de-paraffinized and rehydrated in graded ethanol. After incubation with antibodies targeting rat type II collagen and MMP-13, the samples were treated with secondary antibodies. Finally, the specimens were photographed under a microscope (Olympus, Tokyo, Japan). Percentages of MMP-13-positive cells in cartilage samples were obtained in six fields at 200× magnification. Cell counts were performed by three independent observers, and mean values were used as the final scores.

A modified Mankin histological score was used to assess histological injuries of the articular cartilage as follows. Structural changes were scored on a scale of 0–6: 0 normal; 1 irregular surface, including fissures into the radial layer; 2 pannus; 3 absence of superficial cartilage layers; 4 slight disorganization (absent cell row and small superficial clusters); 5 fissure into the calcified cartilage layer; 6 disorganization (chaotic structure, clusters, and osteoclast activity). Cellular abnormalities were scored on a scale of 0–3: 0 normal; 1 hypercellularity, including small superficial clusters; 2 clusters; 3 hypocellularity. Matrix staining was scored on a scale of 0–4: 0 normal/slight reduction in staining; 1 reduced staining in the radial layer; 2 reduced staining in the interterritorial matrix; 3 staining present only in the pericellular matrix; 4 no staining [15]. Three independent observers assessed cartilage damage in a blinded manner.

2.10. Statistical Analysis

Data are mean ± Standard Error of Mean (SEM) unless otherwise stated. Gene expression, Western blot, MTT assay and immunohistochemistry data were analyzed by one-way Analysis of Variance followed by Dunnett's multiple comparisons. Differences in Mankin scores were assessed by the Kruskal–Wallis test. Dunnett's multiple comparison test was used to assess MMP-13-positive chondrocytes. $p < 0.05$ was considered statistically significant. Statistical analyses were performed with the SPSS software version 16.0 (SPSS Inc., Chicago, IL, USA) or GraphPad Prism (GraphPad Software, San Diego, CA, USA).

3. Results

3.1. Effects of Rg1 on Gene Expression of Extracellular Matrix and Inflammatory Mediators after Induction by IL-1β

Gene expression levels of type II collagen (Figure 1A) and aggrecan (Figure 1B) in the IL-1β group were reduced after treatment. They were increased by 1.7- and 2.1-fold after treatment with 1 μg/mL Rg1, and by 2.1- and 4.1-fold after treatment with 10 μg/mL Rg1 MMP-13 (Figure 1C) and COX-2 (Figure 1D) mRNA amounts in the IL-1β group were increased; they were reduced by 5.6- and 1.6-fold, and 7.5- and 2.2-fold, respectively, after treatment with 1 μg/mL and 10 μg/mL Rg1 (all $p < 0.05$). However, no effect was observed at 0.1 μg/mL Rg1. Thus, Rg1 effects were dose-dependent.

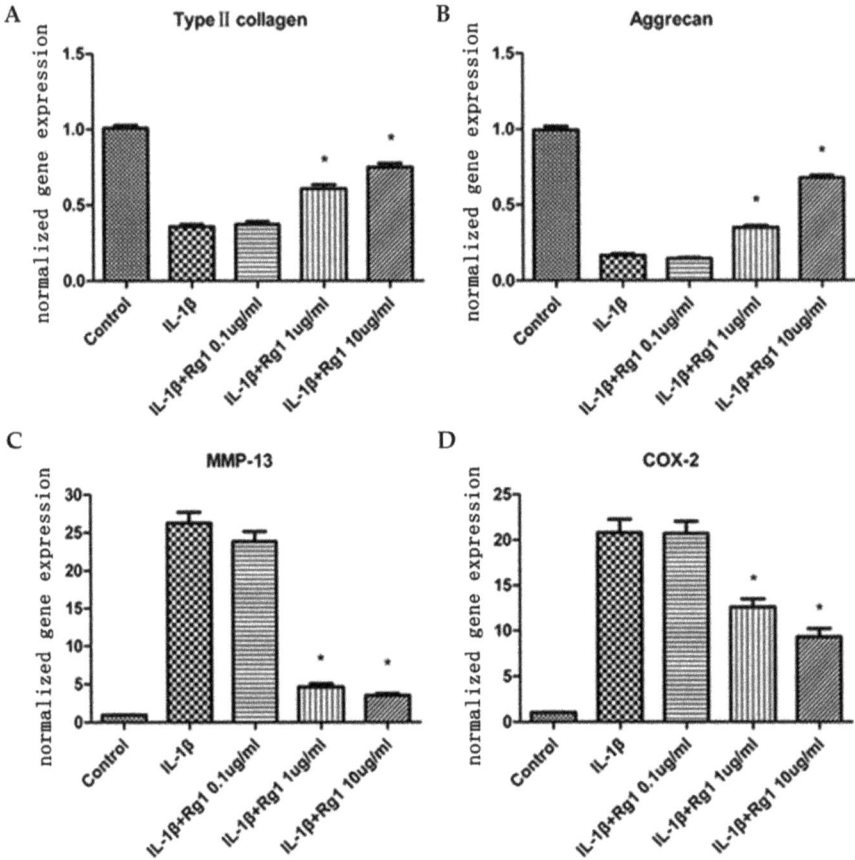

Figure 1. Effect of Ginsenoside Rg1 (Rg1) on gene expression levels of extracellular matrix and inflammatory mediators after induction by Interleukin (IL)-1β. Human osteoarthritis (OA) chondrocytes were treated with the medium (control group), and IL-1β (10 ng/mL) alone or in combination with Rg1 (0.1, 1, or 10 μg/mL). Gene expression levels of type II collagen (**A**), aggrecan (**B**), matrix metalloproteinase (MMP)-13 (**C**) and cyclooxygenase-2 (COX-2) (**D**) were determined by quantitative real-time PCR, normalized to glyceraldehyde-3-phosphatedehydrogenase (GADPH) and expressed as means ± Standard Error of Mean (SEM) of four independent experiments. * $p < 0.05$ compared with cells treated with IL-1β alone.

3.2. Effects of Rg1 on Protein Expression of Extracellular Matrix and Inflammatory Mediators after Induction by IL-1β

Protein levels of type II collagen, aggrecan, MMP-13, and COX-2 were analyzed by Western blotting (Figure 2A), and quantified by densitometry (Figure 2B). Protein levels of both type II collagen and aggrecan were reduced by IL-1β treatment; administration of 1 or 10 μg/mL Rg1 resulted in increased amounts of these proteins. Analysis of MMP-13 and COX-2 by Western blotting, alongside PGE$_2$ amount assessment by ELISA (Figure 2C), revealed that the levels of all three proteins increased significantly in the IL-1β-treatment group, and significantly inhibited by Rg1 at 1 or 10 μg/mL.

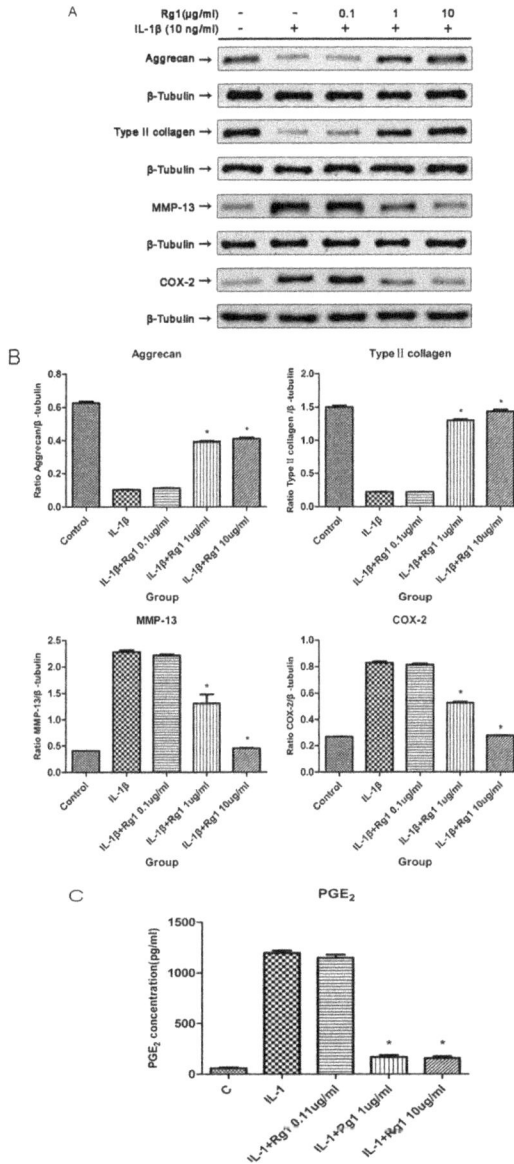

Figure 2. Effect of Rg1 on protein levels of extracellular matrix and inflammatory mediators after induction by IL-1β. Human OA chondrocytes were treated with medium (control group), and IL-1β (10 ng/mL) alone or in combination with Rg1 (0.1, 1, or 10 μg/mL). Total protein was extracted for Western blot analysis. The following proteins were assessed: type II collagen, aggrecan, MMP-13, COX-2 and β-Tubulin (**A**); the relative protein levels of type II collagen, aggrecan, MMP-13 and COX-2 were quantified by densitometric analysis and normalized to β-tubulin (**B**); Prostaglandin E2 (PGE$_2$) concentrations in the corresponding culture media were measured by enzyme-linked immunosorbent assay (ELISA) (**C**). Data are mean ± SEM of four independent experiments. * $p < 0.05$ compared with cells treated with IL-1β alone.

3.3. Rat OA and Gross Morphology

Visible abrasion of the articular surface was detected in the right knee joint of OA rats (Figure 3). Compared with the ACLT group, cartilage destruction was slightly improved in the R1 group, with partial repair on the articular surface. However, slight cartilage erosion was still detected on the tibial plateau, the lateral and medial condyles, and patellar surface (Figure 3). The situation was further improved in the R2 group, which showed smoother and more regular articular surface (Figure 3).

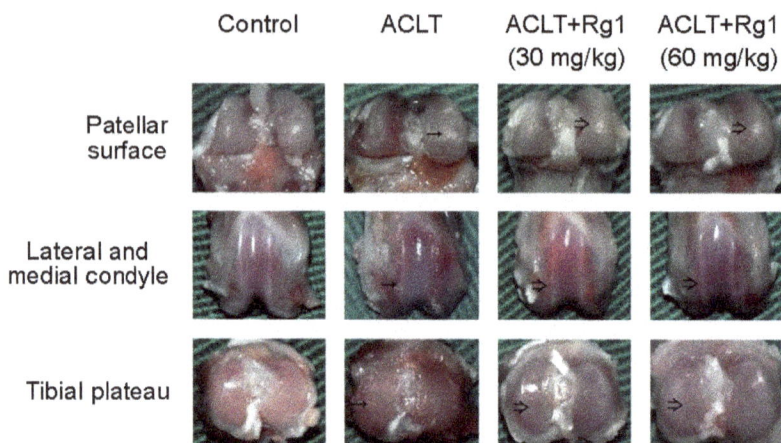

Figure 3. Rat OA and gross morphology. OA was induced by Anterior Cruciate Ligament Transaction (ACLT) of the right knee. Four weeks after ACLT, Rg1 intervention groups received Rg1 (30 or 60 mg/kg/day) by gavage once daily for eight consecutive weeks. Gross morphological changes of femoral condyles and the tibial plateau were photographed to compare cartilage lesions at 12 weeks after surgery. "→" indicates articular surface abrasion; "⇒" indicates articular cartilage repair.

3.4. Histology and Immunohistochemistry Findings

Structural changes in the joints as well as aggrecan content were evaluated histologically (Figure 4A); type II collagen and MMP-13 (Figure 4C) levels were analyzed by immunohistochemistry. The ACLT group showed serious cartilage destruction, irregular abrasions of the cartilage surfaces, chondrocyte disappearance, and aggrecan depletion. Type II collagen showed similar changes as aggrecan, as demonstrated by immunohistochemistry. No Safranin O/fast green or toluidine blue staining was evident, and remarkably more MMP-13-positive chondrocytes (Figure 4D) were found in the OA cartilage specimens. These histo-morphological and immunohistochemical changes were reduced in the R1 group and further decreased in the R2 group. The modified Mankin scores (Figure 4B) were significantly higher in the ACLT group compared with the control values, and lower in the R1 and R2 groups than the ACLT group. Percentages of MMP-13-positive cells were higher in the ACLT group than the control, R1 and R2 groups.

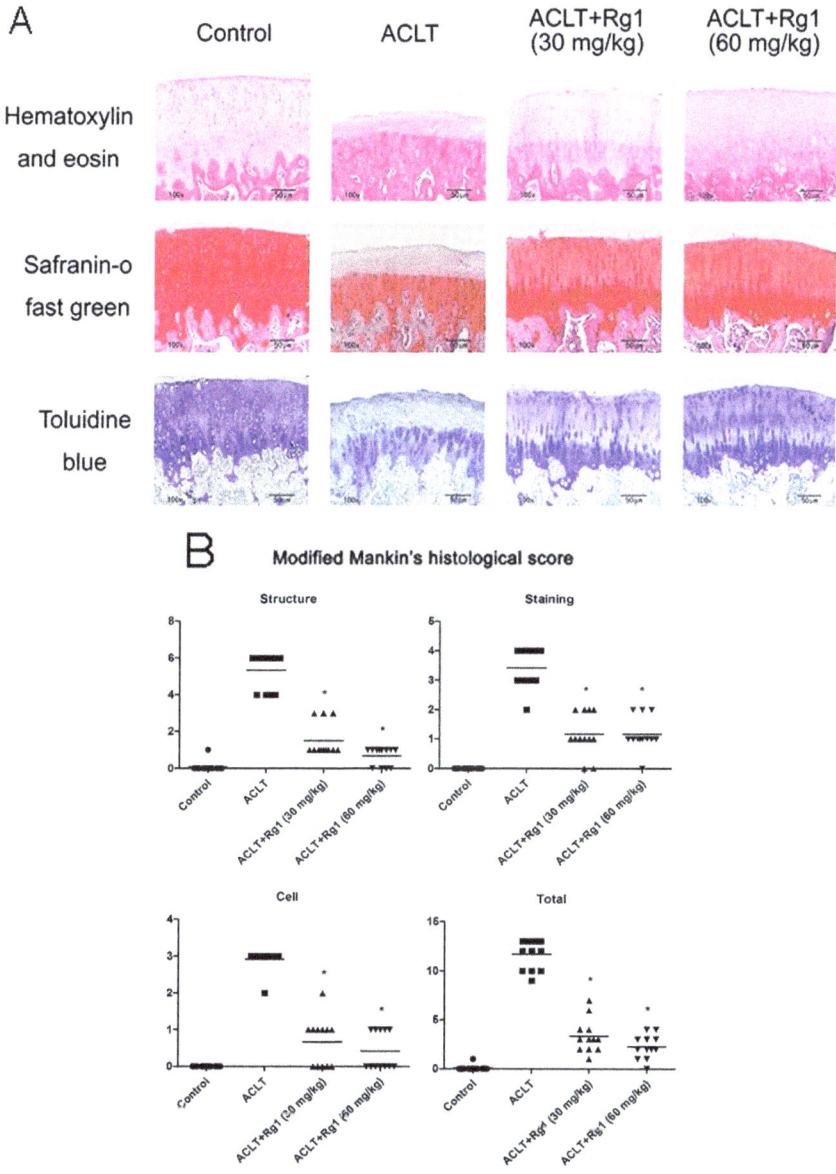

Figure 4. *Cont.*

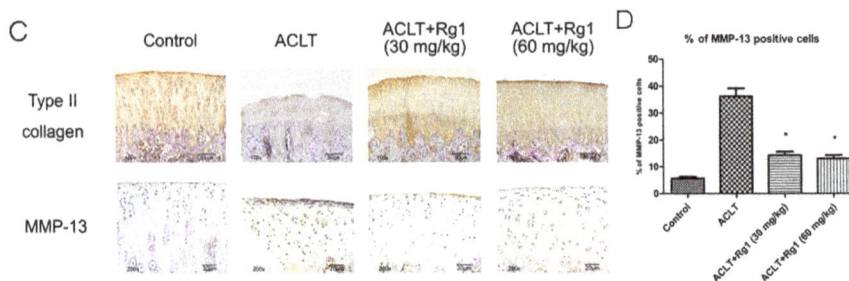

Figure 4. Histological and immunohistochemical findings. Four weeks after ACLT, Rg1 intervention groups received Rg1 (30 or 60 mg/kg/day) by gavage once daily for eight consecutive weeks. (**A**) representative photomicrographs of hematoxylin and eosin (H&E), Safranin O-fast green, and toluidine blue -stained tissue sections of knee joint specimens from OA rats treated with Rg1 or saline (original magnification ×100); (**B**) joint lesions were graded on a scale of 0–13 using the modified Mankin scoring. They were graphed as dot plots with the mean (bar) of 12 rats per group; (**C**) representative photomicrographs of immunohistochemical staining for type II collagen (original magnification ×100) and MMP-13 (original magnification ×200). Yellow staining indicates collagen type II expression in the cartilage. MMP-13 positive cells are stained brown; and (**D**) bar graphs represent mean percentage ± SEM of MMP-13-positive cells in 12 rats per group. * $p < 0.05$ compared with the ACLT group.

4. Discussion

Our previous study demonstrated that Rg1 protects chondrocytes from IL-1β-induced apoptosis. However, inflammatory responses induced by IL-1β in chondrocytes also play a critical role in the initiation and development of OA [7]. The anti-inflammatory effects of Rg1 on chondrocytes and protective effects in osteoarthritis had not been reported.

Hitherto, more than 30 ginsenosides have been identified from ginseng [16]. Ginsenosides are classified into 20 (*S*)-protopanaxadiol and 20 (*S*)-protopanaxatriol groups. Rg1 is one of the most abundant ginsenosides in ginseng and categorized into the 20 (*S*)-protopanaxatriol group [17]. After systematic review, no published article was found on the human use of the bioavailable Rg1 monomer. Several studies focused on its bioavailability in rats in vivo; in vitro studies demonstrated that Rg1 can be absorbed after oral gavage [18]. A preliminary study showed that ginsenoside Rg1 is metabolized into 20 (*S*)-protopanaxatriol via ginsenoside Rh1 and ginsenoside F1 by the gut microbiota of humans and mice. These metabolites may ameliorate inflammatory diseases, such as colitis, by inhibiting the binding of lipopolysaccharide to toll-like receptor 4 on peritoneal macrophages and restoring the Th17/Treg cell balances [19].

MMP-13 plays a pivotal role in initiating the degradation of type II collagen [20]. Previous studies have shown that herbs can decrease MMP-13 expression, when administered as complementary OA treatment [7,21]. Senescent chondrocytes have been detected in damaged OA cartilage samples [22]. The current view is that the local balance of MMP and tissue inhibitor of metalloproteinase activities is pivotal in determining the extent of ECM turnover. Thus, a hypothesis was formed that chondrocytes in aging or diseased cartilage may become senescent, with associated phenotypic changes contributing towards the development or progression of OA [22]. While serving as pro-senescent compounds in cancer cell models [23,24], ginsenoside Rg1 was reported to exhibit protective effects against IL-1β, H_2O_2, and tert-Butyl hydroperoxide-induced senescence in endothelial progenitor cells [25] and fibroblasts [26]. In the current study, Rg1 decreased IL-1β-induced MMP-13 expression. This inhibition may be crucial for inhibiting ECM degradation. The impact of ginsenoside Rg1 on modulating chondrocyte senescence need to be further explored.

PGE$_2$ can influence chondrocyte metabolism, degrade type II collagen, and cause articular cartilage degeneration. Thus, it can be considered a putative therapeutic target for OA [27]. As shown above, Rg1 inhibited COX-2 and PGE$_2$ protein levels, in a similar manner as symptomatic slow-acting drugs of OA, which also ameliorated OA symptoms by inhibiting the expression of COX-2 and PGE$_2$ [28,29]. We predict that these effects of Rg1 could be crucial in slowing OA progression.

The rat model of ACLT has been used previously to assess tissue alterations in OA [30–32]. Reported findings demonstrate similarities between ACLT induced OA in rats and human OA, including type II collagen and aggrecan degradation, and chondrocyte loss, leading to progressive articular cartilage degeneration [33,34]. We found that Rg1 protected the cartilage and ECM. In addition, Rg1 inhibited OA progression association with reduced MMP-13 expression in the OA model. This phenomenon may represent one of the mechanisms, whereby Rg1 mitigates cartilage and ECM damage.

Rg1 is composed of a steroidal skeleton with sugar moieties, as demonstrated by Chan et al., and exhibits estrogen-like effects [35–37]. In addition to bone-inducing effects, estrogen sustains normal chondrocyte metabolism and reduces the sensitization of articular cartilage to OA pathogenic factors [20]. Thus, we suggest that Rg1 may exert estrogen-like effects in OA treatment.

Rg1 can bind glucocorticoid receptors as a functional ligand, thereby exhibiting glucocorticoid effects [8]. Du et al. confirmed that Rg1 exerts dexamethasone-like anti-inflammatory effects on osteoblasts, not promoting hyperglycemia and osteoporosis, indicating less severe side-effects than glucocorticoids [8].

5. Conclusions

In conclusion, Ginsenoside Rg1 inhibited IL-1β-induced human chondrocyte gene and protein expressions of MMP-13, COX-2 and PGE2, and prevented type II collagen and aggrecan degradation. Administration of Ginsenoside Rg1 to OA rats attenuated cartilage degeneration, and reduced type II collagen loss and MMP-13 levels. These findings suggest that Rg1 has potential clinical benefits for OA treatment.

Author Contributions: D.W. and Y.H. designed experiments. W.C., J.J. and Z.W. carried out experiments; W.C. analyzed experimental results; and D.W. and Y.H. wrote the manuscript.

Conflicts of Interest: The authors declare no conflict of interest.

References

1. Cheng, W.; Wu, D.; Zuo, Q.; Wang, Z.; Fan, W. Ginsenoside Rb1 prevents interleukin-1 beta induced inflammation and apoptosis in human articular chondrocytes. *Int. Orthop.* **2013**, *37*, 2065–2070. [CrossRef] [PubMed]
2. Charlier, E.; Relic, B.; Deroyer, C.; Malaise, O.; Neuville, S.; Collee, J.; Malaise, M.G.; De Seny, D. Insights on Molecular Mechanisms of Chondrocytes Death in Osteoarthritis. *Int. J. Mol. Sci.* **2016**, *17*, 2146. [CrossRef] [PubMed]
3. Schuerwegh, A.J.; Dombrecht, E.J.; Stevens, W.J.; Van Offel, J.F.; Bridts, C.H.; De Clerck, L.S. Influence of pro-inflammatory (IL-1 alpha, IL-6, TNF-alpha, IFN-gamma) and anti-inflammatory (IL-4) cytokines on chondrocyte function. *Osteoarthr. Cartil.* **2003**, *11*, 681–687. [CrossRef]
4. Attur, M.; Al-Mussawir, H.E.; Patel, J.; Kitay, A.; Dave, M.; Palmer, G.; Pillinger, M.H.; Abramson, S.B. Abramson, Prostaglandin E2 exerts catabolic effects in osteoarthritis cartilage: Evidence for signaling via the EP4 receptor. *J. Immunol.* **2008**, *181*, 5082–5088. [CrossRef] [PubMed]
5. Zhang, X.H.; Xu, X.X.; Xu, T. Ginsenoside Ro suppresses interleukin-1beta-induced apoptosis and inflammation in rat chondrocytes by inhibiting NF-kappaB. *Chin. J. Nat. Med.* **2015**, *13*, 283–289. [PubMed]
6. Martel-Pelletier, J.; Pelletier, J.P.; Fahmi, H. Cyclooxygenase-2 and prostaglandins in articular tissues. *Semin. Arthritis Rheum.* **2003**, *33*, 155–167. [CrossRef]

7. Jeong, J.W.; Lee, H.H.; Lee, K.W.; Kim, K.Y.; Kim, S.G.; Hong, S.H.; Kim, G.Y.; Park, C.; Kim, H.K.; Choi, Y.W.; et al. Mori folium inhibits interleukin-1beta-induced expression of matrix metalloproteinases and inflammatory mediators by suppressing the activation of NF-kappaB and p38 MAPK in SW1353 human chondrocytes. *Int. J. Mol. Med.* **2016**, *37*, 452–460. [PubMed]
8. Du, J.; Cheng, B.; Zhu, X.; Ling, C. Ginsenoside Rg1, a novel glucocorticoid receptor agonist of plant origin, maintains glucocorticoid efficacy with reduced side effects. *J. Immunol.* **2011**, *187*, 942–950. [CrossRef] [PubMed]
9. Lu, J.M.; Yao, Q.; Chen, C. Ginseng compounds: An update on their molecular mechanisms and medical applications. *Curr. Vasc. Pharmacol.* **2009**, *7*, 293–302. [CrossRef] [PubMed]
10. Radad, K.; Gille, G.; Moldzio, R.; Saito, H.; Rausch, W.D. Ginsenosides Rb1 and Rg1 effects on mesencephalic dopaminergic cells stressed with glutamate. *Brain Res.* **2004**, *1021*, 41–53. [CrossRef] [PubMed]
11. Ye, J.; Yao, J.P.; Wang, X.; Zheng, M.; Li, P.; He, C.; Wan, J.B.; Yao, X.; Su, H. Neuroprotective effects of ginsenosides on neural progenitor cells against oxidative injury. *Mol. Med. Rep.* **2016**, *13*, 3083–3091. [CrossRef] [PubMed]
12. Li, N.; Liu, Y.; Li, W.; Zhou, L.; Li, Q.; Wang, X.; He, P. A UPLC/MS-based metabolomics investigation of the protective effect of ginsenosides Rg1 and Rg2 in mice with Alzheimer's disease. *J. Ginseng. Res.* **2016**, *40*, 9–17. [CrossRef] [PubMed]
13. Huang, Y.; Wu, D.; Fan, W. Protection of ginsenoside Rg1 on chondrocyte from IL-1beta-induced mitochondria-activated apoptosis through PI3K/Akt signaling. *Mol. Cell. Biochem.* **2014**, *392*, 249–257. [CrossRef] [PubMed]
14. Appleton, C.T.; McErlain, D.D.; Pitelka, V.; Schwartz, N.; Bernier, S.M.; Henry, J.L.; Holdsworth, D.W.; Beier, F. Forced mobilization accelerates pathogenesis: Characterization of a preclinical surgical model of osteoarthritis. *Arthritis Res. Ther.* **2007**, *9*, R13. [CrossRef] [PubMed]
15. Moon, S.J.; Woo, Y.J.; Jeong, J.H.; Park, M.K.; Oh, H.J.; Park, J.S.; Kim, E.K.; Cho, M.L.; Park, S.H.; Kim, H.Y.; et al. Rebamipide attenuates pain severity and cartilage degeneration in a rat model of osteoarthritis by downregulating oxidative damage and catabolic activity in chondrocytes. *Osteoarthr. Cartil.* **2012**, *20*, 1426–1438. [CrossRef] [PubMed]
16. Liu, C.X.; Xiao, P.G. Recent advances on ginseng research in China. *J. Ethnopharmacol.* **1992**, *36*, 27–38. [PubMed]
17. Wang, A.; Wang, C.Z.; Wu, J.A.; Osinski, J.; Yuan, C.S. Determination of major ginsenosides in Panax quinquefolius (American ginseng) using high-performance liquid chromatography. *Phytochem. Anal.* **2005**, *16*, 272–277. [CrossRef] [PubMed]
18. Xiong, J.; Sun, M.; Guo, J.; Huang, L.; Wang, S.; Meng, B.; Ping, Q. Active absorption of ginsenoside Rg1 in vitro and in vivo: The role of sodium-dependent glucose co-transporter 1. *J. Pharm. Pharmacol.* **2009**, *61*, 381–386. [CrossRef] [PubMed]
19. Lee, S.Y.; Jeong, J.J.; Eun, S.H.; Kim, D.H. Anti-inflammatory effects of ginsenoside Rg1 and its metabolites ginsenoside Rh1 and 20(S)-protopanaxatriol in mice with TNBS-induced colitis. *Eur. J. Pharmacol.* **2015**, *762*, 333–343. [CrossRef] [PubMed]
20. Liang, Y.; Duan, L.; Xiong, J.; Zhu, W.; Liu, Q.; Wang, D.; Liu, W.; Li, Z.; Wang, D. E2 regulates MMP-13 via targeting miR-140 in IL-1beta-induced extracellular matrix degradation in human chondrocytes. *Arthritis Res. Ther.* **2016**, *18*, 105. [CrossRef] [PubMed]
21. Kwon, H.O.; Lee, M.; Kim, O.K.; Ha, Y.; Jun, W.; Lee, J. Effect of Hijikia fusiforme extracts on degenerative osteoarthritis in vitro and in vivo models. *Nutr. Res. Pract.* **2016**, *10*, 265–273. [CrossRef] [PubMed]
22. Price, J.S.; Waters, J.G.; Darrah, C.; Pennington, C.; Edwards, D.R.; Donell, S.T.; Clark, I.M. The role of chondrocyte senescence in osteoarthritis. *Aging Cell* **2002**, *1*, 57–65. [CrossRef] [PubMed]
23. Liu, J.; Cai, S.Z.; Zhou, Y.; Zhang, X.P.; Liu, D.F.; Jiang, R.; Wang, Y.P. Senescence as a consequence of ginsenoside rg1 response on k562 human leukemia cell line. *Asian Pac. J. Cancer Prev.* **2012**, *13*, 6191–6196. [CrossRef] [PubMed]
24. Sin, S.; Kim, S.Y.; Kim, S.S. Chronic treatment with ginsenoside Rg3 induces Akt-dependent senescence in human glioma cells. *Int. J. Oncol.* **2012**, *41*, 1669–1674. [PubMed]
25. Shi, A.W.; Gu, N.; Liu, X.M.; Wang, X.; Peng, Y.Z. Ginsenoside Rg1 enhances endothelial progenitor cell angiogenic potency and prevents senescence in vitro. *J. Int. Med. Res.* **2011**, *39*, 1306–1318. [CrossRef] [PubMed]

26. Chen, X.; Zhang, J.; Fang, Y.; Zhao, C.; Zhu, Y. Ginsenoside Rg1 delays tert-butyl hydroperoxide-induced premature senescence in human WI-38 diploid fibroblast cells. *J. Gerontol. A Biol. Sci. Med. Sci.* **2008**, *63*, 253–264. [CrossRef] [PubMed]

27. Sun, T.W.; Wu, Z.H.; Weng, X.S. Celecoxib can suppress expression of genes associated with PGE2 pathway in chondrocytes under inflammatory conditions. *Int. J. Clin. Exp. Med.* **2015**, *8*, 10902–10910. [PubMed]

28. Heinecke, L.F.; Grzanna, M.W.; Au, A.Y.; Mochal, C.A.; Rashmir-Raven, A.; Frondoza, C.G. Inhibition of cyclooxygenase-2 expression and prostaglandin E2 production in chondrocytes by avocado soybean unsaponifiables and epigallocatechin gallate. *Osteoarthr. Cartil.* **2010**, *18*, 220–227. [CrossRef] [PubMed]

29. Nakamura, H.; Shibakawa, A.; Tanaka, M.; Kato, T.; Nishioka, K. Effects of glucosamine hydrochloride on the production of prostaglandin E2, nitric oxide and metalloproteases by chondrocytes and synoviocytes in osteoarthritis. *Clin. Exp. Rheumatol.* **2004**, *22*, 293–299. [PubMed]

30. Afara, I.; Prasadam, I.; Crawford, R.; Xiao, Y.; Oloyede, A. Non-destructive evaluation of articular cartilage defects using near-infrared (NIR) spectroscopy in osteoarthritic rat models and its direct relation to Mankin score. *Osteoarthr. Cartil.* **2012**, *20*, 1367–1373. [CrossRef] [PubMed]

31. Wen, Z.H.; Tang, C.C.; Chang, Y.C.; Huang, S.Y.; Hsieh, S.P.; Lee, C.H.; Huang, G.S.; Ng, H.F.; Neoh, C.A.; Hsieh, C.S.; et al. Glucosamine sulfate reduces experimental osteoarthritis and nociception in rats: Association with changes of mitogen-activated protein kinase in chondrocytes. *Osteoarthr. Cartil.* **2010**, *18*, 1192–1202. [CrossRef] [PubMed]

32. Naito, K.; Watari, T.; Furuhata, A.; Yomogida, S.; Sakamoto, K.; Kurosawa, H.; Kaneko, K.; Nagaoka, I. Evaluation of the effect of glucosamine on an experimental rat osteoarthritis model. *Life Sci.* **2010**, *86*, 538–543. [CrossRef] [PubMed]

33. Hayami, T.; Pickarski, M.; Zhuo, Y.; Wesolowski, G.A.; Rodan, G.A.; Duong, L.T. Characterization of articular cartilage and subchondral bone changes in the rat anterior cruciate ligament transection and meniscectomized models of osteoarthritis. *Bone* **2006**, *38*, 234–243. [CrossRef] [PubMed]

34. Stoop, R.; Buma, P.; van der Kraan, P.M.; Hollander, A.P.; Billinghurst, R.C.; Meijers, T.H.M; Poole, A.R.; van den Berg, W.B. Type II collagen degradation in articular cartilage fibrillation after anterior cruciate ligament transection in rats. *Osteoarthr. Cartil.* **2001**, *9*, 308–315. [CrossRef] [PubMed]

35. Lau, W.S.; Chan, R.Y.; Guo, D.A.; Wong, M.S. Ginsenoside Rg1 exerts estrogen-like activities via ligand-independent activation of ERalpha pathway. *J Steroid Biochem. Mol. Biol.* **2008**, *108*, 64–71. [CrossRef] [PubMed]

36. Ling, C.; Li, Y.; Zhu, X.; Zhang, C.; Li, M. Ginsenosides may reverse the dexamethasone-induced down-regulation of glucocorticoid receptor. *Gen. Comp. Endocrinol.* **2005**, *140*, 203–209. [CrossRef] [PubMed]

37. Chan, R.Y.; Chen, W.F.; Dong, A.; Guo, D.; Wong, M.S. Estrogen-like activity of ginsenoside Rg1 derived from Panax notoginseng. *J. Clin. Endocrinol. Metab.* **2002**, *87*, 3691–3695. [CrossRef] [PubMed]

nutrients

MDPI

Review

Microbiota and Probiotics in Health and HIV Infection

Chiara D'Angelo, Marcella Reale * and Erica Costantini

Unit of Immunodiagnostic and Molecular Pathology, Department of Medical, Oral and Biotechnological Sciences, University "G. d'Annunzio" Chieti-Pescara, 66100 Chieti, Italy; chiara.dangelo@unich.it (C.D.); erica.costantini@unich.it (E.C.)
* Correspondence: mreale@unich.it; Tel.: +39-0871-3554029

Received: 26 January 2017; Accepted: 12 June 2017; Published: 16 June 2017

Abstract: Microbiota play a key role in various body functions, as well as in physiological, metabolic, and immunological processes, through different mechanisms such as the regulation of the development and/or functions of different types of immune cells in the intestines. Evidence indicates that alteration in the gut microbiota can influence infectious and non-infectious diseases. Bacteria that reside on the mucosal surface or within the mucus layer interact with the host immune system, thus, a healthy gut microbiota is essential for the development of mucosal immunity. In patients with human immunodeficiency virus (HIV), including those who control their disease with antiretroviral drugs (ART), the gut microbiome is very different than the microbiome of those not infected with HIV. Recent data suggests that, for these patients, dysbiosis may lead to a breakdown in the gut's immunologic activity, causing systemic bacteria diffusion and inflammation. Since in HIV-infected patients in this state, including those in ART therapy, the treatment of gastrointestinal tract disorders is frustrating, many studies are in progress to investigate the ability of probiotics to modulate epithelial barrier functions, microbiota composition, and microbial translocation. This mini-review analyzed the use of probiotics to prevent and attenuate several gastrointestinal manifestations and to improve gut-associated lymphoid tissue (GALT) immunity in HIV infection.

Keywords: microbiome; probiotics; dietary supplements; nutrition; HIV; inflammation

---

1. Introduction

Over the past 20 years, the increasing interest in the health effects of probiotic consumption has erupted in studies both in food and pharmaceutical companies, and studies have been conducted to understand the effects of probiotics on the regulation of the immune response and potential applications for disease prevention. Probiotic benefits are not a recent discovery: they were already present a long time ago in traditional foods, such as cheese, yogurt, milk, and salty fishes, and used for nutritional purposes. Subsequently, people noted the beneficial health effects of eating fermented foods.

Over the years, probiotics have been described as "organisms and substances which contribute to intestinal microbial balance" [1–3]. For the Food and Agriculture Organization/World Health Organization (FAO/WHO), the term probiotic is defined as "live microorganisms which, when administered in adequate amounts, confer a health benefit on the host".

Improving health could be a useful strategy for protecting us from several illnesses, and probiotics are able to enrich our digestive system with good microbes that are able to neutralize the harmful ones and restore the balance between bacteria such as lactobacilli, streptococci, clostridia, coliform, and bacteroides. Thus, probiotics may confer a health benefit on the host by the modulation of the immune system [4,5], limiting pathogen colonization [6,7], and controlling inflammatory gut disorders [8] and metabolic disorders [9]. Probiotics are also helpful during antibiotic administration—reducing antibiotic-associated diarrhea—and in restoring normal gut permeability, mechanical integrity, and homeostasis [10].

Some effects attributed to probiotics have been proved by clinical trials, and the effectiveness of probiotics has been demonstrated in disorders, such as inflammatory bowel diseases (IBD), diarrhea, allergies, and the prevention of upper respiratory tract infections [11–13], and also in unbalanced conditions of intestinal flora induced by stress, lifestyle, host genetics, inadequate food, and exposure to environmental toxins [14–16].

Many studies have demonstrated that the human immunodeficiency virus (HIV) has harmful effects on the human immune system, mainly on the cluster differentiation (CD4)$^+$ T-cells, and that HIV infection is characterized by gut microbiota dysbiosis, an altered intestinal barrier, and systemic inflammation [17–20]. The mucosal immune system can be modulated by gut-resident bacteria, and alteration of the mucosal innate immune system can result in the outgrowth of a dysbiotic pro-inflammatory group accountable for chronic inflammation in the mucosa and the periphery [21,22]. HIV infection significantly alters total microbial colonization as well as the microbiota composition in the oral cavity, and decreased CD4 cell counts have been associated with the presence of oral lesions [23].

Progressive HIV infection is characterized by the dysregulation of intestinal immunity that may also persist during highly-active antiretroviral therapy, and the extent of the gut and oral microbiota dysbiosis correlates with markers of disease progression [24,25].

Thus, interventions in HIV-positive patients are necessary to restore the integrity of the immune system of gut-associated lymphoid tissue (GALT), and the use of probiotics may recover gut barrier functions, remodel the microbiome, and aid to decrease bacterial translocation and pro-inflammatory cytokine production, thereby improving immune functions in HIV-infected subjects, including during short-term antiretroviral therapy (ART) [26–28].

Mechanisms by which probiotics may exert their effects are strain-related and include the host's microbiota modulation, improvement of mucosal barrier functions, and modulation of the immune system [29,30]. As all the implicated mechanisms are not completely known, probiotic clinical use needs to be related to probiotic strain and dosage, in order to identify their efficacy under specific conditions [31]. Studies have been conducted, and others are in progress, with the aim of understanding probiotic-specific mechanisms and selecting probiotic strains in relation to the target patient's specific pathogenic and clinical defects [32,33].

2. The Intestinal Microbiota Functions

The total human body surface, lung, oral and vaginal mucosa, and the gastrointestinal (GI) tract host over 10^{14} microorganisms—starting from birth—which form the microbiota. About 99% of the microbiota is present in the GI, achieving a configuration during human evolution, and has a major impact on the gastrointestinal tract and mucosal immune functions, and significantly affects the health of their host. For this reason, the gastrointestinal microbiome is the best-investigated microbiome and serves as a model for understanding host–microbiota interactions and disease. The development of next-generation DNA sequencing platforms has clarified the composition of the intestinal microbiota, that is, a complex microbial ecosystem. Under healthy conditions, it includes different species of bacteria, each of which contains many functionally different strains with significant genetic diversity. The majority of strains are strictly anaerobes, even if facultative anaerobes and aerobes are present. Some bacterial strains are prevalent: fermenting bacteria (such as *Lactobacillus* and *Bifidobacteria*) represent 80% of the gut microbiota, while the remaining 20% includes *Escherichia*, *Bacteroides*, *Eubacteria*, and *Clostridium*. Lactic acid bacteria (LAB) are considered a major group of probiotic bacteria and have been isolated directly from humans. To date, different bacterial genera are known, including *Bifidobacterium* and *Lactobacillus*: they survive stomach acid pH and intestinal bile salts, reach sites of action, and their ingestion does not cause any risk for the host. It is known that a healthy gut flora is largely responsible of the overall health of the host, while gut microbiota alteration is associated with several human diseases, such as bowel diseases, metabolic and allergic diseases, or neurodevelopmental illnesses [34–36]. Thus, researchers are beginning to consider intestinal microbiota as another organ of the human body with different functions, such as maintenance of the epithelial barrier, inhibition of pathogen adhesion to intestinal surfaces, and modulation of the immune system [37].

2.1. Function and Preservation of the Intestinal Barrier

The GI mucosal surface is the largest area of the body in contact with the external environment; it plays a key role in blocking the access of potentially harmful substances. The epithelium and the mucus layer, lining the gut, represent the host's first line of defense and the essential mechanical barrier that avoids contact between the internal and the external environments by blocking the passage of antigens, toxins, and microbial products, thus acting as a component of innate immunity [38].

The intestinal barrier is equipped with several levels of defense mechanisms to limit luminal antigen translocation. In a normal gut, the epithelial barrier consists of a layer of enterocyte tight junctions, anchoring junctions, and desmosomes—which hinder microbe passage—goblet cells producing mucus, and Paneth cells. Intestinal epithelial cells (IECs) can sense and respond to microbial stimuli, support barrier functions, and participate in immune responses [39,40]. The function of the epithelial barrier depends on junctional complexes formed by transmembrane proteins, such as claudins that form paracellular channels for small cations and water. Yuan et al. [41] showed the changes in expression and distribution of claudin proteins, which are essential for the formation and the integrity of tight junctions, which regulate the flow of water ions and small molecules, and their relationships with barrier dysfunction.

Paneth cells may limit bacterial penetration through pattern recognition receptors (PRR) and secretion of mucins and antimicrobial proteins (AMPs), establishing a physical and biochemical barrier to microbial penetration and underlying immune cells [42]. Intestinal epithelial cells produce immunoregulatory signals for tolerizing immune cells, limiting steady-state inflammation, and directing innate and adaptive immune cell responses against pathogens and commensal bacteria. Specialized epithelial cells, called M-cells, mediate the constant sampling of luminal antigens, and both microorganisms and macromolecules can gain entry through the M-cells [43].

Commensal bacteria induce cytokine production by IECs via PRR signaling, promoting the development of dendritic cells (DC) and macrophages with tolerogenic properties [44,45]. Commensal microorganisms may regulate barrier functions, controlling mucus production by goblet cells [46] or the expression of AMPs. Intestinal epithelial cells secrete Immunoglobulin (Ig)A in the lamina propria and express microbial recognition receptors, such as Toll-like receptors (TLR), that can recognize both antigens derived from the microbiota or invading pathogens. Under homeostatic conditions, IECs are unresponsive to TLR stimuli, while increased TLR expression was observed under inflammatory conditions. TLRs act as a link between microbiota alterations and immune homeostasis [47]. TLRs promote epithelial cell proliferation, secretion of IgA into the gut lumen, the expression of antimicrobial peptides [48], and play a role in intestinal barrier homeostasis [49]. The expression of tight junction proteins was modulated by TLR activation [50], and during inflammatory disorders epithelial tight junctions are impaired and result in increased bacterial translocation into the lamina propria, supporting the inflammatory response.

Many factors can alter the intestinal permeability and GI infections may be responsible for altered nutrient absorption, depleted levels of micronutrients, and waste secretion. As a consequence of microbe activity and the release of soluble peptides or toxins, there are alterations in enterocyte components and their metabolism, leading to a breakdown of the epithelial barrier and to microbial translocation in the gut [51]. Moreover, lifestyle and dietetic factors, including alcohol and energy-dense foods, can increase intestinal permeability [16]. The resulting increased permeability does facilitate chronic intestinal inflammation, strictly connected to the immune system, as observed in the existing association between inflammation and barrier dysfunction in several GI diseases. The proper defense activity of the epithelial barrier is supported by the microbiota, which influences cell metabolism and proliferation, maintenance and repair of barrier integrity, nutrient acquisition and energy regulation, inflammatory response, and angiogenesis [52–54].

The intake of probiotics can reduce the risk of diseases associated with intestinal barrier dysfunction. The mechanisms by which probiotics can influence the barrier function are also an area of interest, although many studies have shown that probiotics increase the barrier function by increasing mucus, antimicrobial peptides, and secretory IgA production, as well as increasing competitive adherence for pathogens, and the tight junctions (TJ) integrity of epithelial cells [55–57]. It is known that certain lactobacilli adhere to mucosal surfaces, inhibiting the attachment of pathogenic bacteria and enhancing the secretion of mucin.

2.2. Resistance to Pathogenic Colonization

One of the major functions of the intestinal microbiota is the protection of the host from colonization and overgrowth of ingested invading bacteria, a phenomenon known as resistance to colonization [41]. Endogenous microbial populations act via several mechanisms, including the modification of the pH in the environment and ecological niches, the release of antimicrobial substances, and the direct competition for the adhesion sites on the epithelium and for nutritive substrates.

After ingestion, pathogens penetrate the highly-colonized mucus layer, where they compete with the resident microbiota for adhesion to the intestinal epithelial cell receptors. In healthy subjects, the direct competition for nutrients limits the possibilities for exogenous pathogenic microbes to colonize and replicate within the gut lumen and invading deeper tissues [58]. Additionally, the production of pathogen growth inhibitors or the resistance to colonization, due to the induction of immune responses and to metabolic products of beneficial bacterial, makes the host resistant to pathogenic infections. In addition, in the GI tract, the microbiota affects biosynthesis and the availability of neurotransmitters that modulate peristalsis, the flow of blood, and the secretion of ions [35,36].

Traditional probiotic approaches to maintain colonization resistance are designed to modulate the competition for nutritious substrates and adhesion sites, as well as the prevention of microorganism translocation and stimulation of the immune system.

2.3. Development and Stimulation of GALT

The presence of the microbiota is crucial for the normal development of GALT. Already from birth, the presence of intestinal microorganisms stimulates GALT to recognize the conserved microbial structures, ensuring an appropriate immune activity. GALT composition is modified immediately after microbial colonization of the GI tract, with a number of intraepithelial lymphocytes and immunoglobulin-producing cells in follicles and in the lamina propria. Bacterial antigen detection is performed by the resident cells of the innate and adaptive immune system. Signals from bacteria can be transmitted to macrophages, dendritic cells, and lymphocytes through molecules expressed on the epithelial cell surface, such as molecules of the major histocompatibility complex I and II, Toll-like receptors, and nucleotide oligomerization domain (NOD)-like receptors or nucleotide-binding domain leucine-rich repeat-containing (NLRs) proteins [59]. Antigen-presenting cells (APCs) provide processed antigens to naïve lymphocytes within distinct T- and B-cell zones.

Mucosal effector sites consist of T lymphocytes, primarily CD8+, located in the epithelium and in the lamina propria, and CD4+ T-cells and plasma cells that heavily populate the large and small intestines, beneath the lamina propria.

The CD4+ T lymphocytes can differentiate into T helper (Th)1, Th2, Th17, and regulatory T (Treg) cells. CD4+ Th17 cells share differentiation pathways and a reciprocal relationship with antigen-induced cells and CD4+ Treg cells, which are both able to maintain the balance between inflammation and tolerance. Th17 cells, characterized by the production of cytokines interleukin (IL)-17A, IL-17F, and IL-22, which have their receptors on epithelial cells [60,61], are specialized in maintaining mucosal integrity, stimulating the proliferation of epithelial cells, producing tight junction proteins (claudins), and modulating a robust antimicrobial inflammatory response by neutrophil and macrophage recruitment via chemokine, antimicrobial defensins, and mucin production [62–65].

Treg cells, maintaining immune homeostasis, have anti-inflammatory activity and prevent autoimmunity, inducing tolerance against self-antigens. Without an inflammatory stimulus, commensal microorganisms induce tolerogenic maturation of DCs, leading to the induction of various types of Treg or hypo-responsive T-cells [66].

Humoral immune response represents the main mechanism of protection given by GALT, mediated also by B cells secreting IgA, of which the intestinal DCs are potent inducers. It has anti-pathogenic effects and prevents commensal bacteria penetration in the host [67].

Epithelial cells, APCs, and lymphocytes can secrete cytokines, chemokines, and other factors that can be tuned to promote tolerance, inflammation, or specific immunity.

The dualistic effect that the microbiota exerts on GALT consists in maintaining tolerance and preventing inflammation through β-defensins and IgA production in the epithelium, whose integrity is enhanced through TLR signaling and Treg induction [45]. The equilibrium between microbiota, immune response, and tolerance mechanisms is important for a healthy intestine, and an aberrant colonization may drive mucosal inflammation, which plays a pivotal function in the development of feeding intolerance. The constant interplay between the microbiota, the intestinal barrier, and the mucosal immune system ensures the balance between permissive or tolerogenic responses to pathogens or food antigens [68].

Probiotics may induce a tolerogenic situation by modulating anti-inflammatory/regulatory cytokines, such as IL-10 and transforming growth factor (TGF)-β, and DC functionality. The supplementation with specific probiotics can promote the restoration of the intestinal CD4+ T-cell population in many immunological diseases, while the anti-inflammatory effects of probiotics in Th17-related diseases might be a consequence of the downregulation of pro-inflammatory IL-17 production [38].

3. Bowel Conditions in People Living with HIV

The GI tract is a major site of HIV replication, and its disorders are among the most frequent complaints in patients with HIV infection. Patients with HIV infection are susceptible to gastric hypoacidity, which may be responsible for a greater risk of opportunistic infection. Additionally, delayed gastric emptying may contribute to the increased bacterial colonization of the upper digestive tract, playing a key role in chronic diarrhea and weight loss, and dysphagia and odynophagia, in which nausea, vomiting, and abdominal pain are the most frequent symptoms [69,70]. HIV infection has an unfavorable effect on the interaction between the commensal microbiota and the immune system, with progressive immune decline associated with inefficient epithelial repair and enhanced epithelial permeability responsible for GI disorders [69]. In people with HIV infection or acquired immune deficiency syndrome (AIDS), the wall of the small intestine is impaired, the crypts are enlarged, and the atrophy of the microvilli decreases their surface area. These modifications are responsible for malabsorption, digestive discomfort, or decreased intake of nutrients.

HIV infection causes a breakdown of the GI barrier, alters the homeostatic balance between GI bacteria and gut immunity, and induces a compositional shift of gut microbiota [71] with the enrichment of either pro-inflammatory or potentially pathogenic bacterial populations [72], such as *Pseudomonas aeruginosa* and *Candida albicans*, and the reduction of *Bifidobacteria* levels and *Lactobacillus* species. These bacterial populations are associated with damage and loss of mucosal barrier functions [73,74] that are correlated with immune status [25,75]. In HIV infection, the increased translocation of microbes and bacterial products from the intestinal tract may induce a systemic immune activation, which causes further damage to the gut barrier function, augmenting bacterial translocation and subsequently increasing systemic inflammation and, in turn, HIV progression [76,77].

Throughout the initial stage of HIV infection, the immune system is unprepared for the attack of the virus, which therefore reproduces at very high levels in the lamina propria, spreading throughout the body. HIV causes a disruption of gut microbiota and 50% of lamina propria CD4 cells are depleted in early and acute HIV infection [19], as these cells may be more susceptible to HIV infection due to high levels of activation and expression of C-C chemokine receptor (CCR)5 receptors [78], in particular the CD4 cells that produce IL-17 and IL-22. The mechanism of this depletion is likely cell death of productively infected cells via apoptosis as well as of bystander cells via pyroptosis and the direct killing of infected cells by natural killer (NK) cells or cytotoxic T-cells [79,80]. The combination of these mechanisms may contribute to $CD4^+$ T-cells loss, mucosal barrier damage, and chronic systemic inflammation. The consequences of reduced CD4 cells is the failure of gut mucosal barrier to protect against invading pathogens as well as the loss of cytokines necessary to support normal barrier function. Usually, with <100 $CD4^+$ T-cells/mL, opportunistic infections of pathogenic bacteria and/or fungi drive GI dysfunctions, and HIV-1 directly drives mucosal inflammation, causing HIV-related enteropathies [81]. Poor CD4 recovery is linked to microbial translocation, and in HIV-infected persons with poor CD4 recovery, intestinal barrier dysfunction and mortality has been linked to elevated plasma kynurenine/tryptophan ratio [82].

The existence of HIV-specific IL-17-producing $CD4^+$ T-cells, named Th17, have been reported [83,84], but it was not completely determined whether Th17 cells have direct anti-viral functions during HIV infection. Th17 and Th22 cells could play a role in amplifying the innate responses to HIV infection by enhancing the production of IL-22, a critical cytokine for epithelial barrier maintenance, which enhances epithelial regeneration inducing stem cell–mediated epithelial cell proliferation [85], and the expression of anti-microbial peptides [65].

During HIV infection, high levels of viremia are associated with an important Th17 reduction in the gut; the loss of mucosal Th17 cells may be related to a decrease in mucosal restoration and an increase of microbial translocation from the gut lumen to the systemic circulation and immune hyperactivation, contributing to the exacerbation of the infection and to opportunistic infections [86–88]. The loss of Th17 cells was accompanied by a concomitant rise of Treg cells, resulting in an imbalanced Th17/Treg ratio during HIV progression [89–91]. A low Th17/Treg ratio in HIV-infected individuals correlates with microbial translocation and with a higher frequency of activated $CD8^+$ T-cells, which is one of the strongest predictors of mortality. Treg cells may have both a beneficial and a detrimental role; the first is by limiting immune activation, while the second is based on the ability of Treg cells to suppress virus-specific immune responses. Thus, the role of Treg cells in regulating T-cell activation in HIV infection is still debated [92] (Figure 1).

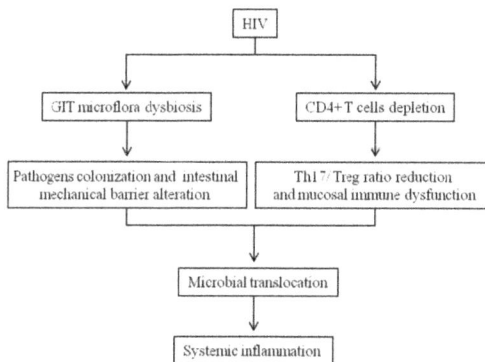

Figure 1. Gastrointestinal tract dysfunctions in HIV-infected patients. HIV: human immunodeficiency virus; GIT: gastro intestinal tract; Treg: T regulatory cells.

HIV infection is associated with an inflammatory state, as evidenced by high levels of Tumor necrosis Factor (TNF) and Tumor necrosis Factor Receptors (TNFRs) 1 and 2, IL-6, and Interferon (IFN)α [93] that may also lead to tight junction destruction. These changes may lead to impaired barrier function [94] and intestinal permeability with an increase of markers for microbial translocation/monocyte activation, such as lipopolysaccharide (LPS) and soluble CD14 into the plasma. Brenchley et al. [76] reported that plasma LPS levels and bacterial ribosomal DNA were elevated in patients with HIV infection compared with healthy controls, and circulating microbial products have been appointed as a possible cause of HIV-related systemic immune activation, HIV progression promotion, and suboptimal response to therapy and co-morbidity. Chronic TLR activation in HIV disease, through recognition of translocated bacterial products and/or viral products, can cause the dysregulation of immune responses.

4. Probiotics as a New Therapeutic Approach That Might Improve Life in HIV-Positive Subjects

Although ART and other pharmacological therapies are life-saving in HIV-positive subjects, due to the suppression of plasma viremia, the number of mucosal CD4 cells does not always fully recover, and microbial translocation is still not under full control and remains associated with systemic immune activation and inflammation, characterized by elevated pro-inflammatory cytokine levels, as well as T and B cell activation and tight junction dysfunction between the epithelial cells of the mucosal barrier.

Epithelial barrier dysfunction, measured by peripheral blood levels of intestinal fatty acid-binding protein and zonulin-1, predicted mortality in HIV infection, even after adjustment for CD4 count [95].

Several HIV-affected patients may be effectively managed by controlling the HIV infection with high-efficacy and improved ART, while other HIV-positive patients have many side effects, such as diarrhea and other GI symptoms associated with a worse quality of life, leading to a discontinuation of treatment and the requirement of more complex approaches [96–99].

The hypothesis that probiotic administration protects the gut surface and can delay the progression of HIV infection to AIDS was proposed some years ago. The use of probiotics may be inexpensive and potentially useful to reduce HIV-related morbidity and mortality [100].

There are many possible mechanisms by which probiotics may interfere with HIV (Figure 2). Probiotics can compete for nutrients and epithelial and mucosal adherence, inhibit epithelial invasion, counteract the inflammatory process by stabilizing and strengthening the gut microbiota responsible for the intestinal barrier integrity, prevent microbial translocation, lower mucosal and systemic inflammation, stimulate production of antimicrobial substances [101–103], and promote intestinal immunoglobulin A responses to improve the immunological barrier function [104–107]. The effectiveness of diet supplementation with different probiotic strains has been shown in people with HIV and has especially been shown as an additional strategy in patients on ART, in order to improve antioxidant defenses and aid in the reconstitution of the immune function.

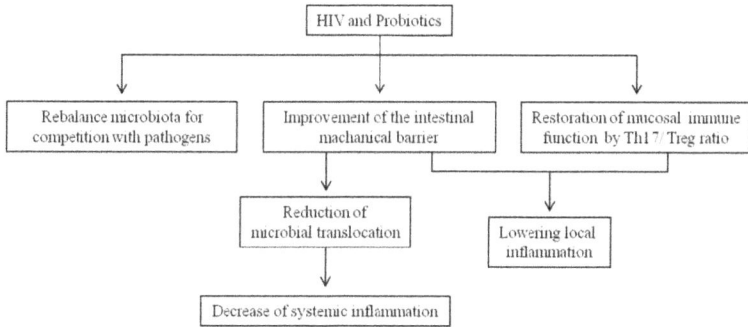

Figure 2. Probiotics use and beneficial effects in the gastrointestinal tract of HIV-1-infected patients.

Gut reconditioning through probiotic administration could be protective of the gut surface and delay the progression to AIDS [108]. Probiotics, by altering intestinal flora, may induce epithelial healing, and by preventing the decline in CD4$^+$ cell counts may lower the risk of virus transmission and reduce hospitalization for co-infections. ART-treated patients who fail to have an immunologic response (CD4 < 200) have lower levels of lactobacilli, elevated levels of LPS and sCD14, and increased inflammatory markers, such as IL-6 and sCD14 [109,110]. In 2010, Irvine et al. ran an observational retrospective study to assess the effect of a *Lactobacillus rhamnosus* Fiti yogurt on CD4$^+$ cell counts in HIV subjects; the study showed an increased CD4$^+$ cell average count over a period of three years in yogurt consumers [111]. Gori et al. reported that, in Highly Active Anti-Retroviral Therapy (HAART)-naïve HIV-infected patients, dietary supplementation with a prebiotic mixture results in the improvement of gut microbiota composition, the reduction of sCD14, CD4$^+$ T-cell activation (CD25), and improved NK cell activity [28]. The study of Kim et al. evaluated the ability of probiotics, provided during combined antiretroviral therapy (cART), to reduce inflammation and improve gut immune health in HIV-positive treatment-naïve individuals (PROOV IT I) and in individuals with suboptimal CD4 recovery on cART (PROOV IT II) [108].

A combination of probiotic bacteria upregulates Treg cell activation and suppresses pro-inflammatory immune responses in models of autoimmunity, including IBD, thus providing a rationale for the use of probiotics in HIV infection.

In addition to the ability of probiotics to improve barrier function and intestinal homeostasis, specific probiotic strains may be able to revert the HIV-induced Th-2 polarization [112]. The study carried out by d'Ettorre et al. in 2015, where HIV-infected patients on ART were supplemented with probiotics, showed that inflammation and markers of microbial translocation were significantly reduced [101]. In HIV-infected subjects, diet supplementation for four weeks with *Lactobacillus casei* Shirota were virologically, bacteriologically, and immunologically beneficial, leading to increased levels of CD56$^+$ cells and to a reduction of inflammatory status with significantly increased IL-23 serum levels. In addition, probiotic supplementation could be useful in the reduction of risk factors for cardiovascular diseases, such as hypercholesterolemia, as well as in the improvement of quality of life by improving the nutritional status, alleviating GI manifestations, and stimulating mucosal immune function [103].

Bacterial vaginosis may increase the risk of transmission or acquisition of HIV, increasing proinflammatory cytokines and disrupting the mucosal barrier function [113], and probiotic intervention may be prophylactic for bacterial vaginosis [114].

In HIV-affected patients, a periodontal disease, an extensive dysbiosis in the oral microbiome is a comorbidity that could act as a font of chronic inflammation, or a risk of various systemic diseases such as diabetes, hyperlipedemia, chronic kidney diseases. Recently, various studies have reported the lactic acid inhibition of oral bacteria, suggesting a promising role in combating periodontal diseases.

Thus, in HIV-affected patients, probiotics may be a low-cost and accessible treatment approach to periodontal diseases that confer benefits upon host well-being, improving the quality of life [115–118].

5. Conclusions

The helpful effects of probiotics to maintain our body in health are well-known, and several clinical and in vitro studies have shown a large field of application for probiotic supplementation related to benefits that occur in infections and diseases [119–121]. Probiotics reduce gastrointestinal discomfort and reinforce the various lines of gut defense: immune exclusion, immune elimination, and immune regulation. Probiotics also stimulate non-specific host resistance to microbial pathogens and thereby aid in their eradication, maintaining a 'healthy' microbiota [122].

The intestinal microbiome has been proposed as a novel therapeutic target for reducing chronic inflammation [78,123], and probiotics have been proposed to improve the resident gut microbiome [27,28,124,125]. In HIV-infected patients, probiotics may provide a beneficial effect [109,110] by restoring the balance of commensals, pathobionts, and pathogens resident at a mucosal surface, as well as by inducing improvements in the epithelial barrier function, to improve CD4 counts and to impact markers of bacterial translocation, inflammation, and immune activation [126–128]

There are evidences that beneficial effects of probiotics are strain-dependent and not all interventions are equally effective. It is likely that some probiotic strains adhere better to the small intestine, while others bind specifically to different parts of the large intestine, as well as different strains adhering differently in healthy or injured mucosa. Strictly-related probiotics have shown different in vitro properties, which may mirror differences in clinical effects.

Thus, immunomodulatory properties of all probiotic bacteria should be characterized in order to develop clinical applications in different target populations [120,121,129]. The recent expansion in the sale and use of probiotics has resulted in an increase in the standards required to scientifically substantiate their claimed beneficial effects.

Many studies reported that probiotics were "well-tolerated" without side effects, or no statistically significantly increased relative risk of overall number of adverse events. In conclusion: "Across studies, there was no indication that critically ill and high-risk participants taking probiotics were more likely to experience adverse events than control participants with the same health status" [130].

However, additional investigations may provide a full clarification of the mechanism of action by which probiotics can be used as innovative tools to alleviate intestinal inflammation, normalize gut mucosal dysfunction, and downregulate hypersensitivity reactions, with the aim of improving the quality of life during HIV infection, and underlining the economic advantages of probiotic diet supplementation.

Acknowledgments: The authors are grateful to Arianna Rolandi and Paola Rocca for their generosity in providing time for reading this manuscript and improving its scientific quality. No funding was received for this project.

Author Contributions: Marcella Reale conceived the project. All authors contributed to researching, writing, and editing the manuscript. All authors reviewed and approved the final manuscript.

Conflicts of Interest: The authors declare no conflict of interest.

References

1. Parker, R.B. Probiotics, the other half of the antibiotics story. *Anim. Nutr. Health* **1974**, *29*, 4–8.
2. Fuller, R. Probiotics in man and animals. *J. Appl. Bacteriol.* **1989**, *66*, 365–378. [PubMed]
3. Salminen, S.; von Wright, A.; Morelli, L.; Marteau, P.; Brassart, D.; de Vos, W.M.; Fonden, R.; Saxelin, M.; Collins, K.; Mogensen, G.; et al. Demonstration of safety of probiotics—A review. *Int. J. Food Microbiol.* **1998**, *44*, 93–106. [CrossRef]
4. Yan, F.; Polk, D.B. Probiotics and immune health. *Curr. Opin. Gastroenterol.* **2011**, *27*, 496–501. [CrossRef] [PubMed]
5. Kang, H.J.; Im, S.H. Probiotics as an immune modulator. *J. Nutr. Sci. Vitaminol.* **2015**, *61*, 103–105. [CrossRef] [PubMed]

6.   O'Toole, P.W.; Cooney, J.C. Probiotic bacteria influence the composition and function of the intestinal microbiota. *Interdiscip. Perspect. Infect. Dis.* **2008**, *2008*, 175285. [CrossRef] [PubMed]
7.   Sanders, M.E. Impact of probiotics on colonizing microbiota of the gut. *J. Clin. Gastroenterol.* **2011**, *45*, 115–119. [CrossRef] [PubMed]
8.   Ganji-Arjenaki, M.; Rafieian-Kopaei, M. Probiotics are a good choice in remission of inflammatory bowel diseases: A meta-analysis and systematic review. *J. Cell Physiol.* **2017**. [CrossRef] [PubMed]
9.   Yoo, J.Y.; Kim, S.S. Probiotics and prebiotics: Present status and future perspectives on metabolic disorders. *Nutrients* **2016**, *8*, 173. [CrossRef] [PubMed]
10.   Guandalini, S. Probiotics for prevention and treatment of diarrhea. *J. Clin. Gastroenterol.* **2011**, *45*, 149–153. [CrossRef] [PubMed]
11.   Sheil, B.; Shanahan, F.; O'Mahony, L. Probiotic effects on inflammatory bowel disease. *J. Nutr.* **2007**, *137*, 819–824.
12.   Yang, G.; Liu, Z.; Yang, P.C. Treatment of allergic rhinitis with probiotics: An alternative approach. *N. Am. J. Med. Sci.* **2013**, *5*, 465–468. [CrossRef] [PubMed]
13.   Wang, Y.; Li, X.; Ge, T.; Xiao, Y.; Liao, Y.; Cui, Y.; Zhang, Y.; Ho, W.; Yu, G.; Zhang, T. Probiotics for prevention and treatment of respiratory tract infections in children: A systematic review and meta-analysis of randomized controlled trials. *Medicine* **2016**, *95*, 4509. [CrossRef] [PubMed]
14.   Claesson, M.J.; Jeffery, I.B.; Conde, S.; Power, S.E.; O'Connor, E.M.; Cusack, S.; Harris, H.M.; Coakley, M.; Lakshminarayanan, B.; O'Sullivan, O.; et al. Gut microbiota composition correlates with diet and health in the elderly. *Nature* **2012**, *488*, 178–184. [CrossRef] [PubMed]
15.   Bailey, M.T. Exposure to a social stressor alters the structure of the intestinal microbiota: Implications for stressor-induced immunomodulation? *Brain Behav. Immun.* **2011**, *25*, 397. [CrossRef] [PubMed]
16.   Moreira, A.P.; Texeira, T.F.; Ferreira, A.B.; Peluzio Mdo, C.; Alfenas Rde, C. Influence of a high-fat diet on gut microbiota, intestinal permeability and metabolic endotoxaemia. *Br. J. Nutr.* **2012**, *108*, 801–809. [CrossRef] [PubMed]
17.   Brenchley, J.M.; Douek, D.C. HIV infection and the gastrointestinal immune system. *Mucosal Immunol.* **2008**, *1*, 23–30. [CrossRef] [PubMed]
18.   Kotler, D.P. HIV infection and the gastrointestinal tract. *AIDS* **2005**, *19*, 107–117. [CrossRef] [PubMed]
19.   Brenchley, J.M.; Schacker, T.W.; Ruff, L.E.; Price, D.A.; Taylor, J.H.; Beilman, G.J.; Nguyen, P.L.; Khoruts, A.; Larson, M.; Haase, A.T.; et al. CD4+ T-cell depletion during all stages of HIV disease occurs predominantly in the gastrointestinal tract. *J. Exp. Med.* **2004**, *200*, 749–759. [CrossRef] [PubMed]
20.   Dillon, S.M.; Frank, D.N.; Wilson, C.C. The gut microbiome and HIV-1 pathogenesis: A two-way street. *AIDS* **2016**, *30*, 2737–2751. [CrossRef] [PubMed]
21.   Liu, J.; Williams, B.; Frank, D.; Dillon, S.M.; Wilson, C.C.; Landay, A.L. Inside Out: HIV, the gut microbiome, and the mucosal immune system. *J. Immunol.* **2017**, *198*, 605–614. [CrossRef] [PubMed]
22.   Mudd, J.C.; Brenchley, J.M. Gut mucosal barrier dysfunction, microbial dysbiosis, and their role in HIV-1 disease progression. *J. Infect. Dis.* **2016**, *214*, 58–66. [CrossRef] [PubMed]
23.   Hamza, O.J.; Matee, M.I.; Simon, E.N.; Kikwilu, E.; Moshi, M.J.; Mugusi, F.; Mikx, F.H.; Verweij, P.E.; van der Ven, A.J. Oral manifestations of HIV infection in children and adults receiving highly active anti-retroviral therapy [HAART] in Dar es Salaam, Tanzania. *BMC Oral Health* **2006**, *6*, 12–20. [CrossRef] [PubMed]
24.   Vujkovic-Cvijin, I.; Dunham, R.M.; Iwai, S.; Maher, M.C.; Albright, R.G.; Broadhurst, M.J.; Hernandez, R.D.; Lederman, M.M.; Huang, Y.; Somsouk, M.; et al. Dysbiosis of the gut microbiota is associated with HIV disease progression and tryptophan catabolism. *Sci. Transl. Med.* **2013**, *5*, 193ra91. [CrossRef] [PubMed]
25.   Nowak, P.; Troseid, M.; Avershina, E.; Barqasho, B.; Neogi, U.; Holm, K.; Hov, J.R.; Noyan, K.; Vesterbacka, J.; Svärd, J.; et al. Gut microbiota diversity predicts immune status in HIV-1 infection. *AIDS* **2015**, *29*, 2409–2418. [CrossRef] [PubMed]
26.   Carter, G.M.; Esmaeili, A.; Shah, H.; Indyk, D.; Johnson, M.; Andreae, M.; Sacks, H.S. Probiotics in Human Immunodeficiency Virus Infection: A systematic review and evidence synthesis of benefits and risks. *Open Forum Infect. Dis.* **2016**, *3*, ofw164. [CrossRef] [PubMed]

27. Klatt, N.R.; Canary, L.A.; Canary, L.A.; Sun, X.; Vinton, C.L.; Funderburg, N.T.; Morcock, D.R.; Quiñones, M.; Deming, C.B.; Perkins, M.; et al. Probiotic/prebiotic supplementation of antiretrovirals improves gastrointestinal immunity in SIV-infected macaques. *J. Clin. Investig.* **2013**, *123*, 903–907. [CrossRef] [PubMed]

28. Gori, A.; Rizzardini, G.; Van't Land, B.; Amor, K.B.; van Schaik, J.; Torti, C.; Quirino, T.; Tincati, C.; Bandera, A.; Knol, J.; et al. Specific prebiotics modulate gut microbiota and immune activation in HAART-naïve HIV-infected adults: Results of the "COPA" pilot randomized trial. *Nature* **2011**, *4*, 554–563. [CrossRef] [PubMed]

29. Ohland, C.L.; MacNaughton, W.K. Probiotic bacteria and intestinal epithelial barrier function. *Am. J. Physiol. Gastrointest. Liver Physiol.* **2010**, *298*, G807–G819. [CrossRef] [PubMed]

30. Delcenserie, V.; Martel, D.; Lamoureux, M.; Amiot, J.; Boutin, Y.; Roy, D. Immunomodulatory effects of probiotics in the intestinal tract. *Curr. Issues Mol. Biol.* **2008**, *10*, 37–54. [PubMed]

31. Rijkers, G.T.; Bengmark, S.; Enck, P.; Haller, D.; Herz, U.; Kalliomaki, M.; Kudo, S.; Lenoir-Wijnkoop, I.; Mercenier, A.; Myllyluoma, E.; et al. Guidance for substantiating the evidence for beneficial effects of probiotics: Current status and recommendations for future research. *J. Nutr.* **2010**, *140*, 671S–676S. [CrossRef] [PubMed]

32. Cinque, B.; La Torre, C.; Lombardi, F.; Palumbo, P.; Evtoski, Z.; Santini, S.J.; Falone, S.; Cimini, A.; Amicarelli, F.; Cifone, M.G. Vsl#3 probiotic differently influence iec-6 intestinal epithelial cell status and function. *J. Cell. Physiol.* **2017**. [CrossRef]

33. Allen, S.J.; Martinez, E.G.; Gregorio, G.V.; Dans, L.F. Probiotics for treating acute infectious diarrhoea. *Cochrane Database Syst. Rev.* **2010**, *11*, CD003048. [CrossRef]

34. Hooper, L.V.; Littman, D.R.; Macpherson, A.J. Interactions between the microbiota and the immune system. *Science* **2012**, *336*, 1268–1273. [CrossRef] [PubMed]

35. Buffie, C.G.; Pamer, E.G. Microbiota-mediated colonization resistance against intestinal pathogens. *Nat. Rev. Immunol.* **2013**, *13*, 790–801. [CrossRef] [PubMed]

36. Bäumler, A.J.; Sperandio, V. Interactions between the microbiota and pathogenic bacteria in the gut. *Nature* **2016**, *535*, 85–93. [CrossRef] [PubMed]

37. Baquero, F.; Nombela, C. The microbiome as a human organ. *Clin. Microbiol. Infect.* **2012**, *18*, 2–4. [CrossRef] [PubMed]

38. Nishio, J.; Honda, K. Immunoregulation by the gut microbiota. *Cell. Mol. Life Sci.* **2012**, *69*, 3635–3650. [CrossRef] [PubMed]

39. Peterson, L.W.; Artis, D. Intestinal epithelial cells: Regulators of barrier function and immune homeostasis. *Nat. Rev. Immunol.* **2014**, *14*, 141–153. [CrossRef] [PubMed]

40. Garrett, W.S.; Gordon, J.I.; Glimcher, L.H. Homeostasis and inflammation in the intestine. *Cell* **2010**, *140*, 859–870. [CrossRef] [PubMed]

41. Yuan, B.; Zhou, S.; Lu, Y.; Liu, J.; Jin, X.; Wan, H.; Wang, F. Changes in the expression and distribution of claudins, increased epithelial apoptosis, and a mannan-binding lectin-associated immune response lead to barrier dysfunction in dextran sodium sulfate-induced rat colitis. *Gut Liver* **2015**, *9*, 734–740. [CrossRef] [PubMed]

42. Nakamura, K.; Sakuragi, N.; Takakuwa, A.; Ayabe, T. Paneth cell α-defensins and enteric microbiota in health and disease. *Biosci. Microbiota Food Health* **2016**, *35*, 57–67. [CrossRef] [PubMed]

43. Kucharzik, T.; Lugering, N.; Rautenberg, K.; Lugering, A.; Schmidt, M.A.; Stoll, R.; Domschke, W. Role of M-cells in intestinal barrier function. *Ann. N. Y. Acad. Sci.* **2000**, *915*, 171–183. [CrossRef] [PubMed]

44. Zeuthen, L. H.; Fink, L. N.; Frokiaer, H. Epithelial cells prime the immune response to an array of gut-derived commensals towards a tolerogenic phenotype through distinct actions of thymic stromal lymphopoietin and transforming growth factor β. *Immunology* **2008**, *123*, 197–208. [CrossRef] [PubMed]

45. Baba, N.; Samson, S.; Bourdet-Sicard, R.; Rubio, M.; Sarfati, M. Commensal bacteria trigger a full dendritic cell maturation program that promotes the expansion of non-Tr1 suppressor T-cells. *J. Leukoc. Biol.* **2008**, *84*, 468–476. [CrossRef] [PubMed]

46. Kim, Y.S.; Ho, S.B. Intestinal goblet cells and mucins in health and disease: Recent insights and progress. *Curr. Gastroenterol. Rep.* **2010**, *12*, 319–330. [CrossRef] [PubMed]

47. Rogier, R.; Koenders, M.I.; Abdollahi-Roodsaz, S. Toll-like receptor mediated modulation of T-cell response by commensal intestinal microbiota as a trigger for autoimmune. *Arthritis* **2015**, *2015*, 527696. [CrossRef] [PubMed]

48. Abreu, M.T. Toll-like receptor signalling in the intestinal epithelium: How bacterial recognition shapes intestinal function. *Nat. Rev. Immunol.* **2010**, *11*, 215. [CrossRef]

49. Cario, E.; Gerken, G.; Podolsky, D. Toll-like receptor 2 controls mucosal inflammation by regulating epithelial barrier function. *Gastroenerology* **2004**, *127*, 224–238. [CrossRef]

50. Melmed, G.; Thomas, L.S.; Lee, N.; Tesfay, S.Y.; Lukasek, K.; Michelsen, K.S.; Hou, Y.; Hu, B.; Arditi, M.; Abreu, M.T. Human intestinal epithelial cells are broadly unresponsive to Toll-like receptor 2-dependent bacterial ligands: Implications for host-microbial interactions in the gut. *J. Immunol.* **2003**, *170*, 1406–1415. [CrossRef] [PubMed]

51. Ashida, H.; Ogawa, M.; Kim, M.; Mimuro, H.; Sasakawa, C. Bacteria and host interactions in the gut epithelial barrier. *Nat. Chem. Biol.* **2012**, *8*, 36–45. [CrossRef] [PubMed]

52. Caricilli, A.M.; Castoldi, A.; Câmara, N.O. Intestinal barrier: A gentlemen's agreement between microbiota and immunity. *World J. Gastrointest. Pathophysiol.* **2014**, *5*, 18–32. [CrossRef] [PubMed]

53. Ahrne, S.; Hagslatt, M.L. Effect of lactobacilli on paracellular permeability in the gut. *Nutrients* **2011**, *3*, 104–117. [CrossRef] [PubMed]

54. Ferreira, C.M.; Vieira, A.T.; Vinolo, M.A.; Oliveira, F.A.; Curi, R.; Martins, F.S. The central role of the gut microbiota in chronic inflammatory diseases. *J. Immunol. Res.* **2014**, 689492. [CrossRef] [PubMed]

55. Rao, R.K.; Samak, G. Protection and Restitution of gut barrier by probiotics: Nutritional and clinical implications. *Curr. Nutr. Food Sci.* **2013**, *9*, 99–107. [PubMed]

56. Bron, P.A.; Kleerebezem, M.; Brummer, R.J.; Cani, P.D.; Mercenier, A.; MacDonald, T.T.; Garcia-Ródenas, C.L.; Wells, J.M. Can probiotics modulate human disease by impacting intestinal barrier function? *Br. J. Nutr.* **2017**, *117*, 93–107. [CrossRef] [PubMed]

57. Abedi, D.; Feizizadeh, S.; Akbari, S.; Jafarian-Dehkordi, A. In vitro anti-bacterial and anti-adherence effects of *Lactobacillus delbrueckii* subsp *bulgaricus* on *Escherichia coli*. *Res. Pharm. Sci.* **2013**, *8*, 260–268. [PubMed]

58. Ribet, D.; Cossart, P. How bacterial pathogens colonize their hosts and invade deeper tissues. *Microbes Infect.* **2015**, *17*, 173–183. [CrossRef] [PubMed]

59. Coombes, J.L.; Powrie, F. Dendritic cells in intestinal immune regulation. *Nat. Rev. Immunol.* **2008**, *8*, 435–446. [CrossRef] [PubMed]

60. Van de Veerdonk, F.L.; Gresnigt, M.S.; Kullberg, B.J.; van der Meer, J.W.; Joosten, L.A.; Netea, M.G. Th17 responses and host defense against microorganisms: An overview. *BMB Rep.* **2009**, *42*, 776–787. [CrossRef] [PubMed]

61. Donkor, O.N.; Ravikumar, M.; Proudfoot, O.; Day, S.L.; Apostolopoulos, V.; Paukovics, G.; Vasiljevic, T.; Nutt, S.L.; Gill, H. Cytokine profile and induction of T helper type 17 and regulatory T-cells by human peripheral mononuclear cells after microbial exposure. *Clin. Exp. Immunol.* **2012**, *167*, 282–295. [CrossRef] [PubMed]

62. Symons, A.; Budelsky, A.L.; Towne, J.E. Are Th17 cells in the gut pathogenic or protective? *Mucosal Immunol.* **2012**, *5*, 4–6. [CrossRef] [PubMed]

63. Ouyang, W.; Kolls, J.K.; Zheng, Y. The biological functions of T helper 17 cell effector cytokines in inflammation. *Immunity* **2008**, *28*, 454–467. [CrossRef] [PubMed]

64. Bettelli, E.; Korn, T.; Oukka, M.; Kuchroo, V.K. Induction and effector functions of T(H)17 cells. *Nature* **2008**, *453*, 1051–1057. [CrossRef] [PubMed]

65. Liang, S.C.; Tan, X.Y.; Luxenberg, D.P.; Karim, R.; Dunussi-Joannopoulos, K.; Collins, M.; Fouser, L.A. Interleukin (IL)-22 and IL-17 are coexpressed by Th17 cells and cooperatively enhance expression of antimicrobial peptides. *J. Exp. Med.* **2006**, *203*, 2271–2279. [CrossRef] [PubMed]

66. Sakaguchi, S.; Yamaguchi, T.; Nomura, T.; Ono, M. Regulatory T-cells and immune tolerance. *Cell* **2008**, *133*, 775–787. [CrossRef] [PubMed]

67. Mathias, A.; Pais, B.; Favre, L.; Benyacoub, J.; Corthésy, B. Role of secretory IgA in the mucosal sensing of commensal bacteria. *Gut Microbes* **2014**, *5*, 688–695. [CrossRef] [PubMed]

68. Sharma, R.; Young, C.; Neu, J. Molecular modulation of intestinal epithelial barrier: Contribution of microbiota. *J. Biomed. Biotechnol.* **2010**, 305879. [CrossRef] [PubMed]

69. Tincati, C.; Douek, D.C.; Marchetti, G. Gut barrier structure, mucosal immunity and intestinal microbiota in the pathogenesis and treatment of HIV infection. *AIDS Res. Ther.* **2016**, *13*, 19. [CrossRef] [PubMed]

70. Neild, P.J.; Nijran, K.S.; Yazaki, E.; Evans, D.F.; Wingate, D.L.; Jewkes, R.; Gazzard, B.G. Delayed gastric emptying in human immunodeficiency virus infection: Correlation with symptoms, autonomic function, and intestinal motility. *Dig. Dis. Sci.* **2000**, *45*, 1491–1499. [CrossRef] [PubMed]

71. Mutlu, E.A.; Keshavarzian, A.; Losurdo, J.; Swanson, G.; Siewe, B.; Forsyth, C.; French, A.; Demarais, P.; Sun, Y.; Koenig, L.; et al. A compositional look at the human gastrointestinal microbiome and immune activation parameters in HIV infected subjects. *PLoS Pathog.* **2014**, *10*, e1003829. [CrossRef] [PubMed]

72. Lozupone, C.A.; Li, M.; Campbell, T.B.; Flores, S.C.; Linderman, D.; Gebert, M.J.; Knight, R.; Fontenot, A.P.; Palmer, B.E. Alterations in the gut microbiota associated with HIV-1 infection. *Cell Host Microbe* **2013**, *14*, 329–339. [CrossRef] [PubMed]

73. Dillon, S.M.; Lee, E.J.; Kotter, C.V.; Austin, G.L.; Dong, Z.; Hecht, D.K.; Gianella, S.; Siewe, B.; Smith, D.M.; Landay, A.L.; et al. An altered intestinal mucosal microbiome in HIV-1 infection is associated with mucosal and systemic immune activation and endotoxemia. *Mucosal Immunol.* **2014**, *7*, 983–994. [CrossRef] [PubMed]

74. Perez-Santiago, J.; Gianella, S.; Massanella, M.; Spina, C.A.; Karris, M.Y.; Var, S.R.; Patel, D.; Jordan, P.S.; Young, J.A.; Little, S.J.; et al. Gut *Lactobacillales* are associated with higher CD4 and less microbial translocation during HIV infection. *AIDS* **2013**, *27*, 1921–1931. [CrossRef] [PubMed]

75. Nwosu, F.C.; Avershina, E.; Wilson, R.; Rudi, K. Gut microbiota in HIV Infection: Implication for disease progression and management. *Gastroenterol. Res. Pract.* **2014**, *2014*, 803185. [CrossRef] [PubMed]

76. Brenchley, J.M.; Price, D.A.; Schacker, T.W.; Asher, T.E.; Silvestri, G.; Rao, S.; Kazzaz, Z.; Bornstein, E.; Lambotte, O.; Altmann, D.; et al. Microbial translocation is a cause of systemic immune activation in chronic HIV infection. *Nat. Med.* **2006**, *12*, 1365–1371. [CrossRef] [PubMed]

77. Assimakopoulos, S.F.; Dimitropoulou, D.; Marangos, M.; Gogos, C.A. Intestinal barrier dysfunction in HIV infection: Pathophysiology, clinical implications and potential therapies. *Infection* **2014**, *42*, 951–959. [CrossRef] [PubMed]

78. Lapenta, C.; Boirivant, M.; Marini, M.; Santini, S.M.; Logozzi, M.; Viora, M.; Belardelli, F.; Fais, S. Human intestinal lamina propria lymphocytes are naturally permissive to HIV-1 infection. *Eur. J. Immunol.* **1999**, *29*, 1202–1208. [CrossRef]

79. Steele, A.K.; Lee, E.J.; Manuzak, J.A.; Dillon, S.M.; Beckham, J.D.; McCarter, M.D.; Santiago, M.L.; Wilson, C.C. Microbial exposure alters HIV-1-induced mucosal CD4$^+$ T-cell death pathways ex vivo. *Retrovirology* **2014**, *11*, 14. [CrossRef] [PubMed]

80. Doitsh, G.; Galloway, N.L.; Geng, X.; Yang, Z.; Monroe, K.M.; Zepeda, O.; Hunt, P.W.; Hatano, H.; Sowinski, S.; Munoz-Arias, I.; et al. Cell death by pyroptosis drives CD4 T-cell depletion in HIV-1 infection. *Nature* **2014**, *505*, 509–514. [CrossRef] [PubMed]

81. French, M.; Keane, N.; McKinnon, E.; Phung, S.; Price, P. Susceptibility to opportunistic infections in HIV-infected patients with increased CD4 T-cell counts on antiretroviral therapy may be predicted by markers of dysfunctional effector memory CD4 T-cells and B cells. *HIV Med.* **2007**, *8*, 148–155. [CrossRef] [PubMed]

82. Singh, A.; Vajpayee, M.; Ali, S.A.; Mojumdar, K.; Chauhan, N.K.; Singh, R. HIV-1 diseases progression associated with loss of Th17 cells in subtype 'C' infection. *Cytokine* **2012**, *60*, 55–63. [CrossRef] [PubMed]

83. Yue, F.Y.; Merchant, A.; Kovacs, C.M.; Loutfy, M.; Persad, D.; Ostrowski, M.A. Virus-specific interleukin-17-producing CD4$^+$ T-cells are detectable in early human immunodeficiency virus type 1 infection. *J. Virol.* **2008**, *82*, 6767–6771. [CrossRef] [PubMed]

84. Ancuta, P.; Monteiro, P.; Sekaly, R.P. Th17 lineage commitment and HIV-1 pathogenesis. *Curr. Opin. HIV AIDS* **2010**, *5*, 158–165. [CrossRef] [PubMed]

85. Lindemans, C.A.; Calafiore, M.; Mertelsmann, A.M.; O'Connor, M.H.; Dudakov, J.A.; Jenq, R.R.; Velardi, E.; Young, L.F.; Smith, O.M.; Lawrence, G.; et al. Interleukin-22 promotes intestinal-stem-cell-mediated epithelial regeneration. *Nature* **2015**, *528*, 560–564. [CrossRef] [PubMed]

86. Dandekar, S.; George, M.D.; Baumler, A.J. Th17 cells, HIV and the gut mucosal barrier. *Curr. Opin. HIV AIDS* **2010**, *5*, 173–178. [CrossRef] [PubMed]

87. Chege, D.; Sheth, P.M.; Kain, T.; Kim, C.J.; Kovacs, C.; Loutfy, M.; Halpenny, R.; Kandel, G.; Chun, T.W.; Ostrowski, M.; et al. Sigmoid Th17 populations, the HIV latent reservoir, and microbial translocation in men on long-term antiretroviral therapy. *AIDS* **2011**, *25*, 741–749. [CrossRef] [PubMed]

88. Hunt, P.W. Th17, gut, and HIV: Therapeutic implications. *Curr. Opin. HIV AIDS* **2010**, *5*, 189–193. [CrossRef] [PubMed]
89. Hartigan-O'Connor, D.J.; Hirao, L.A.; McCune, J.M.; Dandekar, S. Th17 cells and regulatory T-cells in elite control over HIV and SIV. *Curr. Opin. HIV AIDS* **2011**, *6*, 221–227. [CrossRef] [PubMed]
90. Kanwar, B.; Favre, D.; McCune, J.M. Th17 and regulatory T-cells: Implications for AIDS pathogenesis. *Curr. Opin. HIV AIDS* **2010**, *5*, 151–157. [CrossRef] [PubMed]
91. Brandt, L.; Benfield, T.; Mens, H.; Clausen, L.N.; Katzenstein, T.L.; Fomsgaard, A.; Karlsson, I. Low level of regulatory T-cells and maintenance of balance between regulatory T-cells and TH17 cells in HIV-1-infected elite controllers. *J. Acquir. Immune Defic. Syndr.* **2011**, *57*, 101–108. [CrossRef] [PubMed]
92. Chevalier, M.F.; Weiss, L. The split personality of regulatory T-cells in HIV infection. *Blood* **2013**, *121*, 29–37. [CrossRef] [PubMed]
93. Neuhaus, J.; Jacobs, D.R., Jr.; Baker, J.V.; Calmy, A.; Duprez, D.; La Rosa, A.; Kuller, L.H.; Pett, S.L.; Ristola, M.; Ross, M.J.; et al. Markers of inflammation, coagulation, and renal function are elevated in adults with HIV infection. *J. Infect. Dis.* **2010**, *201*, 1788–1795. [CrossRef] [PubMed]
94. Sankaran, S.; George, M.D.; Reay, E.; Guadalupe, M.; Flamm, J.; Prindiville, T.; Dandekar, S. Rapid onset of intestinal epithelial barrier dysfunction in primary human immunodeficiency virus infection is driven by an imbalance between immune response and mucosal repair and regeneration. *J. Virol.* **2008**, *82*, 538–545. [CrossRef] [PubMed]
95. Hunt, P.W.; Sinclair, E.; Rodriguez, B.; Shive, C.; Clagett, B.; Funderburg, N.; Robinson, J.; Huang, Y.; Epling, L.; Martin, J.N.; et al. Gut epithelial barrier dysfunction and innate immune activation predict mortality in treated HIV infection. *J. Infect. Dis.* **2014**, *210*, 1228–1238. [CrossRef] [PubMed]
96. Rajasuriar, R.; Wright, E.; Lewin, S.R. Impact of antiretroviral therapy (ART) timing on chronic immune activation/inflammation and end-organ damage. *Curr. Opin. HIV AIDS* **2015**, *10*, 35–42. [CrossRef] [PubMed]
97. Bolsewicz, K.; Debattista, J.; Vallely, A.; Whittaker, A.; Fitzgerald, L. Factors associated with antiretroviral treatment uptake and adherence: A review. Perspectives from Australia, Canada, and the United Kingdom. *AIDS Care* **2015**, *27*, 1429–1438. [CrossRef] [PubMed]
98. Pham, M.D.; Romero, L.; Parnell, B.; Anderson, D.A.; Crowe, S.M.; Luchters, S. Feasibility of antiretroviral treatment monitoring in the era of decentralized HIV care: A systematic. *AIDS Res. Ther.* **2017**, *14*, 3. [CrossRef] [PubMed]
99. Katlama, C.; Deeks, S.G.; Autran, B.; Martinez-Picado, J.; van Lunzen, J.; Rouzioux, C.; Miller, M.; Vella, S.; Schmitz, J.E.; Ahlers, J.; et al. Barriers to a cure for HIV: New ways to target and eradicate HIV-1 reservoirs. *Lancet* **2013**, *381*, 2109–2117. [CrossRef]
100. Hummelen, R.; Vos, A.P.; van't Land, B.; van Norren, K.; Reid, G. Altered host-microbe interaction in HIV: A target for intervention with pro- and pre-biotics. *Int. Rev. Immunol.* **2010**, *29*, 485–513. [CrossRef] [PubMed]
101. D'Ettorre, G.; Ceccarelli, G.; Giustini, N.; Serafino, S.; Calantone, N.; De Girolamo, G.; Bianchi, L.; Bellelli, V.; Ascoli-Bartoli, T.; Marcellini, S.; et al. Probiotics reduce inflammation in antiretroviral treated, HIV-infected individuals: Results of the "Probio-HIV" clinical trial. *PLoS ONE* **2015**, *10*. [CrossRef] [PubMed]
102. Surendran Nair, M.; Amalaradjou, M.A.; Venkitanarayanan, K. Antivirulence properties of probiotics in combating microbial pathogenesis. *Adv. Appl. Microbiol.* **2017**, *98*, 1–29. [PubMed]
103. Falasca, K.; Vecchiet, J.; Ucciferri, C.; Di Nicola, M.; D'Angelo, C.; Reale, M. Effect of probiotic supplement on cytokine levels in HIV-infected individuals: A preliminary study. *Nutrients* **2015**, *7*, 8335–8347. [CrossRef] [PubMed]
104. Hardy, H.; Harris, J.; Lyon, E.; Beal, J.; Foey, A.D. Probiotics, prebiotics and immunomodulation of gut mucosal defences: Homeostasis and immunopathology. *Nutrients* **2013**, *5*, 1869–1912. [CrossRef] [PubMed]
105. Tanabe, S. The effect of probiotics and gut microbiota on Th17 cells. *Int. Rev. Immunol.* **2013**, *32*, 511–525. [CrossRef] [PubMed]
106. Kwon, H.K.; Lee, C.G.; So, J.S.; Chae, C.S.; Hwang, J.S.; Sahoo, A.; Nam, J.H.; Rhee, J.H.; Hwang, K.C.; Im, S.H. Generation of regulatory dendritic cells and CD4+ Foxp3+ T-cells by probiotics administration suppresses immune disorders. *Proc. Natl. Acad. Sci. USA* **2010**, *107*, 2159–2164. [CrossRef] [PubMed]
107. López, P.; González-Rodríguez, I.; Gueimonde, M.; Margolles, A.; Suárez, A. Immune response to Bifidobacterium bifidum strains support Treg/Th17 plasticity. *PLoS ONE* **2011**, *6*, e24776. [CrossRef] [PubMed]

108. Kim, C.J.; Walmsley, S.L.; Raboud, J.M.; Kovacs, C.; Coburn, B.; Rousseau, R.; Reinhard, R.; Rosenes, R.; Kaul, R. Can probiotics reduce inflammation and enhance gut immune health in people living with HIV: Study designs for the Probiotic Visbiome for Inflammation and Translocation (PROOV IT) Pilot Trials. *HIV Clin. Trials* **2016**, *17*, 147–157. [CrossRef] [PubMed]

109. Cunningham-Rundles, S.; Ahrne, S.; Johann-Liang, R.; Abuav, R.; Dunn-Navarra, A.M.; Grassey, C.; Bengmark, S.; Cervia, J.S. Effect of probiotic bacteria on microbial host defense, growth, and immune function in human immunodeficiency virus type-1 infection. *Nutrients* **2011**, *3*, 1042–1070. [CrossRef] [PubMed]

110. Yang, O.O.; Kelesidis, T.; Cordova, R.; Khanlou, H. Immunomodulation of antiretroviral drug-suppressed chronic HIV-1 infection in an oral probiotic double-blind placebo-controlled trial. *AIDS Res. Hum. Retrovir.* **2014**, *30*, 988–995. [CrossRef] [PubMed]

111. Irvine, S.L.; Hummelen, R.; Hekmat, S.; Looman, C.W.; Habbema, J.D.; Reid, G. Probiotic yogurt consumption is associated with an increase of CD4 count among people living with HIV/AIDS. *J. Clin. Gastroenterol.* **2010**, *44*, 201–205. [CrossRef] [PubMed]

112. Iwabuchi, N.; Takahashi, N.; Xiao, J.Z. Suppressive effects of *Bifidobacterium longum* on the production of Th2-attracting chemokines induced with T-cell-antigen-presenting cell interactions. *FEMS Immunol. Med. Microbiol.* **2009**, *55*, 324–334. [CrossRef] [PubMed]

113. Mirmonsef, P.; Krass, L.; Landay, A.; Spear, G.T. The role of bacterial vaginosis and trichomonas in HIV transmission across the female genital tract. *Curr. HIV Res.* **2012**, *10*, 202–210. [CrossRef] [PubMed]

114. Hummelen, R.; Changalucha, J.; Butamanya, N.L.; Cook, A.; Habbema, J.D.; Reid, G. Lactobacillus rhamnosus GR-1 and *L. reuteri* RC-14 to prevent or cure bacterial vaginosis among women with HIV. *Int. J. Gynaecol. Obstet.* **2010**, *111*, 245–248. [CrossRef] [PubMed]

115. Staab, B.; Eick, S.; Knöfler, G.; Jentsch, H. The influence of a probiotic milk drink on the development of gingivitis: A pilot study. *J. Clin. Periodontol.* **2009**, *36*, 850–856. [CrossRef] [PubMed]

116. Ryder, M.I.; Nittayananta, W.; Coogan, M.; Greenspan, D.; Greenspan, J.S. Periodontal disease in HIV/AIDS. *Periodontology 2000* **2012**, *60*, 78–97. [CrossRef] [PubMed]

117. Gupta, G. Probiotics and periodontal health. *J. Med. Life* **2011**, *4*, 387–394. [PubMed]

118. Noguera-Julian, M.; Guillén, Y.; Peterson, J.; Reznik, D.; Harris, E.V.; Joseph, S.J.; Rivera., J.; Kannanganat, S.; Amara, R.; Nguyen, M.L.; et al. Oral microbiome in HIV-associated periodontitis. *Medicine* **2017**, *96*, e5821. [CrossRef] [PubMed]

119. Martinez, R.C.; Bedani, R.; Saad, S.M. Scientific evidence for health effects attributed to the consumption of probiotics and prebiotics: an update for current perspectives and future challenges. *Br. J. Nutr.* **2015**, *114*, 1993–2015. [CrossRef] [PubMed]

120. Forsberg, A.; West, C.E.; Prescott, S.L.; Jenmalm, M.C. Pre- and probiotics for allergy prevention: Time to revisit recommendations? *Clin. Exp. Allergy* **2016**, *46*, 1506–1521. [CrossRef] [PubMed]

121. Parvez, S.; Malik, K.A.; Ah Kang, S.; Kim, H.Y. Probiotics and their fermented foods products are beneficial for health. *J. Appl. Microbiol.* **2006**, *100*, 1171–1185. [CrossRef] [PubMed]

122. Isolauri, E. Probiotics in human disease. *Am. J. Clin. Nutr.* **2001**, *73*, 1142S–1146S. [PubMed]

123. Rajasuriar, R.; Khoury, G.; Kamarulzaman, A.; French, M.A.; Cameron, P.U.; Lewin, S.R. Persistent immune activation in chronic HIV infection: Do any interventions work? *AIDS* **2013**, *27*, 1199–1208. [CrossRef] [PubMed]

124. González-Hernández, L.A.; Jave-Suarez, L.F.; Fafutis-Morris, M.; Montes-Salcedo, K.E.; Valle-Gutierrez, L.G.; Campos-Loza, A.E.; Enciso-Gómez, L.F.; Andrade-Villanueva, J.F. Synbiotic therapy decreases microbial translocation and inflammation and improves immunological status in HIV-infected patients: A double-blind randomized controlled pilot trial. *Nutr. J.* **2012**, *11*, 90. [CrossRef] [PubMed]

125. Amara, A.A.; Shibl, A. Role of Probiotics in health improvement, infection control and disease treatment and management. *Saudi Pharm. J.* **2015**, *23*, 107–114. [CrossRef] [PubMed]

126. Connolly, N.C.; Riddler, S.A.; Rinaldo, C.R. Proinflammatory cytokines in HIV disease—A review and rationale for new therapeutic approaches. *AIDS Rev.* **2005**, *7*, 168–180. [PubMed]

127. Merlini, E.; Bai, F.; Bellistri, G.M.; Tincati, C.; d'Arminio Monforte, A.; Marchetti, G. Evidence for polymicrobic flora translocating in peripheral blood of HIV-infected patients with poor immune response to antiretroviral therapy. *PLoS ONE* **2011**, *6*, e18580. [CrossRef] [PubMed]

128. Gad, M.; Ravn, P.; Søborg, D.A.; Lund-Jensen, K.; Ouwehand, A.C.; Jensen, S.S. Regulation of the IL-10/IL-12 axis in human dendritic cells with probiotic bacteria. *FEMS Immunol. Med. Microbiol.* **2011**, *63*, 93–107. [CrossRef] [PubMed]

129. Sánchez, B.; Delgado, S.; Blanco-Míguez, A.; Lourenço, A.; Gueimonde, M.; Margolles, A. Probiotics, gut microbiota, and their influence on host health and disease. *Mol. Nutr. Food Res.* **2017**, *61.* [CrossRef] [PubMed]

130. Hempel, S.; Newberry, S.; Ruelaz, A.; Wang, Z.; Miles, J.; Suttorp, M.J.; Johnsen, B.; Shanman, R.; Slusser, W.; Fu, N.; et al. Safety of probiotics to reduce risk and prevent or treat disease. *Evid. Rep. Technol. Assess* **2011**, *200*, 1–645.

nutrients

MDPI

Article

The Anti-Inflammatory and Antioxidant Potential of Pistachios (*Pistacia vera* L.) In Vitro and In Vivo

Irene Paterniti [1], Daniela Impellizzeri [1], Marika Cordaro [1], Rosalba Siracusa [1], Carlo Bisignano [1], Enrico Gugliandolo [1], Arianna Carughi [2], Emanuela Esposito [1], Giuseppina Mandalari [1] and Salvatore Cuzzocrea [1,3,*]

[1] Department of Chemical, Biological, Pharmaceutical and Environmental Science, University of Messina, Viale Ferdinando Stagno D'Alcontres 31, 98166 Messina, Italy; ipaterniti@unime.it (I.P.); dimpellizzeri@unime.it (D.I.); cordarom@unime.it (M.C.); rosiracusa@gmail.com (R.S.); cbisignano@unime.it (C.B.); egugliandolo@unime.it (E.G.); eesposito@unime.it (E.E.); gmandalari@unime.it (G.M.)
[2] American Pistachio Growers, 9 River Park Pl E, Fresno, CA 93720, USA; carughia@smccd.edu
[3] Department of Pharmacological and Physiological Science, Saint Louis University School of Medicine, 1402 South Grand Blvd, St. Louis, MO 63104, USA
* Correspondence: salvator@unime.it

Received: 12 May 2017; Accepted: 10 August 2017; Published: 22 August 2017

Abstract: Several reports have demonstrated the effectiveness of pistachio against oxidative stress and inflammation. In this study, we investigate if polyphenols extracts from natural raw shelled pistachios (NP) or roasted salted pistachio (RP) kernels have anti-inflammatory and antioxidant properties at lower doses than reported previously, in both in vitro and in vivo models. The monocyte/macrophage cell line J774 was used to assess the extent of protection by NP and RP pistachios against lipopolysaccharide (LPS)-induced inflammation. Moreover, antioxidant activity of NP and RP was assessed in an in vivo model of paw edema in rats induced by carrageenan (CAR) injection in the paw. Results from the in vitro study demonstrated that pre-treatment with NP (0.01, 0.1 and 0.5 mg/mL) and RP (0.01 and 0.1 mg/mL) exerted a significant protection against LPS induced inflammation. Western blot analysis showed NP reduced the degradation of IκB-α, although not significantly, whereas both NP and RP decreased the TNF-α and IL-1β production in a dose-dependent way. A significant reduction of CAR-induced histological paw damage, neutrophil infiltration and nitrotyrosine formation was observed in the rats treated with NP. These data demonstrated that, at lower doses, polyphenols present in pistachios possess antioxidant and anti-inflammatory properties. This may contribute toward a better understanding of the beneficial health effects associated with consumption of pistachios.

Keywords: pistachio; oxidative stress; inflammation; paw edema

1. Introduction

Over the past decade, numerous studies have shown that classes of natural substances derived from higher plants are potentially interesting for therapeutic interventions in various inflammatory diseases [1]. Frequently, treating inflammation with analgesics, non-steroidal anti-inflammatory drugs, and corticosteroid leads to side effects such as gastric discomfort, hypersensitivity reactions, gastric erosion, diabetes mellitus and increased susceptibility to infection [2]. Therefore, it is time to consider plants as possible remedies. Most members of the pistachio genus have chemical and therapeutic similarities. The fruits, nuts, resin and leaves of *Pistacia lentiscus* are used for the treatment of eczema, throat infections, asthma, kidney stones, diarrhea and stomach ache, with astringent, antipyretic, anti-inflammatory, antibacterial, antiviral, pectoral and stimulating properties [3–6].

In addition, pistachio tree nut has been reported to cause IgE-mediated allergic reactions, comparing three different extracts from raw, roasted, and steam-roasted pistachio nut treatments. The most significant finding of this study was the successful reduction of IgE-binding by pistachio extracts using steam-roast processing without any significant changes in sensory quality of product [7]. The extracts from galls of *P. integerrima* are known to have expectorant, bronchodilator, antiemetic, appetizer, diuretic and antirheumatic effects [8]. The galls of *P. terebinthus*, a small tree from Mediterranean countries, has been used for hip pain gout and rheumatisms [9]. *P. vera* L. (Anacardiaceae family) is a high value product, widely consumed globally because of its nutritional characteristics and health benefits [10]. The United States is considered the second largest producer of pistachios after Iran [11]. There are currently several pistachio bioavailability studies available in the literature [12–16]. One of our previous studies has shown that polyphenols from Natural raw shelled pistachios (*Pistacia vera* L.) (NP) and roasted salted pistachio (RP) kernels were bioaccessible in the upper gastrointestinal tract during simulated human digestion: more than 90% of the total polyphenols were released in the gastric compartment, with virtual total release in the duodenal phase [17]. Anti-inflammatory effects of pistachio nut and anti-inflammatory activity of its components have been the subject of numerous studies in recent years. These effects have been demonstrated in various animal models of acute inflammation such as paw edema [4,8,18,19], LPS-induced inflammation [20], and chronic inflammation models such as colitis [21–24].

Based of this evidence, the purpose of the present study was to investigate if polyphenols extracted from natural raw shelled pistachios (NP) and from roasted salted pistachio (RP) kernels had anti-inflammatory and antioxidant properties, even at lower doses compared to that observed in the literature. We used two different models where inflammation and oxidative stress play a crucial role. In particular, we have induced the inflammation process in both an in vitro model using cultured LPS-stimulated macrophage cells and an in vivo model of carrageenan-induced fist edema in rats, which is a useful model of acute inflammation.

2. Materials and Methods

2.1. Pistachios

Californian natural raw, shelled pistachios (NP) and roasted salted pistachio (RP) kernels (*Pistacia vera* L.) were kindly provided by the American Pistachio Growers (Fresno, CA, USA). Pistachio polyphenolic extracts were prepared as previously reported [17,25]. Briefly, NPs or RPs (10 g) were extracted five times with *n*-hexane (100 mL) to remove lipids, after which the residues were mixed with 100 mL of methanol/HCl 0.1% (*v/v*), extracted and centrifuged. After four extractions, the residues were dissolved in distilled water (40 mL) and extracted five times with ethyl acetate (40 mL). Polyphenols analysis was performed using an Ascentis Express C18 column (150 × 4.6 mm, 2.7 μm, Ascentis Express, Supelco, Bellefonte, PA, USA). Polyphenols identification in samples of NP and RP is reported in Table 1. As previously reported [20], NP showed higher amounts of total polyphenols (6.7 mg/100 g) compared with RP (6.0 mg/100 g), with significant differences ($p < 0.05$) in the concentration of gallic acid, catechin, epicatechin and isoquercetin. Experimental research on pistachios complied with the Convention on Biological Diversity and the Convention on the Trade in Endangered Species of Wild Fauna and Flora.

Table 1. Flavonoids and phenolic acids in NP and RP.

Compound	NP	RP
Gallic acid	1.18 ± 0.12 *	2.05 ± 0.24 *
Protocatechuic acid	0.88 ± 0.04	0.96 ± 0.16
Chlorogenic acid	-	0.18 ± 0.02
Catechin	2.19 ± 0.20 *	0.95 ± 0.06 *
Epicatechin	0.15 ± 0.01 *	0.08 ± 0.02 *

Table 1. *Cont.*

Compound	NP	RP
Eriodictyol-7-*O*-glucoside	0.01 ± 0.00	0.03 ± 0.00
Quercetin-3-*O*-rutinoside	0.58 ± 0.04	0.55 ± 0.04
Isoquercetin	1.52 ± 0.22 *	0.81 ± 0.10 *
Daidzein	-	0.15 ± 0.01
Eriodictyol	0.06 ± 0.02	0.05 ± 0.01
Luteolin	0.18 ± 0.03	0.22 ± 0.11

-, trace of detected compound; NP, natural raw pistachio extract; RP, roasted salted pistachio extract. Values are expressed as mg per 100 g and represent the average of triplicate measurements ± SD. Differences among concentration of polyphenols in NP and RP were assessed by analysis of variance followed by the Tukey pairwise comparison. Two-sample *t* tests (two-tailed) were used. The regression values were considered statistically significant at $p < 0.05$. * indicates significant differences.

2.2. In Vitro Study

2.2.1. Cell Culture and Experimental Groups

The monocyte/macrophage cell-line J774-A1 was cultured and a preliminary analysis involved the study of cell viability: 4×10^4 cells were plated (in a volume of 150 μL) in 96-well plates and allowed to adhere for 4 h at 37 °C. Thereafter, the medium was replaced with fresh medium and cells were treated with 4 different concentrations (0.01 mg/mL, 0.1 mg/mL, 0.5 mg/mL, and 1.0 mg/mL) of both NP and RP, to determine the high concentrations with less toxicity on cell viability. Once the high concentrations with less toxicity were determined, we stimulated the cells with LPS (from *Escherichia coli* 1.0 μg/mL) for 24 h [26].

2.2.2. Vital Staining

To assess viability of cell cultures, cells were incubated at 37 °C with 0.2 mg/mL MTT (3-[4,5-dimethylthiazol-2-yl]-2,5 diphenyl tetrazolium bromide) for 1 h. Cell viability was quantified by measurement of optical density at 550 nm (OD550) using a microplate reader [27].

2.2.3. Western Blot Analysis

Extracts of macrophages stimulated for 24 h with LPS were prepared as previous described [28]. Specific primary antibodies anti-iNOS (1:500 Trasduction), anti-COX2 (1:500; Cayman Chemical, Ann Arbor, MI, USA) and anti-IκB-α (1:500; Santa Cruz Biotechnology, Dallas, TX, USA) were used. Membranes were then incubated secondary antibody (1:2000, Jackson ImmunoResearch, West Grove, PA, USA) for 1 h at room temperature. To ascertain that blots were loaded with equal amounts of proteic lysates, they were also incubated in the presence of the antibody β-actin (1:500; Santa Cruz Biotechnology). Signals were detected with an enhanced chemiluminescence detection system reagent. Relative expression of protein bands was quantified by densitometry (optical density [OD] per mm^2) with ChemiDoc™ XRS+ (Image Lab version 5.2.1 build 11, Bio-Rad Laboratories, Hercules, CA, USA) and standardized to β-actin levels.

2.2.4. Measurement of Nitrite Levels

Total nitrite levels, as an indicator of nitric oxide (NO) synthesis, were measured in the supernatant as previously described [28].

2.2.5. Measurement of Cytokine Production

The medium samples were mixed prior to their use in TNF-α and IL-1β ELISA assays, according to manufacturer's details. Absorbance was read at 450 nm and background wavelength correction set at 540 nm or 570 nm.

2.2.6. Determination of Intracellular ROS

Intracellular ROS was detected using the total ROS detection kit as previously shown [29]. After various treatments, the monocyte/macrophage cell-line J774 were trypsinized and then washed twice with 1× washing buffer. Subsequently, the cells were incubated with 5-(and-6)-carboxy-2′,7′-dichlorodihydrofluorescein diacetate (carboxy-H2DCFDA; 10 μM final concentration) at 37 °C in the dark for 30 min. The fluorescence microplate reader detected the light emission. The level of intracellular ROS was expressed as the percentage of the control (nmol/mL).

2.2.7. Determination of Malondialdehyde (MDA) Levels

The monocyte/macrophage cell-line J774 (1×105 cells/ well) was seeded in poly-L-lysine-coated six-well plates. The cells were harvested to detect the levels of malondialdehyde (MDA) using the MDA assay kit as previously described.

2.3. In Vivo Study

2.3.1. Animals

The study was carried out on Sprague–Dawley male rats (200–230 g, Harlan, Nossan, Italy). Food and water were available ad libitum; the animals were fed with a standard diet. The study was approved by the University of Messina Review Board for the care of animals. Animal care was in compliance with Italian regulations on protection of animals used for experimental and other scientific purposes (D.M.116192) as well as with the EEC regulations (O.J. of E.C. L 358/1 12/18/1986). This study conforms to the "ARRIVE Guidelines for Reporting Animal Research". Authors declare that the research complies with the commonly-accepted "3Rs": Replacement, Reduction and Refinement.

2.3.2. Carrageenan-Induced Paw Edema

Paw edema was induced as previously described by subplantar injection of CAR (0.1 mL of a 1% suspension in 0.85% saline) into the right hind paw on rats [30,31]. At the end of the experiment, animals were killed under anesthesia and hind paws were fixed in 10% neutral buffered formalin and embedded in paraffin for both histological and immunohistochemical examinations or stored at −70 °C and used for further analyses.

The volume of paw edema was measured by a plethysmometer (Ugo Basile, Comerio, Varese, Italy) prior to car injection and every hour for 6 h. Edema was expressed as the increase in paw volume (mL) after carrageenan injection relative to the pre-injection value for all animal. Scores are expressed as paw volume difference (mL).

2.3.3. Experimental Groups

First, rats were randomly allocated into the following groups:

(i)   CAR group, rats were injected with CAR to induced paw edema (n = 10);
(ii)   CAR + NP, same as the CAR group and NP (30 mg/kg) was orally administered 30 min before CAR (n = 10); and
(iii)   CAR + RP, same as the CAR group and RP (30 mg/kg) was orally administered 30 min before CAR (n = 10).

The sham-operated group received saline, a vehicle of pistachios, instead of carrageenan (n = 10 for all experimental groups).

The dose of 30 mg/kg was chosen based on a previous dose–response experiment that we did in our laboratories, in which rats were treated with 10, 30 and 100 mg/kg of NP or RP and we observed that the dose of 30 mg/kg was the highest dose without toxicity.

2.3.4. Histological Examination of the CAR-Inflamed Hind Paw

Seven-micrometer-thick sections stained with haematoxylin and eosin (H&E) were examined using light microscopy associated to an Imaging system (AxioVision, Zeiss, Milan, Italy) and scored by two investigators in a blind fashion. The sections were stained with H&E to allow a complete histological analysis that identified the morphological characteristics of the muscle fibers, from 0 to 5, defined as follows: 0 = no inflammation; 1 = mild inflammation; 2 = mild/moderate inflammation; 3 = moderate inflammation; 4 = moderate/severe inflammation; and 5 = severe inflammation [32].

2.3.5. Myeloperoxidase Activity

MPO activity, an index of polymorphonuclear cell accumulation, was determined as previously described [33] in the palm of hind paw tissues. The rate of change in absorbance was measured spectrophotometrically at 650 nm. MPO activity was measured as the quantity of enzyme degrading 1 mM of peroxide 1 minute at 37 °C, and was expressed in units per gram weight of wet tissue.

2.3.6. Immunohistochemistry for Nitroyrosine

Immunohistochemical analysis for nitroyrosine was performed in the palm of hind paw sections as described in previous studies [34]. At the end of the experiment, the tissues were fixed in 10% (w/v) PBS-buffered formaldehyde, and 7-μm sections were prepared from paraffin- embedded tissues. After de-paraffinization, endogenous peroxidase was quenched with 0.3% (v/v) hydrogen peroxide in 60% (v/v) methanol for 30 min. The sections were permeabilized with 0.1% (w/v) Triton X-100 in PBS for 20 min. Non-specific adsorption was minimized by incubating the sections in 2% (v/v) normal goat serum in PBS for 20 min. Endogenous biotin- or avidin-binding sites were blocked by sequential incubation for 15 min with biotin and avidin (DBA), respectively. Sections were incubated overnight with anti-nitrotyrosine polyclonal antibody (1:500 in PBS (v/v)). Sections were washed in PBS and incubated with secondary antibody. Specific labeling was detected with a biotin conjugated goat anti-rabbit IgG and avidin–biotin peroxidase complex (Vector) (D.B.A s.r.l, Milan, Italy). The counter stain was developed with diaminobenzidine (brown color) and nuclear fast red (red background). The photographs obtained (n = 5 photos from each sample collected from all animals in each experimental group) were assessed by densitometry using Leica QWin (software version V3, Leica Microsystems, Cambridge, UK). The percentage area of immunoreactivity was expressed as percent of total tissue area.

2.4. Materials

Unless otherwise stated, all compounds were obtained from Sigma-Aldrich (St. Louis, MO, USA). All other chemicals were of the highest commercial grade available. All stock solutions were prepared in non-pyrogenic saline (0.9% NaCl, Baxter, Milan, Italy) or 10% dimethyl sulfoxide.

2.5. Statistical Analysis

All values are expressed as mean ± SEM. The results were analyzed by one-way ANOVA followed by a Bonferroni post-hoc test for multiple comparisons. A value of $p \leq 0.05$ was pre-determined as the criterion of significance. The number of animals used for in vivo studies was carried out by G * Power 3 software (Die Heinrich-Heine-Universität Düsseldorf, Düsseldorf, Germany).

3. Results

3.1. Effect of NP and RP on Cell Viability

To test the effect on cell viability, J774 cells were incubated with increasing concentrations of NP and RP (from 0.01 mg/mL to 1.0 mg/mL). NP and RP at the high concentration of 1.0 mg/mL reduced cell viability by 55% and 33%, respectively (Figure 1), whereas the concentrations of 0.5 mg/mL, 0.1 mg/mL, and 0.01 mg/mL showed no toxic effects on cell viability, which is around 82% and 77%.

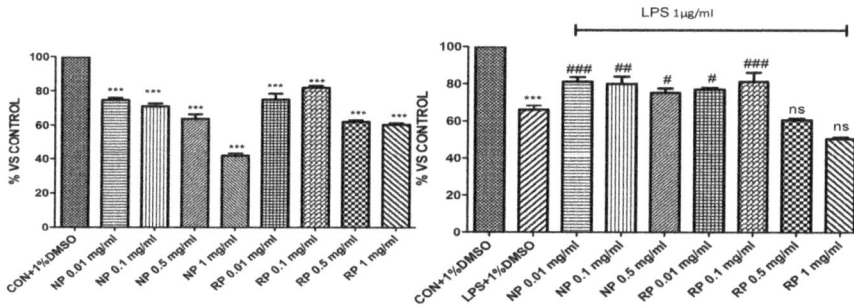

Figure 1. Effect of the pistachios on cell viability. Cell viability was assessed 24 h after treatment with the indicated concentrations (1.0 mg/mL, 0.5 mg/mL, 0.1 mg/mL, and 0.01 mg/mL) of NP and RP, respectively; cell viability was significantly reduced with NP at the highest concentration of 1 mg/mL. NP and RP at 0.5, 0.1, and 0.01 mg/mL lacked cytotoxicity. Moreover, incubation of cells with LPS significantly reduced cell viability compared to the control group, whereas pretreatment with NP at the concentrations of 0.01, 0.1 and 0.5 mg/mL and RP at the concentrations of 0.0 and 0.1 mg/mL significantly limited reduction of cell viability. Data are representative of at least three independent experiments; *** $p < 0.001$ vs. Ctr; ### $p < 0.001$ vs. LPS; ## $p < 0.01$ vs. LPS; # $p < 0.05$ vs. LPS.

Since the concentration of NP at 1.0 mg/mL induced a high reduction of the cell viability, this concentration was not used in further experiments.

Further, we stimulated cells with LPS to induce inflammatory response, and pre-treated cells with NP and RP. The results obtained showed a significant protective effect against LPS induced inflammatory process in cells pre-treated with NP at the three concentrations used (0.5 mg/mL, 0.1 mg/mL and 0.01 mg/mL), whereas pre-treatment with RP exerted significant protection only at the concentrations of 0.01 mg/mL and 0.1 mg/mL.

3.2. Effect of NP and RP on IκB-α Expression

To investigate how NP and RP could attenuate the inflammatory process induced by LPS stimulation, we evaluated IκB-α expression. The results obtained showed a basal expression of IκB-α in the cytoplasmic fraction of the control cells, while IκB-α levels significantly decreased after stimulation with LPS (Figure 2). Pre-treatment with NP at the highest tested concentration 0.5 mg/mL and RP at 0.1 reduced IκB-α degradation, although not significantly. No effect was observed with RP and NP at the concentrations of 0.01 mg/mL.

3.3. Effect of NP and RP on TNF-α and IL-1β Expression

The levels of the pro-inflammatory cytokines TNF-α and IL-1β were evaluated by Elisa kit. An increase in the production of both TNF-α and IL-1β was recorded after LPS stimulation (Figure 3a,b). However, pre-treatment with both NP and RP significantly decreased the levels of TNF-α and IL-1β in a concentration-dependent manner (Figure 3a,b).

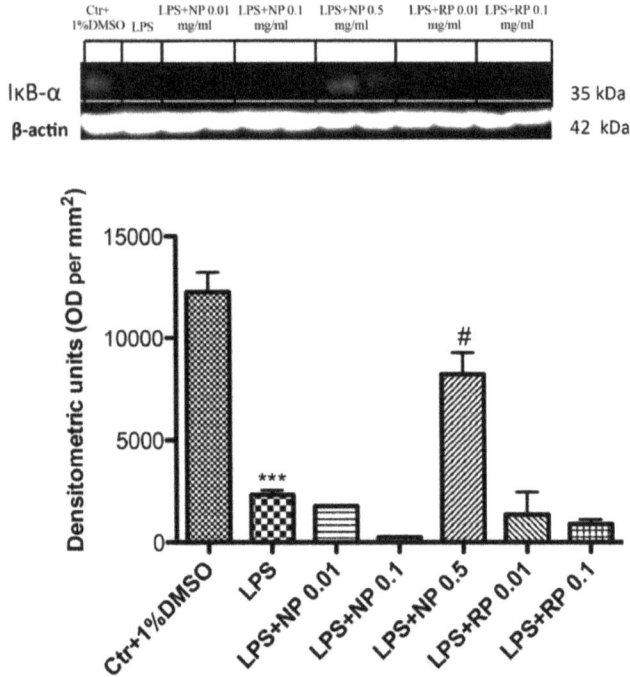

Figure 2. Effect of NP and RP on IκB-α expression. Western blot analysis demonstrated basal levels for IκB-α in the control group, whereas stimulation with LPS significantly induced the degradation of IκB-α levels. Treatments with NP at the concentration of 0.5 mg/mL and RP at 0.1 mg/mL increased the levels of IκB-α, but this protection was not significant. Data are representative of at least three independent experiments; *** $p < 0.001$ vs. Ctr; # $p < 0.05$ vs. LPS.

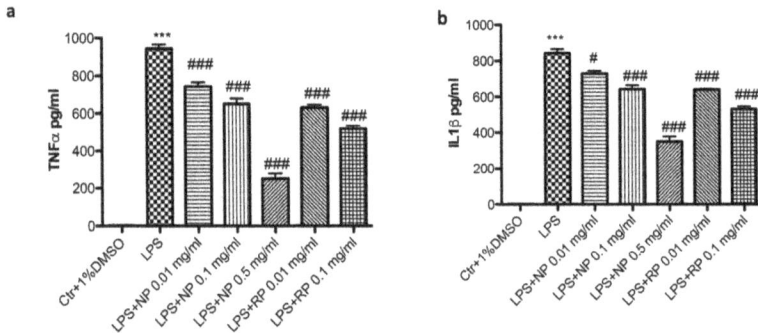

Figure 3. Effect of NP and RP of pro-inflammatory cytokines production. LPS stimulation significantly increased the levels of: TNF-α (**a**); and IL-1β (**b**). Treatments with NP and RP decreased the levels of: TNF-α (**a**); and IL-1β (**b**) in a concentration dependent manner. *** $p < 0.001$ vs. Ctr; ### $p < 0.001$ vs. LPS; # $p < 0.05$ vs. LPS.

3.4. Effect of NP and RP on iNOS, COX-2 and Nitrite Expression

To evaluate the nitrosative stress induced by LPS stimulation and the protective role played by pistachios, we performed Western blots for iNOS and COX-2. Basal levels of iNOS were observed in the control groups, whereas LPS stimulation induced a significant increase in iNOS expression (Figure 4a). Pre-treatment with NP at the concentration of 0.1 and 0.5 mg/mL significantly reduced the expression of iNOS, whereas the concentration of 0.01 mg/mL had no significant effect. Pre-treatment with RP at the concentration of 0.1 significantly reduced the expression of iNOS, whereas the concentration of 0.01 mg/mL had no significant effect (Figure 4a).

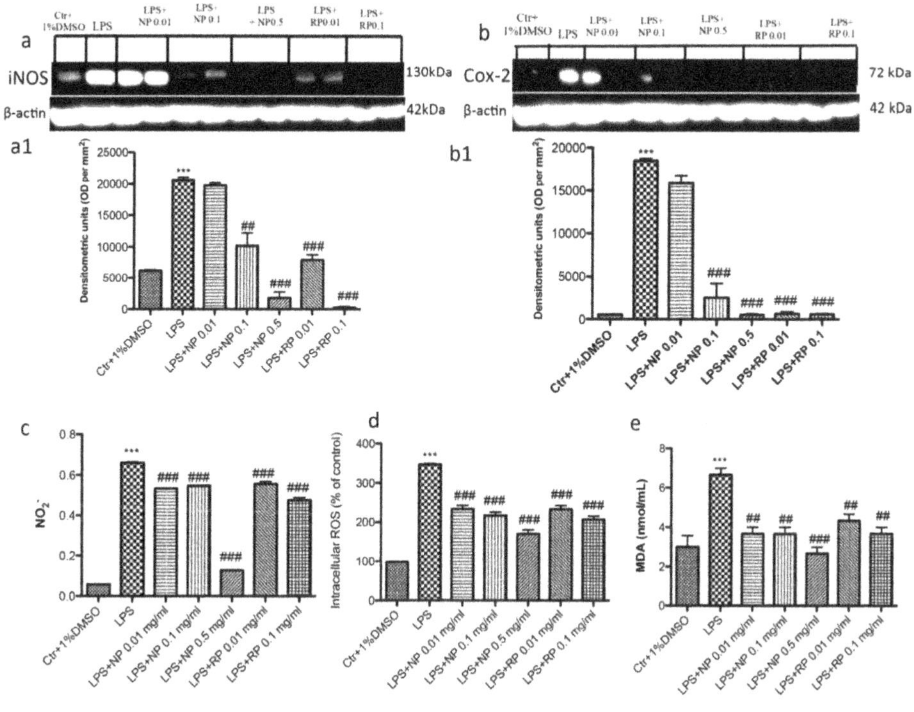

Figure 4. Effect of NP and RP on the expression of iNOS, COX-2 and nitrite levels. Western blot analysis for iNOS and COX-2 demonstrated a significant increased levels after LPS stimulation (**a,b**); treatments with NP only at concentrations of 0.1 and 0.5 mg/mL and RP at 0.01 and 0.1 mg/mL significantly reduced iNOS expressions. No protection was observed with NP 0.01 mg/mL. The levels of COX-2 were also reduced with NP treatment at concentrations of 0.1 and 0.5 mg/mL and RP at 0.01 and 0.1 mg/mL. Less protection, but significant, was observed with NP 0.01 mg/mL. Moreover, we analyzed the levels of nitrite production and we observed an increase of nitrite levels after LPS stimulation (**c**), whereas treatments with NP, only at 0.5 mg/mL, significantly reduced nitrite production. Moreover, we determinate the levels of ROS content and MDA production (**d,e**), and observed an increase of ROS content and MDA levels after LPS stimulation, whereas treatments with NP and RP reduced ROS and MDA levels at concentration dependent manner. Data are representative of at least three independent experiments; *** $p < 0.001$ vs. Ctr; ### $p < 0.001$ vs. LPS; ## $p < 0.01$ vs. LPS.

The COX-2 levels were significantly increased after LPS stimulation, whereas pre-treatment with both NP and RP at all concentrations tested significantly reduced COX-2 levels (Figure 4b).

Moreover, we investigated the levels of nitrite released into the culture medium by Griess reagent. The untreated control group released low levels of NO^{2-}, whereas LPS stimulation significantly increased the levels of NO^{2-} production (Figure 4c). Pre-treatment with both NP and RP extracts decreased NO production in a concentration-dependent manner (Figure 4c).

Moreover, to better investigate the antioxidant capacity of NP and RP, we measure the ROS content and the MDA levels (Figure 4d,e, respectively).

The untreated control group released low levels of ROS and MDA, whereas LPS stimulation significantly increased ROS content and the MDA levels (Figure 4d,e, respectively). Pre-treatment with both NP and RP extracts significantly decreased ROS and MDA production in a concentration-dependent manner (Figure 4d,e, respectively).

3.5. Effect of NP and RP on the Time-Course of Carrageenan-Induced Paw Edema in Rats

Injection of CAR into the sub-plantar region of the right hind paw rapidly induced paw edema in rats, which was maximal after 5 h in CAR injected rats (Figure 5a,b). A significant reduction of the paw edema volume was observed in rats treated with NP at 30 and 100 mg/kg (Figure 5a) compared to the sham group, whereas treatment RP did not significantly affect the paw edema (Figure 5b). Moreover, MPO activity was measured in the palm of hind paw tissues as a marker of neutrophilic infiltration: an increase of MPO activity was found in CAR injected rats (Figure 5c). Administration of NP 30 mg/kg significantly reduced MPO activity, whereas RP at 30 mg/kg did not produce a reduction in neutrophil infiltration in the paw tissues (Figure 5c).

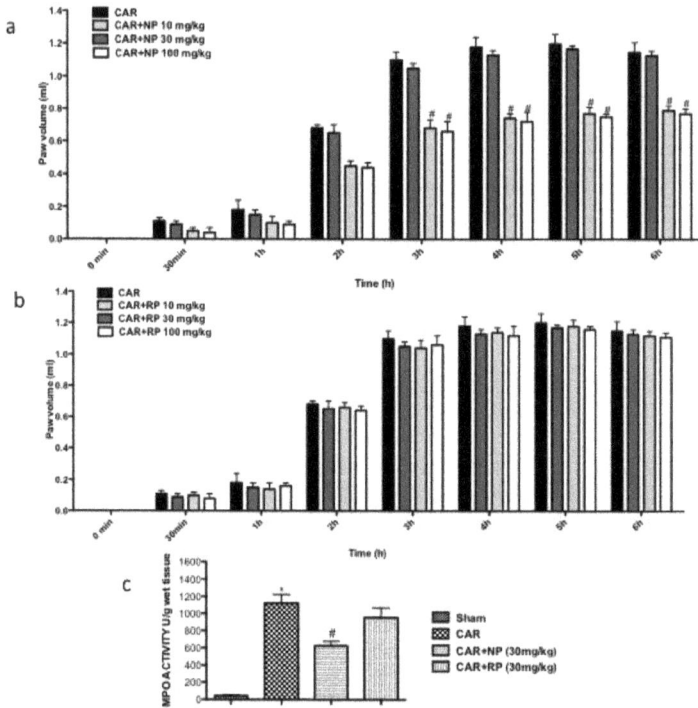

Figure 5. Effect of NP and RP on the time course of carrageenan-induced paw edema. NP and RP were administered orally 30 min before CAR injection. Paw edema was assessed at the time points indicated (**a,b**). NP produced significant improvements in the paw edema (measured as paw volume) in comparison to RP administered at the same time point and at the same doses. Moreover, we observed increased levels of MPO after CAR injections and treatments with NP, but no RP significantly reduced levels of MPO (**c**). Values are means ± SEM. * $p < 0.05$ vs. Sham; [#] $p < 0.05$ vs. CAR.

3.6. NP Inhibited CAR-Induced Histological Paw Damage and Neutrophil Infiltration

To assess the anti-inflammatory and antioxidant effects of NP and RP, the palm of hind paw tissues were examined by hematoxylin and eosin staining. While tissue from sham-treated rats showed no histologic alteration and normal fibers (Figure 6a and insert Figure 6a1, see histological score Figure 6e, a disorganized muscle fibers of various shapes and sizes with irregular contours, important amassing of infiltrating inflammatory cells, edema, loss of normal muscle paw architecture, and increased inter-fiber space were evident after CAR injection into the right hind paw (Figure 6b and insert Figure 6b1, see histological score Figure 6e). Muscle fibers of normal appearance, exhibiting some infiltrating inflammatory cells, was observed after NP treatment (Figure 6c and insert Figure 6c1, see histological score Figure 6e). However, treatment with RP failed to ameliorate this damage, in which accumulation of infiltrating inflammatory cells, edema, increased inter-fiber space and disorganization of normal muscle paw morphology were still observed (Figure 6d and insert Figure 6d1, see histological score Figure 6e).

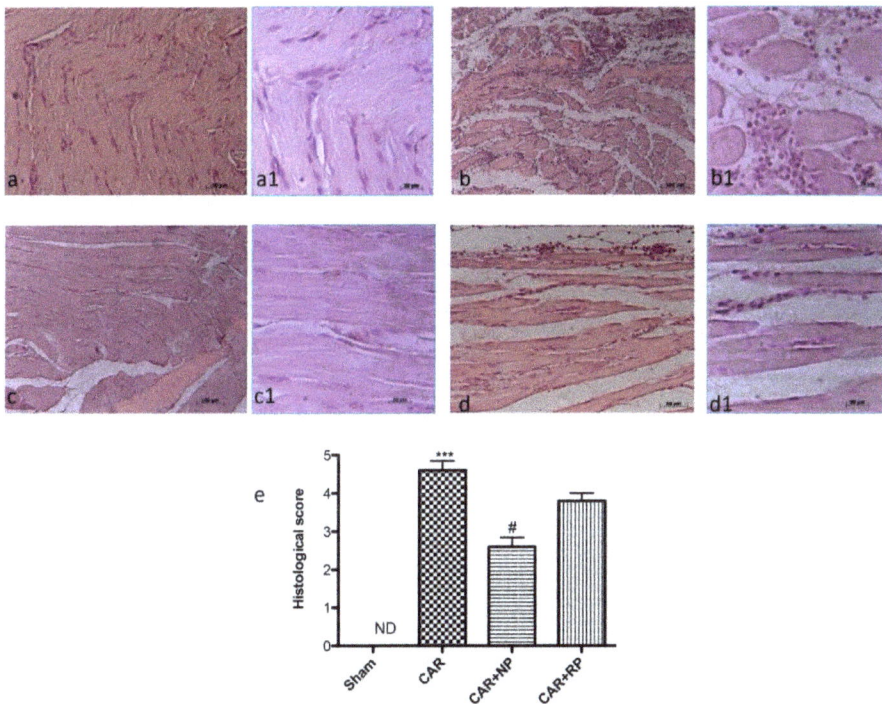

Figure 6. Anti-inflammatory effects of NP histological analysis. Histological evaluation was performed by hematoxylin and eosin staining: (**a**) the control group; (**b**) the intraplantar CAR injection; (**c,d**) CAR with NP treatment and CAR with RP treatment, respectively; (**a1–d1**) low magnification of the respective panels; and (**e**) histological score for the various treatment groups. The figures are representative of at least three independent experiments. *** $p < 0.001$ vs. Sham; # $p < 0.05$ vs. CAR.

3.7. NP Inhibited CAR-Induced Nitrotyrosine Formation

The possible participation of peroxynitrite in reactive oxygen species (ROS)-mediated nociception was evaluated by immunohistochemical detection of nitrated proteins (nitrotyrosine formation). Nitrotyrosine expression was clearly detectable after CAR injection into the hind paw tissue (Figure 7b and insert Figure 7b1, see densitometry analysis Figure 7e). The formation of nitrated proteins was blocked by NP but not by RP (Figure 7c and insert Figure 7c1, see densitometry analysis Figure 7e). This effect better explains the antioxidant effect of NP.

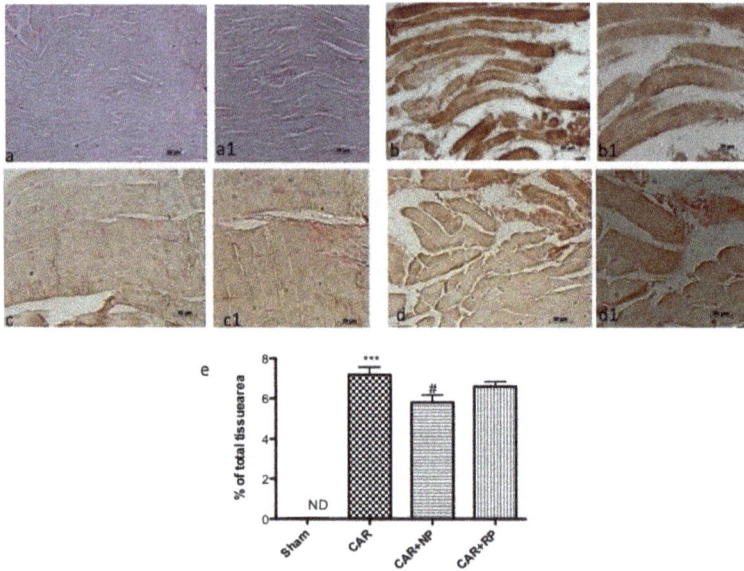

Figure 7. Peroxynitrite production following intraplantar injection of CAR in rat hind paw. Immunohistochemistry for nitrotyrosine showed positive staining in paw tissue sections from CAR-injected rats (**b**). The intensity of nitrotyrosine staining was significantly reduced in the paw from NP-treated rats (**c**); compared to RP-treated rats (**d**). (**a1–d1**) low magnification of the respective panels. The figures are representative of at least three independent experiments. *** $p < 0.001$ vs. Sham; # $p < 0.05$ vs. CAR.

4. Discussion

In the literature, there are several studies regarding the beneficial effects of pistachio. In particular, there are studies on carrageenan or LPS-induced acute inflammatory response [4,16], inflammatory bowel disease and colitis [21,35–37], cancer [38–40], allergic inflammation in asthmatic model [41] and many other experimental models. Furthermore, the antimicrobial properties of polyphenolic fractions obtained from natural raw and roasted salted pistachios have also been evaluated [42,43]. Several studies have reported the potent antioxidant, anti-inflammatory and anti-apoptotic potential of pistachio [44–48].

Therefore, the purpose of our research was to demonstrate for the first time that polyphenols extracted from natural raw shelled pistachios (NP) and from roasted salted pistachio (RP) kernels possessed antioxidant and anti-inflammatory properties at doses lower than those found in the literature.

In the in vivo study, we have demonstrated that the polyphenols-rich extract obtained from NP significantly reduced the paw edema in rats, with a decrease in MPO activity that a marker of neutrophilic infiltration. No significant effect was detected when using the RP polyphenols-rich extract. This could be due to the higher levels of bioactive compounds identified in NP compared with RP [17]. Furthermore, the synergistic interaction amongst the polyphenols identified in NP could enhance its bioactivity. The strong antioxidant effect of catechin, whose concentration is double in NP compared to RP (Table 1), has been widely reported [49]. The concentration of epicatechin and isoquercetin is also significantly higher in NP compared to RP (Table 1): we believe that these compounds contribute to the strong antioxidant and anti-inflammatory properties of the extract.

Various studies on pistachio have clearly indicated a crucial role played by NF-κB in the gene regulation associated to proteins or mediators of inflammation [4,50]. For example, it has been previously shown that NP is able to inhibit the degradation of IKB-alpha and the consequent NFKB translocation in the nucleus [16]. Furthermore, it has been shown that the hydrophilic extract from Sicilian *Pistacia* L. is capable of influencing redox-sensitive signal transduction pathways thus modulating NF-κB activity and finally decreasing the regulation of iNOS expression, COX-2 and TNF-α [16,20,44,45].

Therefore, we sought to evaluate where NP and RP polyphenols-rich extracts had effects on the expression of these proteins, even at lower doses than reported previously. Treatment with NP and RP significantly reduced TNF-α and IL-1β levels, and attenuated the production of iNOS. These observations are in agreement with previous studies evaluating the anti-inflammatory properties of plant materials and molecules of bioactives [51–54]. Thus, the antioxidant and anti-inflammatory properties of polyphenols from pistachios could be attributed to the reduction of the nitrosative stress and subsequent formation of NO. This was reported earlier, but at doses much higher than those used in the present study.

Taken together, our data demonstrated that polyphenols from pistachios, at lower doses that reported in literature, were able to protect from oxidative stress reducing the expression of markers of nitrosative stress such as iNOS, COX2 and NO formation.

5. Conclusions

In conclusion, we have demonstrated that the bioactives present in pistachios exhibit some antioxidants and anti-inflammatory properties at lower doses in vitro and in vivo, suggesting a potential therapeutic use of these natural products. However, in-depth studies and more appropriate models are warranted on the mechanisms involved.

Acknowledgments: The pistachios used in this study were generously provided by the American Pistachios Growers (Fresno, CA, USA).

Author Contributions: S.C. and G.M. planned experiments. D.I., M.C. and E.G. performed in vivo experiments. I.P. and R.S. performed in vitro experiments. E.E. and A.C. analyzed the results and prepared the manuscript. C.B. prepared pistachios compounds. All authors read and approved the final manuscript.

Financial Support: This research received no specific grant from any funding agency, commercial or not-for-profit sectors.

Conflicts of Interest: The authors declare no conflict of interest. Author disclosures: AC is scientific advisor for American Pistachio Growers.

References

1   Calixto, J.B.; Otuki, M.F.; Santos, A.R. Anti-inflammatory compounds of plant origin. Part I. Action on arachidonic acid pathway, nitric oxide and nuclear factor kappa b (nf kappab). *Planta Med.* **2003**, *69*, 973–983. [PubMed]

2.   Ong, C.K.; Lirk, P.; Tan, C.H.; Seymour, R.A. An evidence-based update on nonsteroidal anti-inflammatory drugs. *Clin. Med. Res.* **2007**, *5*, 19–34. [CrossRef] [PubMed]

3.   Lev, E.; Amar, Z. Ethnopharmacological survey of traditional drugs sold in israel at the end of the 20th century. *J. Ethnopharmacol.* **2000**, *72*, 191–205. [CrossRef]

4.   Ben Khedir, S.; Mzid, M.; Bardaa, S.; Moalla, D.; Sahnoun, Z.; Rebai, T. In vivo evaluation of the anti-inflammatory effect of pistacia lentiscus fruit oil and its effects on oxidative stress. *Evid.-Based Complement. Altern. Med. eCAM* **2016**, *2016*, 6108203. [CrossRef] [PubMed]

5.   Said, O.; Khalil, K.; Fulder, S.; Azaizeh, H. Ethnopharmacological survey of medicinal herbs in israel, the golan heights and the west bank region. *J. Ethnopharmacol.* **2002**, *83*, 251–265. [CrossRef]

6. Ali-Shtayeh, M.S.; Yaniv, Z.; Mahajna, J. Ethnobotanical survey in the palestinian area: A classification of the healing potential of medicinal plants. *J. Ethnopharmacol.* **2000**, *73*, 221–232. [CrossRef]

7. Noorbakhsh, R.; Mortazavi, S.A.; Sankian, M.; Shahidi, F.; Maleki, S.J.; Nasiraii, L.R.; Falak, R.; Sima, H.R.; Varasteh, A. Influence of processing on the allergenic properties of pistachio nut assessed in vitro. *J. Agric. Food Chem.* **2010**, *58*, 10231–10235. [CrossRef] [PubMed]

8. Ahmad, N.S.; Waheed, A.; Farman, M.; Qayyum, A. Analgesic and anti-inflammatory effects of pistacia integerrima extracts in mice. *J. Ethnopharmacol.* **2010**, *129*, 250–253. [CrossRef] [PubMed]

9. Adams, M.; Berset, C.; Kessler, M.; Hamburger, M. Medicinal herbs for the treatment of rheumatic disorders—A survey of european herbals from the 16th and 17th century. *J. Ethnopharmacol.* **2009**, *121*, 343–359. [CrossRef] [PubMed]

10. Dreher, M.L. Pistachio nuts: Composition and potential health benefits. *Nutr. Rev.* **2012**, *70*, 234–240. [CrossRef] [PubMed]

11. Grace, M.H.; Esposito, D.; Timmers, M.A.; Xiong, J.; Yousef, G.; Komarnytsky, S.; Lila, M.A. Chemical composition, antioxidant and anti-inflammatory properties of pistachio hull extracts. *Food Chem.* **2016**, *210*, 85–95. [CrossRef] [PubMed]

12. Eisenhauer, B.; Natoli, S.; Liew, G.; Flood, V.M. Lutein and zeaxanthin-food sources, bioavailability and dietary variety in age-related macular degeneration protection. *Nutrients* **2017**, *9*, 120. [CrossRef] [PubMed]

13. Rafiee, Z.; Barzegar, M.; Sahari, M.A.; Maherani, B. Nanoliposomal carriers for improvement the bioavailability of high-valued phenolic compounds of pistachio green hull extract. *Food Chem.* **2017**, *220*, 115–122. [CrossRef] [PubMed]

14. De Giudici, G.; Medas, D.; Meneghini, C.; Casu, M.A.; Gianoncelli, A.; Iadecola, A.; Podda, S.; Lattanzi, P. Microscopic biomineralization processes and zn bioavailability: A synchrotron-based investigation of *Pistacia lentiscus* L. Roots. *Environ. Sci. Pollut. Res. Int.* **2015**, *22*, 19352–19361. [CrossRef] [PubMed]

15. Attoub, S.; Karam, S.M.; Nemmar, A.; Arafat, K.; John, A.; Al-Dhaheri, W.; Al Sultan, M.A.; Raza, H. Short-term effects of oral administration of pistacia lentiscus oil on tissue-specific toxicity and drug metabolizing enzymes in mice. *Cell. Physiol. Biochem.* **2014**, *33*, 1400–1410. [CrossRef] [PubMed]

16. Gentile, C.; Allegra, M.; Angileri, F.; Pintaudi, A.M.; Livrea, M.A.; Tesoriere, L. Polymeric proanthocyanidins from sicilian pistachio (*Pistacia vera* L.) nut extract inhibit lipopolysaccharide-induced inflammatory response in raw 264.7 cells. *Eur. J. Nutr.* **2012**, *51*, 353–363. [CrossRef] [PubMed]

17. Mandalari, G.; Bisignano, C.; Filocamo, A.; Chessa, S.; Saro, M.; Torre, G.; Faulks, R.M.; Dugo, P. Bioaccessibility of pistachio polyphenols, xanthophylls, and tocopherols during simulated human digestion. *Nutrition* **2013**, *29*, 338–344. [CrossRef] [PubMed]

18. Esmat, A.; Al-Abbasi, F.A.; Algandaby, M.M.; Moussa, A.Y.; Labib, R.M.; Ayoub, N.A.; Abdel-Naim, A.B. Anti-inflammatory activity of pistacia khinjuk in different experimental models: Isolation and characterization of its flavonoids and galloylated sugars. *J. Med. Food* **2012**, *15*, 278–287. [CrossRef] [PubMed]

19. Giner-Larza, E.M.; Manez, S.; Giner, R.M.; Recio, M.C.; Prieto, J.M.; Cerda-Nicolas, M.; Rios, J.L. Anti-inflammatory triterpenes from pistacia terebinthus galls. *Planta Med.* **2002**, *68*, 311–315. [CrossRef] [PubMed]

20. Grace, M.H.; Esposito, D.; Timmers, M.A.; Xiong, J.; Yousef, G.; Komarnytsky, S.; Lila, M.A. In vitro lipolytic, antioxidant and anti-inflammatory activities of roasted pistachio kernel and skin constituents. *Food Funct.* **2016**, *7*, 4285–4298. [CrossRef] [PubMed]

21. Naouar, M.S.; Mekki, L.Z.; Charfi, L.; Boubaker, J.; Filali, A. Preventive and curative effect of pistacia lentiscus oil in experimental colitis. *Biomed. Pharmacother.* **2016**, *83*, 577–583. [CrossRef] [PubMed]

22. Tanideh, N.; Jamshidzadeh, A.; Sepehrimanesh, M.; Hosseinzadeh, M.; Koohi-Hosseinabadi, O.; Najibi, A.; Raam, M.; Daneshi, S.; Asadi-Yousefabad, S.L. Healing acceleration of acetic acid-induced colitis by marigold (calendula officinalis) in male rats. *Saudi J. Gastroenterol.* **2016**, *22*, 50–56. [CrossRef] [PubMed]

23. Gholami, M.; Ghasemi-Niri, S.F.; Maqbool, F.; Baeeri, M.; Memariani, Z.; Pousti, I.; Abdollahi, M. Experimental and pathalogical study of pistacia atlantica, butyrate, lactobacillus casei and their combination on rat ulcerative colitis model. *Pathol. Res. Pract.* **2016**, *212*, 500–508. [CrossRef] [PubMed]

24. Gioxari, A.; Kaliora, A.C.; Papalois, A.; Agrogiannis, G.; Triantafillidis, J.K.; Andrikopoulos, N.K. Pistacia lentiscus resin regulates intestinal damage and inflammation in trinitrobenzene sulfonic acid-induced colitis. *J. Med. Food* **2011**, *14*, 1403–1411. [CrossRef] [PubMed]

25. Mandalari, G.; Bisignano, C.; D'Arrigo, M.; Ginestra, G.; Arena, A.; Tomaino, A.; Wickham, M.S. Antimicrobial potential of polyphenols extracted from almond skins. *Lett. Appl. Microbiol.* **2010**, *51*, 83–89. [CrossRef] [PubMed]

26. Esposito, E.; Dal Toso, R.; Pressi, G.; Bramanti, P.; Meli, R.; Cuzzocrea, S. Protective effect of verbascoside in activated c6 glioma cells: Possible molecular mechanisms. *Naunyn Schmiedebergs Arch. Pharmacol.* **2010**, *381*, 93–105. [CrossRef] [PubMed]

27. Abe, K.; Matsuki, N. Measurement of cellular 3-(4,5-dimethylthiazol-2-yl)-2,5-diphenyltetrazolium bromide (mtt) reduction activity and lactate dehydrogenase release using mtt. *Neurosci. Res.* **2000**, *38*, 325–329. [CrossRef]

28. Paterniti, I.; Cordaro, M.; Campolo, M.; Siracusa, R.; Cornelius, C.; Navarra, M.; Cuzzocrea, S.; Esposito, E. Neuroprotection by association of palmitoylethanolamide with luteolin in experimental alzheimer's disease models: The control of neuroinflammation. *CNS Neurol. Disord. Drug Targets* **2014**, *13*, 1530–1541. [CrossRef] [PubMed]

29. Perez, A.P.; Casasco, A.; Schilrreff, P.; Tesoriero, M.V.; Duempelmann, L.; Pappalardo, J.S.; Altube, M.J.; Higa, L.; Morilla, M.J.; Petray, P.; et al. Enhanced photodynamic leishmanicidal activity of hydrophobic zinc phthalocyanine within archaeolipids containing liposomes. *Int. J. Nanomed.* **2014**, *9*, 3335–3345.

30. Salvemini, D.; Wang, Z.Q.; Wyatt, P.S.; Bourdon, D.M.; Marino, M.H.; Manning, P.T.; Currie, M.G. Nitric oxide: A key mediator in the early and late phase of carrageenan-induced rat paw inflammation. *Br. J. Pharmacol.* **1996**, *118*, 829–838. [CrossRef] [PubMed]

31. Impellizzeri, D.; Bruschetta, G.; Cordaro, M.; Crupi, R.; Siracusa, R.; Esposito, E.; Cuzzocrea, S. Micronized/ultramicronized palmitoylethanolamide displays superior oral efficacy compared to nonmicronized palmitoylethanolamide in a rat model of inflammatory pain. *J. Neuroinflamm.* **2014**, *11*, 136. [CrossRef] [PubMed]

32. Petrosino, S.; Campolo, M.; Impellizzeri, D.; Paterniti, I.; Allara, M.; Gugliandolo, E.; D'Amico, R.; Siracusa, R.; Cordaro, M.; Esposito, E.; et al. 2-pentadecyl-2-oxazoline, the oxazoline of pea, modulates carrageenan-induced acute inflammation. *Front. Pharmacol.* **2017**, *8*, 308. [CrossRef] [PubMed]

33. Casili, G.; Cordaro, M.; Impellizzeri, D.; Bruschetta, G.; Paterniti, I.; Cuzzocrea, S.; Esposito, E. Dimethyl fumarate reduces inflammatory responses in experimental colitis. *J. Crohns Colitis* **2016**, *10*, 472–483. [CrossRef] [PubMed]

34. Esposito, E.; Impellizzeri, D.; Cordaro, M.; Siracusa, R.; Gugliandolo, E.; Crupi, R.; Cuzzocrea, S. A new co-micronized composite containing palmitoylethanolamide and polydatin shows superior oral efficacy compared to their association in a rat paw model of carrageenan-induced inflammation. *Eur. J. Pharmacol.* **2016**, *782*, 107–118. [CrossRef] [PubMed]

35. Kim, H.J.; Neophytou, C. Natural anti-inflammatory compounds for the management and adjuvant therapy of inflammatory bowel disease and its drug delivery system. *Arch. Pharm. Res.* **2009**, *32*, 997–1004. [CrossRef] [PubMed]

36. Papalois, A.; Gioxari, A.; Kaliora, A.C.; Lymperopoulou, A.; Agrogiannis, G.; Papada, E.; Andrikopoulos, N.K. Chios mastic fractions in experimental colitis: Implication of the nuclear factor kappab pathway in cultured ht29 cells. *J. Med. Food* **2012**, *15*, 974–983. [CrossRef] [PubMed]

37. Tafti, L.D.; Sharialpanahi, S.M.; Damghani, M.M.; Javadi, B. Traditional persian topical medications for gastrointestinal diseases. *Iran J. Basic Med. Sci.* **2017**, *20*, 222–241. [PubMed]

38. Catalani, S.; Palma, F.; Battistelli, S.; Nuvoli, B.; Galati, R.; Benedetti, S. Reduced cell viability and apoptosis induction in human thyroid carcinoma and mesothelioma cells exposed to cidofovir. *Toxicol. In Vitro* **2017**, *41*, 49–55. [CrossRef] [PubMed]

39. Spyridopoulou, K.; Tiptiri-Kourpeti, A.; Lampri, E.; Fitsiou, E.; Vasileiadis, S.; Vamvakias, M.; Bardouki, H.; Goussia, A.; Malamou-Mitsi, V.; Panayiotidis, M.I.; et al. Dietary mastic oil extracted from pistacia lentiscus var. Chia suppresses tumor growth in experimental colon cancer models. *Sci. Rep.* **2017**, *7*, 3782. [CrossRef] [PubMed]

40. Balan, K.V.; Prince, J.; Han, Z.; Dimas, K.; Cladaras, M.; Wyche, J.H.; Sitaras, N.M.; Pantazis, P. Antiproliferative activity and induction of apoptosis in human colon cancer cells treated in vitro with constituents of a product derived from pistacia lentiscus l. Var. Chia. *Phytomedicine* **2007**, *14*, 263–272. [CrossRef] [PubMed]

41. Qiao, J.; Li, A.; Jin, X.; Wang, J. Mastic alleviates allergic inflammation in asthmatic model mice by inhibiting recruitment of eosinophils. *Am. J. Respir. Cell Mol. Biol.* **2011**, *45*, 95–100. [CrossRef] [PubMed]

42. Smeriglio, A.; Denaro, M.; Barreca, D.; Calderaro, A.; Bisignano, C.; Ginestra, G.; Bellocco, E.; Trombetta, D. In vitro evaluation of the antioxidant, cytoprotective, and antimicrobial properties of essential oil from *Pistacia vera* L. Variety bronte hull. *Int. J. Mol. Sci.* **2017**, *18*, 1212. [CrossRef] [PubMed]

43. Bisignano, C.; Filocamo, A.; Faulks, R.M.; Mandalari, G. In vitro antimicrobial activity of pistachio (*Pistacia vera* L.) polyphenols. *FEMS Microbiol. Lett.* **2013**, *341*, 62–67. [CrossRef] [PubMed]

44. Rauf, A.; Maione, F.; Uddin, G.; Raza, M.; Siddiqui, B.S.; Muhammad, N.; Shah, S.U.; Khan, H.; De Feo, V.; Mascolo, N. Biological evaluation and docking analysis of daturaolone as potential cyclooxygenase inhibitor. *Evid.-Based Complement. Altern. Med. eCAM* **2016**, *2016*, 4098686. [CrossRef] [PubMed]

45. Yayeh, T.; Hong, M.; Jia, Q.; Lee, Y.C.; Kim, H.J.; Hyun, E.; Kim, T.W.; Rhee, M.H. Pistacia chinensis inhibits no production and upregulates ho-1 induction via pi-3k/akt pathway in lps stimulated macrophage cells. *Am. J. Chin. Med.* **2012**, *40*, 1085–1097. [CrossRef] [PubMed]

46. Mehla, K.; Balwani, S.; Kulshreshtha, A.; Nandi, D.; Jaisankar, P.; Ghosh, B. Ethyl gallate isolated from pistacia integerrima linn. Inhibits cell adhesion molecules by blocking ap-1 transcription factor. *J. Ethnopharmacol.* **2011**, *137*, 1345–1352. [CrossRef] [PubMed]

47. Zhang, J.; Kris-Etherton, P.M.; Thompson, J.T.; Vanden Heuvel, J.P. Effect of pistachio oil on gene expression of ifn-induced protein with tetratricopeptide repeats 2: A biomarker of inflammatory response. *Mol. Nutr. Food Res.* **2010**, *54* (Suppl. 1), S83–S92. [CrossRef] [PubMed]

48. Zhou, L.; Satoh, K.; Takahashi, K.; Watanabe, S.; Nakamura, W.; Maki, J.; Hatano, H.; Takekawa, F.; Shimada, C.; Sakagami, H. Re-evaluation of anti-inflammatory activity of mastic using activated macrophages. *In Vivo* **2009**, *23*, 583–589. [PubMed]

49. Gahruie, H.H.; Niakousari, M. Antioxidant, antimicrobial, cell viability and enzymatic inhibitory of antioxidant polymers as biological macromolecules. *Int. J. Biol. Macromol.* **2017**, *104*, 606–617. [CrossRef] [PubMed]

50. Beg, S.; Swain, S.; Hasan, H.; Barkat, M.A.; Hussain, M.S. Systematic review of herbals as potential anti-inflammatory agents: Recent advances, current clinical status and future perspectives. *Pharm. Rev.* **2011**, *5*, 120–137. [CrossRef] [PubMed]

51. Mandalari, G.; Bisignano, C.; Genovese, T.; Mazzon, E.; Wickham, M.S.; Paterniti, I.; Cuzzocrea, S. Natural almond skin reduced oxidative stress and inflammation in an experimental model of inflammatory bowel disease. *Int. Immunopharmacol.* **2011**, *11*, 915–924. [CrossRef] [PubMed]

52. Mandalari, G.; Genovese, T.; Bisignano, C.; Mazzon, E.; Wickham, M.S.; Di Paola, R.; Bisignano, G.; Cuzzocrea, S. Neuroprotective effects of almond skins in experimental spinal cord injury. *Clin. Nutr.* **2011**, *30*, 221–233. [CrossRef] [PubMed]

53. Impellizzeri, D.; Bruschetta, G.; Di Paola, R.; Ahmad, A.; Campolo, M.; Cuzzocrea, S.; Esposito, E.; Navarra, M. The anti-inflammatory and antioxidant effects of bergamot juice extract (bje) in an experimental model of inflammatory bowel disease. *Clin. Nutr.* **2015**, *34*, 1146–1154. [CrossRef] [PubMed]

54. Impellizzeri, D.; Talero, E.; Siracusa, R.; Alcaide, A.; Cordaro, M.; Maria Zubelia, J.; Bruschetta, G.; Crupi, R.; Esposito, E.; Cuzzocrea, S.; et al. Protective effect of polyphenols in an inflammatory process associated with experimental pulmonary fibrosis in mice. *Br. J. Nutr.* **2015**, *114*, 853–865. [CrossRef] [PubMed]

![nutrients logo] *nutrients*

MDPI

Article

Anti-Inflammatory Mechanism Involved in Pomegranate-Mediated Prevention of Breast Cancer: the Role of NF-κB and Nrf2 Signaling Pathways

Animesh Mandal [1], Deepak Bhatia [2] and Anupam Bishayee [3,*

[1] Cancer Therapeutics and Chemoprevention Group, Department of Pharmaceutical Sciences, College of Pharmacy, Northeast Ohio Medical University, Rootstown, OH 44272, USA; animandal0@gmail.com

[2] Department of Pharmacogenomics, Bernard J. Dunn School of Pharmacy, Shenandoah University, Ashburn, VA 20147, USA; dbhatia@su.edu

[3] Department of Pharmaceutical Sciences, College of Pharmacy, Larkin University, Miami, FL 33169, USA

* Correspondence: abishayee@ULarkin.org or abishayee@gmail.com; Tel.: +1-305-760-75211

Received: 1 February 2017; Accepted: 5 April 2017; Published: 28 April 2017

Abstract: Pomegranate (*Punica granatum* L.), a nutrient-rich unique fruit, has been used for centuries for the prevention and treatment of various inflammation-driven diseases. Based on our previous study, a characterized pomegranate emulsion (PE) exhibited a striking inhibition of dimethylbenz(a)anthracene (DMBA)-initiated rat mammary tumorigenesis via antiproliferative and apoptosis-inducing mechanisms. The objective of the present work is to investigate the anti-inflammatory mechanism of action of PE during DMBA rat mammary carcinogenesis by evaluating the expression of cyclooxygenase-2 (COX-2), heat shock protein 90 (HSP90), nuclear factor-κB (NF-κB) and nuclear factor erythroid 2p45 (NF-E2)-related factor 2 (Nrf2). Mammary tumor samples were harvested from our previous chemopreventive study in which PE (0.2–5.0 g/kg) was found to reduce mammary tumorigenesis in a dose-dependent manner. The expressions of COX-2, HSP90, NF-κB, inhibitory κBα (IκBα) and Nrf2 were detected by immunohistochemical techniques. PE decreased the expression of COX-2 and HSP90, prevented the degradation of IκBα, hindered the translocation of NF-κB from cytosol to nucleus and increased the expression and nuclear translocation of Nrf2 during DMBA-induced mammary tumorigenesis. These findings, together with our previous results, indicate that PE-mediated prevention of DMBA-evoked mammary carcinogenesis may involve anti-inflammatory mechanisms through concurrent but differential regulation of two interrelated molecular pathways, namely NF-κB and Nrf2 signaling.

Keywords: anti-inflammatory effects; breast tumor; COX-2; DMBA; HSP90; NF-κB; Nrf2; *Punica granatum*

1. Introduction

Pomegranate (*Punica granatum* L.) is a nutrient-rich fruit which represents a reservoir of bioactive phytochemicals with exceptional medicinal values. The pomegranate is a plant native from the Himalayas to Iran and has been cultivated and naturalized throughout the world and in the United States, including Arizona, California and Texas. Pomegranate, known as *"a pharmacy unto itself"* has been used for centuries in various traditional and folk medicine for the treatment of a large number of ailments [1–3]. During the last decade, pomegranate fruit has been gaining a widespread reputation as a dietary supplement as well as a functional food due to emerging scientific evidence on potential health benefits, including prevention and/or treatment of cardiovascular ailments, neurological disorders, oncologic diseases, dental problems, inflammation, ulcer, arthritis, microbial infection, obesity, diabetes, acquired immune deficiency syndrome and erectile dysfunction [4–9]. Pomegranate fruit contains phytochemicals, including flavonoids (e.g., anthocyanins and catechins), flavonols (e.g., kaempferol and quercetin), flavones (e.g., apigenin and luteolin), conjugated fatty acids, hydrolyzable tannins and related compounds which are thought to be responsible for various biological and pharmacological activities [4,10–14]. Based on preclinical and clinical studies conducted by various laboratories worldwide, pomegranate-derived substances, such as juice, extracts and phytoconstituents exhibited cancer preventive and therapeutic effects against colon, liver, lung, prostate and skin cancer [4,15–18]. Various extracts, fractions and phytochemicals from pomegranate fruit, peel, seed and flower demonstrated cytotoxic, antiproliferative, proapoptotic, antiangiogenic, anti-invasive, and antimetastatic properties against estrogen receptor-positive and -negative breast cancer cells [19–32]. Pomegranate seed oil and fermented juice concentrate were found to suppress 7,12-dimethyl benz(a)anthracene (DMBA)-induced preneoplastic mammary gland lesions ex vivo in a mouse mammary organ culture model [33]. Oral administration of pomegranate juice concentrate reduced the volume and weight of xenografted BT-474 tumors in female athymic nude mice [26].

Recently, we have documented the novel finding that oral feeding of a pomegranate emulsion (PE), containing most bioactive phytochemicals present in the whole fruit, exerted a significant chemopreventive activity against DMBA-initiated mammary tumorigenesis in female rats [34]. PE reduced the incidence, total burden and average weight of mammary tumors with a concomitant inhibition of intratumor cell proliferation, induction of apoptosis, and it altered the expression of Bax, Bcl2, Bad, caspase-3, caspase-7, caspase-9, poly (ADP ribose) polymerase and cytochrome c [34]. We have also observed that PE diminished the expression of estrogen receptor-α (ER-α), ER-β and cyclin D1 and abrogated the expression, cytoplasmic accumulation and nuclear translocation of β-catenin, an essential transcriptional cofactor for Wnt/β-catenin signaling, during DMBA mammary carcinogenesis [35]. Emerging studies indicate that chronic inflammation is involved in the development and progression of mammary carcinoma [36–40] and pomegranate phytochemicals are endowed with anti-inflammatory properties [4,8,11,12]. Accordingly, this study was conducted to investigate the anti-inflammatory mechanisms of PE administration by analyzing various proinflammatory and stress markers, such as cyclooxygenase-2 (COX-2) and heat shock protein 90 (HSP90) as well as several inflammation-regulatory pathways, namely nuclear factor-κB (NF-κB) and nuclear factor erythroid 2p45 (NF-E2)-related factor 2 (Nrf2) signaling, during DMBA-inflicted mammary gland tumorigenesis in rats.

2. Materials and Methods

2.1. Test Materials, Chemicals and Antibodies

PE, a proprietary formulation containing pomegranate aqueous extract and seed oil, was purchased from Rimonest Ltd. (Haifa, Israel). A detailed description of the preparation of this emulsion has been published previously [15]. The chemical analyses of this formulation revealed the presence of caffeic acid, corilagin, ellagic acid, ferulic acid, gallic acid, 5-hydroxymethylfurfural, protocatechuic acid, punicalagins (A and B) and *trans*-p-coumaric acid in the aqueous phase and mixed octadecatrienoic acids, sterols and steroids (e.g., 17-α-estradiol), tocol and γ-tocopherol in the lipid phase [15]. The mammary carcinogen DMBA was procured from Sigma-Aldrich (St. Louis, MO, USA). Paraformaldehyde was obtained from Ted Pella (Redding, CA, USA). Primary antibodies, such as COX-2, inhibitory κBα (IκBα), NF-κB-p65, Nrf2 as well as the ABC staining kit were purchased from Santa Cruz Biotechnology (Santa Cruz, CA, USA). HSP90 was a product of Enzo Life Sciences (Farmingale, NY, USA).

2.2. Experimental Protocol and Tumor Tissue Harvesting

Breast tumor sections for this work were collected from our previously completed chemopreventive study [34] based on an animal protocol approved by the Institutional Animal

Care and Use Committee of Northeast Ohio Medical University (Rootstown, OH, USA). In short, female Sprague-Dawley rats (Harlan Laboratories, Indianapolis, IN, USA) were divided into six separate groups. All animals were fed a basal diet (LabDiet, St. Louis, MO, USA) ad libitum. Group A ($n = 12$) and group B ($n = 11$) were kept untreated. The remaining rats were orally administered (gavaged) with PE three times per week as follow: 0.2 g/kg (group C, $n = 8$), 1.0 g/kg (group D, $n = 8$) and 5.0 g/kg (group E, $n = 7$ and group F, $n = 5$) for a total of 18 weeks. Two weeks following the commencement of the study, animals from groups B, C, D and E were orally administered with a single dose of DMBA (50 mg/kg body weight) to induce mammary tumorigenesis. Eighteen weeks after initiation of the study, all animals were sacrificed and mammary glands were harvested and fixed in 4% paraformaldehyde for further analysis.

2.3. Immunohistochemical Analysis

The intratumor protein expressions of COX-2, HSP90, IκBα, NF-κB-p65 and Nrf2 were analyzed by methods described previously [41]. In brief, tumor tissue sections were first hydrated using phosphate-buffered saline followed by incubation in sodium citrate buffer for antigen retrieval. Endogenous peroxidases were blocked by treating the sections with 1% H_2O_2. Tissue sections were then treated with blocking solution followed by washing with phosphate-buffered saline and incubating overnight at 4 °C with primary antibodies (1:100 dilution). After several washes, tissue sections were treated with horseradish peroxidase-conjugated secondary antibody (1:200 dilution) for 30 min at room temperature, and then with 3,3'-diaminobenzidine tetrahydrochloride solution to visualize brown antigen-antibody complexes. Finally, the sections were counterstained with Gill's hematoxylin solution. The immunohistochemical slides were viewed under a light microscope (BX43, Olympus, Center Valley, PA, USA) and 1000 tumor cells/rat were analyzed.

2.4. Statistical Analyses

All data are expressed as mean ± standard error of mean (SEM). Statistical analyses were performed by using a commercial software (SigmaPlot 11.0, Systat Software, Inc., San Jose, CA, USA). One-way analysis of variance with least significant difference post hoc analysis was employed to compare various parameters among different treatment and control groups. A *p*-value less than 0.05 was considered statistically significant.

3. Results

3.1. PE Abrogated Elevated COX-2 Expression during DMBA-Induced Mammary Tumorigenesis

A substantial expression of COX-2 was observed predominantly in the cytoplasm of tumor cells of DMBA control animals (Figure 1(Aa)). PE at a dose of 0.2 g/kg slightly reduced intratumor COX-2 immunopositivity compared to the DMBA control (Figure 1(Ab)). On the other hand, a moderate and drastic inhibition of COX-2 expression was noticed following PE treatment at a dose of 1 and 5 g/kg, respectively (Figure 1(Ac,Ad)). Although PE at 0.2 g/kg slightly decreased the percentage of COX-2-positive cells, this result was statistically insignificant (Figure 1B). On the other hand, there was a significant ($p < 0.01$ or 0.001) inhibition of the percentage of intratumor COX-2-positive cells in animals administered with 1 or 5 g/kg PE compared to the DMBA control, respectively.

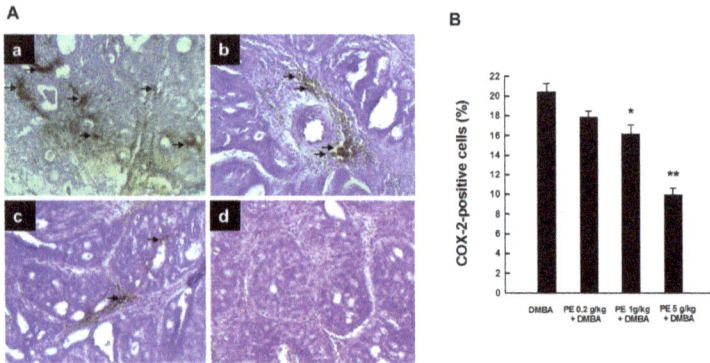

Figure 1. COX-2 expression during 7,12-dimethyl benz(a)anthracene (DMBA)-induced breast tumorigenesis in rats in the presence or absence of pomegranate emulsion (PE) treatment. (**A**) Immunohistochemical localization of COX-2-positive cells (arrows) in tumor sections (magnification: ×200). The various treatment groups are: (**a**) DMBA control; (**b**) PE (0.2 g/kg) plus DMBA; (**c**) PE (1 g/kg) plus DMBA; and (**d**) PE (5 g/kg) plus DMBA. (**B**) Quantitative analysis of COX-2-immunopositive cells from representative images. Results (mean ± SEM) are based on 1000 cells per animal and four animals per group. * $p < 0.01$ and ** $p < 0.001$ compared to the DMBA control.

3.2. PE Suppressed HSP90 Expression during DMBA Mammary Tumorigenesis

Tumor sections from the DMBA control animals showed a considerable expression of HSP90 (Figure 2(Aa)). PE treatment at 0.2 g/kg in conjunction with DMBA exposure did not change intratumor HSP90 expression (Figure 2(Ab)). PE at 1 g/kg substantially reduced the expression of HSP90 in tumor samples compared to the DMBA control (Figure 2(Ac)). A further decrease of HSP90 was achieved with PE at 5 g/kg (Figure 2(Ad)). The quantitative data on HSP90-immunopositivity revealed dose-dependent and statistically significant ($p < 0.001$) suppression of this protein expression in tumor samples from DMBA-treated rats that received PE treatment at 1 or 5 g/kg (Figure 2B).

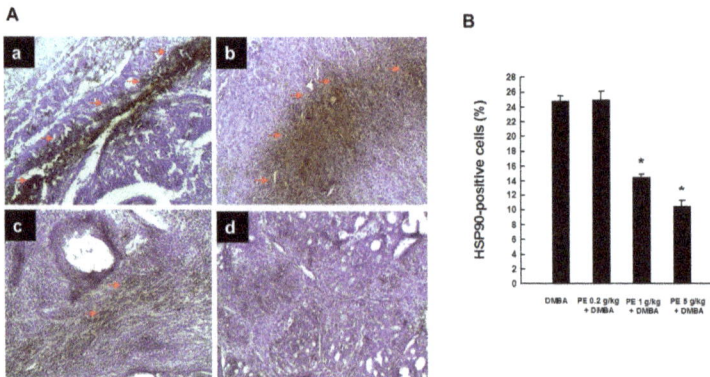

Figure 2. HSP90 expression during DMBA-induced breast tumorigenesis in rats in the presence or absence of PE treatment. (**A**) Immunohistochemical localization of HSP90-positive cells (arrows) in tumor sections (magnification: ×200). The various treatment groups are: (**a**) DMBA control; (**b**) PE (0.2 g/kg) plus DMBA; (**c**) PE (1 g/kg) plus DMBA; and (**d**) PE (5 g/kg) plus DMBA. (**B**) Quantitative analysis of HSP90-immunopositive cells from representative images. Results (mean ± SEM) are based on 1000 cells per animal and four animals per group. * $p < 0.001$ compared to the DMBA control.

3.3. PE Inhibited Activation of NF-κB during DMBA Mammary Tumorigenesis

We observed an extensive expression of NF-κB p65 in the nucleus and very limited expression of this protein in the cytoplasm in tumor sections harvested from the DMBA control animals, suggesting activation and subsequent translocation of NF-κB p65 from cytosol to nucleus (Figure 3(Aa)). Almost similar expression of nuclear and cytoplasmic NF-κB p65 was noticed following the treatment with PE at 0.2 g/kg (Figure 3(Ab)). Conversely, PE at 1 or 5 g/kg attenuated nuclear NF-κB p65 expression and elevated the expression of this protein in cytosol (Figure 3(Ac,Ad), respectively). The quantitative analysis of NF-κB p65-immunopositive cells displays a significant ($p < 0.001$) decrease in nuclear NF-κB p65-positive cells (Figure 3B) and a significant ($p < 0.001$) increase in cytoplasmic NF-κB p65-positive cells (Figure 3C) in two PE-treated groups (1 and 5 g/kg) compared to the DMBA control.

We also noticed very limited expression of cytosolic IκBα in the DMBA control animals (Figure 4(Aa)), indicating the possible degradation of IκBα protein. PE treatment inhibited DMBA-induced degradation of IκBα protein in cytosol in a dose-dependent fashion. Our results showed a marginal increase in cytosolic IκBα expression by 0.2 g/kg PE (Figure 4(Ab)) and sizable upregulation of this protein by 1 or 5 g/kg PE (Figure 4(Ac,Ad), respectively). These results are supported by the quantitative analysis of IκBα-positive cells that demonstrated a significant ($p < 0.01$ or 0.001) increase of immuno-positive cells in the mammary tumor sections harvested from rats treated with 1 or 5 mg/kg PE, respectively (Figure 4B).

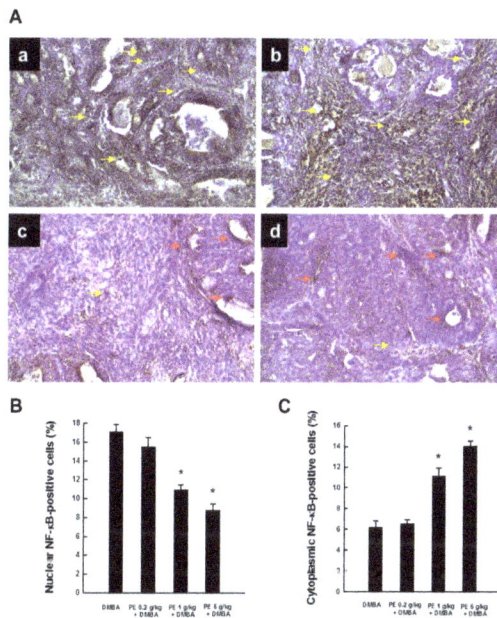

Figure 3. NF-κB p65 expression during DMBA-induced breast tumorigenesis in rats in the presence or absence of PE treatment. (**A**) Immunohistochemical localization of NF-κB p65 in nucleus (yellow arrows) and cytoplasm (red arrows) (magnification: ×200). The various treatment groups are: (**a**) DMBA control; (**b**) PE (0.2 g/kg) plus DMBA; (**c**) PE (1 g/kg) plus DMBA; and (**d**) PE (5 g/kg) plus DMBA. Quantitative analysis of (**B**) nuclear and (**C**) cytoplasmic NF-κB-immunopositive cells from representative images. Results (mean ± SEM) are based on 1000 cells per animal and four animals per group. * $p < 0.001$ compared to the DMBA control.

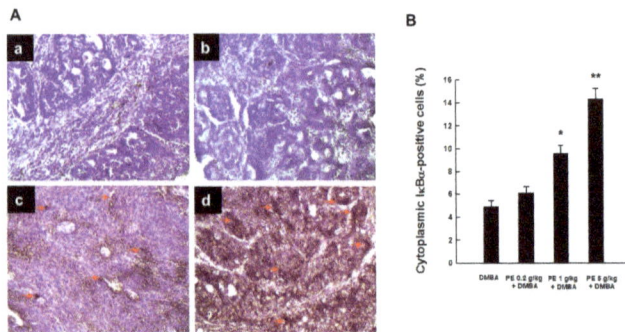

Figure 4. IκBα expression during DMBA-induced breast tumorigenesis in rats in the presence or absence of PE treatment. (**A**) Immunohistochemical localization of IκBα-positive cells (arrows) in the cytoplasm of tumor sections (magnification: ×200). The various treatment groups are: (**a**) DMBA control; (**b**) PE (0.2 g/kg) plus DMBA; (**c**) PE (1 g/kg) plus DMBA; and (**d**) PE (5 g/kg) plus DMBA. (**B**) Quantitative analysis of IκBα-immunopositive cells from representative images (**A**). Results (mean ± SEM) are based on 1000 cells per animal and four animals per group. * $p < 0.01$ and ** $p < 0.001$ compared to the DMBA control.

3.4. PE Induced Nrf2 Expression during DMBA Mammary Tumorigenesis

Figure 5A shows the intratumor immunohistochemical profiles of Nrf2 expression of various animal groups. The DMBA control group exhibited a marginal expression of Nrf2 (Figure 5(Aa)). PE treatment at 0.2 g/kg showed a similar expression of Nrf2 (Figure 5(Ab)). In contrast, tumor sections from DMBA-exposed animals treated with 1 or 5 g/kg PE showed a strong upregulation of Nrf2 (Figure 5(Ac,Ad), respectively). The majority of Nrf2-immunopositivity was noticed in the nucleus, suggesting the activation and subsequent translocation of Nrf2 from the cytoplasm to the nucleus. Our quantitative analysis revealed a statistically significant ($p < 0.001$) increase in the percentage of Nrf2-positive cells following PE treatment at 1 or 5 g/kg compared to the DMBA control (Figure 5B)

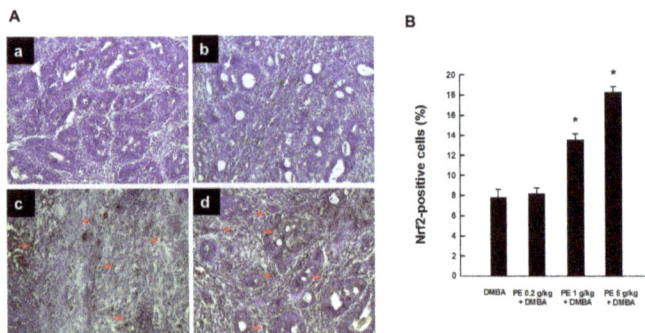

Figure 5. Expression of Nrf2 from DMBA-induced breast tumors in rats treated with or without PE. (**A**) Immunohistochemical localization of Nrf2-positive cells (arrows) in the nucleus of tumor sections (magnification: ×200). The treatment groups are: (**a**) DMBA control; (**b**) PE (0.2 g/kg) plus DMBA; (**c**) PE (1 g/kg) plus DMBA; and (**d**) PE (5 g/kg) plus DMBA. (**B**) Quantitative analysis of Nrf2-immunopositive cells from representative images. Results (mean ± SEM) are based on 1000 cells per animal and four animals per group. * $p < 0.001$ compared to the DMBA control.

4. Discussion

Breast cancer is the second leading cause of death in women worldwide. Globally, more than one million women are diagnosed with breast tumor. In the United States, breast cancer is the second most frequent female cancer. Approximately 246,660 new cases of breast cancer and 40,450 breast cancer-related deaths were estimated to occur in women in the United States in 2016 [42]. Emerging evidence suggests that chronic inflammation contributes to breast cancer development and progression [36–40]. Experimental studies have provided convincing evidence that various inflammatory molecules and signaling pathways are involved in the proliferation, survival, epithelial-mesenchymal transition, invasion and metastasis of breast cancer cells [43]. Accordingly, agents that inhibit chronic inflammation may be effective in the prevention and therapy of mammary carcinoma. Numerous natural products, phytochemicals and dietary agents with anti-inflammatory activities have shown promise in the prevention and treatment of breast cancer [44–49]. Recently, we have published the novel finding that a new formulation (PE) consisting of pomegranate phytoconstituents exerts a striking chemoprevention of DMBA-initiated rat mammary tumorigenesis though the molecular mechanisms of action of such beneficial activity is not completely elucidated [34]. Accordingly, the present work was designed to investigate the ability of PE to interfere with DMBA-mediated inflammatory signaling cascades by analyzing breast tumor sections harvested from our earlier study [34].

The COX family of enzymes catalyzes the rate-limiting step in the synthesis of prostaglandins (PGs), including PGE2, mainly from arachidonic acid. There are two isoforms of COX enzymes, namely COX-1 and COX-2. COX-1 is constitutively expressed in most cells, whereas the expression of COX-2 is induced by various stimuli, including shear stress, cytokines, growth factors and oncogenes [50]. PGs produced by COX-2 are involved in various critical steps in oncogenesis, including proliferation, mutagenesis, apoptosis evasion, immune suppression, angiogenesis and invasion [51,52]. Interestingly, the inhibition of COX-2 has been shown to reduce breast tumor cell proliferation [53,54]. COX-2 protein expression in ductal carcinoma in situ and invasive breast carcinoma indicates the crucial role of COX-2 in early stages of mammary carcinogenesis [55]. Our results clearly indicate the PE-mediated suppression of COX-2 in chemically-induced mammary carcinogenesis in rats. We suggest that our previously reported mammary tumor inhibitory effect of PE in DMBA carcinogenesis [34] could be, at least in part, due to the inhibition of COX-2 expression. Other investigators reported that cold-pressed pomegranate seed oil inhibited sheep cyclooxygenases by around 30–40% [56]. In addition, plasma isolated from rabbits following oral ingestion of pomegranate fruit extract inhibited both COX-1 and COX-2 enzymes ex vivo, and the effect was more pronounced for COX-2 [57]. Finally, several pomegranate constituents, e.g., ellagic acid, gallic acid and punicalagin A and B, inhibited lipopolysaccharide-induced PGE2, nitric oxide and interleukin-16 (IL-6) production [58].

Heat shock proteins (HSPs) represent stress-inducible proteins which are known to play important roles in cellular stress response. There are several core HSP families in humans, such as DNAJ (HSP40), HSPA (HSP70), HSPB (small HSP), HSPC (HSP90), HSPD/E (HSP60/HSP10), HSPH (HSP110) and CCT (TRiC). HSP90, a highly conserved and abundant 90 kDa protein, is involved in chaperoning the structures of over 200 client proteins, many of which are involved in growth control as well as mammary tumor cell proliferation [59]. Based on scientific reports, HSP90 may regulate inflammatory events through the modulation of various cytokines and cell signaling pathways [60–62]. The efficacy of various HSP90 inhibitors in the treatment of several oncologic diseases, including breast cancer, is currently underway [63–65]. In our study, an elevated expression of HSP90 in tumor sections suggests that DMBA may exert heat shock response, perhaps due to inflammatory stress, leading to the irregular proliferation and evasion of apoptosis, as we have previously observed [34]. Our present results show the substantial reduction of HSP90 expression in the PE plus DMBA group, suggesting the capability of PE to abrogate mammary tumor cell growth and survival by the downregulation of HSP90, which may be associated with a lesser magnitude of inflammatory stress. A recent study shows that pomegranate exerts anti-inflammatory effects by upregulating HSP70, transforming growth factor-β1 (TGF-β1) and IL-10 expression during chemically-induced hepatotoxicity in rats [66].

Various signaling pathways activated in oncogenesis are likely to be networked through the activation of NF-κB, a proinflammatory transcription factor [67]. The major inactive form of NF-κB complex is a p50–p65 heterodimer which binds to the inhibitory protein IκBα and predominantly resides in the cytoplasm. In the classical (canonical) pathway initiated by proinflammatory cytokines, such as tumor necrosis factor-α and IL-1β, IκBα undergoes degradation by IκB kinase (IKK)γ-containing IKK complex via the TGF-β1-dependent pathway [68]. Subsequently, the free p50–p65 dimer translocates into the nucleus, binds to its cognate response element in the DNA, and induces the transcription of a battery of target genes involved in proliferation, survival, inflammation, angiogenesis, invasion and metastasis [69,70]. A critical association between NF-κB-mediated inflammatory pathways and breast cancer development and progression has been well established [37,71]. Abrogation of constitutive activation of NF-κB diminishes the oncogenic potential of transformed cells, and the suppression of this proinflammatory pathway may be valuable for cancer prevention and therapy [72–74]. In our present study, the limited expression of NF-κB p65 and IκBα in cytosol and the substantial expression of NF-κB p65 in the nucleus of mammary tumor cells from animals exposed to DMBA only indicate the conceivable degradation of IκBα, release of activated NF-κB p65, and its subsequent translocation to the nucleus. These results are supported by an earlier study which documented that the activation of NF-κB is implicated in DMBA-initiated early malignant transformation in rat mammary glands [75]. Our results of PE-mediated protection against IκBα degradation and hindrance with the translocation of activated NF-κB p65 to the nucleus suggest that PE possibly impedes early events in DMBA-induced mammary carcinogenesis in rats. Consistent with our results, Khan and colleagues [21] observed that pomegranate fruit extract inhibited constitutive NF-κB activation and NF-κB-dependent reported gene expression associated with proliferation, invasion and motility in two aggressive breast carcinoma cells, namely MDA-MB-231 and SUM 149. Banerjee and coinvestigators [26] also reported that pomegranate juice concentrate suppressed the growth of MDA-MB-231 cells concomitant with reduced transcriptional and translational expression of NF-κB. Similar results were observed in nude mice xenografted with BT474 breast carcinoma cells [26].

Nrf2, a basic-leucine zipper transcription factor, is known to play a critical role in protecting mammalian cells against injury inflicted by inflammation and oxidative stress [76]. Under normal cellular conditions, Nrf2 is anchored in the cytoplasm by its association with Kelch-like erythroid Cap-N-Collar homolog-associated protein 1 (Keap1). Upon cellular stress as well as pharmacological stimuli, Nrf2 is liberated from Keap1, translocates to the nucleus, recognizes and binds to *cis*-acting enhancer known as an antioxidant response element or electrophile response element, and eventually facilitates the transcription of a plethora of genes that encode antioxidant and detoxifying enzymes [77]. The Nrf2-mediated signaling pathway is profoundly involved in the regulation of inflammation and inflammation-associated carcinogenesis, and accordingly this pathway embodies an important target for the prevention of inflammation-related cancer [78–80]. Interestingly, a potential cross-talk between Nrf2 and NF-κB transcription factors modulated by mitogen-activated protein kinase to influence inflammation-associated etiopathogenesis of cancer has been proposed [81,82]. Singh and colleague [83–85] elegantly showed decreased protein and mRNA expression of Nrf2 and Nrf2-regulated genes in estrogen-exposed mammary tissue and mammary tumors in rats. In line with the aforementioned results, we have also found the limited expression of Nrf2 in rat mammary tumors induced by DMBA. However, PE treatment was effective in upregulating the expression as well as the nuclear translocation of Nrf2 which may, in turn, relieve NF-κB-mediated inflammatory action and ultimately lead to breast cancer prevention. However, it is noteworthy to mention another aspect of Nfr2 induction. Stable expression of Nfr2 results in the enhanced resistance of cancer cells to chemotherapeutic agents, including cisplatin, doxorubicin and etoposide. The increment of Nrf2, induced by PE, can be useful for the prevention of inflammation associated with carcinogenesis, but it can compromise the therapeutic outcome of patients treated with chemotherapeutic drugs.

The identification of specific phytochemicals of PE responsible for the observed anti-inflammatory activities during experimental mammary tumorigenesis may not be possible at this time, and requires further investigations. Phytochemical profiling of pomegranate reveals the presence of several constituents, including ellagic acid, ellagitannins, punicic acid, and punicalagin, which are capable of the modulation of inflammatory cell signaling [86–88]. In view of the notion that pomegranate phytochemicals exhibit maximum beneficial effects when they are used in combination rather than individually [89], it is reasonable to propose that various phytochemicals present in this unique fruit may regulate various inflammatory molecules and cascades during mammary carcinogenesis through a synergistic effect.

5. Conclusions

The present results coupled with those we previously reported possibly indicate that the prevention of DMBA-initiated mammary tumorigenesis by PE may involve anti-inflammatory mechanisms propelled by synchronized and differential regulation of two interrelated molecular pathways, such as NF-κB and Nrf2 signaling. Further in-depth studies using appropriate techniques are required to confirm the results presented here, in order to facilitate the development of a pomegranate emulsion for the prevention and treatment of cancers linked to inflammation.

Acknowledgments: This work was financially supported by a new faculty start-up research grant from the Northeast Ohio Medical University (NEOMED, Rootstown, OH) awarded to Anupam Bishayee. The authors express sincere gratitude to personnel from NEOMED Comparative Medicine Unit for assistance with animal care and maintenance.

Author Contributions: Anupam Bishayee conceived and designed the experiments. Animesh Mandal and Deepak Bhatia performed the experiments. Anupam Bishayee wrote the manuscript.

Conflicts of Interest: The authors declare no conflicts of interest.

References

1. Cáceres, A.; Girón, L.M.; Alvarado, S.R.; Torres, M.F. Screening of antimicrobial activity of plants popularly used in Guatemala for the treatment of dermatomucosal diseases. *J. Ethnopharmacol.* **1987**, *20*, 223–237.
2. Naqvi, S.A.; Khan, M.S.; Vohra, S.B. Antibacterial, antifungal, and antihelmintic investigations on Indian medicinal plants. *Fitoterapia* **1991**, *62*, 221–228.
3. Saxena, A.; Vikram, N.K. Role of selected Indian plants in management of type 2 diabetes: A review. *J. Altern. Complement. Med.* **2004**, *10*, 369–378. [CrossRef] [PubMed]
4. Lansky, E.P.; Newman, R.A. *Punica granatum* (pomegranate) and its potential for prevention and treatment of inflammation and cancer. *J. Ethnopharmacol.* **2007**, *109*, 177–206. [CrossRef] [PubMed]
5. Jurenka, J.S. Therapeutic applications of pomegranate (*Punica granatum* L.): A review. *Altern. Med. Rev.* **2008**, *13*, 128–144. [PubMed]
6. Basu, A.; Penugonda, K. Pomegranate: A health-healthy fruit juice. *Nutr. Rev.* **2009**, *67*, 49–56. [CrossRef]
7. Johanningsmeier, S.D.; Harris, G.K. Pomegranate as a functional food and nutraceutical source. *Annu. Rev. Food Sci. Technol.* **2011**, *2*, 181–201. [CrossRef] [PubMed]
8. Faria, A.; Calhau, C. The bioactivity of pomegranate: Impact on health and disease. *Crit. Rev. Food Sci. Nutr.* **2011**, *51*, 626–634. [CrossRef] [PubMed]
9. Sreekumar, S.; Sithul, H.; Muraleedharan, P.; Azeez, J.M.; Sreeharshan, S. Pomegranate fruit as a rich source of biologically active compounds. *Biomed. Res. Int.* **2014**, *2014*. [CrossRef] [PubMed]
10. Heber, D. Pomegranate ellagitannins. In *Herbal Medicine: Biomolecular and Clinical Aspects*, 2nd ed.; Benzie, I.F.F., Wachtel-Galor, S., Eds.; CRC Press: Boca Raton, FL, USA, 2011; pp. 201–210.
11. Rahimi, H.R.; Arastoo, M.; Ostad, S.N. A Comprehensive Review of *Punica granatum* (Pomegranate) Properties in Toxicological, Pharmacological, Cellular and Molecular Biology Researches. *Iran J. Pharm. Res.* **2012**, *11*, 385–400. [PubMed]
12. Ismail, T.; Sestili, P.; Akhtar, S. Pomegranate peel and fruit extracts: A review of potential anti-inflammatory and anti-infective effects. *J. Ethnopharmacol.* **2012**, *143*, 397–405. [CrossRef] [PubMed]

13. Viladomiu, M.; Hontecillas, R.; Lu, P.; Bassaganya-Riera, J. Preventive and prophylactic mechanisms of action of pomegranate bioactive constituents. *Evid.-Based Complement. Altern. Med.* **2013**, *2013*. [CrossRef] [PubMed]

14. Shaygannia, E.; Bahmani, M.; Zamanzad, B.; Rafieian-Kopaei, M. A Review Study on *Punica granatum* L. *J. Evid.-Based Complement. Altern. Med.* **2016**, *21*, 221–227. [CrossRef] [PubMed]

15. Bishayee, A.; Bhatia, D.; Thoppil, R.J.; Darvesh, A.S.; Nevo, E.; Lansky, E.P. Pomegranate-mediated chemoprevention of experimental hepatocarcinogenesis involves Nrf2-regulated antioxidant mechanisms. *Carcinogenesis* **2011**, *32*, 888–896. [CrossRef] [PubMed]

16. Syed, D.N.; Chamcheu, J.C.; Adhami, V.M.; Mukhtar, H. Pomegranate extracts and cancer prevention: Molecular and cellular activities. *Anticancer Agents Med. Chem.* **2013**, *13*, 1149–1161. [CrossRef] [PubMed]

17. Vlachojannis, C.; Zimmermann, B.F.; Chrubasik-Hausmann, S. Efficacy and safety of pomegranate medicinal products for cancer. *Evid.-Based Complement. Altern. Med.* **2015**, *2015*, 258598. [CrossRef] [PubMed]

18. Turrini, E.; Ferruzzi, L.; Fimognari, C. Potential Effects of Pomegranate Polyphenols in Cancer Prevention and Therapy. *Oxid. Med. Cell. Longev.* **2015**, *2015*, 938475. [CrossRef] [PubMed]

19. Kim, N.D.; Mehta, R.; Yu, W.; Neeman, I.; Livney, T.; Amichay, A.; Poirier, D.; Nicholls, P.; Kirby, A.; Jiang, W.; et al. Chemopreventive and adjuvant therapeutic potential of pomegranate (*Punica granatum*) for human breast cancer. *Breast Cancer Res. Treat.* **2002**, *71*, 203–217. [CrossRef] [PubMed]

20. Toi, M.; Bando, H.; Ramachandran, C.; Melnick, S.J.; Imai, A.; Fife, R.S.; Carr, R.E.; Oikawa, T.; Lansky, E.P. Preliminary studies on the anti-angiogenic potential of pomegranate fractions in vitro and in vivo. *Angiogenesis* **2003**, *6*, 121–128. [CrossRef] [PubMed]

21. Khan, G.N.; Gorin, M.A.; Rosenthal, D.; Pan, Q.; Bao, L.W.; Wu, Z.F.; Newman, R.A.; Pawlus, A.D.; Yang, P.; Lansky, E.P.; et al. Pomegranate fruit extract impairs invasion and motility in human breast cancer. *Integr. Cancer Ther.* **2009**, *8*, 242–253. [CrossRef] [PubMed]

22. Adams, L.S.; Zhang, Y.; Seeram, N.P.; Heber, D.; Chen, S. Pomegranate ellagitannin-derived compounds exhibit antiproliferative and antiaromatase activity in breast cancer cells in vitro. *Cancer Prev. Res.* **2010**, *3*, 108–113. [CrossRef] [PubMed]

23. Dai, Z.; Nair, V.; Khan, M.; Ciolino, H.P. Pomegranate extract inhibits the proliferation and viability of MMTV-Wnt-1 mouse mammary cancer stem cells in vitro. *Oncol. Rep.* **2010**, *24*, 1087–1091. [PubMed]

24. Dikmen, M.; Ozturk, N.; Ozturk, Y. The antioxidant potency of *Punica granatum* L. fruit peel reduces cell proliferation and induces apoptosis on breast cancer. *J. Med. Food* **2011**, *14*, 1638–1646. [CrossRef] [PubMed]

25. Joseph, M.M.; Aravind, S.R.; Varghese, S.; Mini, S.; Sreelekha, T.T. Evaluation of antioxidant, antitumor and immunomodulatory properties of polysaccharide isolated from fruit rind of *Punica granatum*. *Mol. Med. Rep.* **2012**, *5*, 489–496. [PubMed]

26. Banerjee, N.; Talcott, S.; Safe, S.; Mertens-Talcott, S.U. Cytotoxicity of pomegranate polyphenolics in breast cancer cells in vitro and vivo: Potential role of miRNA-27a and miRNA-155 in cell survival and inflammation. *Breast Cancer Res. Treat.* **2012**, *136*, 21–34. [CrossRef]

27. Rocha, A.; Wang, L.; Penichet, M.; Martins-Green, M. Pomegranate juice and specific components inhibit cell and molecular processes critical for metastasis of breast cancer. *Breast Cancer Res. Treat.* **2012**, *136*, 647–658. [CrossRef]

28. Sreeja, S.; Santhosh Kumar, T.R.; Lakshmi, B.S.; Sreeja, S. Pomegranate extract demonstrate a selective estrogen receptor modulator profile in human tumor cell lines and in vivo models of estrogen deprivation. *J. Nutr. Biochem.* **2012**, *23*, 725–732. [CrossRef] [PubMed]

29. Bekir, J.; Mars, M.; Vicendo, P.; Fterrich, A.; Bouajila, J. Chemical composition and antioxidant, anti-inflammatory, and antiproliferation activities of pomegranate (*Punica granatum*) flowers. *J. Med. Food* **2013**, *16*, 544–550. [CrossRef] [PubMed]

30. Shirode, A.B.; Kovvuru, P.; Chittur, S.V.; Henning, S.M.; Heber, D.; Reliene, R. Antiproliferative effects of pomegranate extract in MCF-7 breast cancer cells are associated with reduced DNA repair gene expression and induction of double strand breaks. *Mol. Carcinog.* **2014**, *53*, 458–470. [CrossRef] [PubMed]

31. Lucci, P.; Pacetti, D.; Loizzo, M.R.; Frega, N.G. *Punica granatum* cv. Dente di Cavallo seed ethanolic extract: Antioxidant and antiproliferative activities. *Food Chem.* **2015**, *167*, 475–483. [CrossRef] [PubMed]

32. Modaeinama, S.; Abasi, M.; Abbasi, M.M.; Jahanban-Esfahlan, R. Anti Tumoral Properties of *Punica Granatum* (Pomegranate) Peel Extract on Different Human Cancer Cells. *Asian Pac. J. Cancer Prev.* **2015**, *16*, 5697–56701. [CrossRef] [PubMed]

33. Mehta, R.; Lansky, E.P. Breast cancer chemopreventive properties of pomegranate (*Punica granatum*) fruit extracts in a mouse mammary organ culture. *Eur. J. Cancer Prev.* **2004**, *13*, 345–348. [CrossRef] [PubMed]

34. Bishayee, A.; Mandal, A.; Bhattacharyya, P.; Bhatia, D. Pomegranate exerts chemoprevention of experimentally induced mammary tumorigenesis by suppression of cell proliferation and induction of apoptosis. *Nutr. Cancer* **2016**, *68*, 120–130. [CrossRef] [PubMed]

35. Mandal, A.; Bishayee, A. Mechanism of Breast Cancer Preventive Action of Pomegranate: Disruption of Estrogen Receptor and Wnt/β-Catenin Signaling Pathways. *Molecules* **2015**, *20*, 22315–22328. [CrossRef] [PubMed]

36. Macciò, A.; Madeddu, C. Obesity, inflammation, and postmenopausal breast cancer: Therapeutic implications. *Sci. World J.* **2011**, *11*, 2020–2036.

37. Shostak, K.; Chariot, A. NF-κB, stem cells and breast cancer: The links get stronger. *Breast Cancer Res.* **2011**, *13*, 214. [CrossRef] [PubMed]

38. Baumgarten, S.C.; Frasor, J. Minireview: Inflammation: An instigator of more aggressive estrogen receptor (ER) positive breast cancers. *Mol. Endocrinol.* **2012**, *26*, 360–371. [CrossRef] [PubMed]

39. Harris, R.E.; Casto, B.C.; Harris, Z.M. Cyclooxygenase-2 and the inflammogenesis of breast cancer. *World J. Clin. Oncol.* **2014**, *5*, 677–692. [CrossRef] [PubMed]

40. Brown, K.A. Impact of obesity on mammary gland inflammation and local estrogen production. *J. Mammary Gland Biol. Neoplasia* **2014**, *19*, 183–189. [CrossRef] [PubMed]

41. Mandal, A.; Bhatia, D.; Bishayee, A. Suppression of inflammatory cascade is implicated in methyl amooranin-mediated inhibition of experimental mammary carcinogenesis. *Mol. Carcinog.* **2014**, *53*, 999–1010. [CrossRef] [PubMed]

42. Siegel, R.L.; Miller, K.D.; Jemal, A. Cancer statistics, 2016. *CA Cancer J. Clin.* **2016**, *66*, 7–30. [CrossRef] [PubMed]

43. Fouad, T.M.; Kogawa, T.; Reuben, J.M.; Ueno, N.T. The role of inflammation in inflammatory breast cancer. *Adv. Exp. Med. Biol.* **2014**, *816*, 53–73. [PubMed]

44. Johnson, J.J. Carnosol: A promising anti-cancer and anti-inflammatory agent. *Cancer Lett.* **2011**, *305*, 1–7. [CrossRef] [PubMed]

45. Sinha, D.; Biswas, J.; Sung, B.; Aggarwal, B.B.; Bishayee, A. Chemopreventive and chemotherapeutic potential of curcumin in breast cancer. *Curr. Drug Targets* **2012**, *13*, 1799–1819. [CrossRef] [PubMed]

46. Howe, L.R.; Subbaramaiah, K.; Hudis, C.A.; Dannenberg, A.J. Molecular pathways: Adipose inflammation as a mediator of obesity-associated cancer. *Clin. Cancer Res.* **2013**, *19*, 6074–6083. [CrossRef] [PubMed]

47. Subbaramaiah, K.; Sue, E.; Bhardwaj, P.; Du, B.; Hudis, C.A.; Giri, D.; Kopelovich, L.; Zhou, X.K.; Dannenberg, A.J. Dietary polyphenols suppress elevated levels of proinflammatory mediators and aromatase in the mammary gland of obese mice. *Cancer Prev. Res.* **2013**, *6*, 886–897. [CrossRef] [PubMed]

48. Crawford, S. Anti-inflammatory/antioxidant use in long-term maintenance cancer therapy: A new therapeutic approach to disease progression and recurrence. *Ther. Adv. Med. Oncol.* **2014**, *6*, 52–68. [CrossRef] [PubMed]

49. Sinha, D.; Sarkar, N.; Biswas, J.; Bishayee, A. Resveratrol for breast cancer prevention and therapy: Preclinical evidence and molecular mechanisms. *Semin. Cancer Biol.* **2016**, *40–41*, 209–232. [CrossRef] [PubMed]

50. Williams, C.S.; Mann, M.; DuBois, R.N. The role of cyclooxygenases in inflammation, cancer, and development. *Oncogene* **1999**, *18*, 7908–7916. [CrossRef] [PubMed]

51. Harris, R.E. Cyclooxygenase-2 (cox-2) blockade in the chemoprevention of cancers of the colon, breast, prostate, and lung. *Inflammopharmacology* **2009**, *17*, 55–67. [CrossRef] [PubMed]

52. Bocca, C.; Ievolella, M.; Autelli, R.; Motta, M.; Mosso, L.; Torchio, B.; Bozzo, F.; Cannito, S.; Paternostro, C.; Colombatto, S.; et al. Expression of Cox-2 in human breast cancer cells as a critical determinant of epithelial-to-mesenchymal transition and invasiveness. *Expert Opin. Ther. Targets* **2014**, *18*, 121–135. [CrossRef] [PubMed]

53. Trimboli, A.J.; Waite, B.M.; Atsumi, G.; Fonteh, A.N.; Namen, A.M.; Clay, C.E.; Kute, T.E.; High, K.P.; Willingham, M.C.; Chilton, F.H. Influence of coenzyme A-independent transacylase and cyclooxygenase inhibitors on the proliferation of breast cancer cells. *Cancer Res.* **1999**, *59*, 6171–6177. [PubMed]

54. Alshafie, G.A.; Abou-Issa, H.M.; Seibert, K.; Harris, R.E. Chemotherapeutic evaluation of Celecoxib, a cyclooxygenase-2 inhibitor, in a rat mammary tumor model. *Oncol. Rep.* **2000**, *7*, 1377–1381. [CrossRef] [PubMed]

55. Glover, J.A.; Hughes, C.M.; Cantwell, M.M.; Murray, L.J. A systematic review to establish the frequency of cyclooxygenase-2 expression in normal breast epithelium, ductal carcinoma in situ, microinvasive carcinoma of the breast and invasive breast cancer. *Br. J. Cancer* **2011**, *105*, 13–17. [CrossRef] [PubMed]

56. Schubert, S.Y.; Lansky, E.P.; Neeman, I. Antioxidant and eicosanoid enzyme inhibition properties of pomegranate seed oil and fermented juice flavonoids. *J. Ethnopharmacol.* **1999**, *66*, 11–17. [CrossRef]

57. Shukla, M.; Gupta, K.; Rasheed, Z.; Khan, K.A.; Haqqi, T.M. Bioavailable constituents/metabolites of pomegranate (*Punica granatum* L.) preferentially inhibit COX2 activity ex vivo and IL-1beta-induced PGE2 production in human chondrocytes in vitro. *J. Inflamm.* **2008**, *5*, 9. [CrossRef] [PubMed]

58. BenSaad, L.A.; Kim, K.H.; Quah, C.C.; Kim, W.R.; Shahimi, M. Anti-inflammatory potential of ellagic acid, gallic acid and punicalagin A&B isolated from *Punica granatum*. *BMC Complement. Altern. Med.* **2017**, *17*, 47.

59. Calderwood, S.K.; Gong, J. Molecular chaperons in mammary cancer growth and breast tumor therapy. *J. Cell. Biochem.* **2012**, *113*, 1096–1103. [CrossRef] [PubMed]

60. Shimp, S.K., 3rd; Parson, C.D.; Regna, N.L.; Thomas, A.N.; Chafin, C.B.; Reilly, C.M.; Nichole Rylander, M. HSP90 inhibition by 17-DMAG reduces inflammation in J774 macrophages through suppression of Akt and nuclear factor-κB pathways. *Inflamm. Res.* **2012**, *61*, 521–533. [CrossRef] [PubMed]

61. Tamura, Y.; Torigoe, T.; Kutomi, G.; Hirata, K.; Sato, N. New paradigm for intrinsic function of heat shock proteins as endogenous ligands in inflammation and innate immunity. *Curr. Mol. Med.* **2012**, *12*, 1198–1206. [CrossRef] [PubMed]

62. Kumar, P.; Kolls, J.K. Act1-hsp90 heats up TH17 inflammation. *Nat. Immunol.* **2013**, *14*, 16–17. [CrossRef] [PubMed]

63. Trepel, J.; Mollapour, M.; Giaccone, G.; Neckers, L. Targeting the dynamic HSP90 complex in cancer. *Nat. Rev. Cancer* **2010**, *10*, 537–549. [CrossRef] [PubMed]

64. Sankhala, K.K.; Mita, M.M.; Mita, A.C.; Takimoto, C.H. Heat shock proteins: A potential anticancer target. *Curr. Drug Targets* **2011**, *12*, 2001–2008. [CrossRef] [PubMed]

65. Barrott, J.J.; Haystead, T.A. Hsp90, an unlikely ally in the war on cancer. *FEBS J.* **2013**, *280*, 1381–1396. [CrossRef] [PubMed]

66. Ibrahim, Z.S.; Nassan, M.A.; Soliman, M.M. Ameliorative effects of pomegranate on carbon tetrachloride hepatotoxicity in rats: A molecular and histopathological study. *Mol. Med. Rep.* **2016**, *13*, 3653–3660. [CrossRef] [PubMed]

67. Grivennikov, S.I.; Greten, F.R.; Karin, M. Immunity, inflammation, and cancer. *Cell* **2010**, *140*, 883–899. [CrossRef] [PubMed]

68. Hayden, M.S.; Ghosh, S. Shared principles in NF-κB signaling. *Cell* **2008**, *132*, 344–362. [CrossRef] [PubMed]

69. Sethi, G.; Sung, B.; Aggarwal, B.B. Nuclear factor-κB activation: From bench to bedside. *Exp. Biol. Med.* **2008**, *233*, 21–31. [CrossRef] [PubMed]

70. Viatour, P.; Merville, M.P.; Bours, V.; Chariot, A. Phosphorylation of NF-kappaB and IkappaB proteins: Implications in cancer and inflammation. *Trends Biochem. Sci.* **2005**, *30*, 43–52. [CrossRef] [PubMed]

71. Zubair, A.; Frieri, M. Role of nuclear factor-κB in breast and colorectal cancer. *Curr. Allergy Asthma Rep.* **2013**, *13*, 44–49. [CrossRef] [PubMed]

72. Ralhan, R.; Pandey, M.K.; Aggarwal, B.B. Nuclear factor-kappa B links carcinogenic and chemopreventive agents. *Front. Biosci.* **2009**, *1*, 45–60. [CrossRef]

73. Gupta, S.C.; Sundaram, C.; Reuter, S.; Aggarwal, B.B. Inhibiting NF-κB activation by small molecules as a therapeutic strategy. *Biochim. Biophys. Acta* **2010**, *1799*, 775–787. [CrossRef] [PubMed]

74. Zhu, Z.; Zhong, S.; Shen, Z. Targeting the inflammatory pathways to enhance chemotherapy of cancer. *Cancer Biol. Ther.* **2011**, *12*, 95–105. [CrossRef] [PubMed]

75. Kim, D.W.; Sovak, M.A.; Zanieski, G.; Nonet, G.; Romieu-Mourez, R.; Lau, A.W.; Hafer, L.J.; Yaswen, P.; Stampfer, M.; Rogers, A.E.; et al. Activation of NF-kappaB/Rel occurs early during neoplastic transformation of mammary cells. *Carcinogenesis* **2000**, *21*, 871–879. [CrossRef] [PubMed]

76. Baird, L.; Dinkova-Kostova, A.T. The cytoprotective role of the Keap1–Nrf2 pathway. *Arch. Toxicol.* **2011**, *85*, 241–272. [CrossRef] [PubMed]

77. Okawa, H.; Motohashi, H.; Kobayashi, A.; Aburatani, H.; Kensler, T.W.; Yamamoto, M. Hepatocyte-specific deletion of the keap1 gene activates Nrf2 and confers potent resistance against acute drug toxicity. *Biochem. Biophys. Res. Commun.* **2006**, *339*, 79–88. [CrossRef] [PubMed]

78. Hu, R.; Saw, C.L.; Yu, R.; Kong, A.N. Regulation of NF-E2-related factor 2 signaling for cancer chemoprevention: Antioxidant coupled with antiinflammatory. *Antioxid. Redox Signal.* **2010**, *13*, 1679–1698. [CrossRef] [PubMed]

79. Kundu, J.K.; Surh, Y.J. Nrf2-Keap1 signaling as a potential target for chemoprevention of inflammation-associated carcinogenesis. *Pharm. Res.* **2010**, *27*, 999–1013. [CrossRef] [PubMed]

80. Slocum, S.L.; Kensler, T.W. Nrf2: Control of sensitivity to carcinogens. *Arch. Toxicol.* **2011**, *85*, 273–284. [CrossRef] [PubMed]

81. Nair, S.; Doh, S.T.; Chan, J.Y.; Kong, A.N.; Cai, L. Regulatory potential for concerted modulation of Nrf2- and Nfkb1-mediated gene expression in inflammation and carcinogenesis. *Br. J. Cancer* **2008**, *99*, 2070–2082. [CrossRef] [PubMed]

82. Bellezza, I.; Mierla, A.L.; Minelli, A. Nrf2 and NF-κB and their concerted modulation in cancer pathogenesis and progression. *Cancers* **2010**, *2*, 483–497. [CrossRef] [PubMed]

83. Singh, B.; Bhat, N.K.; Bhat, H.K. Induction of NAD(P)H-quinone oxidoreductase 1 by antioxidants in female ACI rats is associated with decrease in oxidative DNA damage and inhibition of estrogen-induced breast cancer. *Carcinogenesis* **2012**, *33*, 156–163. [CrossRef] [PubMed]

84. Singh, B.; Bhat, H.K. Superoxide dismutase 3 is induced by antioxidants, inhibits oxidative DNA damage and is associated with inhibition of estrogen-induced breast cancer. *Carcinogenesis* **2012**, *33*, 2601–2610. [CrossRef] [PubMed]

85. Singh, B.; Chatterjee, A.; Ronghe, A.M.; Bhat, N.K.; Bhat, H.K. Antioxidant-mediated up-regulation of OGG1 via NRF2 induction is associated with inhibition of oxidative DNA damage in estrogen-induced breast cancer. *BMC Cancer* **2013**, *13*, 253. [CrossRef] [PubMed]

86. Adams, L.S.; Seeram, N.P.; Aggarwal, B.B.; Takada, Y.; Sand, D.; Heber, D. Pomegranate juice, total pomegranate ellagitannins, and punicalagin suppress inflammatory cell signaling in colon cancer cells. *J. Agric. Food Chem.* **2006**, *54*, 980–985. [CrossRef] [PubMed]

87. Boussetta, T.; Raad, H.; Lettéron, P.; Gougerot-Pocidalo, M.A.; Marie, J.C.; Driss, F.; El-Benna, J. Punicic acid a conjugated linolenic acid inhibits TNFα-induced neutrophil hyperactivation and protects from experimental colon inflammation in rats. *PLoS ONE* **2009**, *4*, e6458. [CrossRef] [PubMed]

88. Rosillo, M.A.; Sanchez-Hidalgo, M.; Cárdeno, A.; de la Lastra, C.A. Protective effect of ellagic acid, a natural polyphenolic compound, in a murine model of Crohn's disease. *Biochem. Pharmacol.* **2011**, *82*, 737–745. [CrossRef] [PubMed]

89. Lansky, E.P. Beware of pomegranates bearing 40% ellagic acid. *J. Med. Food* **2006**, *9*, 119–122. [CrossRef] [PubMed]

![nutrients logo] *nutrients*

MDPI

Article

Intravenous Arginine Administration Promotes Proangiogenic Cells Mobilization and Attenuates Lung Injury in Mice with Polymicrobial Sepsis

Chiu-Li Yeh [1,2], Man-Hui Pai [3], Yao-Ming Shih [4], Juey-Ming Shih [4] and Sung-Ling Yeh [2,5,*]

[1] Department of Nutrition and Health Sciences, Chinese Culture University, Taipei 11114, Taiwan; m8707008@hotmail.com
[2] School of Nutrition and Health Sciences, College of Nutrition, Taipei Medical University, Taipei 11031, Taiwan
[3] Department of Anatomy and Cell Biology, School of Medicine, College of Medicine, Taipei Medical University, Taipei 11031, Taiwan; pai0507@tmu.edu.tw
[4] Department of Surgery, Cathay General Hospital, Taipei 11073, Taiwan; yaomingshih2@gmail.com (Y.-M.S.); shihcvs@gmail.com (J.-M.S.)
[5] Nutrition Research Center, Taipei Medical University Hospital, Taipei 11031, Taiwan
* Correspondence: sangling@tmu.edu.tw; Tel.: +886-2-2736-1661 (ext. 6547)

Received: 13 February 2017; Accepted: 15 May 2017; Published: 17 May 2017

Abstract: This study investigated the influence of intravenous arginine (Arg) administration on alteration of circulating proangiogenic cells and remote lung injury in a model of polymicrobial sepsis. Mice were assigned to one normal control group (NC) and two sepsis groups that were induced by cecal ligation and puncture (CLP). One of the sepsis groups was injected with saline (SS), whereas the other (SA) was administered with a single bolus of 300 mg Arg/kg body weight via the tail vein 1 h after CLP. Septic mice were sacrificed at either 24 or 48 h after CLP, with their blood and lung tissues collected for analysis. Results showed that septic groups had higher proangiogenic cells releasing factors and proangiogenic cells percentage in blood. Also, concentration of inflammatory cytokines and expression of angiopoietin (Angpt)/Tie-2 genes in lung tissues were upregulated. Arg administration promoted mobilization of circulating proangiogenic cells while it downregulated the production of inflammatory cytokines and expression of Angpt/Tie-2 genes in the lung. The results of this investigation suggested that intravenous administration of Arg shortly after the onset of sepsis enhanced the mobilization of circulating proangiogenic cells, maintained the homeostasis of the Angpt/Tie-2 axis, and attenuated remote organ injury in polymicrobial sepsis.

Keywords: arginine; sepsis; proangiogenic cells; Angpt/Tie-2; lung injury

---

1. Introduction

Sepsis is a systemic inflammation induced by severe infection that commonly occurs in critically ill patients. A recent finding suggested that during the clinical progression of a severe, uncontrolled infection, a continuum of severe sepsis to septic shock and ultimately to multiple organ failure (MOF) exists [1]. It has been speculated that the dysfunction of endothelial cells (ECs) may play a major role in the pathophysiology of sepsis-related MOF and that the degree of dysfunction has been suggested to be a predictor of mortality in sepsis [2]. Under physiological conditions, ECs provide an endothelial barrier to prevent vascular leakage and leukocyte migration into extravascular compartments. However, the excessive activation of the inflammatory responses during sepsis can alter the function of ECs, which consequently lead to an increase in vascular permeability, pro-coagulability, leukocyte adhesion and impairment of microcirculatory flow [3]. Thus, maintaining or restoring the normal function of ECs may be a novel approach for managing patients with severe infection and represents an attractive therapeutic target in cases of uncontrolled/refractory sepsis or septic shock [4,5].

Cumulative evidence has demonstrated that the repair of damaged vascular endothelium does not depend solely on the proliferation of local ECs. Bone marrow-derived endothelial progenitor cells (EPCs) may also play important roles in the process of angiogenesis and the re-endothelialization after vascular injury [6]. The percentage of circulating EPCs is low in normal physiological condition. However, EPCs increase significantly under conditions of metabolic stress, vascular injury, and inflammation [7–9]. A previous study revealed that an increased number of circulating EPCs was predictive of survival outcome in septic patients, which suggested the importance of EPCs in maintaining the integrity of vascular endothelium and restoring microcirculation during sepsis [10]. EPCs encompass two categories of cell types: the proangiogenic hematopoietic cells and the endothelial colony forming cells, both of which are involved in the process of angiogenesis [11]. Endothelial colony forming cells proliferate to form new blood vessels, whereas proangiogenic (hematopoietic) cells interact with local ECs to favor angiogenesis [11].

Although classified as a non-essential amino acid, L-Arginine (Arg) is an indispensable nitrogen carrier for the synthesis of urea, polyamines, proline, and other proteins in healthy adults [12]. In fact, it is considered semi-essential under stressed conditions such as trauma and sepsis [13,14]. Arg supplementation was originally discovered to have immune-regulatory properties [13] and was later demonstrated to be capable of attenuating inflammation [15]; thus, Arg is often used as a supplement for critical patients under high catabolic stress. Additionally, Arg is the substrate of nitric oxide (NO) synthase and the sole precursor of endogenous NO production [16]. A previous study found that Arg supplementation enhanced the expression of vascular endothelial growth factor (VEGF) and transforming growth factor (TGF)-β mRNAs in wound tissue of rats [17]. Finally, a study performed by Evrard et al. has shown that NO bioavailability and VEGF expression were critical for EPC mobilization and that TGF-β expression was found to increase the angiogenic property of EPCs [18]. The imbalance of angiopoietin (Angpt)/Tie-2 signaling is an accepted marker of vascular dyshomeostasis [19]. In this study, by identifying proangiogenic cells in mononuclear cells using specific markers, we hypothesized that Arg supplementation may promote proangiogenic cell mobilization and vascular endothelium maturation and thus alleviate remote organ damage. Since remote lung injury is the most frequently encountered complication in sepsis-associated MOF, the effects of Arg on systemic inflammatory response, Angpt/Tie-2 balance, and the histopathology of lungs were evaluated in a mouse model of polymicrobial sepsis.

2. Materials and Methods

2.1. Animals

Eight-week-old male C57BL/6J mice weighing 20–22 g were purchased from National Laboratory Animal Center, Taipei, Taiwan at the beginning of the experiment. Mice were maintained on a 12-h day/night cycle in a temperature- and humidity-controlled room. Animal care and all experiments protocols were approved by the Institutional Animal Care and Use Committee at Taipei Medical University (LAC-2015-0018).

2.2. Experimental Procedures

Mice were randomly assigned to a normal group (NC, $n = 8$), a septic saline group (SS, $n = 20$), and a septic Arg group (SA, $n = 20$). There were no differences in the initial body weight (BW) among the three groups (data not shown). Sepsis was induced by cecal ligation and puncture (CLP) as described previously [19]. Briefly, mice were anesthetized with intraperitoneal (IP) injection of Zoletil (25 mg/kg BW) and Rumpon (10 mg/kg BW). A 1-cm midline abdominal incision was made with subsequent opening of the underlying peritoneum. The cecum was fully extracted from the peritoneal cavity and then ligated with 3-0 silk (Ethicon, Somerville, NJ, USA) at a level approximately 50% below the ileocecal valve. The distal cecum was punctured in a "through and through" manner using a 23-gauge needle. A small amount of fecal content was squeezed out and smeared onto the serosa of cecum. The punctured, fecal-coated cecum was then placed back into the peritoneal cavity and the laparotomy wound was closed in layers using 3.0 silk. Immediately after surgery, each mouse was resuscitated with sterile saline (40 mL/kg of BW) subcutaneously. One hour after CLP procedure, the SS group was injected with saline, while the SA group was treated with a single bolus of 300 mg Arg/kg BW given intravenously via tail vein. Mice were given buprenorphine (0.05 mg/kg BW) subcutaneously every 12 h for pain control and were euthanatized at either 24 or 48 h after CLP by cardiac puncture under anesthesia. Blood sample from each mouse was collected in heparinized tubes. Part of the whole blood collected was used for analyzing percentage of EPCs. The rest was centrifuged at $3000\times g$ at 4 °C for 10 min to obtain the plasma, which was stored at −80 °C for further analysis. Lung tissues were removed and frozen at −80 °C for gene expression assays, but the right middle lobe of the lung from each animal was used specifically for histological analysis.

2.3. Flow Cytometric Analysis Of Proangiogenic Cells in Blood

One hundred microliters of fresh blood were incubated with fluorescein isothiocyanate (FITC)-conjugated anti-mouse CD34 (RAM34, eBioscience, San Diego, CA, USA), allophycocyanin (APC)-conjugated anti-mouse CD309 (Avas12a1, eBioscience, San Diego, CA, USA), and phycoerythrin (PE)-conjugated anti-mouse CD133 (13A4, eBioscience, San Diego, CA, USA). After thirty minutes, lysing buffer (PharmLyse; BD Pharmingen, San Diego, CA, USA) was added to lyse the red blood cells (RBCs). Then the isolated proangiogenic cells were fixed using 2% paraformaldehyde before cytometric analysis. Mononuclear cells were first identified and CD34$^+$/CD133$^+$/CD309$^+$-cells were gated. Flow cytometric analysis was carried out in accordance to standard settings on a multicolor BD FACS CantoII flow cytometer (BD Biosciences, San Diego, CA, USA), and data were analyzed with BD FACSDiva™ v6.1.3 software (BD Biosciences, San Diego, CA, USA) as described in the previous report [20]. We presented the value of proangiogenic cells in percentage instead of the absolute number among mononuclear cells because the number of proangiogenic cells in plasma are relatively low. In addition, cell loss may occur during the standard staining process. Therefore, in order to minimize the discrepancies between samples due to possible cell loss during the staining process, percentage of proangiogenic cells was calculated based on the number of mononuclear cells obtained from the same sample.

2.4. Measurements of Proangiogenic Cell-Mobilizing Factors in Plasma

The concentrations of C-X-C motif chemokine (CXCL) 12, matrix metallopeptidase (MMP)-9, VEGF, and tumor necrosis factor (TNF)-α were measured by Enzyme-Linked ImmunoSorbent Assay (ELISA) kits (eBioscience, San Diego, CA, USA). Known to be unstable in solution, NO was converted to stable nitrite and nitrate ions in aqueous solution. Using nitrate reductase, nitrate in the solution mixture was converted to nitrite of which the concentrations were measured with the Griess reagent. Plasma nitrite/nitrate concentrations were determined with a commercial kit (R&D Systems, Minneapolis, MN, USA) according to the manufacturers' instructions.

2.5. Measurement of Cytokines in Lung Tissue

For each animal, a 20% lung homogenate was prepared by grinding 20 milligrams of lung tissue together with 100 microliters of ice-cold phosphate-buffered saline (PBS) using a homogenizer. The homogenate was centrifuged at 15,000 rpm for 20 min, and the supernatant was used for the analysis of cytokines. Concentrations of interleukin (IL)-1β, IL-6, TNF-α, IL-10, and TGF-β1 in lung tissue extracted supernatant were measured by ELISA kits (eBioscience, San Diego, CA, USA) according to the manufacturer's instructions.

2.6. Angpt1, Angpt2, and Tie-2 Messenger (m)RNA Extraction and Analysis by Quantitative Real-Time Reverse-Transcription (RT) Polymerase Chain Reaction (PCR)

Total RNA was extracted from homogenized lung tissues using the Trizol reagent (Invitrogen, Carlsbad, CA, USA). The RNA pellet obtained was dissolved in RNase-free water and stored at −80 °C for further analysis. The concentration of RNA was quantified by a spectrophotometer with absorbance wavelengths set at 260 and 280 nm. Complementary (c)DNA was synthesized from total RNA using a RevertAid™ first-strand cDNA synthesis kit (Fermentas, Vilnius, Lithuania) according to standard protocols. Reverse transcription was performed by sequential incubation of the RNA and cDNA strand for 5 min at 65 °C, 60 min at 42 °C, and 5 min at 70 °C. Specific genes were amplified by a real-time RT-PCR using the 7300 Real-Time PCR System (Applied Biosystems, Foster City, CA, USA) with SYBR Green as the detection format. Primers that were used included the following: β-actin forward, ACC CAC ACT GTG CCC ATC TAC, and reverse, TCG GTG AGG ATC TTC ATG AGG TA; Angpt1 forward, GGG CAC ACT CAT GCA TTC CT, and reverse, GCG TCA GCT GCG AGT ACA TA; Angpt2 forward, CCA ACT CCA AGA GCT CGG TT, and reverse, CGG TGT TGG ATG ACT GTC CA; and Tie-2 forward, AGG CTA GTT CCA GGC TTG CTA A, and reverse, TGG AGA ACT TGG CAC AGG AAG A. All primers were purchased from Mission Biotech (Taipei, Taiwan) based on deposited cDNA sequences (GenBank database, NCBI). A total volume of 25 μL containing 1× Power SYBR Green PCR Master Mix (Applied Biosystems, Foster City, CA, USA), 400 nM of each primer, and 100 ng of cDNA were used in the amplification procedure. The proceeding reaction included: one cycle of 2 min at 50 °C and 10 min at 95 °C, followed by forty cycles of 15 s at 95 °C and 1 min at 60 °C, with a final dissociation curve (DC) analysis. Expression levels were quantified in duplicate by means of a real-time RT-PCR. Cycle threshold (CT) values for genes were normalized to β-actin and were used to calculate the relative quantity of mRNA expression.

2.7. Histopathology of Lung Tissue

From each animal, lung tissue of the right middle lobe was fixed with a 4% phosphate-buffered paraformaldehyde solution, dehydrated with a graded ethanol series, and embedded in paraffin. In order to determine the morphology of the tissue, series of 5-μm-thick sections were stained with hematoxylin and eosin (H&E). Digital images at 100× magnification were captured for each section and five fields per section were analyzed for morphological lesions. Images were assessed using an Image-Pro Plus software (Media Cybernetics, Silver Springs, MD, USA). A scoring system was used to grade the severity of lung injury [21] with some modifications. The score calculated for each animal was based on the following histological features: (1) atelectasis and interstitial edema; (2) alveolar edema; (3) perivascular hemorrhage and overdistension. Since specific immunostaining was not used to stain the lung tissues, the simplified histological parameters used for identifying the extent of lung injury among groups was less ambiguous and more objective when calculating the injury scores. Each feature was graded (0 = normal; 1 = mild; 2 = moderate; 3 = severe), and a total summed score can range from 0 to 9.

2.8. Statistical Analysis

Data are expressed as the mean ± standard deviation (SD). All analyses were conducted using GraphPad Prism 5 (GraphPad Software, La Jolla, CA, USA). Differences among groups were analyzed by one-way analysis of variance (ANOVA) followed by Newman–Keuls multiple-comparison test. Differences within the same group at 24 and 48 h were analyzed by Student's *t*-test. A *p* value of <0.05 was considered statistically significant.

3. Results

There were no differences in the survival rate between the two septic groups at 24 h and 48 h after CLP (survival rate of 77% in both the SS and SA groups).

3.1. Differences in Proangiogenic Cell Population in Blood after CLP

The proportion of circulating proangiogenic cells increased significantly in all the sepsis groups. Moreover, the Arg-treated groups had even higher percentages of proangiogenic cells at both 24 and 48 h post-CLP when compared to that of the saline-treated groups at corresponding time points (Figure 1).

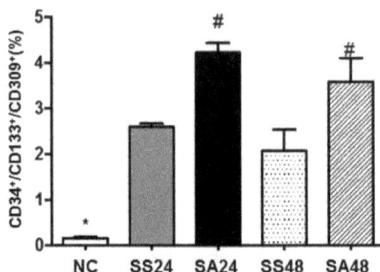

Figure 1. Distribution of circulating proangiogenic cells (CD34+/CD103+/CD309+) among the normal control (NC), sepsis with saline 24 h post-cecal ligation and puncture (CLP) (SS24), sepsis with arginine 24 h post-CLP (SA24), sepsis with saline 48 h post-CLP (SS48), and sepsis with arginine 48 h post-CLP (SA48) groups. All data are representative of duplicate measurements ($n = 8$). Data are presented as the mean $\pm$ SD. * Significantly differs from the sepsis group. # Significantly differs from the SS group at the same time point ($p < 0.05$).

3.2. Expression of Proangiogenic Cell-Related Proteins in Plasma

Compared to the NC group, protein concentrations of CXCL-12, MMP-9, VEGF, TNF-α, and NO were significantly higher in the sepsis groups. Both SA groups (SA24 and SA48) had higher levels of VEGF and NO whereas significant CXCL-12 elevation was observed only in the SA48 group. By contrast, levels of MMP-9 and TNF-α were significantly lowered when compared to those in the SS group at corresponding time points after CLP (Table 1).

Table 1. The expression of endothelial progenitor cell (EPC) mobilizing factors in plasma.

	CXCL-12	MMP-9	VEGF	TNF-α	NO
	ng/mL		pg/mL		µmol/L
NC	4.6 $\pm$ 0.3 *	22.0 $\pm$ 1.4 *	65.0 $\pm$ 6.8 *	1.8 $\pm$ 0.6 *	4.1 $\pm$ 0.5 *
SS24	18.0 $\pm$ 1.2	48.8 $\pm$ 4.7	92.1 $\pm$ 2.4	16.3 $\pm$ 1.9	11.3 $\pm$ 0.5
SA24	17.1 $\pm$ 1.1	33.4 $\pm$ 4.0 #	108.8 $\pm$ 1.0 #	10.5 $\pm$ 0.8 #	38.8 $\pm$ 1.2 #
SS48	21.2 $\pm$ 1.1	46.3 $\pm$ 2.0	112.6 $\pm$ 6.3	10.1 $\pm$ 1.9	8.9 $\pm$ 0.2
SA48	27.9 $\pm$ 1.6 #	30.6 $\pm$ 1.4 #	130.1 $\pm$ 3.4 #	4.96 $\pm$ 0.6 #	15.7 $\pm$ 0.7 #

NC, normal control group; SS24, sepsis at 24 h with saline; SA24, sepsis at 24 h with arginine; SS48, sepsis at 48 h with saline; SA48 sepsis at 48 h with arginine. All data are representative of duplicate measurements ($n = 8$). Data are presented as the mean $\pm$ SD. * Significantly differs from the sepsis group. # Significantly differs from the SS group at the same time point ($p < 0.05$).

3.3. Concentration of Cytokines in Lung Homogenates

The concentrations of IL-1β, IL-6, and TNF-α were significantly higher in the sepsis groups than those in the NC group. These pro-inflammatory cytokines were lower in the SA groups when compared to that of the SS groups. In contrast, the levels of anti-inflammatory cytokines, including IL-10 and TGF-β1, were significantly higher in the SA group than in the SS group, which was significant only at 24 h post-CLP (Table 2).

Table 2. Quantitative level of cytokines protein expression in lung homogenates.

	IL-1	IL-6	TNF-α	IL-10	TGF-β1
			pg/mg		
NC	10.3 ± 1.2 *	12.5 ± 2.1 *	20.7 ± 2.3 *	265.8 ± 17.9	16.4 ± 1.8
SS24	92.6 ± 4.5	1275.6 ± 18.3	80.4 ± 3.4	235.8 ± 4.1	12.8 ± 1.6
SA24	44.3 ± 2.6 #	662.3 ± 13.5 #	42.9 ± 1.9 #	367.8 ± 9.2 #	55.3 ± 2.6 #
SS48	83.2 ± 5.4	834.3 ± 17.3	67.8 ± 3.6	214.2 ± 5.6	16.6 ± 5.6
SA48	32.3 ± 2.3 #	385.4 ± 12.8 #	38.8 ± 3.2 #	229.4 ± 7.7	22.4 ± 3.4

NC, normal control group; SS24, sepsis at 24 h with saline; SA24, sepsis at 24 h with arginine; SS48, sepsis at 48 h with saline; SA48, sepsis at 48 h with arginine. All data are representative of duplicate measurements (n = 8). Data are presented as the mean ± SD. * Significantly differs from the sepsis group. # Significantly differs from the SS group at the same time point ($p < 0.05$).

3.4. Expression of Angpt-1, Angpt-2 and Tie-2 mRNAs in Lung Tissues

The SS groups had higher expression of Angpt-1, Angpt-2, and Tie-2 mRNAs after CLP than those in the NC group. In general, the SA groups had significantly lower expression of the Angpt/Tie-2 genes than that of the SS groups at either time points. Furthermore, the degree of reduced expression of all three genes were comparable to that of the NC group (Figure 2).

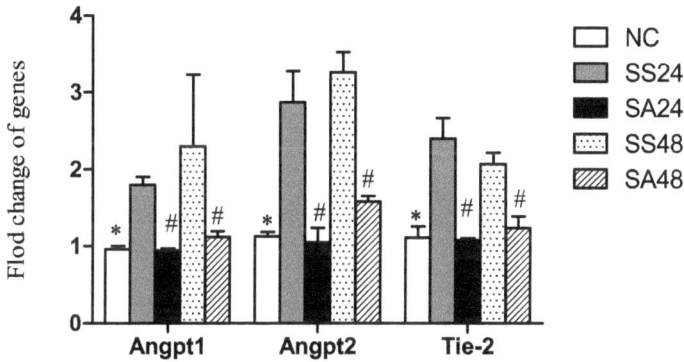

Figure 2. Expression of angiopoietin1 (Angpt1), Angpt2, and Tie-2 mRNAs in lung tissues. NC, normal control group; SS24, sepsis at 24 h with saline; SA24, sepsis at 24 h with arginine; SS48, sepsis at 48 h with saline; SA48, sepsis at 48 h with arginine. Data are presented as the mean ± SD. * Significantly differs from the sepsis group. # Significantly differs from the SS group at the same time point ($p < 0.05$).

3.5. Lung Histology

The histopathological findings showed that the NC group had normal pulmonary architecture, whereas the groups that underwent CLP-induced sepsis resulted in thickening of the septal space. In addition, destruction of alveolar structures, neutrophil infiltration, edema, hyperemia, congestion, intra-alveolar hemorrhage, and scattered parenchymal debris were also observed. However, compared to the SS groups, the extent of inflammatory lesions within the lung alveoli was less severe in the SA groups at both time points (Figure 3A, SA24 and SA48). Semi-quantitative scores calculated according to the histological changes as described in "Methods" showed that the SA groups had significantly lower lung injury scores than that of the SS groups at either time point after CLP (Figure 3B).

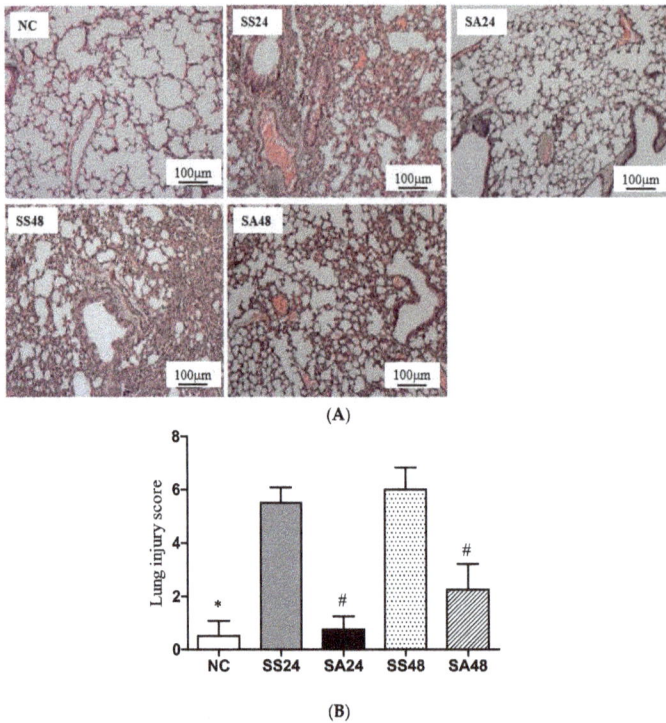

Figure 3. (**A**) Histopathology of the lung tissues; (**B**) Quantification of histological lung injury scores. All data are representative of duplicate measurements (n = 6). Data are presented as the mean ± SD. * Significantly differs from the sepsis group. # Significantly differs from the SS group at the same time point ($p < 0.05$).

4. Discussion

The use of arginine in critically ill patients has always been controversial, especially for septic patients. However, a study by Luiking et al. found that Arg infusion in sepsis stimulated NO production, reduced whole body protein breakdown, and was absent of any negative alterations in hemodynamic parameters [22]. A recent report concluded that Arg is safe and supplemental Arg infusion may be beneficial to septic patients [23]. In this study, we administered intravenously 300 mg Arg/kg BW shortly after onset of CLP-induced peritonitis. A previous study showed that intraperitoneal Arg pretreatment at a dose of 200 mg/kg BW attenuated liver damage in an ischemia/reperfusion rat model [24]. In general, mice have a higher metabolic rate than that of rats. Since we found that a single dose of 300 mg Arg/kg BW was able to reduce the level of inflammatory cytokines in plasma induced by ischemia-reperfusion injury from our preliminary study (unpublished data), this dosage was chosen for the experiment. In this study, the normal control (NC) group was served as the basal reference for comparison to the mice that underwent CLP-induced sepsis. We did not include a normal control group treated with Arg because the metabolism of Arg in sepsis had been previously reported. In a murine model of lipopolysaccharide (LPS) induced sepsis, plasma Arg and citrulline levels dropped significantly at 24 h after LPS induction [25]. In another rat model investigating the effect of LPS-induced sepsis on Arg, citrulline, and ornithine metabolism, arterial levels of Arg were reduced as early as 90 min after the onset of LPS sepsis [26]. Hence, in our study, Arg was administered immediately after the induction of CLP to account for the expected decrease in plasma Arg. In this study, we focused on whether Arg administration promoted EPCs under septic condition.

Findings from the current study showed that Arg administration resulted in higher plasma NO levels, increased the percentage of blood proangiogenic cells, maintained the homeostasis of the Angpt/Tie-2 axis, and attenuated remote organ injury in condition of gut-derived polymicrobial sepsis.

Sepsis is characterized by a systemic inflammatory response, which often leads to prominent disruption of microvascular endothelial structure and function during excessive and prolonged inflammation [27]. EPCs are particularly important in mediating the reparative process of damaged vascular endothelium and in maintaining endothelium homeostasis under catabolic conditions [28]. In this study, we found that sepsis results in a higher percentage of circulating proangiogenic cells. Also, the expression of proangiogenic cells mobilizing-related factors including CXCL-12, MMP-9, VEGF, TNF-α, and NO were all upregulated. CXCL-12, also known as stromal cell-derived factor (SDF)-1, promotes the mobilization and recruitment of EPCs [29]. MMP-9 has been found to induce the proliferation and mobilization of EPCs [30]. VEGF, an angiogenic factor, is responsible for the homing of EPCs [31]. It has been shown that the activation of MMP-9 in vascular endothelial cells involves the stimulation of VEGF receptors expressed on ECs by circulating VEGF in blood [30]. TNF-α is a commonly identified pro-inflammatory cytokine. In an ex vivo study, TNF-α was found to reduce the proliferation and migration of EPCs by decreasing the expression of SDF-1 mRNA and reducing the number of VEGF receptors, respectively [32]. NO has many physiological functions, one of which is involved in the mobilization of EPCs during the process of neovasculogenesis [33]. A study performed by Thum et al. showed that mobilization of EPCs via VEGF stimulation was reduced in an endothelial NO synthase-knockout mice model as compared to the wild type [34]. In this study, the higher proangiogenic cell percentage along with increased levels of NO, CXCL-12, and VEGF and decreased levels of TNF-α in blood after Arg administration suggested that Arg could enhance the release of EPC-mobilizing proteins, thus promoting the mobilization of proangiogenic cells. Arg is the substrate for both arginase and NO synthase (NOS). A pilot study by Tadie et al. [35] evaluated the effect of Arg supplementation on metabolic pathway in medical ICU patients. Their results showed that early enteral Arg administration increased ornithine synthesis, suggesting preferential use of the arginase pathway in these patients. In contrast, in a previous study investigating the differential expression of arginase and inducible NOS (iNOS) in the lungs in a CLP-sepsis rodent model, the authors demonstrated that strong expression of iNOS was found in alveolar, bronchial epithelial cell, endothelial cells, and alveolar macrophages, whereas arginase was almost undetectable after CLP [36]. Since patients in medical ICU are heterogeneous with arrays of underlying co-morbidities, the metabolic changes of these patients may be far more complex than the controlled CLP-induced peritonitis as represented in our model. We speculated that the single bolus administration of Arg may contribute to the observed increase in NO production possibly deriving from the citrulline/NO pathway.

Dysregulation of Angpt/Tie-2 signaling is a phenomenon observed in septic condition. Tie-2 is a transmembrane tyrosine kinase-type receptor specifically expressed on ECs and Angpts are ligands for Tie-2. It has been demonstrated that the activation of Tie-2 signaling in ECs favors blood vessel maturation and sustains the homeostasis of endothelial barrier function [19,37]. There are two types of Angpts: Angpt-1 is largely secreted by peri-endothelial cells, whereas Angpt-2 is mostly synthesized by the ECs [38]. Function of Angpt-2 was determined to act as a competitive antagonist of Angpt-1. A study demonstrated that the expression of Angpt-2 gene was upregulated but Angpt-1 gene was suppressed during sepsis, which correlated to the endothelial barrier dysfunction and consequent vascular leakages and vasculitis [37]. In this study, we found that the expression of Angpt 1 and 2/Tie-2 genes were upregulated in the saline groups, whereas this phenomenon was not observed in the Arg treated groups. Since Tie-2 signaling is known to stabilize endothelium barrier function [19], the elevated expression of Tie-2 gene observed with the saline groups may indicate that a more severe degree of vascular inflammation had occurred; the increased activation of Tie-2 signaling may reflect the severity of the damaged vascular barrier to undergo certain aspects of the reparative process. As expected in the lung tissues, we did find reduced concentrations of pro-inflammatory cytokines and increased concentrations of anti-inflammatory cytokines when Arg was administered. The histopathological findings were also consistent with the inflammatory status, implying that the remote injury to the lungs was less severe in the Arg treated groups. In this study,

Angpt1, Angpt2, and Tie-2 were measured by real-time PCR to detect the expression of mRNA. Future studies are required to measure the actual protein concentrations of Angpt1 and Angpt2 ligands or Tie-2 using western blotting analysis.

Possible mechanisms responsible for the favorable effects of Arg on promoting proangiogenic cells mobilization and reserving Angpt-Tie2 homeostatic signaling axis may involve the participation of several complex regulations. First, a previous study concluded that the favorable effect of Arg may be a result of increased NO bioavailability [39]. An in vitro study by Lu et al. showed that the increased activity of eNOS and the amount of NO production had direct effects on the migration and proliferation of EPCs and their subsequent tube formation in the process of neoangiogenesis [40]. Second, an association between inflammatory status and EPC mobilization has been demonstrated. A previous study revealed that the presence of inflammatory cytokines promoted the apoptosis of EPCs; therefore, excessive pro-inflammatory response may be another crucial factor in reducing the number of circulating EPCs [32]. Arg was known to have an alleviating effect on inflammation; thus, its direct effect on reducing the production of pro-inflammatory cytokines may indirectly prevent apoptosis and impaired mobilization from occurring in the EPCs. The elevated NO production and reduced inflammatory response as observed in the Arg groups may be partly responsible for the enhanced mobilization of proangiogenic cells. We speculated that the attenuated inflammation might also provide a beneficial effect on maintaining more balanced Angpt-Tie2 signaling in remotely injured lung tissues when administered with Arg. However, further investigation is required to clarify whether NO plays a role and what mechanisms are involved in modulating the Angpt-Tie2 signaling axis in sepsis.

In this study, we did not observe a difference in mortality among groups. There are different factors that determine the mortality of CLP: the species of the animal, resuscitation, length of the cecal ligation, and the needle size used. In a previous study with 50% cecum ligated below the ileocecal valve, punctured with 23-gauge needle, and deprived of antibiotics treatment, the survival rate was estimated 77% at 24 h [41], which is identical to this study. According to the classification of sepsis, 50% cecal ligation and the use of a 22-gauge needle puncture was classified as "mid-grade" [42]. With this grade of sepsis, comparing the mortality between groups may not be easy. There are limitations in this study. Since an Arg control group was not included, the influence of Arg on proangiogenic cell alteration at baseline versus in sepsis model cannot be found. Additionally, the possible impact of IV Arg used prophylactically and different Arg dosages on sepsis may also be worthy of further exploration.

5. Conclusions

To our knowledge, this is the first study to investigate the effect of Arg administration on the mobilization of proangiogenic cells and the alteration of Angpt/Tie2 homeostasis in sepsis. The findings of this study revealed that a single intravenous injection of Arg after CLP enhanced the release of CXCL-12, VEGF, and NO, which consequently resulted in the increased mobilization of proangiogenic cells. Also, the magnitude of attenuation on the Angpt/Tie-2 signaling axis after Arg administration in septic condition was comparable to the physiologic level of the normal control, which is suggestive that this signaling axis may play a crucial role in effectively resolving the inflamed vascular endothelium and attenuating remote lung injury. Whether the observed effect of Arg on enhanced proangiogenic cell mobilization and attenuated Angpt-Tie 2 signaling is beneficial in the amelioration of remote lung injury under condition of gut-derived polymicrobial sepsis may need to be clarified.

Acknowledgments: This study were supported by research grant 104-6202-031-112 from Taipei Medical University and MOST-103-2320-B-038-016-MY3 from Ministry of Science and Technology, Taipei, Taiwan.

Author Contributions: Chiu-Li Yeh and Sung-Ling Yeh conceived of and designed the study. Chiu-Li Yeh and Man-Hui Pai performed the study and analyzed the data. Yao-Ming Shih and Juey-Ming Shih did part of the analysis and helped interpret the data. Chiu-Li Yeh and Sung-Ling Yeh prepared the manuscript. All authors read and approved the final submitted manuscript.

Conflicts of Interest: The authors declare no conflict of interest.

References

1. Gustot, T. Multiple organ failure in sepsis: Prognosis and role of systemic inflammatory response. *Curr. Opin. Crit. Care* **2011**, *17*, 153–159. [CrossRef] [PubMed]
2. Boisrame-Helms, J.; Kremer, H.; Schini-Kerth, V.; Meziani, F. Endothelial dysfunction in sepsis. *Curr. Vasc. Pharmacol.* **2013**, *11*, 150–160. [CrossRef] [PubMed]
3. Trzeciak, S.; Cinel, I.; Phillip Dellinger, R.; Shapiro, N.I.; Arnold, R.C.; Parrillo, J.E.; Hollenberg, S.M. Microcirculatory alterations in resuscitation and shock (mars) investigators. Resuscitating the microcirculation in sepsis: The central role of nitric oxide, emerging concepts for novel therapies, and challenges for clinical trials. *Acad. Emerg. Med.* **2008**, *15*, 399–413. [CrossRef] [PubMed]
4. Skibsted, S.; Jones, A.E.; Puskarich, M.A.; Arnold, R.; Sherwin, R.; Trzeciak, S.; Schuetz, P.; Aird, W.C.; Shapiro, N.I. Biomarkers of endothelial cell activation in early sepsis. *Shock* **2013**, *39*, 427–432. [CrossRef] [PubMed]
5. Volk, T.; Kox, W.J. Endothelium function in sepsis. *Inflamm. Res.* **2000**, *49*, 185–198. [CrossRef] [PubMed]
6. Xu, X.; Yang, J.; Li, N.; Wu, R.; Tian, H.; Song, H.; Wang, H. Role of endothelial progenitor cell transplantation in rats with sepsis. *Transplant. Proc.* **2015**, *47*, 2991–3001. [CrossRef] [PubMed]
7. Adams, V.; Lenk, K.; Linke, A.; Lenz, D.; Erbs, S.; Sandri, M.; Tarnok, A.; Gielen, S.; Emmrich, F.; Schuler, G.; et al. Increase of circulating endothelial progenitor cells in patients with coronary artery disease after exercise-induced ischemia. *Arterioscler. Thromb. Vasc. Biol.* **2004**, *24*, 684–690. [CrossRef] [PubMed]
8. Malli, F.; Koutsokera, A.; Paraskeva, E.; Zakynthinos, E.; Papagianni, M.; Makris, D.; Tsilioni, I.; Molyvdas, P.A.; Gourgoulianis, K.I.; Daniil, Z. Endothelial progenitor cells in the pathogenesis of idiopathic pulmonary fibrosis: an evolving concept. *PLoS ONE* **2013**, *8*, e53658. [CrossRef] [PubMed]
9. Marrotte, E.J.; Chen, D.D.; Hakim, J.S.; Chen, A.F. Manganese superoxide dismutase expression in endothelial progenitor cells accelerates wound healing in diabetic mice. *J. Clin. Investig.* **2010**, *120*, 4207–4219. [CrossRef] [PubMed]
10. Rafat, N.; Hanusch, C.; Brinkkoetter, P.T.; Schulte, J.; Brade, J.; Zijlstra, J.G.; van der Woude, F.J.; van Ackern, K.; Yard, B.A.; Beck, G.C. Increased circulating endothelial progenitor cells in septic patients: Correlation with survival. *Crit. Care Med.* **2007**, *35*, 1677–1684. [CrossRef] [PubMed]
11. Duong, H.; Erzurum, S.; Asosingh, K. Pro-angiogenic hematopoietic progenitor cells and endothelial colony forming cells in pathological angiogenesis of bronchial and pulmonary circulation. *Angiogenesis* **2011**, *14*, 411–422. [CrossRef] [PubMed]
12. Patel, J.J.; Miller, K.R.; Rosenthal, C.; Rosenthal, M.D. When is it appropriate to use arginine in critical illness? *Nutr. Clin. Pract.* **2016**, *31*, 438–444. [CrossRef] [PubMed]
13. Evoy, D.; Lieberman, M.D.; Fahey, T.J., 3rd; Daly, J.M. Immunonutrition: The role of arginine. *Nutrition* **1998**, *14*, 611–617. [CrossRef]
14. Casas-Rodera, P.; Gomez-Candela, C.; Benitez, S.; Mateo, R.; Armero, M.; Castillo, R.; Culebras, J.M. Immunoenhanced enteral nutrition formulas in head and neck cancer surgery: A prospective, randomized clinical trial. *Nutr. Hosp.* **2008**, *23*, 105–110. [PubMed]
15. Acquaviva, R.; Lanteri, R.; Li Destri, G.; Caltabiano, R.; Vanella, L.; Lanzafame, S.; Di Cataldo, A.; Li Volti, G.; Di Giacomo, C. Beneficial effects of rutin and L-arginine coadministration in a rat model of liver ischemia-reperfusion injury. *Am. J. Physiol. Gastrointest. Liver Physiol.* **2009**, *296*, G664–G670. [CrossRef] [PubMed]
16. Wu, G.; Morris, S.M., Jr. Arginine metabolism: Nitric oxide and beyond. *Biochem. J.* **1998**, *336 Pt 1*, 1–17. [CrossRef] [PubMed]
17. Ge, K.; Lu, S.L.; Qing, C.; Xie, T.; Kong, L.; Niu, Y.W.; Wang, M.J.; Liao, Z.J.; Shi, J.X. The influence of l-arginine on the angiogenesis in burn wounds in diabetic rats. *Chin. J. Burns* **2004**, *20*, 210–213.
18. Evrard, S.M.; d'Audigier, C.; Mauge, L.; Israel-Biet, D.; Guerin, C.L.; Bieche, I.; Kovacic, J.C.; Fischer, A.M.; Gaussem, P.; Smadja, D.M. The profibrotic cytokine transforming growth factor-beta1 increases endothelial progenitor cell angiogenic properties. *J. Thromb. Haemost.* **2012**, *10*, 670–679. [CrossRef] [PubMed]
19. Moss, A. The angiopoietin: Tie 2 interaction: a potential target for future therapies in human vascular disease. *Cytokine Growth Factor Rev.* **2013**, *24*, 579–592. [CrossRef] [PubMed]
20. Pai, M.H.; Shih, Y.M.; Shih, J.M.; Yeh, C.L. Glutamine administration modulates endothelial progenitor cell and lung injury in septic mice. *Shock* **2016**, *46*, 587–592. [CrossRef] [PubMed]
21. Vaschetto, R.; Kuiper, J.W.; Chiang, S.R.; Haitsma, J.J.; Juco, J.W.; Uhlig, S.; Plotz, F.B.; Corte, F.D.; Zhang, H.; Slutsky, A.S. Inhibition of poly(adenosine diphosphateribose) polymerase attenuates ventilator-induced lung injury. *Anesthesiology* **2008**, *108*, 261–268. [CrossRef] [PubMed]

22. Luiking, Y.C.; Poeze, M.; Deutz, N.E. Arginine infusion in patients with septic shock increases nitric oxide production without haemodynamic instability. *Clin. Sci.* **2015**, *128*, 57–67. [CrossRef] [PubMed]

23. Rosenthal, M.D.; Carrott, P.W.; Patel, J.; Kiraly, L.; Martindale, R.G. Parenteral or enteral arginine supplementation safety and efficacy. *J. Nutr.* **2016**, *146*, 2594S–2600S. [CrossRef] [PubMed]

24. Lucas, M.L.; Rhoden, C.R.; Rhoden, E.L.; Zettler, C.G.; Mattos, A.A. Effects of L-Arginine and L-NAME on ischemia-reperfusion in rat liver. *Acta Cir. Bras.* **2015**, *30*, 345–352. [CrossRef] [PubMed]

25. Braulio, V.B.; Ten Have, G.A.; Vissers, Y.L.; Deutz, N.E. Time course of nitric oxide production after endotoxin challenge in mice. *Am. J. Physiol. Endocrinol. Metab.* **2004**, *287*, E912–E918. [CrossRef] [PubMed]

26. Lortie, M.J.; Satriano, J.; Gabbai, F.B.; Thareau, S.; Khang, S.; Deng, A; Pizzo, D.P.; Thomson, S.C.; Blantz, R.C.; Munger, K.A. Production of arginine by the kidney is impaired in a model of sepsis: Early events following LPS. *Am. J. Physiol. Regul. Integr. Comp. Physiol.* **2004**, *287*, R1434–R1440. [CrossRef] [PubMed]

27. Cepinskas, G.; Wilson, J.X. Inflammatory response in microvascular endothelium in sepsis: Role of oxidants. *J. Clin. Biochem. Nutr.* **2008**, *42*, 175–184. [CrossRef] [PubMed]

28. Patschan, S.A.; Patschan, D.; Temme, J.; Korsten, P.; Wessels, J.T.; Koziolek, M.; Henze, E.; Muller, G.A. Endothelial progenitor cells (EPC) in sepsis with acute renal dysfunction (ARD). *Crit. Care* **2011**, *15*, R94. [CrossRef] [PubMed]

29. George, A.L.; Bangalore-Prakash, P.; Rajoria, S.; Suriano, R.; Shanmugam, A.; Mittelman, A.; Tiwari, R.K. Endothelial progenitor cell biology in disease and tissue regeneration. *J. Hematol. Oncol.* **2011**, *4*, 24–32. [CrossRef] [PubMed]

30. Dery, M.A.; Michaud, M.D.; Richard, D.E. Hypoxia-inducible factor 1: Regulation by hypoxic and non-hypoxic activators. *Int. J. Biochem. Cell Biol.* **2005**, *37*, 535–540. [CrossRef] [PubMed]

31. Asahara, T.; Takahashi, T.; Masuda, H.; Kalka, C.; Chen, D.; Iwaguro, H.; Inai, Y.; Silver, M.; Isner, J.M. VEGF contributes to postnatal neovascularization by mobilizing bone marrow-derived endothelial progenitor cells. *EMBO J.* **1999**, *18*, 3964–3972. [CrossRef] [PubMed]

32. Chen, T.G.; Zhong, Z.Y.; Sun, G.F.; Zhou, Y.X.; Zhao, Y. Effects of tumour necrosis factor-alpha on activity and nitric oxide synthase of endothelial progenitor cells from peripheral blood. *Cell Prolif.* **2011**, *44*, 352–359. [CrossRef] [PubMed]

33. Duda, D.G.; Fukumura, D.; Jain, R.K. Role of eNOS in neovascularization: NO for endothelial progenitor cells. *Trends Mol. Med.* **2004**, *10*, 143–145. [CrossRef] [PubMed]

34. Thum, T.; Fraccarollo, D.; Schultheiss, M.; Froese, S.; Galuppo, P.; Widder, J.D.; Tsikas, D.; Ertl, G.; Bauersachs, J. Endothelial nitric oxide synthase uncoupling impairs endothelial progenitor cell mobilization and function in diabetes. *Diabetes* **2007**, *56*, 666–674. [CrossRef] [PubMed]

35. Tadié, J.M.; Cynober, L.; Peigne, V.; Caumont-Prim, A.; Neveux, N.; Gey, A.; Guerot, E.; Diehl, J.L.; Fagon, J.Y.; Tartour, E.; et al. Arginine administration to critically ill patients with a low nitric oxide fraction in the airways: A pilot study. *Intensive Care Med.* **2013**, *39*, 1663–1665. [CrossRef] [PubMed]

36. Carraway, M.S.; Piantadosi, C.A.; Jenkinson, C.P.; Huang, Y.C. Differential expression of arginine and iNOS in the lung in sepsis. *Exp. Lung. Res.* **1998**, *24*, 253–268. [CrossRef] [PubMed]

37. Parikh, S.M. Dysregulation of the angiopoietin-Tie-2 axis in sepsis and ARDS. *Virulence* **2013**, *4*, 517–524. [CrossRef] [PubMed]

38. Fiedler, U.; Scharpfenecker, M.; Koidl, S.; Hegen, A.; Grunow, V.; Schmidt, J.M.; Kriz, W.; Thurston, G.; Augustin, H.G. The Tie-2 ligand angiopoietin-2 is stored in and rapidly released upon stimulation from endothelial cell Weibel-Palade bodies. *Blood* **2004**, *103*, 4150–4156. [CrossRef] [PubMed]

39. Senbel, A.M.; Omar, A.G.; Abdel-Moneim, L.M.; Mohamed, H.F.; Daabees, T.T. Evaluation of L-arginine on kidney function and vascular reactivity following ischemic injury in rats: Protective effects and potential interactions. *Pharmacol. Rep.* **2014**, *66*, 976–983. [CrossRef] [PubMed]

40. Lu, A.; Wang, L.; Qian, L. The role of eNOS in the migration and proliferation of bone-marrow derived endothelial progenitor cells and *in vitro* angiogenesis. *Cell Biol. Int.* **2015**, *39*, 484–490. [CrossRef] [PubMed]

41. Hu, Y.M.; Yeh, C.L.; Pai, M.H.; Lee, W.Y.; Yeh, S.L. Glutamine administration modulates lung gammadelta T lymphocyte expression in mice with polymicrobial sepsis. *Shock* **2014**, *41*, 115–122. [CrossRef] [PubMed]

42. Rittirsch, D.; Huber-Lang, M.S.; Flierl, M.A.; Ward, P.A. Immunodesign of experimental sepsis by cecal ligation and puncture. *Nat. Protoc.* **2009**, *4*, 31–36. [CrossRef] [PubMed]

MDPI

Review

Vitamin D and Infectious Diseases: Simple Bystander or Contributing Factor?

Pedro Henrique França Gois [1,2,*], Daniela Ferreira [1], Simon Olenski [2] and Antonio Carlos Seguro [1]

[1]   Laboratory of Medical Research-LIM12, Nephrology Department, University of São Paulo School of Medicine, São Paulo CEP 01246-903, Brazil; danferreira61@usp.br (D.F.); trulu@usp.br (A.C.S.)
[2]   Nephrology Department, Royal Brisbane and Women's Hospital, Herston QLD 4029, Australia; Simon.olenski@health.qld.gov.au
*   Correspondence: Pedro.FrancaGois@health.qld.gov.au; Tel.: +61-(07)-3646-8576

Received: 31 March 2017; Accepted: 22 June 2017; Published: 24 June 2017

Abstract: Vitamin D (VD) is a fat-soluble steroid essential for life in higher animals. It is technically a pro-hormone present in few food types and produced endogenously in the skin by a photochemical reaction. In recent decades, several studies have suggested that VD contributes to diverse processes extending far beyond mineral homeostasis. The machinery for VD production and its receptor have been reported in multiple tissues, where they have a pivotal role in modulating the immune system. Similarly, vitamin D deficiency (VDD) has been in the spotlight as a major global public healthcare burden. VDD is highly prevalent throughout different regions of the world, including tropical and subtropical countries. Moreover, VDD may affect host immunity leading to an increased incidence and severity of several infectious diseases. In this review, we discuss new insights on VD physiology as well as the relationship between VD status and various infectious diseases such as tuberculosis, respiratory tract infections, human immunodeficiency virus, fungal infections and sepsis. Finally, we critically review the latest evidence on VD monitoring and supplementation in the setting of infectious diseases.

Keywords: vitamin D; vitamin D deficiency; infectious diseases; HIV/AIDS; tuberculosis; sepsis; fungal infections; oxidative stress

---

1. Introduction

Vitamin D (VD) is a fat-soluble steroid essential for life in higher animals. VD is primarily produced in the skin by the direct action of ultraviolet (UV) sunlight, with a much smaller proportion coming from dietary intake [1]. The two major forms of VD are ergocalciferol (VD$_2$) and cholecalciferol (VD$_3$). VD$_2$ is most commonly added to foods given the paucity of naturally occurring VD rich foods, whereas VD$_3$ is mainly synthesized in the skin but can be also found in animal-based foods [1].

Regarding the VD dermal synthesis pathway, cutaneous-derived 7-dehydroxycholesterol undergoes photolytic conversion by UV sunlight to form previtamin D$_3$ [2,3]. Previtamin D$_3$ subsequently passes through an immediate non-enzymatic temperature dependent isomerization to form VD$_3$ [4]. The pathway of VD conversion in the skin, liver and kidney is illustrated in Figure 1.

Regardless of its source, VD requires a number of steps to become calcitriol (1,25(OH)$_2$-VD), the biologically active metabolite of VD [4]. The first step in the bioactivation of VD is its carriage in the serum by VD-binding protein to the liver [4]. VD then undergoes hydroxylation in the liver by 25-hydroxylase (also known as CYP2R1), an enzyme and member of the cytochrome p450 group of enzymes to become 25-hydroxyvitamin D [25(OH)-VD] [2]. 25(OH)-VD is the major circulating form of VD in humans and its plasma levels are routinely measured as a marker of VD status [4].

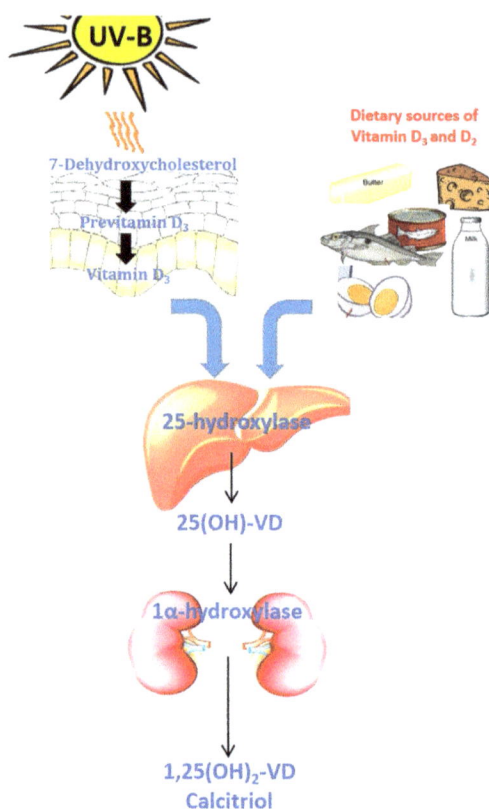

Figure 1. Vitamin D activation and metabolism.

The final activation step of VD occurs primarily in the kidney, involving a second hydroxylation reaction in which 25(OH)-VD is converted to 1,25(OH)$_2$-VD by the enzyme 1α-hydroxylase (also known as CYP27B1) [5]. 1α-hydroxylase is predominantly found in the proximal tubular cells of the kidney but has also been described in other cell types [4].

VD sufficiency is defined as having a serum 25(OH)-VD level equal or greater than 75 nmol/L (30 ng/mL) measured at the end of winter or early spring [6]. VD insufficiency is classified as a serum 25(OH)-VD level between 50 and 74 nmol/L (20–29 ng/mL), whereas VD deficiency (VDD) is defined as 25(OH)-VD levels of less than 50 nmol/L (20 ng/mL) [6]. Despite these widely accepted definitions, prevalence studies from around the world have employed different cut-offs for defining VDD, thus making it difficult to obtain a true estimate of this growing public health problem (Figure 2) [7–9].

VDD is a major public healthcare burden worldwide with estimates that between 20% and 100% of community-dwelling North American and European elderly present with VDD [10,11]. VDD is typically more prevalent among the elderly due to a combination of reduced UV sunlight exposure, reduced dietary VD intake as well as a reduced ability of the skin to produce VD from diminishing amounts of cutaneous 7-dehydroxycholesterol [12]. Moreover, individuals with darker skin need to be exposed to higher levels of UV radiation to produce sufficient levels of 25(OH)-VD, given the fact that pigmentation reduces the production of VD in the skin [13]. However, in the black population this is not always achieved, especially for those living across most latitudes in North America [13,14]. Furthermore, the prevalence of VDD remains high even in tropical and subtropical countries. A cohort study of 603 healthy Brazilian volunteers from the Clinics Hospital of the University of Sao Paulo showed that after winter the prevalence of VDD was as high as 77.4%, with 26.4% of the study subjects also having secondary hyperparathyroidism [9].

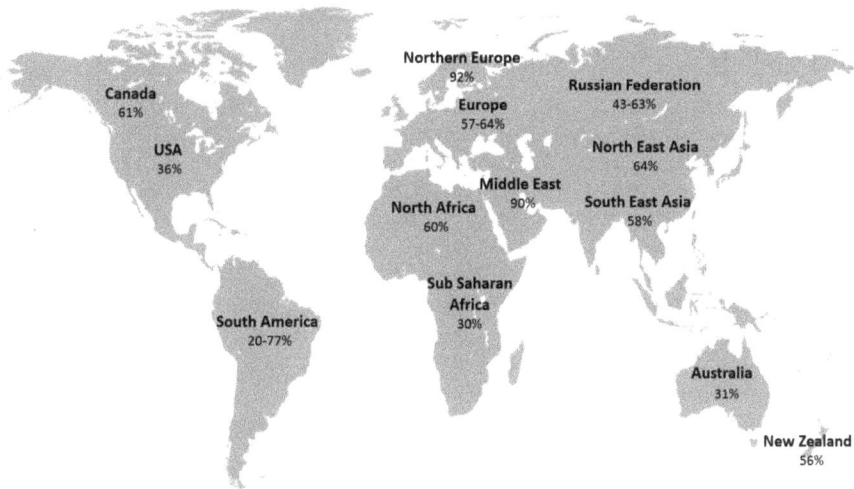

Figure 2. Prevalence of vitamin D deficiency/insufficiency in general population worldwide [7–9].

VDD is directly involved in the pathogenesis of rickets, osteoporosis and osteomalacia. Nevertheless, there is a growing body of clinical and laboratory evidence that supports its association with many other vital biological processes, which is also known as the "non-classical" effects of VD [15]. For example, VDD may affect host immunity leading to an increased incidence and severity of various infectious diseases. In this review, we will further discuss the role of VD in modulating the immune system as well as the relationship between VD status and various infectious diseases such as tuberculosis (TB), respiratory tract infections (RTIs), human immunodeficiency virus (HIV), fungal infections and sepsis. Additionally, we will critically review the latest evidence on VD monitoring and supplementation in the setting of infectious diseases.

2. Vitamin D and the Immune System

The immune response to infections is a complex and dynamic process involving multiple cell types and soluble factors such as cytokines, chemokines and hormones [16]. Numerous studies support the role of VD in both the innate and adaptive immune responses following viral and bacterial infections [17,18]. In addition, several cells of the immune system express the VD receptor (VDR) and respond directly to $1,25(OH)_2$-VD [19,20].

$1,25(OH)_2$-VD has been found to influence cell differentiation within the innate immune system [16]. Previous in vitro studies showed that $1,25(OH)_2$-VD promotes the differentiation of monocytes into macrophages in both mouse and human cells [16]. Lyakh et al. also demonstrated that $1,25(OH)_2$-VD suppresses the differentiation of monocytes into dendritic cells when using in vitro human peripheral blood monocytes exposed to bacterial lipopolysaccharides (LPS) [21]. Furthermore, $1,25(OH)_2$-VD may inhibit the production of interleukin (IL)-12 and IL-10 by LPS-treated monocytes leading to decreased functional capacity of dendritic cells [21]. It is, therefore, proposed that partially mature dendritic cells and low IL-12 levels can induce a tolerogenic state promoting the development of regulatory T-lymphocytes (T cells) with suppressive activity [21,22].

Recent studies have also highlighted the role of VD in enhancing mechanisms of pathogen elimination [15,23]. Apart from dendritic cells, macrophages and their monocyte precursors are the main cells responsible for pathogen phagocytosis and for antigen presentation to T cells [23]. Evidence supporting the concept of an intracrine machinery for VD action within macrophages and monocytes include local activation of VD together with endogenous expression of VDR within these cell types [23]. These findings confirm the study from Koeffler et al. which showed markedly increased $1,25(OH)_2$-VD production after incubating pulmonary alveolar macrophages with interferon gamma (IFN-γ) [24].

Rook et al. also demonstrated that monocytes incubated with VD metabolites have increased anti-tuberculosis activity [25]. In fact, monocytes are able to mediate an immune response against *M. tuberculosis* by phagocytosis or by expressing pathogen-recognition receptors (PRRs) [15]. Among the PRRs, the Toll-like receptors (TLRs) are the most prominent family of receptors triggering anti-microbial activity against *M. tuberculosis* [26]. A heterodimer of TLR1 and TLR2 has been described binding to a *M. tuberculosis* lipoprotein leading to transcriptional induction of VDR and 25(OH)-VD-1-α-hydroxylase [27]. Moreover, activated TLR1-TLR2 heterodimers induce transcription of IL-1 which subsequently increases nuclear factor κB (NFκB) activity via intracrine signaling [15]. Hence, 25(OH)-VD enters the monocyte and after conversion to $1,25(OH)_2$-VD in the mitochondria, binds to the VDR and ultimately acts as a transcription factor for human cathelicidin—an antimicrobial peptide [15,28,29]. It is hypothesized that the transcriptional regulation of cathelicidin by VDR represents an evolutionary change presumably acquired when primates became more exposed to sunlight [23]. Figure 3a,b illustrates VD's immunomodulatory effects in up-regulating the innate defense against *M. tuberculosis*.

Recent research has shown that regulation of the cathelicidin antimicrobial peptide (CAMP) gene by both the VDR and $1,25(OH)_2$-VD is not evolutionary conserved in mice or rats [29,30]. It is proposed that in rodents the CAMP gene is not controlled by the VD pathway due to the absence of a VD response element (VDRE) in its promoter [29,30]. However, the CAMP promoter was found to be present in humans and primates (especially the great apes) suggesting that this could be a specific adaptation of the immune system after the primates [29,31]. CAMP genes can be activated by both TLR ligands and $1,25(OH)_2$-VD, providing specific mechanisms for humans and primates to modulate their immune response and counteract pathogen-mediated suppression [30,32]. Therefore, the absence of VD regulation in the CAMP gene of mice and rats makes rodents limited models to study VD-signaling and the effects on this antimicrobial peptide [30].

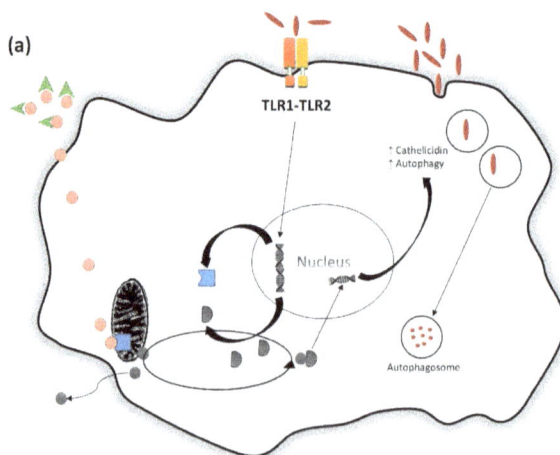

Figure 3. *Cont.*

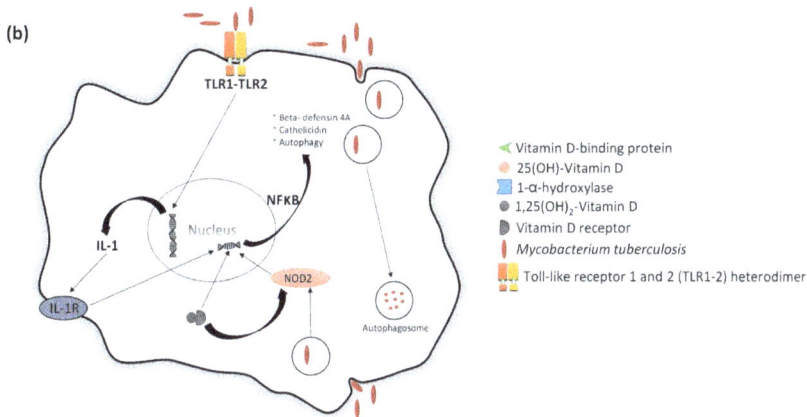

Figure 3. Vitamin D up-regulates the innate defense against *Mycobacterium tuberculosis* (**a**) *M. tuberculosis* binds to the TLR1-TLR2 heterodimer in monocytes leading to the transcriptional induction of VDR and 25(OH)-VD-1-α-hydroxylase; (**b**) Activation of IL-1 and induction of NFκB by TLR1-TLR2. Phagocytosis of *M. tuberculosis* as well as the binding of 1,25(OH)$_2$-VD to the VDR activate the intracellular pathogen-recognition receptor NOD2 increasing NFκB activity. In concert, NFκB enhances the expression of cathelicidin and beta-defensin 4A. This ultimately contributes to bacterial killing. Adapted from Hewison, M [15].

Furthermore, several studies suggest that VD might also have a role in the defense against viral infections [23,33,34]. Hansdottir et al. first demonstrated that in vitro human respiratory epithelial cells expressed higher levels of 1-α-hydroxylase and lower levels of inactivating 24-hydroxylase, thereby increasing overall activation of VD [34]. Similarly, locally activated VD resulted in downstream effects including upregulation of both the cathelicidin peptide gene and TLR co-receptor CD14 [34]. Moreover, 1,25(OH)$_2$-VD bound to the VDR induces transcription of an antimicrobial peptide through a VDRE within the promoter region of human gene DEFB4 for defensins β-defensin 2 and 4A [15,33]. Defensins have multiple roles including acting as antibacterial effectors of the innate immune response and also as antiviral peptides [35]. Defensins can also inhibit viral replication and directly induce viral membrane disruption [36].

1,25(OH)$_2$-VD's anti-inflammatory effect in human T cells is partially mediated by NFκB [37]. NFκB is a transcriptional factor composed of different subunits, e.g., p65 (activating) and p50 (inhibitory), that controls the transcription of inflammatory proteins including cytokines and chemokines [16,37]. Yu et al. illustrated that 1,25(OH)$_2$-VD had an inhibitory effect on NF$_K$B by reducing its protein expression in both the cytosolic and nuclear compartments of T cells [37]. Moreover, 1,25(OH)$_2$-VD has been reported to shift the T helper (Th) cell response from Th1 to Th2 [23]. Multiple in vitro studies revealed that 1,25(OH)$_2$-VD inhibited the development of Th1 mediated immunity, which is required for the induction of cellular responses [16,23,30,38,39]. Likewise, Cantorna et al. demonstrated that 1,25(OH)$_2$-VD treatment suppressed the Th1 response and protected animals from experimental autoimmunity [40]. Therefore, this T cell shift promoted by 1,25(OH)$_2$-VD might act in reducing Th1-mediated tissue damage via inhibition of the Th1 cytokine IFN-γ and an increase in the Th2 cytokines IL-4, IL-5 and IL-10 [23,41]. Overall, while VD can stimulate antibacterial activities in innate immune cells and epithelial cells, VD can simultaneously exert anti-inflammatory effects on adaptive immunity.

Human B cells much like their counterpart T cells also express the VDR [42]. Additionally, B cells also express mRNAs for enzymes involved in VD activity such as 1α-hydroxylase and 24-hydroxylase [41]. Recent studies have suggested that 1,25(OH)$_2$-VD may have direct effects on B cell homeostasis by inhibiting its differentiation into plasma cells and modulating the production of immunoglobulins [23,41].

3. Vitamin D and Tuberculosis

TB is a multi-systemic bacterial infection caused by Mycobacterium tuberculosis [43]. The latest World Health Organization (WHO) global report estimated that 10.4 million people developed TB in 2015, of which 1.8 million died from the disease [44]. The public health burden is even higher, considering that approximately one-third of the world's population has latent TB [44].

TB is an infectious disease known to have a strong association with poverty [45]. Over 95% of cases and deaths are in developing countries [44]. Several reports have shown a linkage between TB and malnutrition, presumed to be due to multiple factors including an increased catabolic state, a lower food intake as well as from treatment side effects [46,47]. Additionally, malnutrition weakens the immune system thus increasing the risk of TB reactivation [46,47]. Overall, nutrient deficiency enhances susceptibility to TB infection, which in turn aggravates undernutrition, creating a vicious cycle [46].

Micronutrients are essential elements to modulate various body reactions such as cell growth and repair, thereby contributing to homeostasis and disease prevention [48]. Previous studies reported lower levels of micronutrients such as vitamin A, D, E, zinc, iron and selenium in TB-infected individuals [46]. Moreover, there has been an increasing amount of literature showing VDD as an independent risk factor for TB infection [49–53]. In recent years, two systematic reviews with meta-analysis have been undertaken to assess this association [52,53]. Nnoaham et al. found a positive correlation between VDD and active TB, although VDD was not uniformly defined among the included studies [52]. Similarly, Zeng et al. reported in a meta-analysis of 15 studies (including 1440 cases and 2558 controls) that 25(OH)-VD levels below 25 nmol/L significantly increased the risk of active TB, while 25(OH)-VD levels between 26 and 50 nmol/L represented only a trend towards an increased risk [53].

The first attempt to use VD supplementation in treating TB was reported by Williams in 1849 [54,55]. He demonstrated that 234 TB cases had an improvement in their symptoms after a few days of treatment with fish liver oil [55]. It was only in 1950 after the identification of VD's structure that Charpy first used pharmacologic doses of VD_2 in the management of cutaneous TB [54].

Over the past decade eight RCTs and one complementary report of an RCT comprising over 1700 patients have been performed aiming to evaluate the effects of adjuvant VD in the setting of pulmonary TB in adults (Table 1) [56–64]. Regarding sputum conversion rates, three studies showed significantly higher sputum conversion in patients that received VD supplementation [54,62,63]. In one study, VD treatment was effective only in individuals with a specific genotype of the VDR, i.e., the tt polymorphism [64]. One study included mortality rate as an outcome but did not find any difference between groups (VD supplementation versus placebo) after 12 months of follow up, however, the authors could not detect increased levels of 25(OH)-VD in the intervention group compared to the placebo group [58]. Four out of eight trials included clinical assessments such as TB score in the outcomes, of which only one study showed beneficial effects of VD supplementation [58–60,62]. Furthermore, VD was generally well tolerated in patients already receiving TB treatment. Adverse events and serum calcium levels were similar between groups (VD supplementation and placebo) in all the studies. Moreover, the frequency of hypercalcemia reported in the trials reviewed here was surprisingly low. Although not directly related to VD supplementation, three cases of paradoxical upgrading reactions were reported in patients allocated to VD groups [57,60]. Finally, three studies demonstrated beneficial immunomodulatory effects of VD supplementation in TB-infected individuals [60,62,64].

Table 1. Overview of the randomized controlled trials evaluating VD supplementation on Tuberculosis.

Author	Study Design	Number of Patients	Dose of Vitamin D	Adverse Events	Primary Outcomes	Conclusion
Coussens et al. [64] #	Double-blind, randomized, placebo-controlled	95	100,000 IU VD₃ PO (4 doses)	Not reported	Sputum smear and culture conversion Circulating immune response	Improved both outcomes
Daley et al. [56]	Double-blind, randomized, placebo-controlled	247	100,000 IU VD₃ PO (4 doses)	Not correlated with intervention	Sputum culture conversion	No difference between groups
Martineau et al. [57]	Double-blind, randomized, placebo-controlled	146	100,000 IU VD₃ PO (4 doses)	Paradoxical upgrading reaction ($n = 2$)	Sputum culture conversion	Improved outcome only for tt genotype (VDR receptor)
Mily et al. [62]	Double-blind, randomized, placebo-controlled	288	5000 IU/day VD₃ PO (2 months)	Not correlated with intervention	Sputum culture conversion Clinical symptoms ¶	Improved only culture conversion
Nursyam et al. [63] ±	Double-blind, randomized, placebo-controlled	67	1000 IU/day VD * PO (2 months)	Not reported	Sputum smear conversion Radiological changes	Improved both outcomes
Ralph et al. [59]	Double-blind, randomized, placebo-controlled	200	50,000 IU VD₃ PO (2 doses)	Similar between groups	Sputum culture conversion Clinical symptoms/lung function test	No difference between groups
Salahuddin et al. [60]	Double-blind, randomized, placebo-controlled	259	600,000 IU VD₃ IM (2 doses)	Paradoxical upgrading reaction ($n = 1$)	Weight gain Radiological changes	Improved both outcomes &
Tukvadze et al. [61]	Double-blind, randomized, placebo-controlled	199	50,000 IU 3×/week VD₃ PO (8 weeks) and 50,000 IU q2week (8 weeks)	Similar between groups	Sputum culture conversion	No difference between groups
Wejse et al. [58]	Double-blind randomized, placebo-controlled	367	100,000 IU VD₃ PO (3 doses)	Similar between groups	Clinical symptoms	No difference between groups

VD: Vitamin D; VD₃: Cholecalciferol; PO: *per os*, oral administration; IM: Intramuscular; VDR: Vitamin D receptor q2week: Every other week; # Coussens et al. reported complementary data from the RCT conducted by Martineau et al.; ¶ TB score: assessment of change in clinical state in patients with TB; * Not specified if VD₂ (ergocalciferol) or VD₃; ± Authors did not assess VD levels; & Sputum smear conversion (secondary outcome): not different between groups.

Taken together, results from these studies must be interpreted with caution since they present several concerns and limitations. Firstly, the doses of VD employed as well as the primary outcomes were strikingly different among the studies. Additionally, four studies did not evaluate VD levels after supplementation and one study did not provide VD baseline levels [56,59,60,63]. Lack of VD measurement in some of these studies makes it impossible to demonstrate that individuals in the supplementation group had significantly higher serum VD compared to controls. Therefore, given the low-cost and high safety profile of VD supplementation together with its well-described pleiotropic effects, there is a need for better quality RCTs in TB-infected subjects. Future RCTs should be adequately powered to detect the effects of adjunctive VD and have more robust outcome measures (including clinical, radiological and laboratory data).

4. Vitamin D and RTI

Evidence supporting the hypothesis that VDD may predispose to influenza infection derives from observational studies, highlighting the seasonal influence of low sun exposure in both conditions [65,66]. Berry et al. described a linear relationship between VD levels and RTI in a cross-sectional study of 6789 British adults. The authors found that for each 10 nmol/L increase in 25(OH)-VD there was a 7% lower risk of RTI [66]. Accordingly, data from the US Third National Health and Nutrition Examination Survey (NHANES) which included 18,883 adults showed an independent association between serum 25(OH)-VD and recent upper RTI [67]. This correlation was even stronger in individuals with underlying lung disease [67]. Furthermore, similar observations have been reported in other studies, some including children as well as different ethnic groups [68–70].

Clinical trials evaluating the effect of VD supplementation in preventing RTIs have reported contrasting protocols and results. Bergman et al. conducted a systematic review with meta-analysis assessing the effect of VD supplementation on the risk of RTI [33]. Pooled data from 11 RCTs comprising 5660 individuals indicated that VD supplementation is safe and might have a beneficial effect in preventing RTIs [33]. More recently, Martineau et al. published a systematic review with meta-analysis of individual participant data from 25 RCTs including 10,933 subjects [71]. Trials were conducted in 14 countries and 19 out of 25 studies assessed baseline 25(OH)-VD levels. All RCTs supplemented VD_3 orally with different protocols of administration: bolus doses every month to every three months (seven studies using doses ranging from 30,000 to 200,000 IU/month), weekly doses (three studies using doses of 1400, 10,000 and 20,000 IU respectively), daily doses (12 studies using doses ranging from 300 to 4000 IU/day) and a combination of bolus and daily doses (three studies using bolus doses from 96,000 to 120,000 IU and daily doses from 400 to 4000 IU/day). High quality evidence from 25 RCTs with moderate heterogeneity revealed that VD_3 supplementation reduced the risk of experiencing at least one RTI (number needed to treat (NNT) to prevent one episode of RTI = 33, 95% CI 20–101). Furthermore, subgroup analysis of 15 RCTs indicated that daily or weekly VD_3 administration protected against RTI (NNT = 20, 95% CI 13–43), while no protective effect was seen in the 10 RCTs that used bolus doses of VD_3. Moreover, VD_3 supplementation was effective in preventing RTIs irrespective of baseline 25(OH)-VD levels, albeit the protective effects were greatest among subjects with more pronounced VDD [71].

Taken together, the RCT data currently available suggest that VD_3 supplementation may represent a novel and safe indication for preventing RTIs [71]. In addition, daily or weekly doses seem to be more efficient than pulse therapy [33,71]. Finally, individuals with lower levels of VD might benefit more from the supplementation [70,71].

5. Vitamin D and Human Immunodeficiency Virus Infection

VDD is a common finding in HIV-infected patients. Coelho et al. found that 63 out of 97 adult patients on antiretroviral therapy presented with insufficient levels of VD [72]. The investigators also showed that VD supplementation over 24 weeks improved CD4+ cell count recovery [72]. Furthermore, a study from the US including over 1700 women found that the prevalence of VDD among HIV+

subjects was 60%, of which African American women had the highest rates of VDD of all the ethnic groups [73]. Similarly, a Brazilian cohort study of adolescents and young adults with perinatally acquired HIV revealed that the prevalence of VD insufficiency was 29.2% [74].

In a prospective study of 398 adult patients with Acquired Immune Deficiency Syndrome (AIDS) on highly active antiretroviral therapy (HAART), those with VDD showed less recovery of CD4+ cells after 18 months of HAART [75]. Coelho et al. also reported a high prevalence of VD insufficiency among HIV-infected subjects with a CD4+ T cells nadir < 50 cells/mm^3 [72]. Hence, the investigators showed a significant positive correlation between CD4+ T cells and 25(OH)-VD levels after 24 weeks of VD supplementation [72]. Therefore, VD supplementation in these patients might be a valuable adjuvant therapy to increase CD4+ cell recovery during HAART.

To date, four cohort studies have investigated the association between VD status and mortality [76–79]. Haug et al. reported lower serum levels of 1,25(OH)$_2$-VD in symptomatic HIV-infected patients when compared to asymptomatic HIV-infected subjects [76]. The investigators also demonstrated that HIV-infected patients with abnormally low 1,25(OH)$_2$-VD levels also had shorter survival times, although serum 1,25(OH)$_2$-VD was not correlated with 25(OH)-VD levels in this study [76]. Moreover, in a cohort study of 884 HIV-infected pregnant women from Tanzania, those with lower 25(OH)-VD levels had a more rapid progression of HIV infection as well as higher all-cause mortality [77]. Sudfeld et al. also showed that VDD may lead to an increased mortality in those on HAART and this association occurred independently of impaired CD4 T cell reconstitution [78]. Finally, VDD was independently related with higher mortality and higher prevalence of AIDS-defining conditions among HIV-infected subjects followed up in the EuroSIDA study [79].

HIV-infected individuals have lower bone mineral densities and are at higher risk for bone fractures when compared to the general population [80–82]. Additionally, a few clinical studies suggested that VD supplementation might reduce bone turnover and increase bone mineral density [83,84]. In the setting of aging HIV patients, where osteopenia and osteoporosis are highly prevalent, assessing and supplementing 25(OH)-VD might be an important adjuvant measure to prevent and treat bone disease [85]. Furthermore, Sudjaritruk et al. demonstrated that VDD was associated with secondary hyperparathyroidism, increased bone turnover and bone loss in perinatally HIV-infected adolescents (aged 10–18 years) on HAART [86]. The authors suggested that VD supplementation might prevent bone loss, particularly if administered before the onset of hyperparathyroidism [86].

Efavirenz, a non-nucleoside reverse transcriptase inhibitor, has also been associated with low bone mineral density [85,87]. In addition, among HIV-infected subjects, those on efavirenz-containing therapy were at a higher risk for VDD [73,88]. The mechanism through which efavirenz lowers VD levels, and therefore may reduce bone mineral density, is through induction of 24-hydroxylase which increases the catabolism of 25(OH)-VD into the inactive form 24,25(OH)$_2$-VD [85].

Similarly, tenofovir disoproxil fumarate (TDF) is another antiretroviral agent from the nucleotide reverse transcriptase inhibitor class and is associated with hypophosphatemia secondary to proximal renal tubular dysfunction, hyperparathyroidism and increased bone loss [85]. Moreover, chronic use of TDF may also lead to a decreased glomerular filtration rate [89]. An experimental study from our laboratory using a rat model demonstrated that VDD aggravated TDF nephrotoxicity and induced hypertension and hyperlipidaemia [90].

VDD may lead to worse outcomes in patients with AIDS through a number of different mechanisms including decreasing bone mineral density, altering immunologic state, as well as increasing the risk for adverse renal and cardiovascular adverse events. Although RCTs are needed, we propose that VD levels should be carefully monitored in AIDS patients, particularly for those on TDF-containing HAART. AIDS patients with VDD might benefit from VD supplementation to improve long-term outcomes.

6. Vitamin D and Fungal Infections

There has been an increase in the incidence and prevalence of fungal infections worldwide over the past three decades [91]. Fungal infections are one of the major causes of human disease, especially among immunocompromised and hospitalized patients with severe underlying diseases [91–93]. There are more than 150 known species of fungi in the world, of which only 15 species have been isolated in patients. 95% of fungal infections are caused by only 5 species: *C. albicans*, *C. glabrata*, *C. parapsilosis*, *C. tropicalis* and *C. krusei* [92,94–96]. Furthermore, *Candida* spp. have been reported as the fourth and seventh most common cause of nosocomial bloodstream infection in the US and Europe respectively [92,97]. According to the Centers for Disease Control and Prevention (CDC), the annual incidence of *Candida* infection was reported as 7.28/100,000 inhabitants between 1992 and 1993 [98]. Another surveillance study found much higher incidence rates in the states of Atlanta and Baltimore (USA) between 2008 and 2011, respectively 13.3 and 26.2 per 100,000 inhabitants [99,100].

Fluconazole is the most commonly prescribed antifungal drug employed to treat *Candida* infections [101]. However, prolonged antifungal prophylaxis and treatment have been associated with increased resistance to antifungal agents [102,103]. Reports from various centers also showed that *C. albicans* and *C. tropicalis* became resistant to fluconazole leading to clinical therapeutic failure [102]. Moreover, there are very few resources available regarding the use of antifungal agents in the setting of antifungal failure. Therefore, in the last decade, antifungal resistance became a matter of great public health concern [97,104].

Despite its high toxicity, amphotericin B (AmB) remains the gold standard antifungal agent for the treatment of systemic mycoses, especially due to its greater activity against the fungal ergosterol membrane [105]. Regarding the different AmB formulations, both liposomal AmB and AmB solubilized in lipid emulsion (AmB/LE) seem to be less toxic, albeit the latter represents a lower cost alternative formulation with similar activity [106]. In a recent experimental study, we investigated whether VDD was a risk factor for AmB-induced nephrotoxicity [107]. We showed that VD deficient rats presented with renal dysfunction associated with increased urinary magnesium excretion following treatment with AmB/LE [107]. On the other hand, VD sufficient animals treated with AmB/LE exhibited normal renal function. Therefore, based on this experimental study we suggest that VDD might be a risk factor for AmB-induced nephrotoxicity regardless of the formulation prescribed [107].

In another experimental study, *Candida*-infected mice treated with low-dose 1,25(OH)$_2$-VD had reduced fungal burden and better survival when compared with untreated mice [108]. However, those animals treated with higher doses of 1,25(OH)$_2$-VD (0.1 and 1 µg/kg) had poorer outcomes [108]. The authors concluded that the beneficial immune response of VD supplementation was only achieved with lower doses of 1,25(OH)$_2$-VD (i.e., 0.001 and 0.01 µg/kg) [108]. Furthermore, data from in vitro *C. albicans* stimulation showed that 1,25(OH)$_2$-VD induced an anti-inflammatory profile of cytokines as well as mediated inhibition of TLR2 and TLR4 [109]. In the clinical setting, Lim et al. showed that a cohort of 28 patients with *Candida* blood stream infection had lower levels of 25(OH)-VD when compared with 78 hospitalized patients and 30 healthy volunteers [108].

Cryptococcal meningitis (CM) has been described as one of the most important opportunistic infections related to HIV and the leading cause of death in AIDS patients living in low-resources areas [110–113]. In a South African study, 25(OH)-VD levels were assessed in 150 patients with CM and 150 HIV-infected matched controls [112]. The authors aimed to evaluate the association between VD levels and disease severity, immune response and microbiological clearance [112]. The prevalence of VDD, defined as plasma 25(OH)-VD ≤50 nmol/L, was 74% in this study. Nevertheless, VDD was not associated with microbiological clearance in this study [112]. Additionally, there was no significant association between 25(OH)-VD levels and fungal load or cytokine profile in the cerebrospinal fluid [112].

At present, many questions regarding the role of VDD as a potential risk factor for fungal infections remain unanswered. Further research could explore the mechanisms of VD in modulating the inflammatory response, the association between VDD and fungal infections, as well as whether VD supplementation improves outcomes in those patients with fungal infections.

7. Vitamin D and Sepsis

Sepsis is a clinical syndrome that has physiological, pathological and biochemical abnormalities induced by infection [114]. Bacterial infections are the most common cause of sepsis, but virtually any infectious organism (e.g., viruses, fungi, and parasites) can precipitate it [115]. According to the third international consensus definitions for sepsis and septic shock (Sepsis-3), sepsis should be defined as life-threatening organ dysfunction caused by a dysregulated host response to infection [114].

Despite significant advances in modern medicine, sepsis remains a major healthcare burden worldwide [116]. It is both an important cause for hospital admission and the leading reason for admission to the intensive care unit (ICU) [117]. Moreover, sepsis is currently the tenth overall leading cause of death in the United States and the commonest cause of non-cardiac death among critically ill patients in a non-coronary ICU [118].

Recent research has shown an increased incidence and mortality from sepsis during the winter months [117]. While the increase in viral RTIs during winter might be a confounding factor regarding the seasonal variation of sepsis, other elements might also contribute to this epidemiological observation [117]. Specifically, VD has been in the spotlight as a potential contributing factor given that it presents with similar seasonal variability [117,119]. Observational studies have revealed correlations between lower levels of VD with the risk of sepsis and increased mortality [120,121]. In fact, two recent systematic reviews with meta-analysis pooled this data together showing an increased risk of sepsis in subjects with lower VD levels [122,123]. However, there are several limitations regarding this evidence. Firstly, both systematic reviews, which were published within one year of each other, included different studies in the meta-analysis. Secondly, the majority of the studies were retrospective in nature. Finally, the definitions used for sepsis and VDD or VD insufficiency were highly heterogeneous among the included studies. Although lower levels of VD might be associated with a higher risk of sepsis, larger prospective studies are needed to confirm this causal relationship.

There is a relatively small body of literature evaluating VD supplementation in the context of sepsis. Currently, seven RCTs have been published on VD levels in the adult ICU setting, of which only two evaluated the effects of supplementation in those patients with sepsis [124]. Quraishi et al. explored the effects of a single dose of VD_3 (200,000 or 400,000 IU orally or via naso/oro-gastric tube) versus placebo in a small sample of subjects (n = 30 in total) within 24 h of new-onset severe sepsis or septic shock [125]. The authors reported no significant difference in clinical outcomes between the study groups [125]. Similarly, Leaf et al. administered a single dose of $1,25(OH)_2$-VD (two mcg intravenously) to 37 critically ill patients with severe sepsis or septic shock versus placebo (n = 31) [126]. Clinical outcome measures were assessed as secondary outcomes, with subjects randomized to $1,25(OH)_2$-VD showing no difference in clinical outcomes (e.g., organ function indexes, length of stay, and mortality) [126]. Additionally, no adverse events related to VD supplementation were reported in the two RCTs available [125,126]. We also identified one ongoing study (ClinicalTrials.gov identifier NCT02684487) of which the results are not yet available.

Further research should be undertaken before introducing VD supplementation into the routine clinical care of sepsis. It is yet to be determined the type of VD metabolite, the appropriate dose and the timing of administration that would improve clinical outcomes in those patients with sepsis. Another challenge might be normalizing VD levels in critically ill patients, as well as maintaining optimum VD levels in the general population, with the aim of preventing sepsis.

8. Conclusions

In summary, the studies reviewed here highlight the role of VD beyond being only a simple bystander in various infectious diseases. VDD might in fact contribute to the pathogenesis of several infectious diseases by negatively modulating vital processes such as the innate and adaptive immune response. Currently the strongest available evidence supports the use of daily or weekly VD$_3$ supplementation as prophylaxis for acute RTIs, especially in individuals with more severe VDD. This benefit may also extend to subjects at higher risk for RTIs, such as those patients with asthma [71]. For the other infectious diseases reviewed here, the RCT data available do not support prophylactic use of VD. While a number of observational studies have evaluated the relationship between VDD and active TB, RCTs on VD supplementation in adults with active TB reported heterogeneous protocols and contrasting results, thus making it difficult to draw conclusions of which supplementation scheme would most benefit TB-infected subjects. Moreover, there is a paucity of studies exploring the association between VDD and HIV, as well as between VDD and fungal infections, and much more research here is warranted. VD toxicity, also known as hypervitaminosis D, is a potentially life threating condition usually caused by excessive VD supplementation. It may lead to hypercalcemia and subsequent renal and cardiovascular damage. Although VD supplementation was generally safe in the studies reviewed here, long-term supplementation above the upper intake level (maximum daily intake without adverse health effects) can increase the risk of toxicity [127]. While symptoms of toxicity are unlikely at daily intakes below 10,000 IU/day, the Food and Nutrition Board at the Institute of Medicine of the National Academies in the US recommended a VD$_3$ dose of 4000 IU as the upper intake level for adults [127,128]. Nevertheless, 25(OH)-VD levels above 125–150 nmol/L should be avoided, as they might be associated with increased cardiovascular events and all-cause mortality [127]. Finally, it remains unclear whether VD supplementation has any benefit for patients with infectious diseases and normal VD levels. A fruitful area for future well-designed RCTs would be to evaluate the role of adjunctive VD therapy in infectious diseases, particularly in those subjects with concomitant VDD.

Author Contributions: All authors contributed substantially to this review article.

Conflicts of Interest: The authors declare no conflict of interest.

References

1. Vitamin, D. The British Dietetic Association (BDA) Food Fact Sheet. Available online: https://www.bda.uk. com/foodfacts/VitaminD.pdf (accessed on 31 March 2017).
2. Deluca, H. History of the discovery of vitamin D and its active metabolites. *Bonekey Rep.* **2014**, *3*, 479. [CrossRef] [PubMed]
3. Holick, M.F.; Frommer, J.E.; McNeill, S.C.; Richtand, N.M.; Henley, J.W.; Potts, J.T. Photometabolism of 7-dehydrocholesterol to previtamin D3 in skin. *Biochem. Biophys. Res. Commun.* **1977**, *76*, 107–114. [CrossRef]
4. Dusso, A.S.; Brown, A.J.; Slatopolsky, E. Vitamin D. *Am. J. Physiol. Ren. Physiol.* **2005**, *289*, F8–F28. [CrossRef] [PubMed]
5. Fraser, D.R.; Kodicek, E. Unique biosynthesis by kidney of a biological active vitamin D metabolite. *Nature* **1970**, *228*, 764–766. [CrossRef] [PubMed]
6. Holick, M.F. Vitamin D status: Measurement, interpretation, and clinical application. *Ann. Epidemiol.* **2009**, *19*, 73–78. [CrossRef] [PubMed]
7. Nowson, C.A.; McGrath, J.J.; Ebeling, P.R.; Haikerwal, A.; Daly, R.M.; Sanders, K.M.; Seibel, M.J.; Mason, R.S. Vitamin D and health in adults in Australia and New Zealand: A position statement. *Med. J. Aust.* **2012**, *196*, 686–687. [CrossRef] [PubMed]
8. Hossein-nezhad, A.; Holick, M.F. Vitamin D for health: A global perspective. *Mayo Clin. Proc.* **2013**, *88*, 720–755. [CrossRef] [PubMed]
9. Unger, M.D.; Cuppari, L.; Titan, S.M.; Magalhães, M.C.; Sassaki, A.L.; dos Reis, L.M.; Jorgetti, V.; Moysés, R.M. Vitamin D status in a sunny country: Where has the sun gone? *Clin. Nutr.* **2010**, *29*, 784–788. [CrossRef] [PubMed]
10. Mithal, A.; Wahl, D.A.; Bonjour, J.P.; Burckhardt, P.; Dawson-Hughes, B.; Eisman, J.A.; El-Hajj Fuleihan, G.; Josse, R.G.; Lips, P.; Morales-Torres, J. Global vitamin D status and determinants of hypovitaminosis D. *Osteoporos. Int.* **2009**, *20*, 1807–1820. [CrossRef] [PubMed]

11. Holick, M.F. High prevalence of vitamin D inadequacy and implications for health. *Mayo Clin. Proc.* **2006**, *81*, 353–373. [CrossRef] [PubMed]

12. Van der Wielen, R.; Lowik, M.; van den Berg, H.; de Groot, L.; Haller, J.; Moreiras, O.; van Staveren, W. Serum vitamin D concentrations among elderly people in Europe. *Lancet* **1995**, *346*, 207–210. [CrossRef]

13. Harris, S.S. Vitamin D and African Americans. *J. Nutr.* **2006**, *136*, 1126–1129. [PubMed]

14. Harris, S.S.; Dawson-Hughes, B. Seasonal changes in plasma 25-hydroxyvitamin D concentrations of young American black and white women. *Am. J. Clin. Nutr.* **1998**, *67*, 1232–1326. [PubMed]

15. Hewison, M. Antibacterial effects of vitamin D. *Nat. Rev. Endocrinol.* **2011**, *7*, 337–345. [CrossRef] [PubMed]

16. Sundaram, M.E.; Coleman, L.A. Vitamin D and influenza. *Adv. Nutr.* **2012**, *3*, 517–525. [CrossRef] [PubMed]

17. Bikle, D.D. Vitamin D and the immune system: Role in protection against bacterial infection. *Curr. Opin. Nephrol. Hypertens.* **2008**, *17*, 348–352. [CrossRef] [PubMed]

18. Beard, J.A.; Bearden, A.; Striker, R. Vitamin D and the anti-viral state. *J. Clin. Virol.* **2011**, *50*, 194–200. [CrossRef] [PubMed]

19. Veldman, C.M.; Cantorna, M.T.; DeLuca, H.F. Expression of 1,25-dihydroxyvitamin D3 receptor in the immune system. *Arch. Biochem. Biophys.* **2000**, *374*, 334–338. [CrossRef] [PubMed]

20. Mahon, B.D.; Wittke, A.; Weaver, V.; Cantorna, M.T. The targets of vitamin D depend on the differentiation and activation status of CD4 positive T cells. *J. Cell. Biochem.* **2003**, *89*, 922–932. [CrossRef] [PubMed]

21. Lyakh, L.A.; Sanford, M.; Chekol, S.; Young, H.A.; Roberts, A.B. TGF-beta and vitamin D3 utilize distinct pathways to suppress IL-12 production and modulate rapid differentiation of human monocytes into CD83+ dendritic cells. *J. Immunol.* **2005**, *174*, 2061–2070. [CrossRef] [PubMed]

22. Adorini, L.; Penna, G. Dendritic cell tolerogenicity: A key mechanism in immunomodulation by vitamin D receptor agonists. *Hum. Immunol.* **2009**, *70*, 345–352. [CrossRef] [PubMed]

23. Hewison, M. Vitamin D and the immune system: New perspectives on an old theme. *Endocrinol. Metab. Clin. N. Am.* **2010**, *39*, 365–379. [CrossRef] [PubMed]

24. Koeffler, H.P.; Reichel, H.; Bishop, J.E.; Norman, A.W. gamma-Interferon stimulates production of 1,25-dihydroxyvitamin D3 by normal human macrophages. *Biochem. Biophys. Res. Commun.* **1985**, *127*, 596–603. [CrossRef]

25. Rook, G.A.; Steele, J.; Fraher, L.; Barker, S.; Karmali, R.; O'Riordan, J.; Stanford, J. Vitamin D3, gamma interferon, and control of proliferation of mycobacterium tuberculosis by human monocytes. *Immunology* **1986**, *57*, 159–163. [PubMed]

26. Medzhitov, R.; Janeway, C., Jr. Innate immune recognition: Mechanisms and pathways. *Immunol. Rev.* **2000**, *173*, 89–97. [CrossRef] [PubMed]

27. Liu, P.T. Toll-Like Receptor Triggering of a vitamin D-mediated human antimicrobial response. *Science* **2006**, *311*, 1770–1773. [CrossRef] [PubMed]

28. Wang, T.-T.; Nestel, F.P.; Bourdeau, V.; Nagai, Y.; Wang, Q.; Liao, J.; Tavera-Mendoza, L.; Lin, R.; Hanrahan, J.H.; Mader, S.; et al. Cutting edge: 1,25-Dihydroxyvitamin D3 is a direct inducer of antimicrobial peptide gene expression. *J. Immunol.* **2004**, *173*, 2909–2912. [CrossRef] [PubMed]

29. Gombart, A.F.; Borregaard, N.; Koeffler, H.P. Human cathelicidin antimicrobial peptide (CAMP) gene is a direct target of the vitamin D receptor and is strongly up-regulated in myeloid cells by 1,25-dihydroxyvitamin D3. *FASEB J.* **2005**, *19*, 1067–1077. [CrossRef] [PubMed]

30. Gombart, A.F. The vitamin D-antimicrobial peptide pathway and its role in protection against infection. *Future Microbiol.* **2009**, *4*, 1151–1165. [CrossRef] [PubMed]

31. Gombart, A.F.; Saito, T.; Koeffler, H.P. Exaptation of an ancient Alu short interspersed element provides a highly conserved vitamin D-mediated innate immune response in humans and primates. *BMC Genom.* **2009**, *10*, 321. [CrossRef] [PubMed]

32. Adams, J.S.; Ren, S.; Liu, P.T.; Chun, R.F.; Lagishetty, V.; Gombart, A.F.; Borregaard, N.; Modlin, R.L.; Hewison, M. Vitamin D-directed rheostatic regulation of monocyte antibacterial responses. *J. Immunol.* **2009**, *182*, 4289–4295. [CrossRef] [PubMed]

33. Bergman, P.; Lindh, Å.U.; Björkhem-Bergman, L.; Lindh, J.D. Vitamin D and respiratory tract infections: A systematic review and meta-analysis of randomized controlled trials. *PLoS ONE* **2013**, *8*, e65835. [CrossRef] [PubMed]

34. Hansdottir, S.; Monick, M.M.; Hinde, S.L.; Lovan, N.; Look, D.C.; Hunninghake, G.W. Respiratory epithelial cells convert inactive vitamin D to its active form: Potential effects on host defense. *J. Immunol.* **2008**, *181*, 7090–7099. [CrossRef] [PubMed]

35. Salvatore, M.; Garcia-Sastre, A.; Ruchala, P.; Lehrer, R.I.; Chang, T.; Klotman, M.E. alpha-Defensin inhibits influenza virus replication by cell-mediated mechanism(s). *J. Infect. Dis.* **2007**, *196*, 835–843. [CrossRef] [PubMed]

36. Chang, T.L.; Vargas, J.; DelPortillo, A.; Klotman, M.E. Dual role of alpha-defensin-1 in anti-HIV-1 innate immunity. *J. Clin. Investig.* **2005**, *115*, 765–773. [CrossRef] [PubMed]

37. Yu, X.P.; Bellido, T.; Manolagas, S.C. Down-regulation of NF-kappa B protein levels in activated human lymphocytes by 1,25-dihydroxyvitamin D3. *Proc. Natl. Acad. Sci. USA* **1995**, *92*, 10990–10994. [CrossRef] [PubMed]

38. Giarratana, N.; Penna, G.; Amuchastegui, S.; Mariani, R.; Daniel, K.C.; Adorini, L. A vitamin D analog down-regulates proinflammatory chemokine production by pancreatic islets inhibiting T cell recruitment and type 1 diabetes development. *J. Immunol.* **2004**, *173*, 2280–2287. [CrossRef] [PubMed]

39. Greiller, C.L.; Martineau, A.R. Modulation of the immune response to respiratory viruses by vitamin D. *Nutrients* **2015**, *7*, 4240–4270. [CrossRef] [PubMed]

40. Cantorna, M.T.; Yu, S.; Bruce, D. The paradoxical effects of vitamin D on type 1 mediated immunity. *Mol. Aspects Med.* **2008**, *29*, 369–375. [CrossRef] [PubMed]

41. Chen, S.; Sims, G.P.; Chen, X.X.; Gu, Y.Y.; Chen, S.; Lipsky, P.E. Modulatory effects of 1,25-dihydroxyvitamin D3 on human B cell differentiation. *J. Immunol.* **2007**, *179*, 1634–1647. [CrossRef] [PubMed]

42. Provvedini, D.M.; Tsoukas, C.D.; Deftos, L.J.; Manolagas, S.C. 1 alpha,25-Dihydroxyvitamin D3-binding macromolecules in human B lymphocytes: Effects on immunoglobulin production. *J. Immunol.* **1986**, *136*, 2734–2740. [PubMed]

43. Smith, I. Mycobacterium tuberculosis pathogenesis and molecular determinants of virulence. *Clin. Microbiol. Rev.* **2003**, *16*, 463–496. [CrossRef] [PubMed]

44. World Health Organization (WHO). Global Tuberculosis Report. Available online: http://www.who.int/tb/publications/global_report/en/ (accessed on 15 February 2017).

45. Barter, D.M.; Agboola, S.O.; Murray, M.B.; Bärnighausen, T. Tuberculosis and poverty: The contribution of patient costs in sub-Saharan Africa—A systematic review. *BMC Public Health* **2012**, *12*, 980. [CrossRef] [PubMed]

46. World Health Organization (WHO). Nutritional Care and Support for Patient with Tuberculosis, GUIDELINE, 2013. Available online: http://www.who.int/nutrition/publications/guidelines/nutcare_support_patients_with_tb/en/ (accessed on 25 March 2017).

47. Cegielski, J.P.; McMurray, D.N. The relationship between malnutrition and tuberculosis: Evidence from studies in humans and experimental animals. *Int. J. Tuberc. Lung Dis.* **2004**, *8*, 286–298. [PubMed]

48. Grobler, L.; Nagpal, S.; Sudarsanam, T.D.; Sinclair, D. Nutritional supplements for people being treated for active tuberculosis. *Cochrane Database Syst. Rev.* **2016**. [CrossRef]

49. Kim, J.H.; Park, J.-S.; Cho, Y.-J.; Yoon, H.-I.; Song, J.H.; Lee, C.-T.; Lee, J.H. Low serum 25-hydroxyvitamin D level: An independent risk factor for tuberculosis? *Clin. Nutr.* **2013**, *33*, 1–6. [CrossRef] [PubMed]

50. Ho-Pham, L.T.; Nguyen, N.D.; Nguyen, T.T.; Nguyen, D.H.; Bui, P.K.; Nguyen, V.N.; Nguyen, T.V. Association between vitamin D insufficiency and tuberculosis in a Vietnamese population. *BMC Infect. Dis.* **2010**, *10*, 306. [CrossRef] [PubMed]

51. Hong, J.Y.; Kim, S.Y.; Chung, K.S.; Kim, E.Y.; Jung, J.Y.; Park, M.S.; Kim, Y.S.; Kim, S.K.; Chang, J.; Kang, Y.A. Association between vitamin D deficiency and tuberculosis in a Korean population. *Int. J. Tuberc. Lung Dis.* **2014**, *18*, 73–78. [CrossRef] [PubMed]

52. Nnoaham, K.E.; Clarke, A. Low serum vitamin D levels and tuberculosis: A systematic review and meta-analysis. *Int. J. Epidemiol.* **2008**, *37*, 113–119. [CrossRef] [PubMed]

53. Zeng, J.; Wu, G.; Yang, W.; Gu, X.; Liang, W.; Yao, Y.; Song, Y. A serum vitamin D level <25 nmol/L pose high tuberculosis risk:A meta-analysis. *PLoS ONE* **2015**, *10*, e0126014.

54. Martineau, A.R.; Honecker, F.U.; Wilkinson, R.J.; Griffiths, C.J. Vitamin D in the treatment of pulmonary tuberculosis. *J. Steroid Biochem. Mol. Biol.* **2007**, *103*, 793–798. [CrossRef] [PubMed]

55. Williams, C.J.B. On the use and administration of cod-liver oil in pulmonary consumption. *Am. J. Med. Sci.* **1849**, *11*, 467–468. [CrossRef]

56. Daley, P.; Jagannathan, V.; John, K.R.; Sarojini, J.; Latha, A.; Vieth, R.; Suzana, S.; Jeyaseelan, L.; Christopher, D.J.; Smieja, M.; et al. Adjunctive vitamin D for treatment of active tuberculosis in India: A randomised, double-blind, placebo-controlled trial. *Lancet Infect. Dis.* **2015**, *15*, 528–534. [CrossRef]

57. Martineau, A.R.; Timms, P.M.; Bothamley, G.H.; Hanifa, Y.; Islam, K.; Claxton, A.P.; Packe, G.E.; Moore-Gillon, J.C.; Darmalingam, M.; Davidson, R.N.; et al. High-dose vitamin D3 during intensive-phase antimicrobial treatment of pulmonary tuberculosis: A double-blind randomised controlled trial. *Lancet* **2011**, *377*, 242–250. [CrossRef]

58. Wejse, C.; Gomes, V.F.; Rabna, P.; Gustafson, P.; Aaby, P.; Lisse, I.M.; Andersen, P.L.; Glerup, H.; Sodemann, M. Vitamin D as supplementary treatment for tuberculosis: A double-blind, randomized, placebo-controlled trial. *Am. J. Respir. Crit. Care Med.* **2009**, *179*, 843–850. [CrossRef] [PubMed]

59. Ralph, A.P.; Waramori, G.; Pontororing, G.J.; Kenangalem, E.; Wiguna, A.; Tjitra, E.; Sandjaja; Lolong, D.B.; Yeo, T.W.; Chatfield, M.D.; et al. L-arginine and vitamin D adjunctive therapies in pulmonary tuberculosis: A randomised, double-blind, placebo-controlled trial. *PLoS ONE* **2013**, *8*, e70032. [CrossRef] [PubMed]

60. Salahuddin, N.; Ali, F.; Hasan, Z.; Rao, N.; Aqeel, M.; Mahmood, F. Vitamin D accelerates clinical recovery from tuberculosis: Results of the SUCCINCT Study [Supplementary Cholecalciferol in recovery from tuberculosis]. A randomized, placebo-controlled, clinical trial of vitamin D supplementation in patients with pulmonar. *BMC Infect. Dis.* **2013**, *13*, 22. [CrossRef] [PubMed]

61. Tukvadze, N.; Sanikidze, E.; Kipiani, M.; Hebbar, G.; Easley, K.A.; Shenvi, N.; Kempker, R.R.; Frediani, J.K.; Mirtskhulava, V.; Alvarez, J.A.; et al. High-dose vitamin D 3 in adults with pulmonary tuberculosis: A double-blind randomized controlled trial. *Am. J. Clin. Nutr.* **2015**, *102*, 1059–1069. [CrossRef] [PubMed]

62. Mily, A.; Rekha, R.S.; Kamal, S.M.M.; Arifuzzaman, A.S.M.; Rahim, Z.; Khan, L.; Haq, M.A.; Zaman, K.; Bergman, P.; Brighenti, S.; et al. Significant effects of oral phenylbutyrate and vitamin D3 adjunctive therapy in pulmonary tuberculosis: A randomized controlled trial. *PLoS ONE* **2015**, *10*, e0138340. [CrossRef] [PubMed]

63. Nursyam, E.W.; Amin, Z.; Rumende, C.M. The effect of vitamin D as supplementary treatment in patients with moderately advanced pulmonary tuberculous lesion. *Acta Med. Indones.* **2006**, *38*, 3–5. [PubMed]

64. Coussens, A.K.; Wilkinson, R.J.; Hanifa, Y.; Nikolayevskyy, V.; Elkington, P.T.; Islam, K.; Timms, P.M.; Venton, T.R.; Bothamley, G.H.; Packe, G.E.; et al. Vitamin D accelerates resolution of inflammatory responses during tuberculosis treatment. *Proc. Natl. Acad. Sci. USA* **2012**, *109*, 15449–15454. [CrossRef] [PubMed]

65. Hope-Simpson, R.E. The role of season in the epidemiology of influenza. *J. Hyg. (Lond.)* **1981**, *86*, 35–47. [CrossRef] [PubMed]

66. Berry, D.J.; Hesketh, K.; Power, C.; Hyppönen, E. Vitamin D status has a linear association with seasonal infections and lung function in British adults. *Br. J. Nutr.* **2011**, *106*, 1433–1440. [CrossRef] [PubMed]

67. Ginde, A.A.; Mansbach, J.M.; Camargo, C.A. Association Between serum 25-hydroxyvitamin D level and upper respiratory tract infection in the Third Naional Health and Nutrition Examination Survey. *Arch. Intern. Med.* **2009**, *169*, 384–390. [CrossRef] [PubMed]

68. Laaksi, I.; Ruohola, J.; Tuohimaa, P.; Auvinen, A.; Haataja, R.; Pihlajama, H. An association of serum vitamin D concentrations <40 nmol/L with acute respiratory tract infection in young Finnish men. *Am. J. Clin. Nutr.* **2007**, *25*, 714–717.

69. Wayse, V.; Yousafzai, A.; Mogale, K.; Filteau, S. Association of subclinical vitamin D deficiency with severe acute lower respiratory infection in Indian children under 5 y. *Eur. J. Clin. Nutr.* **2004**, *58*, 563–567. [CrossRef] [PubMed]

70. Roth, D.E.; Shah, R.; Black, R.E.; Baqui, A.H. Vitamin D status and acute lower respiratory infection in early childhood in Sylhet, Bangladesh. *Acta Paediatr.* **2010**, *99*, 389–393. [CrossRef] [PubMed]

71. Martineau, A.R.; Jolliffe, D.A.; Hooper, R.L.; Greenberg, L.; Aloia, J.F.; Bergman, P.; Dubnov-Raz, G.; Esposito, S.; Ganmaa, D.; Ginde, A.A.; et al. Vitamin D supplementation to prevent acute respiratory tract infections: Systematic review and meta-analysis of individual participant data. *BMJ* **2017**, *356*, i6583. [CrossRef] [PubMed]

72. Coelho, L.; Cardoso, S.W.; Luz, P.M.; Hoffman, R.M.; Mendonça, L.; Veloso, V.G.; Currier, J.S.; Grinsztejn, B.; Lake, J.E. Vitamin D3 supplementation in HIV infection: Effectiveness and associations with antiretroviral therapy. *Nutr. J.* **2015**, *14*, 81. [CrossRef] [PubMed]

73. Adeyemi, O.M.; Agniel, D.; French, A.L.; Tien, P.C.; Weber, K.; Glesby, M.J.; Villacres, M.C.; Sharma, A.; Merenstein, D.; Golub, E.T.; et al. Vitamin D deficiency in HIV-infected and HIV-uninfected women in the United States. *J. Acquir. Immune Defic. Syndr.* **2011**, *57*, 197–204. [CrossRef] [PubMed]

74. Schtscherbyna, A.; Gouveia, C.; Pinheiro, M.F.; Luiz, R.R.; Farias, M.L.; Machado, E.S. Vitamin D status in a Brazilian cohort of adolescents and young adults with perinatally acquired human immunodeficiency virus infection. *Mem. Inst. Oswaldo Cruz* **2016**, *111*, 128–133. [CrossRef] [PubMed]

75. Ezeamama, A.E.; Guwatudde, D.; Wang, M.; Bagenda, D.; Kyeyune, R.; Sudfeld, C.; Manabe, Y.C.; Fawzi, W.W. Vitamin-D deficiency impairs CD4+ T-cell count recovery rate in HIV-positive adults on highly active antiretroviral therapy: A longitudinal study. *Clin. Nutr.* **2016**, *35*, 1110–1117. [CrossRef] [PubMed]

76. Haug, C.; Müller, F.; Aukrust, P.; Frøland, S.S. Subnormal serum concentration of 1,25-vitamin D in human immunodeficiency virus infection: Correlation with degree of immune deficiency and survival. *J. Infect. Dis.* **1994**, *169*, 889–893. [CrossRef] [PubMed]

77. Mehta, S.; Giovannucci, E.; Mugusi, F.M.; Spiegelman, D.; Aboud, S.; Hertzmark, E.; Msamanga, G.I.; Hunter, D.; Fawzi, W.W. Vitamin D status of HIV-infected women and its association with HIV disease progression, anemia, and mortality. *PLoS ONE* **2010**, *5*, e8770. [CrossRef] [PubMed]

78. Sudfeld, C.R.; Wang, M.; Aboud, S.; Giovannucci, E.L.; Mugusi, F.M.; Fawzi, W.W. Vitamin D and HIV progression among Tanzanian adults initiating antiretroviral therapy. *PLoS ONE* **2012**, *7*, e40036. [CrossRef] [PubMed]

79. Viard, J.-P.; Souberbielle, J.-C.; Kirk, O.; Reekie, J.; Knysz, B.; Losso, M.; Gatell, J.; Pedersen, C.; Bogner, J.R.; Lundgren, J.D.; et al. Study Group vitamin D and clinical disease progression in HIV infection: Results from the EuroSIDA study. *AIDS* **2011**, *25*, 1305–1315. [CrossRef] [PubMed]

80. Escota, G.V.; Mondy, K.; Bush, T.; Conley, L.; Brooks, J.T.; Önen, N.; Patel, P.; Kojic, E.M.; Henry, K.; Hammer, J.; et al. High prevalence of low bone mineral density and substantial bone loss over 4 years among HIV-infected persons in the era of modern antiretroviral therapy. *AIDS Res. Hum. Retrovir.* **2015**, *32*, 59–67. [CrossRef] [PubMed]

81. Kooij, K.W.; Wit, F.W.; Bisschop, P.H.; Schouten, J.; Stolte, I.G.; Prins, M.; van der Valk, M.; Prins, J.M.; van Eck-Smit, B.L.; Lips, P.; et al. Low bone mineral density in patients with well-suppressed HIV infection: association with body weight, smoking, and prior advanced HIV disease. *J. Infect. Dis.* **2015**, *211*, 539–548. [CrossRef] [PubMed]

82. Negredo, E.; Domingo, P.; Ferrer, E.; Estrada, V.; Curran, A.; Navarro, A.; Isernia, V.; Rosales, J.; Pérez-Álvarez, N.; Puig, J.; et al. Peak bone mass in young HIV-infected patients compared with healthy controls. *J. Acquir. Immune Defic. Syndr.* **2014**, *65*, 207–212. [CrossRef] [PubMed]

83. Piso, R.J.; Rothen, M.; Rothen, J.P.; Stahl, M.; Fux, C. Per oral substitution with 300,000 IU vitamin D (Cholecalciferol) reduces bone turnover markers in HIV-infected patients. *BMC Infect. Dis.* **2013**, *13*, 577. [CrossRef] [PubMed]

84. McComsey, G.A.; Kendall, M.A.; Tebas, P.; Swindells, S.; Hogg, E.; Alston-Smith, B.; Suckow, C.; Gopalakrishnan, G.; Benson, C.; Wohl, D.A. Alendronate with calcium and vitamin D supplementation is safe and effective for the treatment of decreased bone mineral density in HIV. *AIDS* **2007**, *21*, 2473–2482. [CrossRef] [PubMed]

85. Hileman, C.O.; Overton, E.T.; McComsey, G.A. Vitamin D and bone loss in HIV. *Curr. Opin. HIV AIDS* **2016**, *11*, 277–284. [CrossRef] [PubMed]

86. Sudjaritruk, T.; Bunupuradah, T.; Aurpibul, L.; Kosalaraksa, P.; Kurniati, N.; Prasitsuebsai, W.; Sophonphan, J.; Ananworanich, J.; Puthanakit, T. Hypovitaminosis D and hyperparathyroidism: Effects on bone turnover and bone mineral density among perinatally HIV-infected adolescents. *AIDS* **2016**, *30*, 1059–1067. [CrossRef] [PubMed]

87. Dave, J.A.; Cohen, K.; Micklesfield, L.K.; Maartens, G.; Levitt, N.S. Antiretroviral therapy, especially efavirenz, is associated with low bone mineral density in HIV-infected South Africans. *PLoS ONE* **2015**, *10*, e0144286. [CrossRef] [PubMed]

88. Dao, C.N.; Patel, P.; Overton, E.T.; Rhame, F.; Pals, S.L.; Johnson, C.; Bush, T.; Brooks, J.T. Study to understand the natural history of HIV and AIDS in the era of effective therapy (SUN) investigators low vitamin D among HIV-infected adults: Prevalence of and risk factors for low vitamin D levels in a cohort of HIV-infected adults and comparison to prevalence among adults in the US general population. *Clin. Infect. Dis.* **2011**, *52*, 396–405. [PubMed]

89. Libório, A.B.; Andrade, L.; Pereira, L.V.; Sanches, T.R.; Shimizu, M.H.; Seguro, A.C. Rosiglitazone reverses tenofovir-induced nephrotoxicity. *Kidney Int.* **2008**, *74*, 910–918. [CrossRef] [PubMed]

90. Canale, D.; De Bragança, A.C.; Gonçalves, J.G.; Shimizu, M.H.M.; Sanches, T.R.; Andrade, L.; Volpini, R.A.; Seguro, A.C. Vitamin D deficiency aggravates nephrotoxicity, hypertension and dyslipidemia caused by tenofovir: Role of oxidative stress and renin-angiotensin system. *PLoS ONE* **2014**, *9*, e103055. [CrossRef] [PubMed]

91. Sifuentes-Osornio, J.; Corzo-León, D.E.; Ponce-De-León, L.A. Epidemiology of invasive fungal infections in Latin America. *Curr. Fungal Infect. Rep.* **2012**, *6*, 23–34. [CrossRef] [PubMed]

92. Pfaller, M.A.; Diekema, D.J. Epidemiology of invasive candidiasis: A persistent public health problem. *Clin. Microbiol. Rev.* **2007**, *20*, 133–163. [CrossRef] [PubMed]

93. Hobson, R.P. The global epidemiology of invasive Candida infections—Is the tide turning? *J. Hosp. Infect.* **2003**, *55*, 159–168. [CrossRef] [PubMed]

94. Pappas, P.G. Invasive Candidiasis. *Infect. Dis. Clin. N. Am.* **2006**, *20*, 485–506. [CrossRef] [PubMed]

95. Falagas, M.E.; Roussos, N.; Vardakas, K.Z. Relative frequency of albicans and the various non-albicans Candida spp among candidemia isolates from inpatients in various parts of the world: A systematic review. *Int. J. Infect. Dis.* **2010**, *14*, e954–e966. [CrossRef] [PubMed]

96. Pfaller, M.A.; Jones, R.N.; Messer, S.A.; Edmond, M.B.; Wenzel, R.P. National surveillance of nosocomial blood stream infection due to species of Candida other than Candida albicans: Frequency of occurrence and antifungal susceptibility in the SCOPE program. *Diagn. Microbiol. Infect. Dis.* **1998**, *30*, 121–129. [CrossRef]

97. Wisplinghoff, H.; Bischoff, T.; Tallent, S.M.; Seifert, H.; Wenzel, R.P.; Edmond, M.B. Nosocomial bloodstream infections in US hospitals: Analysis of 24,179 cases form a prospective mnationwide surveillance study. *Clin. Infect. Dis.* **2004**, *39*, 309–317. [CrossRef] [PubMed]

98. Lass-Flörl, C. The changing face of epidemiology of invasive fungal disease in Europe. *Mycoses* **2009**, *52*, 197–205. [CrossRef] [PubMed]

99. Lockhart, S.R.; Iqbal, N.; Cleveland, A.A.; Farley, M.M.; Harrison, L.H.; Bolden, C.B.; Baughman, W.; Stein, B.; Hollick, R.; Park, B.J.; et al. Species Identification and antifungal susceptibility testing of Candida bloodstream isolates from population-based surveillance studies in two US cities from 2008 to 2011. *J. Clin. Microbiol.* **2012**, *50*, 3435–3442. [CrossRef] [PubMed]

100. Cleveland, A.A.; Farley, M.M.; Harrison, L.H.; Stein, B.; Hollick, R.; Lockhart, S.R.; Magill, S.S.; Derado, G.; Park, B.J.; Chiller, T.M. Changes in incidence and antifungal drug resistance in candidemia: Results from population-based laboratory surveillance in Atlanta and Baltimore, 2008–2011. *Clin. Infect. Dis.* **2012**, *55*, 1352–1361. [CrossRef] [PubMed]

101. Ha, Y.E.; Peck, K.R.; Joo, E.-J.; Kim, S.W.; Jung, S.-I.; Chang, H.H.; Park, K.H.; Han, S.H. Impact of first-line antifungal agents on the outcomes and costs of candidemia. *Antimicrob. Agents Chemother.* **2012**, *56*, 3950–3956. [CrossRef] [PubMed]

102. Franz, R.; Kelly, S.L.; Lamb, D.C.; Kelly, D.E.; Ruhnke, M.; Morschhäuser, J. Multiple molecular mechanisms contribute to a stepwise development of fluconazole resistance in clinical Candida albicans strains. *Antimicrob. Agents Chemother.* **1998**, *42*, 3065–3072. [PubMed]

103. Laverdière, M. Systemic antifungal drugs: Are we making any progress? *Can. J. Infect. Dis.* **1994**, *5*, 59–61. [CrossRef] [PubMed]

104. Lockhart, S.R. Current epidemiology of Candida infection. *Clin. Microbiol. Newsl.* **2014**, *36*, 131–136. [CrossRef]

105. Neumann, A.; Wieczor, M.; Zielinska, J.; Baginski, M.; Czub, J. Membrane sterols modulate the binding mode of amphotericin B without affecting its affinity for a lipid bilayer. *Langmuir* **2016**, *32*, 3452–3461. [CrossRef] [PubMed]

106. Dórea, E.L.; Yu, L.; De Castro, I.; Campos, S.B.; Ori, M.; Vaccari, E.M.; Lacaz, C.d.S.; Seguro, A.C. Nephrotoxicity of amphotericin B is attenuated by solubilizing with lipid emulsion. *J. Am. Soc. Nephrol.* **1997**, *8*, 1415–1422.

107. Ferreira, D.; Canale, D.; Volpini, R.A.; Gois, P.H.F.; Shimizu, M.H.M.; Girardi, A.C.C.; Seguro, A.C. Vitamin D deficiency induces acute kidney injury in rats treated with lipid formulation of amphotericin B. In Proceedings of the ASN 2016—American Society of Nephrology Kidney Week, Chicago, IL, USA, 15–20 November 2016; p. 670A.
108. Lim, J.H.; Ravikumar, S.; Wang, Y.-M.; Thamboo, T.P.; Ong, L.; Chen, J.; Goh, J.G.; Tay, S.H.; Chengchen, L.; Win, M.S.; et al. Bimodal influence of vitamin D in host response to systemic Candida infection-vitamin D dose matters. *J. Infect. Dis.* **2015**, *212*, 635–644. [CrossRef] [PubMed]
109. Khoo, A.-L.; Chai, L.Y.; Koenen, H.J.; Kullberg, B.-J.; Joosten, I.; van der Ven, A.J.; Netea, M.G. 1,25-Dihydroxyvitamin D3 modulates cytokine production induced by Candida albicans: Impact of seasonal variation of immune responses. *J. Infect. Dis.* **2011**, *203*, 122–130. [CrossRef] [PubMed]
110. Jarvis, J.N.; Bicanic, T.; Loyse, A.; Meintjes, G.; Hogan, L.; Roberts, C.H.; Shoham, S.; Perfect, J.R.; Govender, N.P.; Harrison, T.S. Very low levels of 25-hydroxyvitamin D are not associated with immunologic changes or clinical outcome in South African patients With HIV-associated cryptococcal meningitis. *Clin. Infect. Dis.* **2014**, *59*, 493–500. [CrossRef] [PubMed]
111. Centers for Disease Control and Prevention (CDC). Cryptococcal Meningitis: A Deadly Fungal Disease Among People Living With HIV/AIDS Cryptococcal Infection. Available online: https://www.cdc.gov/fungal/pdf/at-a-glance-508c.pdf (accessed on 14 March 2017).
112. Jarvis, J.N.; Harrison, T.S. HIV-associated cryptococcal meningitis. *AIDS* **2007**, *21*, 2119–2129. [CrossRef] [PubMed]
113. Park, B.J.; Wannemuehler, K.A.; Marston, B.J.; Govender, N.; Pappas, P.G.; Chiller, T.M. Estimation of the current global burden of cryptococcal meningitis among persons living with HIV/AIDS. *AIDS* **2009**, *23*, 525–530. [CrossRef] [PubMed]
114. Singer, M.; Deutschman, C.S.; Seymour, C.W.; Shankar-Hari, M.; Annane, D.; Bauer, M.; Bellomo, R.; Bernard, G.R.; Chiche, J.-D.; Coopersmith, C.M.; et al. The third international consensus definitions for sepsis and septic shock (sepsis-3). *JAMA* **2016**, *315*, 801. [CrossRef] [PubMed]
115. Remick, D.G. Pathophysiology of sepsis. *Am. J. Pathol.* **2007**, *170*, 1435–1444. [CrossRef] [PubMed]
116. Tiru, B.; DiNino, E.K.; Orenstein, A.; Mailloux, P.T.; Pesaturo, A.; Gupta, A.; McGee, W.T. The economic and humanistic burden of severe sepsis. *Pharmacoeconomics* **2015**, *33*, 925–937. [CrossRef] [PubMed]
117. Danai, P.A.; Sinha, S.; Moss, M.; Haber, M.J.; Martin, G.S. Seasonal variation in the epidemiology of sepsis. *Crit. Care Med.* **2007**, *35*, 410–415. [CrossRef] [PubMed]
118. Mayr, F.B.; Yende, S.; Angus, D.C. Epidemiology of severe sepsis. *Virulence* **2014**, *5*, 4–11. [CrossRef] [PubMed]
119. Klingberg, E.; Oleröd, G.; Konar, J.; Petzold, M.; Hammarsten, O. Seasonal variations in serum 25-hydroxy vitamin D levels in a Swedish cohort. *Endocrine* **2015**, *49*, 800–808. [CrossRef] [PubMed]
120. Moromizato, T.; Litonjua, A.A.; Braun, A.B.; Gibbons, F.K.; Giovannucci, E.; Christopher, K.B. Association of low serum 25-hydroxyvitamin D levels and sepsis in the critically ill. *Crit. Care Med.* **2014**, *42*, 97–107. [CrossRef] [PubMed]
121. Braun, A.B.; Gibbons, F.K.; Litonjua, A.A.; Giovannucci, E.; Christopher, K.B. Low serum 25-hydroxyvitamin D at critical care initiation is associated with increased mortality. *Crit. Care Med.* **2012**, *40*, 63–72. [CrossRef] [PubMed]
122. De Haan, K.; Groeneveld, A.B.; de Geus, H.R.; Egal, M.; Struijs, A. Vitamin D deficiency as a risk factor for infection, sepsis and mortality in the critically ill: Systematic review and meta-analysis. *Crit. Care* **2014**, *18*, 660. [CrossRef] [PubMed]
123. Upala, S.; Sanguankeo, A.; Permpalung, N. Significant association between vitamin D deficiency and sepsis: A systematic review and meta-analysis. *BMC Anesthesiol.* **2015**, *15*, 84. [CrossRef] [PubMed]
124. McNally, J.D.; Ginde, A.A.; Amrein, K. Clarification needed for the systematic review of vitamin D trials in the ICU. *Intensive Care Med.* **2017**, *43*, 595–596. [CrossRef] [PubMed]
125. Quraishi, S.A.; De Pascale, G.; Needleman, J.S.; Nakazawa, H.; Kaneki, M.; Bajwa, E.K.; Camargo, C.A.; Bhan, I.; Bhan, I. Effect of cholecalciferol supplementation on vitamin D status and cathelicidin levels in sepsis: A randomized, placebo-controlled trial. *Crit. Care Med.* **2015**, *43*, 1928–1937. [CrossRef] [PubMed]
126. Leaf, D.E.; Raed, A.; Donnino, M.W.; Ginde, A.A.; Waikar, S.S. Randomized controlled trial of calcitriol in severe sepsis. *Am. J. Respir. Crit. Care Med.* **2014**, *190*, 533–541. [CrossRef] [PubMed]

127. Vitamin D—Health Professional Fact Sheet. Available online: https://ods.od.nih.gov/factsheets/VitaminD-HealthProfessional/ (accessed on 12 June 2017).
128. Institute of Medicine of the National Academices. *Dietary Reference Intakes for Calcium and Vitamin D*; National Academies Press: Washington, DC, USA, 2011.

nutrients

MDPI

Article

Secretory Leukoprotease Inhibitor (Slpi) Expression Is Required for Educating Murine Dendritic Cells Inflammatory Response Following Quercetin Exposure

Stefania De Santis [1,2], Vanessa Galleggiante [1], Letizia Scandiffio [1], Marina Liso [1], Eduardo Sommella [3], Anastasia Sobolewski [4], Vito Spilotro [1], Aldo Pinto [3], Pietro Campiglia [3,5], Grazia Serino [1], Angelo Santino [2], Maria Notarnicola [1,*] and Marcello Chieppa [1,3,*]

[1] National Institute of Gastroenterology "S. de Bellis", Institute of Research, Via Turi, 27, 70013 Castellana Grotte, Italy; stefania.desantis@ispa.cnr.it (S.D.S.); vanessa.galleggiante@libero.it (V.G.); lety.scandiffio@hotmail.it (L.S.); marinaliso@libero.it (M.L.); vito.spilotro@irccsdebellis.it (V.S.); grazia.serino@irccsdebellis.it (G.S.)
[2] Institute of Sciences of Food Production C.N.R., Unit of Lecce, via Monteroni, 73100 Lecce, Italy; angelo.santino@ispa.cnr.it
[3] Department of Pharmacy, School of Pharmacy, University of Salerno, 84084 Fisciano, Italy; esommella@unisa.it (E.S.); pintoal@unisa.it (A.P.); pcampiglia@unisa.it (P.C.)
[4] University of East Anglia, Norwich Research Park, Norwich NR4 7TJ, UK; a.sobolewski@uea.ac.uk
[5] European Biomedical Research Institute of Salerno (EBRIS), Via S. de Renzi, 3, 84125 Salerno, Italy
*   Correspondence: maria.notarnicola@irccsdebellis.it (M.N.); transmed@irccsdebellis.it (M.C.); Tel.: +39-080-4994628 (M.C.)

Received: 21 April 2017; Accepted: 4 July 2017; Published: 6 July 2017

Abstract: Dendritic cells' (DCs) ability to present antigens and initiate the adaptive immune response confers them a pivotal role in immunological defense against hostile infection and, at the same time, immunological tolerance towards harmless components of the microbiota. Food products can modulate the inflammatory status of intestinal DCs. Among nutritionally-derived products, we investigated the ability of quercetin to suppress inflammatory cytokines secretion, antigen presentation, and DCs migration towards the draining lymph nodes. We recently identified the Slpi expression as a crucial checkpoint required for the quercetin-induced inflammatory suppression. Here we demonstrate that Slpi-KO DCs secrete a unique panel of cytokines and chemokines following quercetin exposure. In vivo, quercetin-enriched food is able to induce Slpi expression in the ileum, while little effects are detectable in the duodenum. Furthermore, Slpi expressing cells are more frequent at the tip compared to the base of the intestinal villi, suggesting that quercetin exposure could be more efficient for DCs projecting periscopes in the intestinal lumen. These data suggest that quercetin-enriched nutritional regimes may be efficient for suppressing inflammatory syndromes affecting the ileo-colonic tract.

Keywords: quercetin; dendritic cells; Slpi; inflammation

1. Introduction

Quercetin is among the best known phytochemicals able to impact different aspects of human health [1]. Quercetin also has anti-oxidant [2,3] anti-proliferative [4–6] and anti-inflammatory properties [7–9]. We recently demonstrated that murine dendritic cells, previously exposed to quercetin, suppress inflammatory pathways induced by LPS administration [10–12].

Dendritic cells (DCs) are the most powerful antigen presenting cells and are able to capture antigens, migrate to the draining lymph node and present them to initiate the adaptive immune response [13,14]. DCs were long considered to be pro-inflammatory in the mature state and anti-inflammatory in the immature state [15,16]. Only recently have tolerogenic DCs been characterized [17–19]. In peripheral tissues, the inflammatory abilities of DC progenitors are dynamically regulated by the local milieu. Several factors may help imprint DCs tolerance, particularly in an anatomical compartment exposed to a large variety of antigens [20–22]. In the intestine, incoming progenitors adapt to the temporary need of the tissue, usually becoming inflammatory-impaired DCs; nonetheless, traumatic events and/or hostile infections can change the intestinal milieu that may become inflammatory permissive [23]. DCs polarization in the intestine represents a paradigm for the induction of tolerogenic DCs—a dynamic imprinting mediated by the host secreted factors [24], the microbiome [25,26], and nutritionally-derived factors [27].

In the small intestine, DCs can sample luminal antigen by projecting "periscopes" through the epithelial monolayer into the intestinal lumen [28–31]. Due to the mucus and antimicrobial protein gradient, the tip of the villi are more frequently exposed to luminal antigens, and consequently, DCs periscopes are more frequent in this region [29].

We already demonstrated that Slpi is required for quercetin-mediated immune suppression [32]. Quercetin exposure can induce Slpi expression even in the absence of inflammatory insult, but little is known about the inflammatory pathway regulated by Slpi. Here we further explored the axis between quercetin-exposed DCs, inflammasome secretion, and Slpi expression. We then addressed if a quercetin-enriched diet was able to induce Slpi expression along the ileum and colon. Knowing that DCs extensions into the intestinal lumen were more frequent at the tip of the villi, we investigated if the presence of quercetin in the lumen could determine a gradient of Slpi expression from the tip to the base of the villi. Our results indicate that DCs may better sense (and respond to) the intestinal content if located at the tip of the intestinal villi, highlighting the importance of nutritionally-derived bioactive compounds in educating the immune response.

2. Materials and Methods

2.1. Mice

Ethics Statement: investigation has been conducted according to national and international guidelines and has been approved by the authors' institutional review board (Commission for Animal Wellbeing—OPBA).

Six-to-eight-week-old male mice were purchased from Jackson Laboratories: Wild-type C57BL/6 (Stock No.: 000664; weight: approximately 20 g) and B6; 129-Slpitm1Smw/J (Stock No.: 010926; weight: approximately 20 g). C57BL/6 and B6; 129-Slpitm1Smw/J mice were fed with either a quercetin-enriched food (0.5%) for 4 weeks using aglycone quercetin powder (FARMALABOR SRL Cat n. 1936) or a standard diet (six mice were used in each group). All animal experiments were carried out in accordance with Directive 86/609 EEC enforced by Italian D.L. n. 26/2014, and approved by the Committee on the Ethics of Animal Experiments of Ministero della Salute—Direzione Generale Sanità Animale (Prot. 2012/00000923 A00: Eo_GINRC and 103/2016-PR) and the official RBM veterinarian. Animals were euthanatized if found in severe clinical condition in order to avoid undue suffering.

2.2. Generation and Culture of Murine DCs

Bone marrow derived DCs (BMDCs) were obtained from 6–8 week old male C57BL/6 or Slpi-KO mice. Briefly, a single cell suspension of BMDCs was prepared by flushing the tibiae and femurs with 0.5 mM EDTA followed by hypotonic lysis of red blood cells with ACK (Ammonium-Chloride-Potassium) lysing buffer. Cells were plated in a 10 mL dish (1×10^6 cells/mL) and cultured in RPMI 1640 supplemented with 10% heat-inactivated fetal bovine serum (FBS), 100 U/mL penicillin, 100 mg/mL streptomycin, 25 µg/mL rmGM-CSF and 25 µg/mL rmIL-4 at 37 °C in a humidified 5% CO_2

atmosphere. On day 5, cells were harvested, restimulated with new growth factors, and plated 1×10^6 cells/mL on a 24-well culture plate. Cells were treated with 25 µM of quercetin on day 5 and day 7. On day 8, BMDCs were stimulated with 1 µg/mL of lipopolysaccharide (LPS) for 24 h.

Materials

Cell culture media and antibiotics were obtained from Thermo Fisher Scientific, Waltham, MA, USA. Growth factors were obtained from Miltenyi Biotec, Bergisch Gladbach, Germany. Quercetin and LPS (L6143) were obtained from Sigma-Aldrich, St. Louis, MO, USA.

2.3. Multiplex

Cell culture supernatants were analyzed using the Bead-based Multiplex for the Luminex® platform (LaboSpace srl, Via Ranzato, 12—20128 Milan, Italy).

2.4. Cytofluorimetric Analysis

FoxP3 staining: Spleen, mesenteric lymph nodes (MLNs) and lamina propria (LP) were isolated from 6-to-8-week old mice fed with standard or quercetin-enriched food. Spleen and MLNs were passed through a 40 µm cell strainer (BD Biosciences, Franklin Lakes, NJ, USA) to obtain a single cells suspension that was washed with DPBS 1X (Gibco, NY, USA) + 0.5% bovine serum albumin (BSA, Sigma-Aldrich, St. Louis, MO, USA). LP T-cell analysis was performed starting from single cell suspension of the ileum. Briefly, Peyer's Patches were removed, intestinal segments (1 cm long) were washed with 2.5 mM EDTA to remove the epithelial cells and digested with collagenase and DNase (Sigma Aldrich, St. Louis, MO, USA) using the GentleMacs suggested protocol. Single cells suspensions from spleen, MLNs, and LP were stained with CD4 APC-Vio 700 (Miltenyi Biotec, Bergisch Gladbach, Germany). Cells were then permeabilized with Foxp3 Fixation/Permeabilization Kit (eBioscience, San Diego, CA, USA) and subsequently washed with PERM Buffer (eBioscience, San Diego, CA, USA). Finally, cells were stained with Foxp3 PE-Cy5 (Miltenyi Biotec, Bergisch Gladbach, Germany). Flow Cytometer acquisition was performed using NAVIOS (Beckman Coulter, Brea, CA, USA).

T cells Intracellular Staining: T cells from spleen, MLN, and LP from 8-week-old mice fed with standard or quercetin-enriched food were cultured with a 500X Cell Stimulation Cocktail (eBiosceince, San Diego, CA, USA) for 12 h, washed with DPBS 1X + 0.5% BSA and stained with CD4 APC-Vio 700 (Miltenyi Biotec, Bergisch Gladbach, Germany). After washing, cells were then permeabilized with BD CytoFix/CytoPerm® Fixation/Permeabilization Kit® (BD Biosciences, Franklin Lakes, NJ, USA), washed with PERM Buffer, and stained with: IL-17A FITC, IFNγ-APC, TNFα PE and IL-4 APC as per manufacturer's instructions (Miltenyi Biotec, Bergisch Gladbach, Germany). Flow Cytometer acquisition was performed using NAVIOS.

2.5. Laser Capture Microdissection (LCM)

The lamina propria cells from the tip or the base of the ileum were micro-dissected using a Leica CTR 6000 microscope (Leica Microsystems, Wetzlar, Germany). Tissues were explanted and immediately embedded in OCT (VWR International, Radnor, PA, USA) and frozen at −80 °C. Serial sections (10 µm) were cut by a cryostat (TiEsseLab, Milan, Italy) and placed on 4 µm PEN Frame slides (Cod. 11600289, Leica Microsystems, Wetzlar, Germany). Sections were stained with hematoxylin and eosin (H & E) standard protocol and immediately micro-dissected to collect the mRNA.

2.6. RNA Extraction and Quantitative PCR (qPCR) Analysis

Total RNA was isolated from the ileum and colon. The RNA was extracted using TRIzol® (Thermo Fisher Scientific, Waltham, MA, USA) according to manufacturer's instructions. Five hundred ng of total RNA was reverse transcribed with the High Capacity cDNA Reverse Transcription kit (Thermo Fisher Scientific, Waltham, MA, USA) by using random primers for cDNA synthesis. Gene expression of Slpi and Gapdh were determined by using TaqMan Gene Expression Assays (Thermo Fisher Scientific, Waltham, MA, USA)—murine probes: Mm00441530_g1 and Mm99999915_g1, respectively. Real-time analysis performed on CFX96 System (Biorad, Hercules, CA, USA) and the expression of all target genes were calculated relative to GAPDH expression using $\Delta\Delta$Ct method.

2.7. Quercetin Quantification from Fecal Samples

Stool and intestinal content samples were thawed, accurately weighed and extracted as follows: Samples were solubilized in Methanol and subjected to ultrasonic bath for 15 min. Subsequently samples were centrifuged at 4 °C for 15 min the process was repeated three times. Supernatants were pooled and dried under reduced pressure. All samples were re-solubilized in methanol (1 mL) filtered on 0.45 µm filters and injected. Homogenized tissue samples with METABOPREP-LC kit (Hosmotic, Piano di Sorrento (NA), Italy) according to producer instructions. Liquid chromatography-tandem mass spectrometry (LC-MS/MS) analyses were performed on a Nexera UHPLC coupled to a triple quadrupole LCMS-8050 (Shimadzu, Milan, Italy). The separation was carried out with a 50 × 2.1 mm, 1.7 µm BEH C18 column (Waters, Milford, MA, USA). Mobile phases were: A) 0.1% CH_3COOH in Water v/v, B) ACN plus 0.1% CH_3COOH v/v. Gradient 0–3.00 min, 10–98% B, hold for 30 s, returning at 10%B in 0.1 s. Column equilibration 2 min. Column oven: 45 °C. MS detection was performed in negative ionization (ESI$^-$) in multiple reaction monitoring (MRM): quantifier 301-151, qualifier 301-179. Interface temperature 400 °C, Desolvation line 200 °C, Heat Block 400 °C. Drying gas (N_2) 10 L/min, Heating gas (air) 10 L/min, Nebulizing gas 3 L/min. Probe voltage −3.5 kV, detector voltage 1.80 kV. Quercetin was quantified by external calibration. A stock solution of quercetin 1 mg/1 mL was prepared in methanol and a seven point calibration curve was built (0.025–70 µg/mL) and triplicate analyses of each point were run; this methodology showed excellent linearity (R^2: 0.999). Quercetin quantification refers to stool samples and intestinal content collected from duodenum, ileum, and colon of mice fed for 28 days with quercetin-enriched food as compared to mice fed with standard food. Additionally, the ileum content was assessed for quercetin levels. For this analysis, the tissue was collected in dH_2O and homogenized with GentleMACS® (Miltenyi Biotec, Bergisch Gladbach, Germany) by using the program for protein digestion specific for M tubes® (Miltenyi Biotec, Bergisch Gladbach, Germany).

2.8. Statistical Analysis

Statistical analysis was performed using the Graphpad Prism statistical software release 5.0 for Windows XP. All data were expressed as means ± SEM of data obtained from at least three independent experiments. We evaluated the statistical significance of the grouped analysis with the two-way ANOVA test using the Bonferroni as a post-test. Results were considered statistically significant at $p < 0.05$.

3. Results and Discussion

3.1. Quercetin Fails to Reduce Inflammatory Cytokine Secretion in Slpi-KO DCs

We recently demonstrated that quercetin-exposed DCs from wild-type (WT) mice reduce their ability to release tumor necrosis factor alpha (TNFα), interleukin 1 alpha, 1 beta, 6, 10 and 12 (IL-1α, IL-1β, IL-6, IL-10, and IL-12) following LPS administration [10,11] and that Slpi expression was a non-redundant checkpoint required for quercetin-mediated TNFα secretion suppression [32]. Here we explored the secretion of IL-1α, IL-1β, IL-6, IL-10, and IL-12 from quercetin-exposed Slpi-KO DCs following LPS administration. Slpi-KO DCs respond to LPS by releasing higher amounts of TNFα, IL-1α, IL-1β, and IL-12. However, quercetin administration failed to reduce inflammatory cytokine secretion in Slpi-KO DCs (Figure 1), which was in contrast to previously observed effects on WT DCs [10,11]. Notably, LPS-induced IL-1β secretion from Slpi-KO DCs was increased following pre-exposure to quercetin (Figure 1). Altogether, these analyses of the secretome confirm the role of quercetin as an anti-inflammatory agent [32], and provide a new insight into how quercetin can exert its effects on IL-1β secretion in the absence of Slpi. The anti-inflammatory effects of quercetin were absent or reduced in the absence of Slpi for all cytokines considered, with the exception of IL-1β, suggesting a different regulatory machinery involved in the secretion of this crucial inflammatory cytokine.

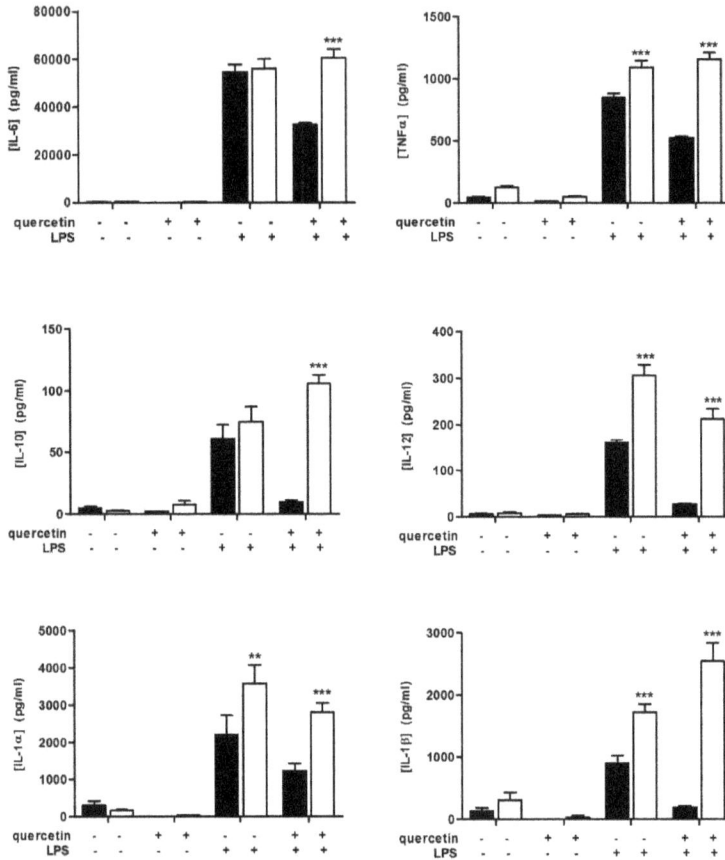

Figure 1. Quercetin reduces inflammatory cytokines secretion in wild-type (WT) dendritic cells (DCs). Bone marrow-derived DCs (BMDCs) were cultured from WT (black bars) and Slpi-KO (white bars) mice, treated with quercetin at day 5 and 7 and exposed to lipopolysaccharide (LPS). Secretion of cytokines was determined 24 h later by Multiplex assay. Bars represent mean concentration of interleukin-6 (IL-6), tumor necrosis factor alpha (TNFα), interleukin-10 (IL-10), interleukin-12 (IL-12), interleukin-1 alpha (IL-1α) and interleukin-1 beta (IL-1β) ± SEM ($n = 4$). *** $p < 0.001$, ** $p < 0.01$.

3.2. Slpi-KO DCs Fail to Secrete CXCL-1 Independently from Quercetin Exposure

The chemokine secretion profile was differentially modulated in WT and Slpi-KO DCs. Slpi-KO cells released higher amounts of CCL-2, CCL-3, and CCL-4 following 24 h LPS stimulation compared to DCs from WT mice. However, quercetin pre-exposure did not affect the secretion of CCL-2, CCL-3, CCL-4, or CCL-11. Of note (and different from other chemokines), CCL-5 secretion was significantly higher in LPS-treated Slpi-KO DCs pre-exposed to quercetin. Strikingly, Slpi-KO DCs failed to release CXCL-1 in response to LPS administration, independent of quercetin exposure (Figure 2). CXCL-1 is a major neutrophil chemoattractant and binds to CXCR2 in mice [33]. This novel observation may be important for future investigation due to the crucial role of CXCL-1 in neutrophil recruitment.

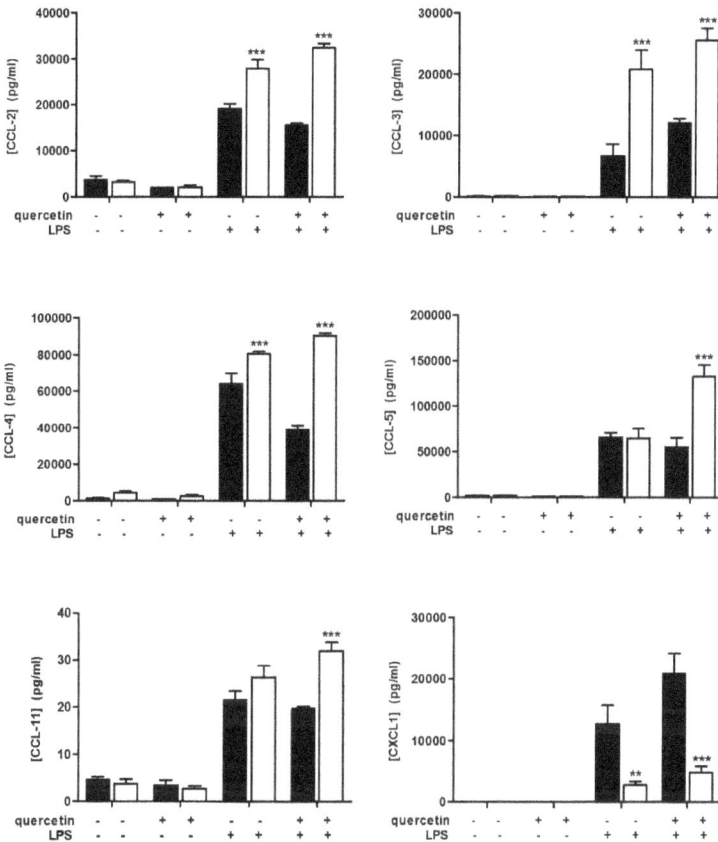

Figure 2. Quercetin exposure and chemokines release. Slpi-KO DCs fail to release the chemokine (C-X-C motif) ligand-1 CXCL-1 following LPS administration independently from quercetin exposure. BMDCs were cultured from WT (black bars) and Slpi-KO (white bars) mice, treated with quercetin at day 5 and 7 and exposed to LPS. Secretion of chemokines was determined after 24 h by Multiplex assay. Bars represent mean concentration of the (C-C motif) ligand-2, -3, -4, -5, -11 (CCL-2, CCL-3, CCL-4, CCL-5, CCL-11) and CXCL-1 $\pm$ SEM ($n = 4$). *** $p < 0.001$, ** $p < 0.01$.

3.3. A Quercetin-Enriched Diet Promotes Slpi Expression in the Ileum and Colon

We already demonstrated that quercetin was able to induce Slpi expression in murine colon if administered by gavage [32]. Quercetin is commonly present in fruits and vegetables; thus, we decided to evaluate whether Slpi expression could be induced by providing a quercetin-enriched diet to wild-type mice. Mice received quercetin-enriched or standard diet for 4 weeks. No difference in food consumption or mice weight was registered (data not shown). To evaluate the quantity of quercetin present in the intestinal lumen, we collected the luminal content of the duodenum, ileum, colon, and mice stools following 4 weeks of enriched-diet and assessed quercetin presence by LC-MS analyses. We also collected 0.5 cm of the ileum and quantified quercetin present in the cellular extract. Quercetin concentration was higher in the duodenum and gradually diminished passing from the ileum to the colon (Table 1). These data confirm that quercetin was able to safely pass the stomach and reach the intestinal lumen. Quercetin was detected at a very low concentration in cell extract from the ileum, suggesting local rather than systemic DCs exposure to quercetin in these experimental conditions. Although not direct evidence of quercetin absorption (serum levels of quercetin were below detection in line with what has previously been demonstrated [34]), the following results showing Slpi induction are consistent with at least some of the ingested quercetin being able to induce biological effects in the host.

Table 1. Quantification of quercetin in stool samples from different regions of the intestine obtained by liquid chromatography-mass spectrometry (LC-MS). Mean values for mice fed with standard and quercetin-enriched food are expressed in μg/mL for fecal sample and ng/10 mg for ileum tissue.

Sample Analysis	Standard Food (μg/mL)	Quercetin-Enriched Food (μg/mL)
Stool	Not Detected	696.47 ± 600.13
Duodenum content	Not Detected	108.44 ± 106.18
Ileum content	Not Detected	76.44 ± 54.21
Colon content	Not Detected	49.43 ± 0.46
Ileum tissue	Not Detected	0.26 ± 0.21

Slpi expression was assessed by qPCR in the duodenum, ileum, and colon. A quercetin-enriched diet induced Slpi expression in the ileum and colon, but failed to induce Slpi in the duodenum (Figure 3). These results suggest that Slpi induction was the result of local rather than systemic DCs exposure to quercetin. To investigate if quercetin-enriched food could change the T cell polarization, we performed the intracellular staining of spleen, MLNs, and LP CD4+ T cells. Figure 4 shows that no significant difference was detected in the TNFα, IFNγ, and Foxp3 populations, likely due to the absence of an inflammatory insult. Additionally, IL-17A and IL-4 CD4+ T cells percentages did not change with the quercetin-enriched diet (data not shown).

Slpi induction observed in the absence of inflammation supports the idea of polyphenol-rich nutritional regimes that may help to prevent chronic inflammatory syndromes.

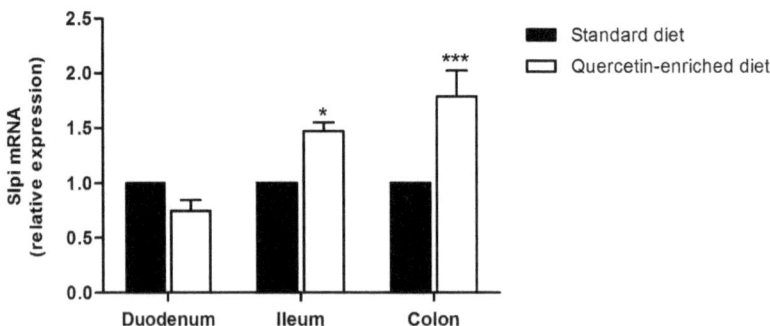

Figure 3. Quercetin induces Slpi expression in the ileum and colon. Quercetin-enriched or standard diet was administered for 4 weeks. Slpi expression was measured by qPCR in the small intestine or colon of standard (black bars) and quercetin (white bars) fed mice. Bars represent mean expression ± SEM ($n = 3$) for each treatment. *** $p < 0.001$, * $p < 0.05$.

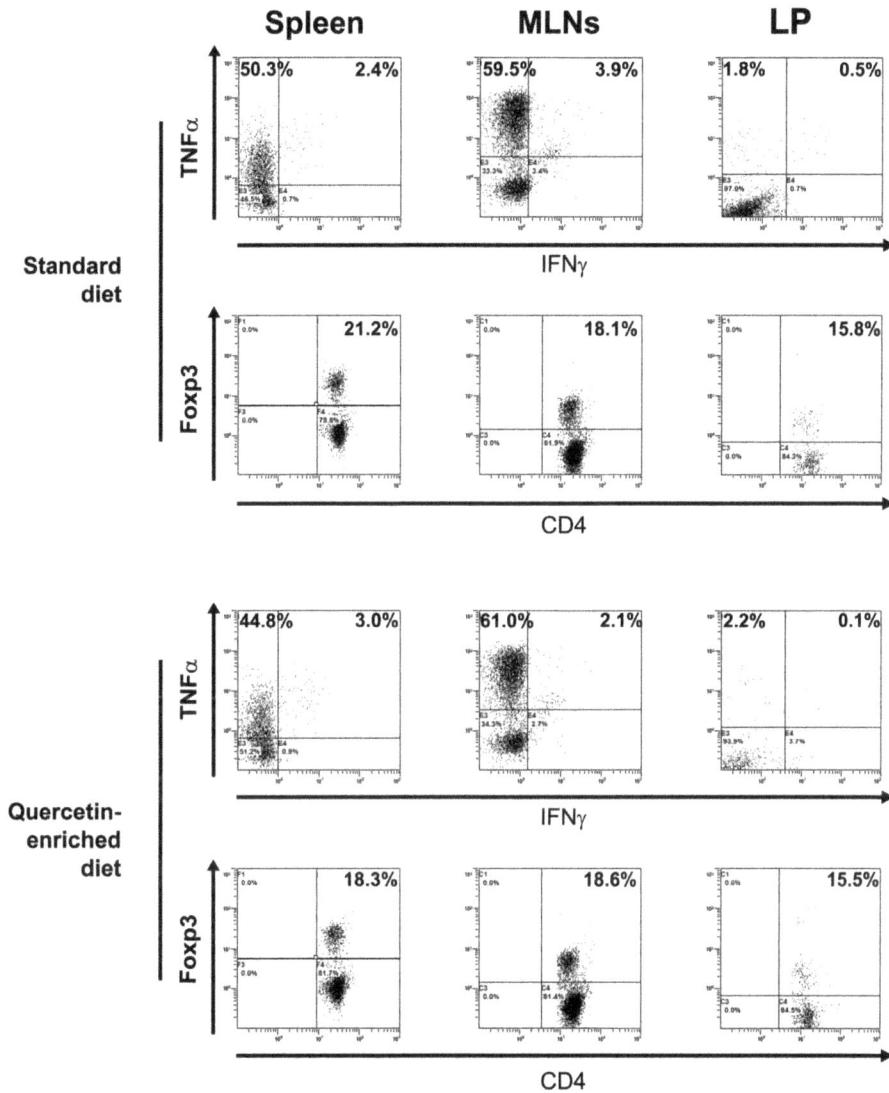

Figure 4. T cell polarization in mesenteric lymph node (MLN), spleen, and lamina propria (LP). Representative intracellular staining of CD4+ T cells. No significant differences could be observed in the percentage of TNFα+, IFNγ+, or Foxp3+ cells in mice treated with standard or quercetin-enriched diet (*n* = 3).

3.4. Quercetin-Mediated Slpi Induction Is Detectable at the Tip of the Villi

In healthy individuals, the mucosal immune system is dynamically educated towards tolerance to luminal antigens, including commensal bacteria and food derived antigens [35]. Several factors contribute to DCs' "maturation" [23,24], including food-derived products [27]. Intestinal dendritic cells resident in the lamina propria of the villi are able to project periscope-like dendrites into the intestinal lumen and sample luminal content [28–30]. As the periscope distribution is mainly concentrated at the tip of the villi, we aimed to detect a correlation between Slpi expression and DCs localization along the villus axis. (Figure 5A). Slpi mRNA expression was increased in samples taken from the tip of the villi in mice fed a quercetin-rich diet (Figure 5B). We acknowledge that this is an indirect indication that sampling DCs encounter quercetin directly in the intestinal lumen and respond to it by up-regulating Slpi. However, it is conceivable that nutritionally-derived factors contribute directly to the intestinal immune cell response by inducing the expression of proteins capable of attenuating the inflammatory response. Indeed, Slpi is also induced by the thymic stromal lymphopoetin (Tslp) [36]—a key factor for intestinal dendritic cell "maturation" [37]. Future studies will further elucidate how nutritional compounds can be used to modulate immune tolerance and mucosal homeostasis through the Slpi–DCs axis, paving the way for the translational use of Slpi-inducing diets for chronic inflammatory syndromes.

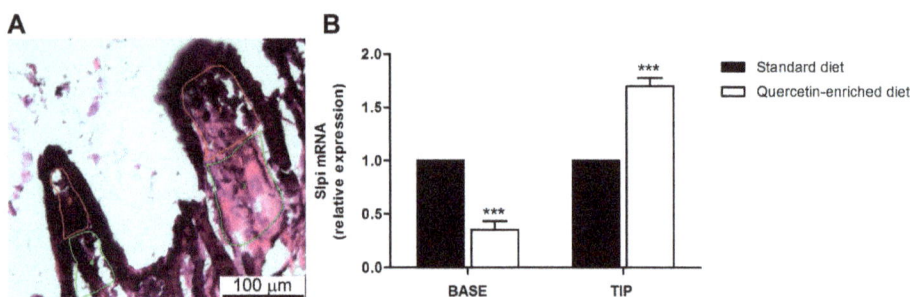

Figure 5. Quercetin induces Slpi expression in the tip of the intestinal villi. Animals were administered quercetin-enriched or standard diet for 4 weeks. (**A**) Representative image demonstrating captured lamina propria cells from the tip (red) or the base (green) of intestinal villi; (**B**) Lamina propria cells from standard (black bars) and quercetin (white bars) diet were laser-captured and Slpi expression measured by qPCR. Bars represent mean expression $\pm$ SEM (n = 3) for each treatment. *** $p < 0.001$.

4. Conclusions

Nutritional regimes enriched in quercetin may create a pro-tolerogenic milieu in the gastrointestinal tract, which could help maintain homeostasis and prevent chronic inflammatory disorders. This work suggests that antigen presenting cells (DCs) residing at the tip of the intestinal villi project dendrites into a quercetin-rich intestinal lumen and subsequently upregulate Slpi expression. Slpi$^+$ DCs are able to suppress the inflammatory cascade even in the presence of LPS, while Slpi-KO DCs fail to respond to quercetin and release high amounts of inflammatory cytokines. Our data indicate that nutritionally-derived quercetin is able to imprint Slpi expression in DCs resident in the tip of the intestinal villi, where the frequency of DC protrusions into the intestinal lumen is far greater. It is important to underline that results obtained within this study address the response of the lamina propria-resident dendritic cells to luminal quercetin, rather than the systemic effects. Additional studies are required to prove nutritional regimes enriched in quercetin can be considered important to sustain intestinal homeostasis and immunological tolerance in an Slpi-dependent manner.

Acknowledgments: This work was supported by the Italian Ministry of Health, "GR-2011-02347991", TomGEM European Union's Horizon 2020 research and innovation programme under grant agreement No. 679796 and by Regione Puglia "NATURE–XUANRO4". We are grateful to all members of LAB-81 and to the I.C. Bregante Volta of Monopoli (BA) for their constructive help and support.

Author Contributions: M.C., M.N. and A.S. conceived and designed the experiments; S.D.S., V.G., L.S. and E.S. performed the experiments; A.S., A.P., S.D.S., G.S. and P.C. analyzed the data; M.L. and V.S. contributed reagents/materials/analysis tools; M.C. and S.D.S. wrote the paper.

Conflicts of Interest: The authors declare no financial or commercial conflict of interest.

References

1. David, A.V.A.; Arulmoli, R.; Parasuraman, S. Overviews of Biological Importance of Quercetin: A Bioactive Flavonoid. *Pharmacogn. Rev.* **2016**, *10*, 84–89. [CrossRef]
2. Kelsey, N.A.; Wilkins, H.M.; Linseman, D.A. Nutraceutical antioxidants as novel neuroprotective agents. *Molecules* **2010**, *15*, 7792–7814. [CrossRef] [PubMed]
3. Jovanovic, S.V.; Simic, M.G. Antioxidants in nutrition. *Ann. N. Y. Acad. Sci.* **2000**, *899*, 326–334. [CrossRef] [PubMed]
4. Niedzwiecki, A.; Roomi, M.W.; Kalinovsky, T.; Rath, M. Anticancer Efficacy of Polyphenols and Their Combinations. *Nutrients* **2016**, *8*, 552. [CrossRef] [PubMed]
5. Harris, Z.; Donovan, M.G.; Branco, G.M.; Limesand, K.H.; Burd, R. Quercetin as an Emerging Anti-Melanoma Agent: A Four-Focus Area Therapeutic Development Strategy. *Front. Nutr.* **2016**, *3*, 48. [CrossRef] [PubMed]
6. Refolo, M.G.; D'Alessandro, R.; Malerba, N.; Laezza, C.; Bifulco, M.; Messa, C.; Caruso, M.G.; Notarnicola, M.; Tutino, V. Anti Proliferative and Pro Apoptotic Effects of Flavonoid Quercetin Are Mediated by CB1 Receptor in Human Colon Cancer Cell Lines. *J. Cell. Physiol.* **2015**, *230*, 2973–2980. [CrossRef] [PubMed]
7. Xue, F.; Nie, X.; Shi, J.; Liu, Q.; Wang, Z.; Li, X.; Zhou, J.; Su, J.; Xue, M.; Chen, W.D.; et al. Quercetin Inhibits LPS-Induced Inflammation and ox-LDL-Induced Lipid Deposition. *Front. Pharmacol.* **2017**, *8*, 40. [CrossRef] [PubMed]
8. Li, Y.; Yao, J.; Han, C.; Yang, J.; Chaudhry, M.T.; Wang, S.; Liu, H.; Yin, Y. Quercetin, Inflammation and Immunity. *Nutrients* **2016**, *8*, 167. [CrossRef] [PubMed]
9. Boots, A.W.; Wilms, L.C.; Swennen, E.L.; Kleinjans, J.C.; Bast, A.; Haenen, G.R. In vitro and ex vivo anti-inflammatory activity of quercetin in healthy volunteers. *Nutrition* **2008**, *24*, 703–710. [CrossRef] [PubMed]
10. Cavalcanti, E.; Vadrucci, E.; Delvecchio, F.R.; Addabbo, F.; Bettini, S.; Liou, R.; Monsurrò, V.; Huang, A.Y.; Pizarro, T.T.; Santino, A.; et al. Administration of reconstituted polyphenol oil bodies efficiently suppresses dendritic cell inflammatory pathways and acute intestinal inflammation. *PLoS ONE* **2014**, *9*, e88898. [CrossRef] [PubMed]
11. Delvecchio, F.R.; Vadrucci, E.; Cavalcanti, E.; De Santis, S.; Kunde, D.; Vacca, M.; Myers, J.; Allen, F.; Bianco, G.; Huang, A.Y.; et al. Polyphenol administration impairs T-cell proliferation by imprinting a distinct dendritic cell maturational profile. *Eur. J. Immunol.* **2015**, *45*, 2638–2649. [CrossRef] [PubMed]
12. Galleggiante, V.; De Santis, S.; Cavalcanti, E.; Scarano, A.; De Benedictis, M.; Serino, G.; Caruso, M.L.; Mastronardi, M.; Pinto, A.; Campiglia, P.; et al. Dendritic Cells Modulate Iron Homeostasis and Inflammatory Abilities Following Quercetin Exposure. *Curr. Pharm. Des.* **2017**, *23*, 2139–2146. [CrossRef]
13. Banchereau, J.; Steinman, R.M. Dendritic cells and the control of immunity. *Nature* **1998**, *392*, 245–252. [CrossRef] [PubMed]
14. Stoll, S.; Delon, J.; Brotz, T.M.; Germain, R.N. Dynamic imaging of T cell-dendritic cell interactions in lymph nodes. *Science* **2002**, *296*, 1873–1876. [CrossRef] [PubMed]
15. Lutz, M.B.; Schuler, G. Immature, semi-mature and fully mature dendritic cells: Which signals induce tolerance or immunity? *Trends Immunol.* **2002**, *23*, 445–449. [CrossRef]
16. Mahnke, K.; Schmitt, E.; Bonifaz, L.; Enk, A.H.; Jonuleit, H. Immature, but not inactive: The tolerogenic function of immature dendritic cells. *Immunol. Cell Biol.* **2002**, *80*, 477–483. [CrossRef] [PubMed]
17. Sim, W.J.; Malinarich, F.; Fairhurst, A.M.; Connolly, J.E. Generation of Immature, Mature and Tolerogenic Dendritic Cells with Differing Metabolic Phenotypes. *J. Vis. Exp.* **2016**. [CrossRef] [PubMed]
18. Raker, V.K.; Domogalla, M.P.; Steinbrink, K. Tolerogenic Dendritic Cells for Regulatory T Cell Induction in Man. *Front. Immunol.* **2015**, *6*, 569. [CrossRef] [PubMed]
19. Longoni, D.; Piemonti, L.; Bernasconi, S.; Mantovani, A.; Allavena, P. Interleukin-10 increases mannose receptor expression and endocytic activity in monocyte-derived dendritic cells. *Int. J. Clin. Lab. Res.* **1998**, *28*, 162–169. [CrossRef] [PubMed]
20. Schmidt, S.V.; Nino-Castro, A.C.; Schultze, J.L. Regulatory dendritic cells: There is more than just immune activation. *Front. Immunol.* **2012**, *3*, 274. [CrossRef] [PubMed]
21. Naik, S.; Bouladoux, N.; Linehan, J.L.; Han, S.J.; Harrison, O.J.; Wilhelm, C.; Conlan, S.; Himmelfarb, S.; Byrd, A.L.; Deming, C.; et al. Commensal-dendritic-cell interaction specifies a unique protective skin immune signature. *Nature* **2015**, *520*, 104–108. [CrossRef] [PubMed]

22. Eri, R.; Chieppa, M. Messages from the Inside. The Dynamic Environment that Favors Intestinal Homeostasis. *Front. Immunol.* **2013**, *4*, 323. [CrossRef] [PubMed]

23. Rescigno, M.; Chieppa, M. Gut-level decisions in peace and war. *Nat. Med.* **2005**, *11*, 254–255. [CrossRef] [PubMed]

24. Rescigno, M.; Lopatin, U.; Chieppa, M. Interactions among dendritic cells, macrophages, and epithelial cells in the gut: Implications for immune tolerance. *Curr. Opin. Immunol.* **2008**, *20*, 669–675. [CrossRef] [PubMed]

25. Thaiss, C.A.; Zmora, N.; Levy, M.; Elinav, E. The microbiome and innate immunity. *Nature* **2016**, *535*, 65–74. [CrossRef] [PubMed]

26. Belkaid, Y.; Hand, T.W. Role of the microbiota in immunity and inflammation. *Cell* **2014**, *157*, 121–141. [CrossRef] [PubMed]

27. De Santis, S.; Cavalcanti, E.; Mastronardi, M.; Jirillo, E.; Chieppa, M. Nutritional Keys for Intestinal Barrier Modulation. *Front. Immunol.* **2015**, *6*, 612. [CrossRef] [PubMed]

28. Rescigno, M.; Urbano, M.; Valzasina, B.; Francolini, M.; Rotta, G.; Bonasio, R.; Granucci, F.; Kraehenbuhl, J.P.; Ricciardi-Castagnoli, P. Dendritic cells express tight junction proteins and penetrate gut epithelial monolayers to sample bacteria. *Nat. Immunol.* **2001**, *2*, 361–367. [CrossRef] [PubMed]

29. Niess, J.H.; Brand, S.; Gu, X.; Landsman, L.; Jung, S.; McCormick, B.A.; Vyas, J.M.; Boes, M.; Ploegh, H.L.; Fox, J.G.; et al. CX3CR1-mediated dendritic cell access to the intestinal lumen and bacterial clearance. *Science* **2005**, *307*, 254–258. [CrossRef] [PubMed]

30. Chieppa, M.; Rescigno, M.; Huang, A.Y.; Germain, R.N. Dynamic imaging of dendritic cell extension into the small bowel lumen in response to epithelial cell TLR engagement. *J. Exp. Med.* **2006**, *203*, 2841–2852. [CrossRef] [PubMed]

31. Mazzini, E.; Massimiliano, L.; Penna, G.; Rescigno, M. Oral tolerance can be established via gap junction transfer of fed antigens from CX3CR1$^+$ macrophages to CD103$^+$ dendritic cells. *Immunity* **2014**, *40*, 248–261. [CrossRef] [PubMed]

32. De Santis, S.; Kunde, D.; Serino, G.; Galleggiante, V.; Caruso, M.L.; Mastronardi, M.; Cavalcanti, E.; Ranson, N.; Pinto, A.; Campiglia, P.; et al. Secretory leukoprotease inhibitor is required for efficient quercetin-mediated suppression of TNFα secretion. *Oncotarget* **2016**, *7*, 75800–75809. [CrossRef] [PubMed]

33. Lida, N.; Grotendorst, G.R. Cloning and sequencing of a new gro transcript from activated human monocytes: Expression in leukocytes and wound tissue. *Mol. Cell. Biol.* **1990**, *10*, 5596–5599.

34. Riches, A.C.; Sharp, J.G.; Thomas, D.B.; Smith, S.V. Blood volume determination in the mouse. *J. Physiol.* **1973**, *228*, 279–284. [CrossRef] [PubMed]

35. Santino, A.; Scarano, A.; Santis, S.; Benedictis, M.; Giovinazzo, G.; Chieppa, M. Gut microbiota modulation and anti-inflammatory properties of dietary polyphenols in IBD: New and consolidated perspectives. *Curr. Pharm. Des.* **2017**. [CrossRef] [PubMed]

36. Reardon, C.; Lechmann, M.; Brüstle, A.; Gareau, M.G.; Shuman, N.; Philpott, D.; Ziegler, S.F.; Mak, T.W. Thymic stromal lymphopoetin-induced expression of the endogenous inhibitory enzyme SLPI mediates recovery from colonic inflammation. *Immunity* **2011**, *35*, 223–235. [CrossRef] [PubMed]

37. Rimoldi, M.; Chieppa, M.; Salucci, V.; Avogadri, F.; Sonzogni, A.; Sampietro, G.M.; Nespoli, A.; Viale, G.; Allavena, P.; Rescigno, M. Intestinal immune homeostasis is regulated by the crosstalk between epithelial cells and dendritic cells. *Nat. Immunol.* **2005**, *6*, 507–514. [CrossRef] [PubMed]

![nutrients logo] *nutrients*

MDPI

Article

Toxoplasma Gondii Moderates the Association between Multiple Folate-Cycle Factors and Cognitive Function in U.S. Adults

Andrew N. Berrett [1,*], **Shawn D. Gale** [1,2], **Lance D. Erickson** [3], **Bruce L. Brown** [1] and **Dawson W. Hedges** [1,2]

[1] Department of Psychology, Brigham Young University, Provo, UT 84602, USA; shawn_gale@byu.edu (S.D.G.); bruce_brown@byu.edu (B.L.B.); dawson_hedges@byu.edu (D.W.H.)
[2] The Neuroscience Center, Brigham Young University, Provo, UT 84602, USA
[3] Department of Sociology, Brigham Young University, Provo, UT 84602, USA; lance_erickson@byu.edu
[*] Correspondence: drew_berrett@byu.edu

Received: 27 March 2017; Accepted: 29 May 2017; Published: 2 June 2017

Abstract: *Toxoplasma gondii* (*T. gondii*) is a microscopic, apicomplexan parasite that can infect muscle or neural tissue, including the brain, in humans. While *T. gondii* infection has been associated with changes in mood, behavior, and cognition, the mechanism remains unclear. Recent evidence suggests that *T. gondii* may harvest folate from host neural cells. Reduced folate availability is associated with an increased risk of neurodevelopmental disorders, neurodegenerative diseases, and cognitive decline. We hypothesized that impairment in cognitive functioning in subjects seropositive for *T. gondii* might be associated with a reduction of folate availability in neural cells. We analyzed data from the third National Health and Nutrition Examination Survey to determine the associations between *T. gondii* infection, multiple folate-cycle factors, and three tests of cognitive functioning in U.S. adults aged 20 to 59 years. In these analyses, *T. gondii* moderated the associations of folate, vitamin B-12, and homocysteine with performance on the Serial Digit Learning task, a measure of learning and memory, as well as the association of folate with reaction time. The results of this study suggest that *T. gondii* might affect brain levels of folate and/or vitamin B-12 enough to affect cognitive functioning.

Keywords: *Toxoplasma gondii*; Folate; Cognition; Memory; Vitamin B-12; Homocysteine

1. Introduction

Toxoplasma gondii (*T. gondii*) is an apicomplexan parasite that infects approximately 12% of the U.S. population as of 2010 [1]. Although the parasite's definitive host is any member of the cat family, humans can become infected with *T. gondii* as an intermediate host via ingestion of contaminated foods or by exposure to cat feces [2]. Upon successful invasion, *T. gondii* infects muscle and neural tissue and encapsulates itself in a cyst that protects it against the host immune system [3]. *T. gondii* biology has unique features that set it apart from other intracellular parasites. For example, using precursors from the infected cell, *T. gondii* can synthesize dopamine [4,5], although the physiological effects of this dopamine production remain unknown.

Beyond dopamine production, *T. gondii* may influence other aspects of host biology. In regards to folate synthesis, *T. gondii* may salvage or harvest folate directly from the infected host cell [6]. It is unknown, however, whether this might affect host health. Further, it is unclear whether *T. gondii* depends on the folate acquired from the host or if the system is only supplementary.

In humans, folate is essential for DNA synthesis and repair and is a key factor in cell division and growth [7]. The brain requires a constant supply of dietary folate for early embryonic neurodevelopment, adult neurogenesis, and the production of neurotransmitters [8,9]. Therefore, reduced availability of folate is associated with neurodevelopmental disorders [10] and may be associated with reduced cognitive function [11–14].

In individuals who do not obtain sufficient dietary folate or who cannot metabolize folate efficiently, neural cells might be generally and persistently deprived of this essential nutrient [15,16]. If *T. gondii* is indeed capable of harvesting folate from host neural cells, then infected cells may be further starved of this important nutrient and thereby magnify potential impairments in cognitive functioning. Based on these factors, we hypothesized that cognitive function may be associated with an interaction between *T. gondii* infection and concentrations of folate and other factors related to folate metabolism such as vitamin B-12 and homocysteine. To test this, we used the National Health and Nutrition Examination Survey III (NHANES III) dataset from the United States' Centers for Disease Control and Prevention (CDC).

2. Materials and Methods

2.1. Study Sample

The NHANES III is a nationally representative survey conducted by the CDC from 1988 to 1994. It includes demographic, examination, laboratory, and dietary information for approximately 34,000 participants recruited from multiple locations around the U.S. Using stratified sampling methods and statistical weighting, the NHANES III was designed to represent the non-institutionalized U.S. population. In the NHANES III, a subsample of subjects aged 20 to 59 years completed tests of cognitive functioning thereby limiting the potential study sample to 4924. Data for serum concentrations of vitamin B-12 and homocysteine were only collected in subjects who were assessed during the second phase (1991–1994) of the NHANES survey, further limiting the potential study sample to 2242 subjects. Finally, we did not include in our analyses subjects who were missing data for any of the remaining variables we used in this study (i.e., serum folate concentration, *T. gondii* infection status, and all controlling variables), resulting in a final sample size of 2037 (Figure 1).

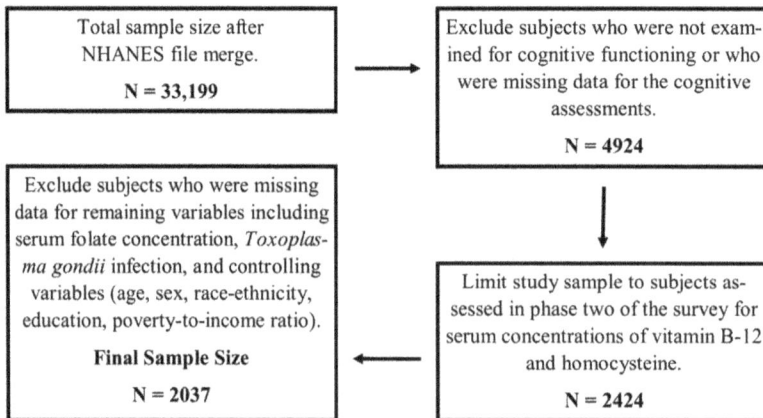

Figure 1. Flow diagram of sampling selection and final number of subjects included in all analyses.

2.2. Toxoplasma Gondii

Detection of *T. gondii*-specific antibodies was accomplished by CDC laboratory technicians via indirect enzyme immunoassay. A full description of the laboratory procedures used is available in the NHANES III laboratory manual [17]. Briefly, test samples were combined with *T. gondii* antigens, peroxidase-labelled human gamma chain immunoglobulin (IgG), peroxidase substrate, and chromogen ortho-phenylene diamine resulting in a color change proportional to the total *T. gondii* antibody concentration in the sample. The color change is read as an optical density via spectrophotometer and compared to a standard curve calibrated to World Health Organization Toxo 60 serum. Possible IgG titer values ranged from 0 IU/mL to 240 IU/mL (higher concentrations were collapsed to 240 IU/mL to reduce the risk of deductive disclosure). Subjects with titer values above 7 IU/mL were considered seropositive for *T. gondii* exposure, whereas subjects below this level were considered negative [17].

Anti-*T. gondii* IgG antibodies indicate past exposure to *T. gondii*. Further, IgG titer levels can vary across the lifespan potentially due to multiple exposures or natural anamnestic decreases in IgG levels [18–20]. However, the procedures and classification system used in the NHANES III remain the industry norm (excepting minor differences in the IgG titer cut-off point for seropositivity depending on the specific assay/manufacturer) in identifying individuals exposed to and likely infected with *T. gondii*.

2.3. B-Vitamin Biomarkers

Serum concentrations of folate and vitamin B-12 were determined by CDC laboratory technicians using a Quantaphase Folate radioassay kit from Bio-Rad Laboratories. The Quantaphase Folate radioassay kit contains a detailed description of the methods used to prepare the folate and B-12 samples [21]. Following sample preparation, concentration is calculated from a standard curve. Serum homocysteine concentration was determined via reverse-phase, high-performance liquid chromatography and fluorescence detection as described in the NHANES III laboratory manual [17]. Importantly, data for the NHANES III was collected before the introduction of mandatory folic acid food fortification [22]. Therefore, reported average serum concentrations of folate and/or vitamin B-12 for subjects surveyed in the NHANES III may be lower than more recent averages.

2.4. Cognitive Functioning

In the NHANES III, computerized versions of the Symbol Digit Substitution (SDS), Serial Digit Learning (SDL), and Reaction Time (RT) tests assessed general cognitive function. In the SDS, subjects were required to match a series of symbols with their respective digits using a key given at the top of the computer screen for four trials. Each trial consisted of nine digit-symbol pairings, and the total latency (in seconds) for each of the nine possible symbol-digit pairings was recorded for each trial. Finally, the average of the two lowest total latencies from the four trials was used as the final score. Therefore, a lower score, or faster pairing, on the SDS indicates better performance. The SDL required subjects to memorize and recall a series of eight random digits over eight trials. Scores on the SDL reflected the number of errors made in each trial. Therefore, a lower score, or fewer errors, indicates better performance on the SDL. Subjects who successfully repeated the eight-digit sequence for two consecutive trials were not required to complete the remaining trials. Finally, RT was measured by pressing a button as quickly as possible whenever a square appeared in the center of the computer screen. The average RT (in milliseconds (ms)) over 50 trials was used as the final score with faster times indicating better RT performance. Krieg et al. [23] provides a detailed description of the three cognitive tests used in the NHANES III.

2.5. Covariates

We included a number of covariates in all analyses to control for potential confounding. Continuous covariates included age in years, poverty-to-income ratio (PIR), and years of education. PIR is a measure of an individual's socioeconomic status computed by dividing the household income by the federal poverty level (at the time of the survey) resulting in a minimum value of zero and a maximum of five (higher values were collapsed to five to reduce the risk of deductive disclosure). Categorical variables included sex and race-ethnicity. For the race-ethnicity variable, the NHANES III survey asked all participants to self-report one of the following race-ethnicity categories: non-Hispanic white, non-Hispanic black, Mexican American, or "other".

2.6. Statistical Analysis

We used Stata 14.2 [24] for all statistical analyses. We computed summary statistics for all variables included in our analyses and computed means and standard errors for continuous variables and proportions and standard errors for categorical variables. Ordinary least squares regression was used to test for differences between *T. gondii* seropositive and seronegative subjects on each of the variables included in this study. Only the *p*-values for these comparisons are reported in Table 1. We used ordinary least squares regression to test for interaction effects between *T. gondii* seropositivity and folate, vitamin B-12, or homocysteine concentrations in the prediction of SDS, SDL, or RT scores for a total of nine separate models. We included age, PIR, education, sex, and race-ethnicity as covariates in each model.

For each analysis, we report main effects for *T. gondii*, a folate-cycle factor, and their interaction. The coefficient for *T. gondii* is the score on the respective cognitive test for subjects seropositive for *T. gondii* infection. The coefficient for the folate-cycle factors represents the relationship between each factor and scores on each of the three cognitive tests for subjects who were seronegative for *T. gondii* infection. The interaction of folate-cycle factors and *T. gondii* is the change in the slope of the folate-cycle factor for those subjects seropositive for *T. gondii* relative to those seronegative for *T. gondii*.

Table 1. Weighted summary statistics of U.S. adults 20 to 59 year olds, National Health and Nutrition Examination Survey III 1988–1994.

	T. Gondii Seropositive *N* = 417		*T. Gondii* Seronegative *N* = 1620		
	Mean or Frequency	SE	Mean or Frequency	SE	*p*
Sociodemographic Factors					
Age (years)	41.54	1.01	36.16	0.63	<0.001
Poverty-to-income Ratio	3.23	0.25	3.38	0.13	0.316
Education (years)	12.36	0.23	13.13	0.12	<0.001
Female	45.63%	2.65%	51.55%	1.08%	0.050
Race-ethnicity					
Non-Hispanic White	72.73%	3.92%	77.98%	1.27%	0.159
Non-Hispanic Black	10.77%	1.64%	10.55%	0.47%	0.896
Hispanic	5.87%	0.81%	5.50%	0.40%	0.691
Other	10.62%	3.25%	5.97%	1.14%	0.097
Biochemistry					
Folate (ng/mL)	7.08	0.51	6.89	0.24	0.807
Vitamin B-12 (pg/mL)	521.83	44.25	474.01	7.08	0.623
Homocysteine (umol/L)	9.36	0.27	9.56	0.25	0.282
Cognitive Testing					
Symbol Digit Substitution Score	2.85	0.05	2.55	0.03	<0.001
Serial Digit Learning Score	5.45	0.31	3.91	0.15	<0.001
Reaction Time (sec)	240.1	3.03	234.47	1.71	0.070

Abbreviations: SE = Standard Error.

3. Results

Of the 2037 subjects included in this study, approximately 20% were seropositive for *T. gondii*, 50% were female, and approximately 75% were non-Hispanic white. Table 1 presents these and other demographic and study-sample characteristics.

We found no significant interactions between *T. gondii* seropositivity and any of the folate-cycle factors in predicting performance on the SDS. In predicting performance on the SDL, *T. gondii* seropositivity interacted with folate ($\beta = -1.02$, (95% CI: -1.88, -0.16), $p = 0.021$; Table 2), vitamin B-12 ($\beta = -1.60$, (95% CI: -2.98, -0.23), $p = 0.023$; Table 3), and homocysteine concentrations ($\beta = 2.40$, (95% CI: 0.94, 3.85), $p = 0.002$; Table 4) (Figure 2). In subjects seronegative for *T. gondii*, SDL performance was relatively constant as folate, vitamin B-12 or homocysteine levels varied. However, in subjects seropositive for *T. gondii*, performance on the SDL worsened as folate and vitamin B-12 levels decreased and as homocysteine levels increased (Figure 2). An interaction between *T. gondii* seropositivity and folate concentration predicted RT (β = 14.88, (95% CI: 5.99, 23.76), *p* = 0.001; Table 2). In contrast to the SDL, subjects seropositive for *T. gondii* performed better on the RT assessment as folate concentration decreased.

Table 2. Two-way interaction effects of serum folate concentration and *Toxoplasma gondii* infection status predicting SDS, SDL, and RT cognitive assessment scores for U.S. 20 to 59 year olds.

	Symbol Digit Substitution			Serial Digit Learning			Reaction Time		
	β	95% CI	*p*	β	95% CI	*p*	β	95% CI	*p*
Folate	0.01	(−0.04, 0.05)	0.786	0.10	(−0.33, 0.53)	0.641	−1.19	(−5.55, 3.16)	0.585
T. gondii	0.15	(−0.08, 0.38)	0.185	2.48	(1.02, 3.94)	0.001	−20.60	(−34.38, −6.83)	0.004
T. gondii × Folate Interaction	−0.07	(−0.20, 0.07)	0.344	−1.13	(−1.97, −0.29)	0.009	13.07	(4.20, 21.93)	0.005
Race-Ethnicity									
Non-Hispanic White (ref)	-	-	-	-	-	-	-	-	-
Non-Hispanic Black	0.30	(0.21, 0.39)	<0.001	1.72	(1.15, 2.29)	<0.001	5.92	(0.14, 11.70)	0.045
Mexican American	0.27	(0.18, 0.37)	<0.001	2.52	(1.95, 3.09)	<0.001	1.23	(−6.86, 9.32)	0.761
Other	0.29	(0.10, 0.47)	0.003	1.88	(0.69, 3.08)	0.003	−4.15	(−14.36, 6.06)	0.418
Female	−0.17	(−0.24, −0.10)	<0.001	0.08	(−0.23, 0.38)	0.611	8.63	(1.60, 15.66)	0.017
Age (years)	0.03	(0.03, 0.03)	<0.001	0.09	(0.07, 0.11)	<0.001	0.36	(0.14, 0.58)	0.002
PIR	−0.03	(−0.04, −0.02)	<0.001	−0.25	(−0.35, −0.14)	<0.001	−3.13	(−4.33, −1.92)	<0.001
Education (years)	−0.10	(−0.11, −0.08)	<0.001	−0.46	(−0.57, −0.35)	<0.001	−2.93	(−4.08, −1.79)	<0.001
Constant	2.91	(2.70, 3.12)	<0.001	6.77	(4.96, 8.57)	<0.001	267.68	(250.01, 285.34)	<0.001

Notes: Age, sex, education, race-ethnicity, and poverty-to-income ratio included as controls in all models. Abbreviations: SDS, the Symbol Digit Substitution; SDL, Serial Digit Learning; RT, Reaction Time; CI = Confidence Interval; ref = Reference category; PIR, poverty-to-income ratio. *N* = 2037.

Table 3. Two-way interaction effects of serum vitamin B-12 concentration and *Toxoplasma gondii* infection status predicting SDS, SDL, and RT cognitive assessment scores for U.S. 20 to 59 year olds.

	Symbol Digit Substitution			Serial Digit Learning			Reaction Time		
	β	95% CI	*p*	β	95% CI	*p*	β	95% CI	*p*
Vitamin B-12	0.05	(−0.00, 0.11)	0.057	0.23	(−0.26, 0.72)	0.346	−3.04	(−8.77, 2.68)	0.291
T. gondii	1.06	(−0.12, 2.25)	0.076	10.24	(2.14, 18.33)	0.014	5.26	(−123.23, 133.74)	0.935
T. gondii Vitamin B-12 Interaction	−0.17	(−0.36, 0.03)	0.087	−1.59	(−2.92, −0.26)	0.020	−0.61	(−21.40, 20.19)	0.954
Race-Ethnicity									
Non-Hispanic White (ref)	-	-	-	-	-	-	-	-	-
Non-Hispanic Black	0.30	(0.21, 0.39)	<0.001	1.78	(1.19, 2.37)	<0.001	6.09	(0.18, 12.01)	0.044
Mexican American	0.27	(0.18, 0.37)	<0.001	2.53	(1.96, 3.10)	<0.001	1.81	(−6.22, 9.84)	0.653
Other	0.29	(0.10, 0.47)	0.003	1.87	(0.67, 3.08)	0.003	−4.59	(−14.98, 5.81)	0.380
Female	−0.17	(−0.24, −0.10)	<0.001	0.08	(−0.21, 0.36)	0.583	8.52	(1.64, 15.40)	0.016
Age (years)	0.03	(0.03, 0.03)	<0.001	0.09	(0.07, 0.11)	<0.001	0.40	(0.17, 0.63)	0.001
PIR	−0.03	(−004, −0.02)	<0.001	−0.24	(−0.34, −0.14)	<0.001	−3.14	(−4.32, −1.97)	<0.001
Education (years)	−0.10	(−0.11, −0.08)	<0.001	−0.47	(−0.58, −0.35)	<0.001	−2.83	(−4.03, −1.64)	<0.001
Constant	2.61	(2.22, 2.99)	<0.001	5.68	(2.42, 8.93)	0.001	281.51	(239.25, 323.77)	<0.001

Notes: Age, sex, education, race-ethnicity, and poverty-to-income ratio included as controls in all models. Abbreviations: CI = Confidence Interval, ref = Reference category. *N* = 2037.

Table 4. Two-way interaction effects of serum homocysteine concentration and *Toxoplasma gondii* infection status predicting SDS, SDL, and RT cognitive assessment scores for U.S. 20 to 59 year olds.

	Symbol Digit Substitution			Serial Digit Learning			Reaction Time		
	β	95% CI	*p*	β	95% CI	*p*	β	95% CI	*p*
Homocysteine	−0.05	(−0.13, 0.02)	0.166	−0.59	(−1.22, 0.03)	0.060	−2.72	(9.05, 3.62)	0.393
T. gondii	−0.35	(−1.17, 0.48)	0.403	−4.87	(−7.88, −1.85)	0.002	−10.87	(−45.43, 23.70)	0.531
T. gondii × Homocysteine Interaction	0.18	(−0.20, 0.56)	0.351	2.49	(1.09, 3.89)	0.001	5.67	(−11.10, 22.43)	0.501
Race-Ethnicity									
Non-Hispanic White (ref)	-	-	-	-	-	-	-	-	-
Non-Hispanic Black	0.31	(0.22, 0.40)	<0.001	1.79	(1.19, 2.38)	<0.001	5.50	(−0.33, 11.34)	0.064
Mexican American	0.27	(0.17, 0.37)	<0.001	2.52	(1.96, 3.08)	<0.001	1.41	(−6.81, 9.62)	0.732
Other	0.29	(0.11, 0.48)	0.002	1.98	(0.81, 3.15)	0.001	−3.92	(−13.95, 6.11)	0.436
Female	−0.18	(−0.25, −0.10)	<0.001	0.02	(−0.31, 0.34)	0.926	8.13	(1.12, 15.13)	0.024
Age (years)	0.03	(0.03, 0.03)	<0.001	0.09	(0.07, 0.11)	<0.001	0.42	(0.18, 0.66)	0.001
PIR	-0.03	(−0.05, −0.02)	<0.001	−0.25	(−0.35, −0.15)	<0.001	−3.21	(−4.45, −1.98)	<0.001
Education (years)	-0.10	(−0.11, −0.08)	<0.001	−0.47	(−0.59, −0.36)	<0.001	−2.83	(−4.02, −1.64)	<0.001
Constant	3.04	(2.78, 3.30)	<0.001	8.36	(6.08, 10.65)	<0.001	268.43	(243.08, 293.78)	<0.001

Notes: Age, sex, education, race-ethnicity, and poverty-to-income ratio included as controls in all models. Abbreviations: CI = Confidence Interval, ref = Reference category. *N* = 2037.

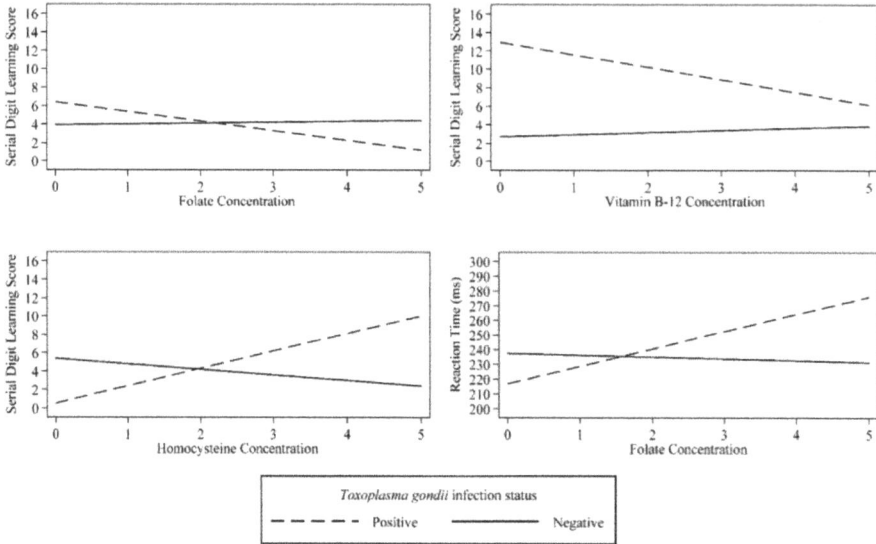

Figure 2. Significant interaction between folate-cycle factors and *Toxoplasma gondii* predicting cognitive performance in U.S. adults. **Notes:** Higher values on each cognitive assessment indicate worse performance. Values for folate, vitamin B-12, and homocysteine were logarithmically transformed.

4. Discussion and Conclusions

Using a large, nationally representative sample of U.S. adults, this study provides evidence of an interaction between *T. gondii* seropositivity and concentrations of several folate-cycle factors in the prediction of general cognitive functioning. In this sample of over 2000 participants, approximately 20% were seropositive for IgG antibodies specific for *T. gondii*. While this prevalence of seropositivity is higher than more recent estimates [1], it is within the expected range for the years in which the NHANES III was conducted (1988–1994) [25,26].

In this study, interactions between *T. gondii* seropositivity and folate, vitamin B-12, and homocysteine concentrations particularly affected performance on the SDL, an assessment of ability in learning and recall [23] that requires attention and intact short-term memory to perform well on the task. Concentrations of folate-cycle factors (including folate, vitamin B-12, and homocysteine) have been found to be associated with memory loss and other cognitive impairments in neurological disorders such as dementia [12,13,27], although these studies did not take into account *T. gondii* seropositivity. Further, deficiency in dietary folate has been linked to reduced genesis of neuroprogenitor cells of the adult hippocampus [8], a brain region that influences both short-term memory and attention [28,29]. Low concentrations of folate and/or vitamin B-12 usually result in an elevation of homocysteine, which can damage neural cells and impair cognitive functioning. Indeed, elevated homocysteine also might increase risk for neurodegenerative disorders in the hippocampus and prefrontal cortex [12,30]. Therefore, reductions in cellular folate concentrations, such as in the case of folate harvesting by *T. gondii*, could limit the bioavailability of this important nutrient and cause elevations in homocysteine in multiple brain regions including the prefrontal cortex or hippocampus [31,32] and thus disrupt learning, attention, or other related cognitive functions. Further, this effect may be magnified in individuals who do not regularly obtain sufficient amounts of dietary folate or who cannot metabolize folate efficiently.

Another potential mechanism by which reduced folate availability might impact cognitive functioning lies in the indirect role of folate metabolism in the generation of serotonin and dopamine.

Purines synthesized during folate metabolism contribute to the biosynthesis of guanosine triphosphate (GTP), a precursor of tetrahydrobiopterin (BH_4). Tyrosine hydroxylase and tryptophan hydroxylase each utilize BH_4 as a co-factor in the production of dopamine and serotonin, respectively. Therefore, reduced availability of dietary folate may indirectly lead to an overall decrease of dopamine and serotonin production [9,33]. In the case of *T. gondii* seropositivity, the combined effects of low concentrations of folate and simultaneous folate harvesting from *T. gondii* might particularly affect infected neural cells. Since dopamine and serotonin both influence various cognitive functions such as memory, attention, and executive function [34–36], this link could potentially offer an alternate explanation of the effects observed in this study.

Beyond performance on the SDL, there was a significant interaction between *T. gondii* and folate concentration in the prediction of RT. However, the direction of the observed relationship suggested that *T. gondii* seropositivity might actually be favorable in regards to RT in cases of low folate concentration and detrimental in cases of high folate concentration. Though not statistically significant, we found a similar relationship in a previous study [37] exploring interactions between the bacterium *Helicobacter pylori* and folate concentration in the prediction of RT. Similar to *T. gondii*, *Helicobacter pylori* might also reduce folate availability [38], though by a different mechanism. Despite these differing mechanisms of folate reduction, in both cases, RT improved, suggesting that folate deficiency associated with *T. gondii* might enable some advantage in RT. We are unaware of a mechanism that might explain how reductions in folate availability by infectious diseases such as *T. gondii* or *Helicobacter pylori* could be favorable towards RT performance. The data available in the NHANES III do not allow for an in-depth exploration of this association. Future research into this finding would likely require animal models or neuroimaging combined with biochemical testing to determine how regions of the brain active during RT tasks might be affected by reductions in folate availability.

Some limitations require consideration when interpreting the results of this study. The NHANES data sets are cross-sectional in design and thus preclude the ability to make casual inferences. Further, as the degree to which *T. gondii* harvests folate from infected host cells is not clear, it is currently unknown whether *T. gondii* harvests enough to affect cognitive functioning. With these data, it is also impossible to determine the time of initial infection by *T. gondii* or the frequency with which individuals have encountered the parasite in their lifetime. It is possible that the effects of *T. gondii* on folate availability and on cognition could vary based on the length of infection with the parasite or based on the number of exposures to the parasite. Finally, folate and vitamin B-12 availability can vary for a multitude of reasons including genetic mutations in key folate-cycle enzymes [7,39,40]. Although the NHANES III does include genetic data, access to the data is restricted and requires funding to obtain it. Therefore, additional research is needed to identify other factors that might modify the effects observed in this study.

This study is strengthened by using multiple controls that limit the number of potential confounds that might explain the observed effects. Specifically, by controlling for demographic and other health factors, we were more confident that the interactions between *T. gondii* and concentrations of folate, vitamin B-12, or homocysteine were associated with the variation in SDL or RT scores. Finally, use of the NHANES III data sets resulted in a large sample size that increases the generalizability of the findings and study power.

This study consisting of 2037 U.S. adults presents evidence of an interaction effect between *T. gondii* seropositivity and concentrations of multiple folate-cycle factors on cognitive functioning in adult humans. Additional research is recommended to explore the specific mechanisms involved in this association and to determine any additional potential consequences to folate and/or vitamin B-12 availability following *T. gondii* infection.

Author Contributions: Dawson W. Hedges and Andrew N. Berrett conceived and designed the study. Andrew N. Berrett collected the data and performed the analyses. Lance D. Erickson provided statistical and methodological support. Andrew N. Berrett authored the manuscript while Shawn D. Gale, Lance D. Erickson, Bruce L. Brown, and Dawson W. Hedges each contributed to and edited the manuscript throughout its preparation. Andrew N. Berrett designed and built the figures and tables with Lance D. Erickson providing support.

Conflicts of Interest: The authors declare no conflict of interest.

References

1. Jones, J.L.; Kruszon-Moran, D.; Rivera, H.N.; Price, C.; Wilkins, P.P. *Toxoplasma gondii* seroprevalence in the United States 2009–2010 and comparison with the past two decades. *Am. J. Trop. Med. Hyg.* **2014**, *90*, 1135–1139. [CrossRef] [PubMed]
2. Elmore, S.A.; Jones, J.L.; Conrad, P.A.; Patton, S.; Lindsay, D.S.; Dubey, J.P. *Toxoplasma gondii*: Epidemiology, feline clinical aspects, and prevention. *Trends Parasitol.* **2010**, *26*, 190–196. [CrossRef] [PubMed]
3. Carruthers, V.B.; Suzuki, Y. Effects of *Toxoplasma gondii* infection on the brain. *Schizophr. Bull.* **2007**, *33*, 745–751. [CrossRef] [PubMed]
4. Martin, H.L.; Alsaady, I.; Howell, G.; Prandovszky, E.; Peers, C.; Robinson, P.; McConkey, G.A. Effect of parasitic infection on dopamine biosynthesis in dopaminergic cells. *Neuroscience* **2015**, *306*, 50–62. [CrossRef] [PubMed]
5. Prandovszky, E.; Gaskell, E.; Martin, H.; Dubey, J.P.; Webster, J.P.; McConkey, G.A. The neurotropic parasite *Toxoplasma gondii* increases dopamine metabolism. *PLoS ONE* **2011**, *6*, e23866. [CrossRef] [PubMed]
6. Massimine, K.M.; Doan, L.T.; Atreya, C.A.; Stedman, T.T.; Anderson, K.S.; Joiner, K.A.; Coppens, I. *Toxoplasma gondii* is capable of exogenous folate transport. A likely expansion of the BT1 family of transmembrane proteins. *Mol. Biochem. Parasitol.* **2005**, *144*, 44–54. [CrossRef] [PubMed]
7. *Folate in Health and Disease*; Marcel Dekker, Inc.: New York, NY, USA, 1995; ISBN: 0824792807.
8. Kruman, I.I.; Mouton, P.R.; Emokpae, R., Jr.; Cutler, R.G.; Mattson, M.P. Folate deficiency inhibits proliferation of adult hippocampal progenitors. *Neuropharmacol. Neurotoxicol.* **2005**, *16*, 1055–1059. [CrossRef]
9. Temudo, T.; Rios, M.; Prior, C.; Carrilho, I.; Santos, M.; Maciel, P.; Sequeiros, J.; Fonseca, M.; Monteiro, J.; Cabral, P.; et al. Evaluation of CSF neurotransmitters and folate in 25 patients with Rett disorder and effects of treatment. *Brain Dev.* **2009**, *31*, 46–51. [CrossRef] [PubMed]
10. Pu, D.; Shen, Y.; Wu, J. Association between *MTHFR* gene polymorphisms and the risk of autism spectrum disorders: A meta-analysis. *Autism Res.* **2013**, *6*, 384–392. [CrossRef] [PubMed]
11. Hultberg, B.; Isaksson, A.; Nilsson, K.; Gustafson, L. Markers for the functional availability of cobalamin/folate and their association with neuropsychiatric symptoms in the elderly. *Int. J. Geriatr. Psychiatry* **2001**, *16*, 873–878. [CrossRef] [PubMed]
12. Mattson, M.P.; Shea, T.B. Folate and homocysteine metabolism in neural plasticity and neurodegenerative disorders. *Trends Neurosci.* **2003**, *26*, 137–146. [CrossRef]
13. Michelakos, T.; Kousoulis, A.A.; Katsiardanis, K.; Dessypris, N.; Anastasiou, A.; Katsiardani, K.P.; Kanavidis, P.; Stefanadis, C.; Papadopoulos, F.C.; Petridou, E.T. Serum folate and B12 levels in association with cognitive impairment among seniors: Results from the Velestino study in Greece and meta-analysis. *J. Aging Health* **2013**, *25*, 589–616. [CrossRef] [PubMed]
14. McGarel, C.; Pentieva, K.; Strain, J.J.; McNulty, H. Emerging roles for folate and related B-vitamins in brain health across the lifecycle. *Proc. Nutr. Soc.* **2015**, *74*, 46–55. [CrossRef] [PubMed]
15. Rai, V. Methylenetetrahydrofolate reductase (MTHFR) C677T polymorphism and Alzheimer disease risk: A meta-analysis. *Mol. Neurobiol.* **2016**, *54*, 1173–1186. [CrossRef] [PubMed]
16. Hua, Y.; Zhao, H.; Kong, Y.; Ye, M. Association between the *MTHFR* gene and Alzheimer's disease: A meta-analysis. *Int. J. Neurosci.* **2011**, *121*, 462–471. [CrossRef] [PubMed]
17. Gunter, E.W.; Lewis, B.G.; Koncikowski, S.M. *Laboratory Procedures Used for the Third National Health and Nutrition Examination Survey (NHANES III), 1988–1994*; U.S. Department of Health and Human Services: Atlanta, GA, USA, 1996.
18. Wyman, C.P.; Gale, S.D.; Hedges-Muncy, A.; Erickson, L.D.; Wilson, E.; Hedges, D.W. Association between *Toxoplasma gondii* seropositivity and memory function in nondemented older adults. *Neurobiol. Aging* **2017**, *53*, 76–82. [CrossRef] [PubMed]
19. Konishi, E. Annual change in immunoglobulin G and M antibody levels to *Toxoplasma gondii* in human sera. *Microbiol. Immunol.* **1989**, *33*, 403–411. [CrossRef] [PubMed]
20. Havlíček, J.; Gašová, Z.; Smith, A.P.; Zvára, K.; Flegr, J. Decrease of psychomotor performance in subjects with latent 'asymptomatic' toxoplasmosis. *Parasitology* **2001**, *122*, 515–520. [CrossRef] [PubMed]
21. Laboratory Procedure Manual. Available online: https://wwwn.cdc.gov/nchs/data/nhanes/2003-2004/labmethods/l06_c_met_folates-b12.pdf (accessed on 2 June 2017).

22. Kessler, D.A.; Shalala, D.E. *Food Standards: Amendment of Standards of Identity for Enriched Grain Products to Require Addition of Folic Acid*; Department of Health and Human Services: Food and Drug Administration; Federal Register: Washington, DC, USA, 1996; Volume 61.
23. Edward, F.; Krieg, J.; Chrislip, D.W.; Letz, R.E.; Otto, D.A.; Crespo, C.J.; Brightwell, W.S.; Ehrenberg, R.L. Neurobehavioral test performance in the third national health and nutrition examination survey. *Neurotoxicol. Teratol.* **2001**, *23*, 569–589. [CrossRef]
24. StataCorp. *Stata Statistical Software: Release 14*, StataCorp LP: College Station, TX, USA, 2015.
25. Jones, J.L.; McAuley, J.B.; Kruszon-Moran, D.; Wilson, M.; McQuillan, G.; Navin, T.; McAuley, J.B. *Toxoplasma gondii* infection in the United States: Seroprevalence and risk factors. *Am. J. Epidemiol.* **2001**, *154*, 357–365. [CrossRef] [PubMed]
26. Pappas, G.; Roussos, N.; Falagas, M.E. Toxoplasmosis snapshots: Global status of *Toxoplasma gondii* seroprevalence and implications for pregnancy and congenital toxoplasmosis. *Int. J. Parasitol.* **2009**, *39*, 1385–1394. [CrossRef] [PubMed]
27. Wahlin, T.-B.R.; Wahlin, A.; Winblad, B.; Backman, L. The influence of serum vitamin B12 and folate status on cognitive functioning in very old age. *Biol. Psychol.* **2001**, *56*, 247–265. [CrossRef]
28. Goldfarb, E.V.; Chun, M.M.; Phelps, E.A. Memory-guided attention: Independent contributions of the hippocampus and striatum. *Neuron* **2016**, *89*, 317–324. [CrossRef] [PubMed]
29. Hannula, D.E.; Ranganath, C. Medial temporal lobe activity predicts successful relational memory binding. *J. Neurosci.* **2008**, *28*, 116–124. [CrossRef] [PubMed]
30. Kruman, I.I.; Kumaravel, T.S.; Lohani, A.; Pedersen, W.A.; Cutler, R.G.; Kruman, Y.; Haughey, N.; Lee, J.; Evans, M.; Mattson, M.P. Folic acid deficiency and homocysteine impair DNA repair in hippocampal neurons and sensitize them to amyloid toxicity in experimental models of Alzheimer's disease. *J. Neurosci.* **2002**, *22*, 1752–1762. [PubMed]
31. Madsen, S.K.; Rajagopalan, P.; Joshi, S.H.; Toga, A.W.; Thompson, P.M.; the Alzheimer's Disease Neuroimaging Initiative (ADNI). Higher homocysteine associated with thinner cortical gray matter in 803 participants from the Alzheimer's disease neuroimaging initiative. *Neurobiol. Aging* **2015**, *36* (Suppl. 1), S203–S210. [CrossRef] [PubMed]
32. Rajagopalan, P.; Hua, X.; Toga, A.W.; Jack, C.R., Jr.; Weiner, M.W.; Thompson, P.M. Homocysteine effects on brain volumes mapped in 732 elderly individuals. *Neuroreport* **2011**, *22*, 391–395. [CrossRef] [PubMed]
33. Ramaekers, V.T.; Thony, B.; Sequeira, J.M.; Ansseau, M.; Philippe, P.; Boemer, F.; Bours, V.; Quadros, E.V. Folinic acid treatment for schizophrenia associated with folate receptor autoantibodies. *Mol. Genet. Metab.* **2014**, *113*, 307–314. [CrossRef] [PubMed]
34. Nakajima, S.; Gerretsen, P.; Takeuchi, H.; Caravaggio, F.; Chow, T.; Le Foll, B.; Mulsant, B.; Pollock, B.; Graff-Guerrero, A. The potential role of dopamine D3 receptor neurotransmission in cognition. *Eur. Neuropsychopharmacol. J. Eur. Coll. Neuropsychopharmacol.* **2013**, *23*, 799–813. [CrossRef] [PubMed]
35. Paul, R.T.; McDonnell, A.P.; Kelly, C.B. Folic acid: Neurochemistry, metabolism and relationship to depression. *Hum. Psychopharmacol.* **2004**, *19*, 477–488. [CrossRef] [PubMed]
36. Terry, A.V., Jr.; Buccafusco, J.J.; Wilson, C. Cognitive dysfunction in neuropsychiatric disorders: Selected serotonin receptor subtypes as therapeutic targets. *Behav. Brain Res.* **2008**, *195*, 30–38. [CrossRef] [PubMed]
37. Berrett, A.N.; Gale, S.D.; Erickson, L.D.; Brown, B.L.; Hedges, D.W. Folate and inflammatory markers moderate the association between *Helicobacter pylori* exposure and cognitive function in us adults. *Helicobacter* **2016**, *21*, 471–480. [CrossRef] [PubMed]
38. Kountouras, J.; Gavalas, E.; Boziki, M.; Zavos, C. *Helicobacter pylori* may be involved in cognitive impairment and dementia development through induction of atrophic gastritis, vitamin B-12–folate deficiency, and hyperhomocysteinemia sequence. *Am. J. Clin. Nutr.* **2007**, *86*, 805–809. [PubMed]
39. Cheng, D.M.; Jiang, Y.G.; Huang, C.Y.; Kong, H.Y.; Pang, W.; Yang, H.P. Polymorphism of MTHFR C677T, serum vitamin levels and cognition in subjects with hyperhomocysteinemia in China. *Nutr. Neurosci.* **2010**, *13*, 175–182. [CrossRef] [PubMed]
40. Ho, V.; Massey, T.E.; King, W.D. Effects of methionine synthase and methylenetetrahydrofolate reductase gene polymorphisms on markers of one-carbon metabolism. *Genes Nutr.* **2013**, *8*, 571–580. [CrossRef] [PubMed]

nutrients

MDPI

Article

The Evolving Interplay among Abundant Adipokines in Patients with Hepatitis C during Viral Clearance

Ming-Ling Chang [1,2,*], Tsung-Hsing Chen [1], Chen-Ming Hsu [1], Cheng-Hui Lin [1], Cheng-Yu Lin [1], Chia-Jung Kuo [1], Shu-Wei Huang [1], Chun-Wei Chen [1], Hao-Tsai Cheng [1], Chau-Ting Yeh [1] and Cheng-Tang Chiu [1]

[1] Liver Research Center, Division of Hepatology, Department of Gastroenterology and Hepatology, Chang Gung Memorial Hospital, Taoyuan 33305, Taiwan; itochenyu@gmail.com (T.-H.C.); hsu3060e@cloud.cgmh.org.tw (C.-M.H.); linchehui@cloud.cgmh.org.tw (C.-H.L.); 8805035@gmail.com (C.-Y.L.); m7011@cloud.cgmh.org.tw (C.-J.K.); huangshuwei@hotmail.com (S.-W.H.); 8902088@cloud.cgmh.org.tw (C.-W.C.); hautai@cloud.cgmh.org.tw (H.-T.C.); chauting@cloud.cgmh.org.tw (C.-T.Y.); ctchiu0508@gmail.com (C.-T.C.)

[2] Department of Medicine, College of Medicine, Chang Gung University, Taoyuan 33305, Taiwan

*   Correspondence: mlchang8210@gmail.com; Tel.: +886-3-3281200-8107; Fax: +886-3-3272236

Received: 4 April 2017; Accepted: 31 May 2017; Published: 2 June 2017

Abstract: How hepatitis C virus (HCV) infection affects the interplay among abundant adipokines in the host remains unclear. A prospective study was conducted with 450 consecutive genotype 1 (G1) and G2 HCV patients who completed a course of anti-HCV therapy and underwent pre-therapy and 24-week post-therapy surveys to assess various profiles and levels of abundant adipokines, including leptin, adiponectin, and plasminogen activator inhibitor-1 (PAI-1). Before anti-HCV therapy, multivariate analyses showed gender to be associated with leptin and adiponectin levels, and BMI with leptin and PAI-1 levels. Among patients with a sustained virological response (SVR, n = 372), associations at 24 weeks post-therapy were as follows: gender and BMI with all adipokine levels; hepatic steatosis and aspartate aminotransferase to platelet ratio index with adiponectin levels; and HOMA-IR and HCV genotype with PAI-1 levels. Paired t-tests revealed increased post-therapeutic PAI-1 levels in G1 SVR patients and decreased adiponectin levels in all SVR patients compared to pre-therapeutic levels. HCV infection may obscure associations between abundant adipokines and metabolic/hepatic profiles. In SVR patients, a higher hierarchical status of PAI-1 versus adiponectin in affecting glucose metabolism was noted at 24 weeks post-therapy. Such genotype-non-specific adiponectin decreases and G1-specific PAI-1 increases warrant careful follow-up of HCV patients after SVR according to viral genotype.

Keywords: HCV; SVR; leptin; adiponectin; PAI-1; HOMA-IR

1. Introduction

Hepatitis C virus (HCV), a human pathogen responsible for acute and chronic liver disease, is classified into seven genotypes [1] and infects an estimated 150 million individuals worldwide [2]. In addition to liver cirrhosis and hepatocellular carcinoma, the virus causes metabolic alterations, including dyslipidemia, hepatic steatosis, obesity, insulin resistance (IR), and type 2 diabetes [2,3]. Much of the HCV life cycle is closely associated with lipid metabolism [2,3]. Moreover, both the direct effects of HCV on hepatocytes and the indirect mechanisms that involve extrahepatic organs influence insulin signaling [2,3].

Adipose tissue has emerged as an important endocrine organ due to its release of adipokines [4], which are biologically-active polypeptides that include the adipose tissue-specific adipokines leptin and adiponectin and non-adipose tissue-specific adipokines, such as plasminogen activator inhibitor-1

(PAI-1) [5]. With circulating concentrations that reach approximately 3–30 µg/mL, adiponectin is the most abundant adipokine [6]. Both adiponectin and leptin are abundant in subcutaneous fat [7], and high levels of PAI-1 are found in the extracellular matrix [8]. Thus, leptin, adiponectin and PAI-1 are abundant adipokines stable in the circulation and may regulate body lipid and glucose metabolism via the adipoinsular axis [9]. For example, leptin plays a crucial role in maintaining glucose metabolism [10], and adiponectin protects against obesity-related complications [11] by enhancing fatty acid oxidation and insulin sensitivity [12]. However, increased visceral adipose tissue reduces the biological activity of leptin and the abundance of circulating adiponectin [13]; the former is pro-inflammatory and positively associated with total body fat [14], whereas the latter is anti-inflammatory and negatively associated with hepatic steatosis and obesity [15]. Overall, circulating leptin levels are increased and adiponectin levels decreased in obese subjects [16]. Both leptin and adiponectin are reliable markers of many diseases, including lung [17] and cardiovascular diseases [18]. In addition, PAI-1 levels are increased in patients with metabolic syndrome [19], obesity, IR [20], non-alcoholic fatty liver disease [5] and inflammatory disease [21]. Moreover, PAI-1 is a central regulator of the fibrinolytic system [22] and can indicate a risk of developing cardiovascular disease [23]. In general, circulating leptin and adiponectin levels are positively associated with the female gender [16], whereas PAI-1 levels are positively associated with the male gender [24]. Together, levels of these adipokines indicate the degree of body fat accumulation, metabolic homeostasis, and sexual dimorphism, as well as the risk of developing the aforementioned diseases, particularly cardiometabolic diseases [10–24].

Since both HCV infection and adipokines are critically involved in metabolism, their potential relationship has attracted much research attention, with conflicting results [2]. In addition to the pleiotropic functions of adipokines described above, the lack of clarity regarding relationships is primarily due to variability among individuals, which is difficult to completely eliminate from case-controlled, retrospective, or prospective studies with a limited number of adjustments for confounders. Thus, our previous studies involved patients with chronic hepatitis C (CHC) who achieved a sustained virological response (SVR) served as their own least biased controls before therapy. Our results showed that after adjusting for crucial confounders, leptin and complement component 3 (C3) possibly maintain immune and metabolic homeostasis through associations with C4 and total cholesterol (TC), with unchanged levels after HCV clearance [25]. Within 24 weeks after anti-HCV therapy, SVR patients showed increasing PAI-1 levels with higher cardiovascular risk [24]. Additionally, we observed an evolving relationship between adiponectin levels and insulin sensitivity in CHC patients during viral clearance [26]. Although HCV infection is treatable using potent direct-acting antiviral agents [2], the interplay among the abundant adipokines produced in CHC patients achieving viral clearance has remained unclear. Furthermore, this process may be crucial for cardiometabolic homeostasis in these patients. Therefore, we sought to elucidate this process by conducting a prospective study to analyze HCV RNA levels as well as hepatic, metabolic, and abundant adipokine profiles in CHC patients before and after anti-HCV therapy.

2. Materials and Methods

2.1. Patients

The study group contained subjects 18 years or older with CHC, which was defined as the presence of documented HCV antibody positivity and detectable HCV RNA levels for at least 24 weeks. Subjects with human immunodeficiency virus or hepatitis B virus infection, hemochromatosis or malignancy and recipients of solid organ transplants were excluded.

2.2. Study Design

A total of 450 CHC (234 genotype 1 (G1) and 216 G2) patients were consecutively recruited at a tertiary referral center between July 2010 and August 2015. According to a response-guided therapeutic protocol, all patients received anti-HCV therapy with pegylated interferon alpha-2b (1.5 μg/kg/week) and ribavirin (800–1400 mg/day) [24–27]. HCV RNA levels were measured using a COBAS Amplicor (ver. 2.0, Roche Diagnostics, Tokyo, Japan), and the HCV genotype was determined using the InoLipa method (COBAS AmpliPrep/COBAS TaqMan HCV Test, Roche Diagnostics, Tokyo, Japan). Single-nucleotide polymorphisms (SNPs) in interferon-λ 3 (IFNL3; rs12979860) were assessed using genomic DNA, as previously described [24–29]. Patients were evaluated under fasting conditions two weeks prior to, and 24 weeks after, the end of therapy to determine the following: body weight; body mass index (BMI); HCV RNA, TC, triglyceride (TG), alanine aminotransferase (ALT), leptin, adiponectin, and PAI-1 levels; homeostasis model assessment-estimated insulin resistance (HOMA-IR)(fasting insulin (μU/mL)×fasting glucose (mmol/L)/22.5) score; and the aspartate aminotransferase (AST)-to-platelet ratio index (APRI) [30]. The leptin/adiponectin (L/A) ratio was also assessed. To monitor hepatic steatosis and cirrhosis, abdominal ultrasound analyses were performed for every patient before therapy and every six months thereafter. SVR was defined as an undetectable HCV RNA level at 24 weeks after therapy completion.

2.3. Adipokine Enzyme-Linked Immunosorbent Assays (ELISAs)

Serum leptin (male: 2.205–11.149 ng/mL; female 3.877–77.273 ng/mL), adiponectin (0.865–21.424 μg/mL), and PAI-1 (0.98–18.7 ng/mL) levels were assayed according to the manufacturer's protocols (R and D Systems, Minneapolis, MN, USA).

2.4. Biochemistry

Serum biochemistry, HCV RNA measurement and genotype evaluation were performed using routine automated techniques in the clinical pathology or liver research laboratories of the hospital.

2.5. Statistical Analysis

All statistical analyses were performed using either Statistical Package for the Social Sciences (SPSS ver. 21.0, SPSS Inc., Chicago, IL, USA) or MedCalc (MedCalc ver. 12.4, MedCalc Software Corp., Brunswick, ME, USA) software. Continuous variables are presented as means ± standard deviations (SDs), and categorical variables are presented as frequencies and percentages. Continuous variables were analyzed using Student's *t*-test and categorical variables using the chi-squared or Fisher exact test, as appropriate, to compare differences in each variable between different groups. Multivariate linear regression models were used to assess the relationship between various dependent and independent variables by adjusting for all independent variables with a *p* value < 0.1 in univariate analyses. Paired *t*-tests were used to compare variables within individuals prior to and 24 weeks after anti-HCV therapy. Levels were logarithmically transformed and used for statistical analyses, as indicated. Statistical significance was defined at the 5% level, based on two-tailed tests of the null hypothesis.

2.6. Informed Consent

Written informed consent was obtained from each patient. The study protocol conformed to the ethical guidelines of the 1975 Declaration of Helsinki and was approved by the hospital institutional review board.

3. Results

3.1. Baseline Characteristics

Compared with those who did not achieve SVR, lower HCV RNA level, HOMA-IR score, and G1 HCV infection and cirrhosis prevalence rates, as well as higher platelet counts and interferon-λ3 (IFNL3) CC genotype frequency were observed in SVR patients. In contrast, pre-therapy leptin, adiponectin and PAI-1 levels and L/A ratios were similar in patients with or without SVR. The pre-therapy demographics of the CHC patients are listed in Table 1.

Table 1. Baseline demographics of patients with CHC prior to anti-HCV therapy.

	Total (*n* = 450)	SVR (+) (*n* = 372)	SVR (−) (*n* = 78)	*p* values Obtained Using Student's *t*-Test or the Chi-Squared Test
Male, *n* (%) #	256 (56.8)	212 (56.9)	44 (56.4)	0.394
Age (years)	54.01 ± 11.51	53.68 ± 11.60	55.66 ± 10.92	0.160
BMI	24.99 ± 3.73	24.84 ± 3.63	25.72 ± 4.09	0.057
HCV RNA (Log IU/mL)	5.95 ± 1.14	5.85 ± 1.18	6.45 ± 0.74	<0.001 *
HCV genotype, G1, *n* (%)	234 (52)	177 (47.5)	57 (73)	0.003 *
ALT (U/L)	94.28 ± 84.95	97.18 ± 88.37	79.80 ± 63.85	0.116
TC (mg/dL)	171.76 ± 32.22	171.53 ± 32.75	172.91 ± 29.62	0.735
TGs (mg/dL)	103.25 ± 50.15	101.21 ± 46.21	113.57 ± 66.05	0.123
Platelet count (10^3 cells/mm)	176.89 ± 64.54	181.27 ± 57.86	156.89 ± 58.72	0.001 *
HOMA-IR	3.14 ± 4.91	2.89 ± 4.70	4.42 ± 5.73	0.043 *
Hepatic steatosis, *n* (%) #	220 (48.8)	186 (50)	34 (43.5)	0.373
Liver cirrhosis, *n* (%) #	117 (26)	83 (22.3)	34 (43.5)	0.001 *
APRI	1.51 ± 1.66	1.47 ± 1.63	1.74 ± 1.83	0.210
Leptin (ng/mL)	9.82 ± 10.1	9.33 ± 9.05	12.2 ± 14.1	0.267
Adiponectin (μg/mL)	9.74 ± 7.15	10.1 ± 7.48	8.04 ± 5.22	0.097
Leptin (ng/mL)/adiponectin (μg/mL) ratio	2.14 ± 3.79	2.06 ± 4.10	2.45 ± 3.00	0.603
PAI-1 (ng/mL)	6.73 ± 3.11	6.76 ± 2.98	6.56 ± 3.69	0.658
rs12979860 (CC) *n* (%) #	379 (84.2)	328 (88.1)	88 (65.3)	0.013 *

SVR: sustained virological response; #: chi-squared test; * $p < 0.05$; BMI: body mass index; G: genotype; Log: logarithmic; ALT: alanine aminotransferase; TC: total cholesterol; TGs: triglycerides; HOMA-IR: homeostasis model assessment-estimated insulin resistance; APRI: aspartate aminotransferase to platelet ratio index; PAI-1: plasminogen activator inhibitor-1.

3.2. Independent Pre-Therapy Factors Associated with Pre-Therapy Leptin, Adiponectin, and PAI-1 Levels in CHC Patients

TC level and HOMA-IR score were independently associated with HCV RNA levels before anti-HCV therapy. Male gender was negatively associated with both leptin and adiponectin levels, and the pre-therapy BMI was positively associated with pre-therapy leptin and PAI-1 levels. Additionally, the TG level was negatively associated with adiponectin level, and the platelet count was positively associated with PAI-1 level. The PAI-1 level exhibited a borderline association with the adiponectin level ($p = 0.05$), and none of the adipokine levels were associated with the HCV RNA level (Table 2 and Figure 1).

Table 2. Univariate and multivariate analyses of associations between pre-therapy factors and pre-therapy levels of hepatitis C virus RNA, leptin, adiponectin, and PAI-1 in all enrolled CHC patients.

Pre-Therapy Factors	Pre-Therapy HCV RNA Level (Log IU/mL)			Pre-Therapy Leptin Level (ng/mL)			Pre-Therapy Adiponectin Level (μg/mL)			Pre-Therapy PAI-1 Level (ng/mL)		
	Univariate	Multivariate		Univariate	Multivariate		Univariate	Multivariate		Univariate	Multivariate	
	p Values	95% CI of Estimated β (Estimated β)	p Values	p Values	95% CI of Estimated β (Estimated β)	p Values	p Values	95% CI of Estimated β (Estimated β)	p Values	p Values	95% CI of Estimated β (Estimated β)	p Values
Sex (Male)	0.322			<0.001*	−12.7–−8.6 (−10.6)	<0.001*	<0.001*	−5.2–−1.5 (−3.4)	<0.001*	0.094	−1.08–1.33 (0.123)	0.84
Age (years)	0.415			0.874			0.012*	−0.03–0.13 (0.05)	0.22	<0.001*	−0.075–0.009 (−0.033)	0.122
BMI	0.795			<0.001*	1.09–1.70 (1.39)	<0.001*	<0.001*	−0.48–−0.038 (−0.222)	0.094	<0.001*	0.041–0.37 (0.205)	0.015*
HCV RNA (Log IU/mL)	NA	NA		0.231			0.887			0.200		
HCV genotype	<0.001*	−0.59–−0.25 (−0.42)	<0.001*	0.356			0.074	−0.1–3.5 (1.7)	0.172	0.064	−0.425–1.44 (0.509)	0.284
ALT (U/L)	0.631			0.468			0.607			0.288		
TC (mg/dL)	0.027*	0.000–0.005 (0.003)	0.036*	0.839			0.312			0.86		
TGs (mg/dL)	0.375			0.341			0.028*	−0.036–−0.003 (−0.017)	0.045*	0.095	−0.002–0.018 (0.008)	0.113
Platelet count (10^3 cells/mm)	0.419			0.659			0.008*	−0.027–−0.005 (−0.011)	0.185	<0.001*	0.005–0.026 (0.015)	0.005*
HOMA-IR	0.063	0.001–0.03 (0.015)	0.04*	0.231			0.03*	−0.254–0.046 (−0.104)	0.171	0.909		
Hepatic steatosis (Yes)	0.837			0.001*	−0.93–3.36 (1.22)	0.265	0.01*	−4.01–−0.014 (−0.302)	0.052	0.009*	−1.14–0.985 (−0.078)	0.885
Cirrhosis (Yes)	0.213			0.577			0.76			0.005*	−1.23–1.56 (0.167)	0.813
APRI	0.496			0.489			0.213			0.012*	−0.163–0.435 (0.136)	0.371
Leptin (ng/mL)	0.231			NA			0.313			0.015*	−0.048–0.181 (0.017)	0.616
Adiponectin (μg/mL)	0.887			0.313			NA	NA		<0.001*	−0.142–0.007 (−0.067)	0.076
Leptin (ng/mL) adiponectin (μg/mL) ratio	0.207			NA#			NA#			<0.001*	NA#	NA#
PAI-1 (ng/mL)	0.2			0.015*	−0.19–0.45 (0.126)	0.442	<0.001*	−0.603–0.000 (−0.302)	0.05	NA#	NA	NA
rs12979860 (CC genotype)	0.246			0.636			0.743			0.236	NA	NA

PAI-1: plasminogen activator inhibitor-1; Log: logarithmic; CI: confidence interval; *: p < 0.05; NA: not accessible; BMI: body mass index; G: genotype; ALT: alanine aminotransferase; TC: total cholesterol; TGs: triglycerides; HOMA-IR: homeostasis model assessment-estimated insulin resistance; APRI: aspartate aminotransferase to platelet ratio index; # these values are not listed as independent factors in univariate or multivariate analyses because they are highly dependent on leptin and adiponectin levels, which are independent factors.

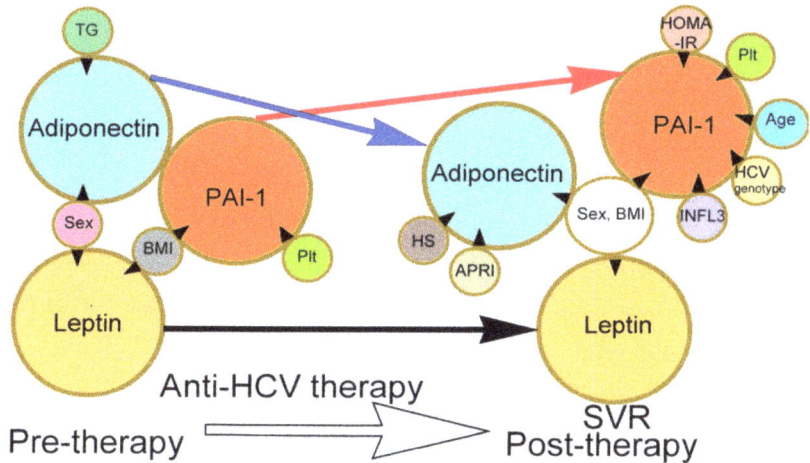

Figure 1. Associations between independent factors and levels of the investigated adipokines, including leptin, adiponectin, and PAI-1, at pre-therapy and 24-week post-HCV therapy stages. The tips of the black arrowheads indicate dependent factors, and the bases of the black arrowheads indicate independent factors. TGs: triglycerides; BMI: body mass index; Plt: platelet; APRI: aspartate aminotransferase to platelet ratio index; HS: hepatic steatosis; HOMA-IR: homeostasis model assessment-estimated insulin resistance; IFNL3: interferon λ3; Tx: anti-HCV therapy. Black arrow: leptin levels were unchanged after anti-HCV therapy; red arrow: PAI-1 levels were increased after anti-HCV therapy; blue arrow: adiponectin levels were decreased after anti-HCV therapy.

3.3. Independent Post-Therapy Factors Associated with 24-Week Post-Therapy Leptin, Adiponectin and PAI-1 Levels in CHC Patients Who Achieved SVR

Gender and BMI were associated with leptin, adiponectin, and PAI-1 levels in patients who achieved SVR at 24 weeks after therapy. Additionally, APRI and steatosis were associated with the adiponectin level; age, HCV, and IFNL3 genotypes, the platelet count and the HOMA-IR score were associated with PAI-1 level (Table 3 and Figure 1).

Table 3. Univariate and multivariate analyses of associations between post-therapy factors and post-therapy levels of leptin, adiponectin, and PAI-1 in CHC patients after SVR.

Post-Therapy Factors	Post-therapy Leptin Level (ng/mL)			Post-therapy Adiponectin Level (μg/mL)			Post-therapy PAI-1 Level (ng/mL)		
	Univariate	Multivariate		Univariate	Multivariate		Univariate	Multivariate	
	p Values	95% CI of Estimated β (Estimated β)	p Values	p Values	95% CI of Estimated β (Estimated β)	p Values	p Values	95% CI of Estimated β (Estimated β)	p Values
Sex (Male)	<0.001 *	−18.9—10.3 (−15.2)	<0.001 *	<0.001 *	−5.7—2.6 (−4.2)	0.001 *	0.008 *	0.8–1.86 (0.97)	0.033 *
Age	0.317			0.223			<0.001 *	−0.12—0.04 (−0.08)	<0.001 *
BMI	<0.001 *	1.31–2.52 (1.88)	<0.001 *	<0.001 *	−0.53—0.05 (−0.29)	0.017 *	<0.001 *	0.06–0.308 (0.186)	0.003 *
HCV genotype	0.193			0.07			0.034 *	−1.6—0.005 (−0.81)	0.049 *
ALT (U/L)	0.958			0.201			0.471		
TC (mg/dL)	0.079			<0.001 *	−0.08–0.04 (−0.02)	0.515	0.246		
TGs (mg/dL)	0.549	−0.003–0.09 (0.05)	0.066	0.448			<0.001 *	−0.005–0.001 (0.002)	0.551
Platelet counts (10³ cells/mm)	0.523			0.006 *	−0.02–0.008 (−0.006)	0.382	<0.001 *	0.014–0.035 (0.025)	<0.001 *
HOMA-IR	0.089	−0.32–0.69 (0.184)	0.473	0.015 *	−0.46—0.003 (−0.22)	0.053	0.039 *	0.05–0.30 (0.182)	0.004 *
APRI	0.564			<0.001 *	0.11–5.59 (2.85)	0.042 *	<0.001 *	−0.77–2.4 (0.814)	0.313
Hepatic steatosis (Yes)	0.069	−2.9–4.6 (0.864)	0.656	0.419	−3.8—0.56 (−2.2)	0.008 *	0.008 *	−0.29–1.41 (0.56)	0.794
Cirrhosis (Yes)	0.315			0.29			<0.001 *	−2.2—0.19 (−0.98)	0.100
Leptin (ng/mL)	NA	NA	NA	NA	NA	NA	0.781		
Adiponectin (μg/mL)	0.29			NA #			0.008 *	−0.05–0.09 (0.02)	0.592
Leptin (ng/mL)/adiponectin (μg/mL) ratio	NA #			0.008 *	−0.18–0.28 (0.053)	0.65	0.044 *	NA #	NA #
PAI-1 (ng/mL)	0.781			0.915			NA #	NA	NA
rs12979860 (CC)	0.853						0.076	0.46–2.4 (1.49)	0.004 *

PAI-1: plasminogen activator inhibitor-1; SVR: sustained virological response; CI: confidence interval; *: p < 0.05, NA: not accessible; BMI: body mass index; ALT: alanine aminotransferase; TC: total cholesterol; TGs: triglycerides; HOMA-IR: homeostasis model assessment–estimated insulin resistance; APRI: aspartate aminotransferase to platelet ratio index; # these values are not listed as independent factors in univariate or multivariate analyses because they are highly dependent on leptin and adiponectin levels, which are independent factors.

3.4. Comparisons between the Pre- and Post-Therapy Levels of Each Variable in CHC Patients

BMI decreased in CHC patients, regardless of whether they achieved SVR. However, post-therapy ALT, APRI and adiponectin levels decreased, whereas TC, TG, and PAI-1 levels increased after therapy only in patients exhibiting SVR. Regardless of the therapeutic response, leptin levels and L/A ratios remained unchanged after therapy (Table 4). Since multivariate analyses showed the HCV genotype to be associated with the post-therapy PAI-1 level in patients who achieved SVR, we stratified these patients by HCV genotype to evaluate alterations in adipokine levels. Only SVR patients with infection by G1 HCV (6.62 ± 2.91 ng/mL versus 8.43 ± 3.66 ng/mL, $p < 0.001$), but not G2 HCV (6.85 ± 3.08 ng/mL vs. 7.30 ± 3.47 ng/mL, $p = 0.13$), exhibited a significantly elevated level of PAI-1 at 24 weeks after therapy compared with the pre-therapy level (Figure 2). In contrast, the leptin level remained unchanged (G1, $p = 0.85$; G2, $p = 0.479$), and the adiponectin level decreased (G1: 9.02 ± 7.52 µg/mL versus 7.27 ± 5.40 µg/mL, $p = 0.003$; G2: 11.10 ± 7.02 µg/mL versus 8.91 ± 6.38 µg/mL, $p < 0.001$) at 24 weeks after therapy, regardless of HCV genotype.

Table 4. Comparison of the pre- and 24-week post-therapy variables in CHC patients stratified by therapeutic response.

Factors	SVR (+) ($n = 372$)		Paired *t*-Test *p* Values	SVR (−) ($n = 78$)		Paired *t*-Test *p* Values
	Pre-Therapy Value	Post-Therapy Value		Pre-Therapy Value	Post-Therapy Value	
BMI	24.84 ± 3.63	24.35 ± 3.51	<0.001 *	25.72 ± 4.09	24.87 ± 3.62	<0.001 *
ALT (U/L)	97.18 ± 88.37	20.0 ± 10.5	<0.001 *	79.80 ± 63.85	63.7 ± 43.3	0.151
TC (mg/dL)	171.53 ± 32.75	184.28 ± 37.39	<0.001 *	172.91 ± 29.62	174.29 ± 36.12	0.7021
TGs (mg/dL)	101.21 ± 46.21	120.54 ± 74.75	<0.001 *	113.57 ± 66.05	102.77 ± 42.88	0.059
Platelet count (10^3 cells/mm)	181.27 ± 57.86	184.10 ± 55.68	0.243	156.89 ± 58.72	149.40 ± 54.55	0.179
HOMA-IR	2.89 ± 4.70	2.83 ± 3.96	0.5493	4.42 ± 5.73	5.40 ± 11.55	0.7332
APRI	1.47 ± 1.63	0.418 ± 0.297	<0.001 *	1.74 ± 1.83	1.28 ± 0.929	0.162
Hepatic steatosis (Yes), n (%)	186 (50)	193 (51.8)	0.499	34 (43.5)	37 (47.4)	0.443
Cirrhosis (Yes), n (%)	83 (22.3)	86 (23)	0.504	34 (43.5)	39 (50)	0.058
Leptin (ng/mL)	9.33 ± 9.05	9.61 ± 8.77	0.15	12.2 ± 14.1	10.1 ± 12.5	0.187
Adiponectin (µg/mL)	10.1 ± 7.48	8.14 ± 5.09	0.003*	8.04 ± 5.16	8.77 ± 6.32	0.473
Leptin (ng/mL)/adiponectin (µg/mL) ratio	1.96 ± 4.07	2.26 ± 3.57	0.188	2.44 ± 2.96	3.02 ± 6.43	0.506
PAI-1 (ng/mL)	6.76 ± 2.98	9.08 ± 4.43	0.003*	6.56 ± 3.69	6.45 ± 4.42	0.9355

SVR: sustained virological response; *: $p < 0.05$; BMI: body mass index; ALT: alanine aminotransferase; TC: total cholesterol; TGs: triglycerides; HOMA-IR: homeostasis model assessment-estimated insulin resistance; APRI: aspartate aminotransferase to platelet ratio index; PAI-1: plasminogen activator inhibitor-1.

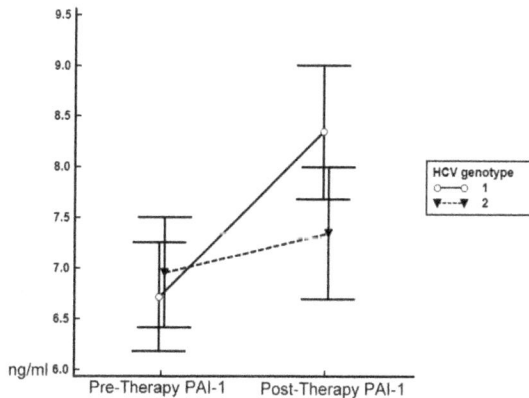

Figure 2. Alterations in patterns of PAI-1 levels at pre- and 24-week post-anti-HCV therapy stages in patients with SVR and stratified by HCV genotype. Pre-therapy PAI-1: pre-anti-HCV therapy levels of PAI-1 (means ± standard deviations); post-therapy PAI-1: 24-week post-anti-HCV therapy PAI-1 level. Blue line: patients with genotype 1 (G1) HCV infection; red line: patients with G2 HCV infection.

Figure 1 summarizes the identified associations between independent factors and the investigated adipokine levels, as well as the altered levels of adipokines observed during viral clearance in CHC patients.

4. Discussion and Conclusions

The most compelling results of this study are as follows. (1) Prior to anti-HCV therapy, HCV genotype, TC level, and HOMA-IR score were independently associated with the HCV RNA level. Moreover, gender was associated with both leptin and adiponectin levels, and BMI was associated with leptin and PAI-1 levels. Additionally, the TG level was associated with that of adiponectin, and the platelet count was associated with the PAI-1 level. (2) Gender and BMI were associated with the levels of all investigated adipokines in patients who achieved SVR at 24 weeks after therapy. Additionally, APRI and steatosis were associated with the adiponectin level. Age, HCV, and IFNL3 genotypes, platelet count and HOMA-IR score were associated with the PAI-1 level. (3) Following anti-HCV therapy, a decrease in the adiponectin level compared to the pre-therapy level was noted in both G1 and G2 SVR patients, whereas an increased PAI-1 level was detected only in the former.

All significantly different pre-therapy variables between SVR and non-SVR patients in the current study have been reported previously [2], confirming the reliability of our data, which did not show differences in pre-therapy levels of any adipokine or the L/A ratio. Moreover, none of the pre-therapy adipokine levels were associated with the pre-therapy HCV RNA level. These results are consistent with our previous studies that focused on one adipokine, with no adjustment for other adipokines [24–26]. The finding of an association between the pre-therapy HCV RNA level and the TC level and HOMA-IR score confirms that HCV is a metabolic virus [2]. In general, associations of gender and BMI with the levels of investigated adipokines appear to be fundamental and more evident after viral clearance. Indeed, except for a link between adiponectin and TG level, no definite hepatic or metabolic profiles were associated with adipokine levels prior to SVR. However, after SVR, APRI (a marker of hepatic fibrosis for chronic hepatitis C [30]) and hepatic steatosis were associated with adiponectin levels, whereas the HOMA-IR score was associated with the PAI-1 level. Based on these results, adipokine levels are, to varying degrees, associated with hepatic or metabolic profiles after viral clearance, even though the effects are not obvious prior to anti-HCV therapy. Thus, HCV infection may mask or overcome interactions between metabolic/hepatic profiles and abundant adipokines.

Regardless of the adjustment for adiponectin and PAI-1 levels [25], leptin levels were consistently associated with gender and BMI and remained unchanged after viral clearance; this finding supports the crucial role of leptin in whole-body homeostasis [25]. In addition, the L/A ratio may represent a marker of adipocyte dysfunction [31] and metabolic syndrome [32] and was unchanged at 24 weeks after therapy, regardless of the therapeutic response. Conversely, although adiponectin levels were consistently associated with TG and gender before viral clearance [26], the direct association of adiponectin level with the HOMA-IR score after SVR noted in our previous study [26] disappeared after adjusting for leptin and PAI-1 levels. Prior to anti-HCV therapy, the PAI-1 level exhibited a borderline association with adiponectin levels; after SVR, the PAI-1 level was associated with the HOMA-IR score, regardless of adjustment for adiponectin and leptin levels [24]. These findings potentially indicate a higher hierarchical status of PAI-1 than adiponectin in affecting glucose metabolism once the patient is free from HCV interference. In line with the results of our previous study [24], the PAI-1 level was consistently associated with the platelet count, a finding that highlights platelets as an important source of PAI-1 [24]. Notably, the positive association between IFNL3-rs12979860 CC (a favorable SNP genotype for anti-HCV therapy [2,24–29]) and the post-therapy PAI-1 level supports the hypothesis that CHC patients who achieve SVR might have an increased risk of cardiovascular events [24]. Unexpectedly, the association of HCV genotype with PAI-1 level after SVR was evident only in the current study, which enrolled patients with both G1 and G2 HCV infections; this was not observed in our previous study, which enrolled patients with pan-genotypic HCV infections, including G1, G2, G3, G6, and mixed genotypes [24]. Based on HCV genotype-stratified analyses, increased PAI-1 after SVR was significant only in patients with G1 HCV infection, whereas altered patterns in leptin (unchanged) [25] and adiponectin (decreased) levels [26] did not change after HCV genotype stratification. The decrease in adiponectin levels observed in CHC patients after SVR was consistent with a previous case-control study showing that CHC is associated with increased serum adiponectin [33]. Our previous studies confirmed the non-hepatic origin of increases in PAI-1 observed after SVR [24] and revealed that the high PAI-1 level is linked to a poor metabolic profile [5]. Accordingly, our current finding of a G1-specific increase in PAI-1 level reiterates the observation that patients infected with G2 HCV benefit more from HCV clearance than those with G1 HCV infection due to a more favorable pattern of lipid alterations and changes in metabolic scores [27].

Since adipose tissue is the major source of adipokines [4], the main limitation of this study is the lack of pathological adipose tissue examination. Moreover, conclusions based on analyzing associated factors is an imperfect yet compromised approach to building a complete picture of adipokine-associated pathways. In addition, patatin-like phospholipase domain-containing 3 (PNPLA3) polymorphism has been shown to affect adiponectin levels in patients with non-alcoholic fatty liver disease [34,35], but not in patients with CHC [34]. PNPLA3 is not associated with the anti-HCV therapeutic response [36], and researchers have not clearly determined whether the PNPLA3 genotype is associated with altered levels of adiponectin in patients with CHC after SVR. Future studies of these adipokines in CHC patients using paired adipose tissue surveys before and at 24 weeks after therapy and an assessment of their correlations with cardiometabolic events and genetics, including the PNPLA3 genotype, as well as associated fundamental cellular or animal models may be required to elucidate the molecular basis and clinical implications of the adipokine alterations observed in patients who achieve viral clearance.

Taken together, the results show that HCV infection possibly masks the effects of abundant adipokines on hepatic or metabolic homeostasis. After viral clearance, adiponectin levels were closely associated with hepatic profiles, whereas PAI-1 was found to be in a higher hierarchical status than adiponectin with regard to affecting glucose metabolism. The genotype non-specific decrease in adiponectin levels and G1-specific increase in PAI-1 levels warrants careful follow-up of co-morbidities in patients with CHC after SVR according to viral genotype. Finally, personalized therapy and follow-up for patients with CHC may become applicable in the near future.

Acknowledgments: This study was funded by grants from the Chang Gung Medical Research Program (CMRPG3F0472, CRRPG3F0012, CMRPG3B1743, and XMRPG3A0525), Research Services Center For Health Information, Chang Gung University, Taoyuan, Taiwan (CIRPD1D0032) and the National Science Council, Taiwan (MOST 105-2314-B-182-023, and MOST 105-2629-B-182-001-). The authors thank Chun-Ming Fan from the Department of Biomedical Sciences, College of Medicine, Chang Gung University for generating the excellent figures presented in this manuscript and Yu-Jr Lin, who was supported by a grant from the Research Services Center for Health Information (CIRPD1D0031) from Chang Gung Memorial Hospital, for the statistical consultation.

Author Contributions: M.-L.C. designed and performed the study and drafted the article and critically revised it for intellectual content. T.-H.C., C.-M.H., C.-H.L, C.-Y.L., C.-J.K., S.-W.H., C.-W.C., H.-T.C., C.-T.Y., and C.-T.C. interpreted the data and wrote the manuscript. All authors approved the final version of the article, including the authorship list.

Conflicts of Interest: The authors declare no conflict of interest.

References

1. Smith, D.B.; Bukh, J.; Kuiken, C.; Muerhoff, A.S.; Rice, C.M.; Stapleton, J.T.; Simmonds, P. Expanded classification of hepatitis C virus into 7 genotypes and 67 subtypes: Updated criteria and genotype assignment web resource. *Hepatology* **2014**, *59*, 318–327. [CrossRef] [PubMed]
2. Chang, M.L. Metabolic alterations and hepatitis C: From bench to bedside. *World J. Gastroenterol.* **2016**, *22*, 1461–1476. [CrossRef] [PubMed]
3. Syed, G.H.; Amako, Y.; Siddiqui, A. Hepatitis C virus hijacks host lipid metabolism. *Trends Endocrinol. Metab.* **2010**, *21*, 33–40. [CrossRef] [PubMed]

4.   Scheja, L.; Heeren, J. Metabolic interplay between white, beige, brown adipocytes and the liver. *J. Hepatol.* **2016**, *64*, 1176–1186. [CrossRef] [PubMed]

5.   Chang, M.L.; Hsu, C.M.; Tseng, J.H.; Tsou, Y.K.; Chen, S.C.; Shiau, S.S.; Chiu, C.T. Plasminogen activator inhibitor-1 is independently associated with non-alcoholic fatty liver disease whereas leptin and adiponectin vary between genders. *J. Gastroenterol. Hepatol.* **2015**, *30*, 329–336. [CrossRef] [PubMed]

6.   Deng, Y.; Scherer, P.E. Adipokines as novel biomarkers and regulators of the metabolic syndrome. *Ann. N. Y. Acad. Sci.* **2010**, *1212*, E1–E19. [CrossRef] [PubMed]

7.   Jung, U.J.; Choi, M.S. Obesity and its metabolic complications: The role of adipokines and the relationship between obesity, inflammation, insulin resistance, dyslipidemia and nonalcoholic fatty liver disease. *Int. J. Mol. Sci.* **2014**, *15*, 6184–6223. [CrossRef] [PubMed]

8.   Podor, T.J.; Loskutoff, D.J. Immunoelectron microscopic localization of type 1 plasminogen activator inhibitor in the extracellular matrix of transforming growth factor-beta-activated endothelial cells. *Ann. N. Y. Acad. Sci.* **1992**, *667*, 46–49. [CrossRef] [PubMed]

9.   Ballantyne, G.H.; Gumbs, A.; Modlin, I.M. Changes in insulin resistance following bariatric surgery and the adipoinsular axis: role of the adipocytokines, leptin, adiponectin and resistin. *Obes. Surg.* **2005**, *15*, 692–699. [CrossRef] [PubMed]

10. Heinrich, G.; Russo, L.; Castaneda, T.R.; Pfeiffer, V.; Ghadieh, H.E.; Ghanem, S.S.; Hill, J.W. Leptin resistance contributes to obesity in mice with null mutation of carcinoembryonic antigen cell adhesion molecule 1. *J. Biol. Chem.* **2016**, *291*, 11124–11132. [CrossRef] [PubMed]

11. Ohashi, K.; Ouchi, N.; Matsuzawa, Y. Adiponectin and hypertension. *Am. J. Hypertens.* **2011**, *24*, 263–269. [CrossRef] [PubMed]

12. Francés, D.E.; Motiño, O.; Agrá, N.; González-Rodríguez, Á.; Fernández-Álvarez, A.; Cucarella, C.; Carnovale, C.E. Hepatic cyclooxygenase-2 expression protects against diet-induced steatosis, obesity, and insulin resistance. *Diabetes* **2015**, *64*, 1522–1531. [CrossRef] [PubMed]

13. Schäffler, A.; Schölmerich, J.; Büchler, C. Mechanisms of disease: adipocytokines and visceral adipose tissue–emerging role in nonalcoholic fatty liver disease. *Nat. Clin. Pract. Gastroenterol. Hepatol.* **2005**, *2*, 273–280. [CrossRef] [PubMed]

14. Janeckova, R. The role of leptin in human physiology and pathophysiology. *Physiol. Res.* **2001**, *50*, 443–459. [PubMed]

15. Pajvani, U.B.; Du, X.; Combs, T.P.; Berg, A.H.; Rajala, M.W.; Schulthess, T.; Scherer, P.E. Structure-function studies of the adipocyte-secreted hormone Acrp30/adiponectin. Implications FPR metabolic regulation and bioactivity. *J. Biol. Chem.* **2003**, *278*, 9073–9085. [CrossRef] [PubMed]

16. Ruscica, M.; Baragetti, A.; Catapano, A.L.; Norata, G.D. Translating the biology of adipokines in atherosclerosis and cardiovascular diseases: Gaps and open questions. *Nutr. Metab. Cardiovasc. Dis.* **2017**, *27*, 379–395. [CrossRef] [PubMed]

17. Tsaroucha, A.; Daniil, Z.; Malli, F.; Georgoulias, P.; Minas, M.; Kostikas, K.; Gourgoulianis, K.I. Leptin, adiponectin, and ghrelin levels in female patients with asthma during stable and exacerbation periods. *J. Asthma* **2013**, *50*, 188–197. [CrossRef] [PubMed]

18. Kappelle, P.J.; Dullaart, R.P.; Van Beek, A.P.; Hillege, H.L.; Wolffenbuttel, B.H. The plasma leptin/adiponectin ratio predicts first cardiovascular event in men: a prospective nested case-control study. *Eur. J. Intern. Med.* **2012**, *23*, 755–759. [CrossRef] [PubMed]

19. Lalić, K.; Jotić, A.; Rajković, N.; Singh, S.; Stošić, L.; Popović, L.; Stanarčić, J. Altered Daytime Fluctuation Pattern of Plasminogen Activator Inhibitor 1 in Type 2 Diabetes Patients with Coronary Artery Disease: A Strong Association with Persistently Elevated Plasma Insulin, Increased Insulin Resistance, and Abdominal Obesity. *Int. J. Endocrinol.* **2015**, *2015*, 390185. [CrossRef] [PubMed]

20. Samad, F.; Ruf, W. Inflammation, obesity, and thrombosis. *Blood* **2013**, *122*, 3415–3422. [CrossRef] [PubMed]

21. Kortlever, R.M.; Bernards, R. Senescence, wound healing and cancer: The PAI-1 connection. *Cell Cycle* **2006**, *5*, 2697–2703. [CrossRef] [PubMed]

22. Eitzman, D.T.; Westrick, R.J.; Nabel, E.G.; Ginsburg, D. Plasminogen activator inhibitor-1 and vitronectin promote vascular thrombosis in mice. *Blood* **2000**, *95*, 577–580. [PubMed]

23. Targher, G.; Marra, F.; Marchesini, G. Increased risk of cardiovascular disease in non-alcoholic fatty liver disease: Causal effect or epiphenomenon? *Diabetologia* **2008**, *51*, 1947–1953. [CrossRef] [PubMed]

24. Chang, M.L.; Lin, Y.S.; Pao, L.H.; Huang, H.C.; Chiu, C.T. Link between plasminogen activator inhibitor-1 and cardiovascular risk in chronic hepatitis C after viral clearance. *Sci. Rep.* **2017**, *7*, 42503. [CrossRef] [PubMed]

25. Chang, M.L.; Kuo, C.J.; Huang, H.C.; Chu, Y.Y.; Chiu, C.T. Association between Leptin and Complement in Hepatitis C Patients with Viral Clearance: Homeostasis of Metabolism and Immunity. *PLoS ONE* **2016**, *11*, e0166712. [CrossRef] [PubMed]

26. Chang, M.L.; Kuo, C.J.; Pao, L.H.; Hsu, C.M.; Chiu, C.T. The evolving relationship between adiponectin and insulin sensitivity in hepatitis C patients during viral clearance. *Virulence* **2017**. [CrossRef] [PubMed]

27. Chang, M.L.; Tsou, Y.K.; Hu, T.H.; Lin, C.H.; Lin, W.R.; Sung, C.M.; Yeh, C.T. Distinct patterns of the lipid alterations between genotype 1 and 2 chronic hepatitis C patients after viral clearance. *PLoS ONE* **2014**, *9*, e104783. [CrossRef] [PubMed]

28. Chang, M.L.; Cheng, M.L.; Chang, S.W.; Tang, H.Y.; Chiu, C.T.; Yeh, C.T.; Shiao, M.S. Recovery of pan-genotypic and genotype-specific amino acid alterations in chronic hepatitis C after viral clearance: Transition at the crossroad of metabolism and immunity. *Amino Acids* **2017**, *49*, 291–302. [CrossRef] [PubMed]

29. Chang, M.L.; Liang, K.H.; Ku, C.L.; Lo, C.C.; Cheng, Y.T.; Hsu, C.M.; Chiu, C.T. Resistin reinforces interferon λ-3 to eliminate hepatitis C virus with fine-tuning from RETN single-nucleotide polymorphisms. *Sci. Rep.* **2016**, *6*, 30799. [CrossRef] [PubMed]

30. Udompap, P.; Mannalithara, A.; Heo, N.Y.; Kim, D.; Kim, W.R. Increasing prevalence of cirrhosis among U.S. adults aware or unaware of their chronic hepatitis C virus infection. *J. Hepatol.* **2016**, *64*, 1027–1032. [CrossRef] [PubMed]

31. Finucane, F.M.; Luan, J.; Wareham, N.J.; Sharp, S.J.; O'Rahilly, S.; Balkau, B.; Savage, D.B. Correlation of the leptin: Adiponectin ratio with measures of insulin resistance in non-diabetic individuals. *Diabetologia* **2009**, *52*, 2345–2349. [CrossRef] [PubMed]

32. Dullaart, R.P.; Kappelle, P.J.; Dallinga-Thie, G.M. Carotid intima media thickness is associated with plasma adiponectin but not with the leptin:adiponectin ratio independently of metabolic syndrome. *Atherosclerosis* **2010**, *211*, 393–396. [CrossRef] [PubMed]

33. Canavesi, E.; Porzio, M.; Ruscica, M.; Rametta, R.; Macchi, C.; Pelusi, S.; Valenti, L. Increased circulating adiponectin in males with chronic HCV hepatitis. *Eur. J. Intern. Med.* **2015**, *26*, 635–639. [CrossRef] [PubMed]

34. Valenti, L.; Rametta, R.; Ruscica, M.; Dongiovanni, P.; Steffani, L.; Motta, B.M.; Magni, P. The I148M PNPLA3 polymorphism influences serum adiponectin in patients with fatty liver and healthy controls. *BMC Gastroenterol.* **2012**, *12*, 111. [CrossRef] [PubMed]

35. Rametta, R.; Ruscica, M.; Dongiovanni, P.; Macchi, C.; Fracanzani, A.L.; Steffani, L.; Valenti, L. Hepatic steatosis and PNPLA3 I148M variant are associated with serum Fetuin-A independently of insulin resistance. *Eur. J. Clin. Investig.* **2014**, *44*, 627–633. [CrossRef] [PubMed]

36. Valenti, L.; Rumi, M.; Galmozzi, E.; Aghemo, A.; Del Menico, B.; De Nicola, S.; Colombo, M. Patatin-like phospholipase domain-containing 3 I148M polymorphism, steatosis, and liver damage in chronic hepatitis C. *Hepatology* **2011**, *53*, 791–799. [CrossRef] [PubMed]

nutrients

MDPI

Review

The Metabolic Response to Stress and Infection in Critically Ill Children: The Opportunity of an Individualized Approach

Valentina De Cosmi [1,2] , Gregorio Paolo Milani [1,3] , Alessandra Mazzocchi [3,4], Veronica D'Oria [5], Marco Silano [6] , Edoardo Calderini [5] and Carlo Agostoni [3,4,*]

1   Pediatric Unit, Fondazione IRCCS Ca' Granda Ospedale Maggiore Policlinico, 20122 Milan, Italy; valentina.decosmi@unimi.it (V.D.C.); gregorio.milani@unimi.it (G.P.M.)
2   Branch of Medical Statistics, Biometry, and Epidemiology "G. A. Maccacaro", Department of Clinical Sciences and Community Health, University of Milan, 20122 Milan, Italy
3   Department of Clinical Sciences and Community Health, Università Degli Studi di Milano, 20122 Milan, Italy; alessandra.mazzocchi@unimi.it
4   Pediatric Intermediate Care Unit, Fondazione IRCCS Ca' Granda Ospedale Maggiore Policlinico, 20122 Milan, Italy
5   Pediatric Intensive Care Unit, Department of Anesthesia and Intensive Care and Emergency, Fondazione IRCCS Ca' Granda, Ospedale Maggiore Policlinico, 20122 Milan, Italy; veronica.doria.vd@gmail.com (V.D.); edoardo.calderini@policlinico.mi.it (E.C.)
6   Unit of Human Nutrition and Health, Department of Food Safety, Nutrition and Veterinary Public Health, Istituto Superiore di Sanità, 00161 Rome, Italy; marco.silano@iss.it
*   Correspondence: carlo.agostoni@unimi.it; Tel.: +39-02-5503-2497

Received: 31 May 2017; Accepted: 14 September 2017; Published: 18 September 2017

Abstract: The metabolic response to stress and infection is closely related to the corresponding requirements of energy and nutrients. On a general level, the response is driven by a complex endocrine network and related to the nature and severity of the insult. On an individual level, the effects of nutritional interventions are highly variable and a possible source of complications. This narrative review aims to discuss the metabolic changes in critically-ill children and the potential of developing personalized nutritional interventions. Through a literature search strategy, we have investigated the importance of blood glucose levels, the nutritional aspects of the different phases of acute stress response, and the reliability of the available tools to assess the energy expenditure. The dynamics of metabolism during stressful events reveal the difficult balance between risk of hypo- or hyperglycemia and under- or overfeeding. Within this context, individualized and accurate measurement of energy expenditure may help in defining the metabolic needs of patients. Given the variability of the metabolic response in critical conditions, randomized clinical studies in ill children are needed to evaluate the effect of individualized nutritional intervention on health outcomes.

Keywords: nutrition; indirect calorimetry; glycemic control; acute phase; ketosis; intensive care

---

1. Introduction

Acute stress conditions, such as sepsis or severe infections, are the major causes of admission into pediatric intensive care units (PICUs) [1]. The metabolic response to these conditions has precise pathophysiological mechanisms, clinical consequences, and therapeutic implications. It is divided into three phases—the acute phase, the stable phase, and the recovery phase—all characterized by specific neuroendocrine, metabolic, and immunologic modifications [2]. From an evolutionary perspective, the pathophysiological mechanisms contribute to the maintenance of body homeostasis, switching nutritional compounds towards different functions and, at a further stage, may facilitate recovery [3,4]. For instance, the translocation of amino acids (AAs) from muscle to liver leads to energy production and facilitates the synthesis of acute-phase proteins [3,5]. On the other hand, the metabolic response to stress also includes catabolic processes that, in many circumstances, may increase physiological instability and resource wasting [6]. The over-activation of inflammatory pathways (e.g., ubiquitin—proteasome system) may cause large protein breakdown, ending with the consumption of muscle tissues [7–9]. Furthermore, prolonged metabolic imbalance may lead to the development of abnormal energy expenditure, mitochondrial failure, and multiple cellular dysfunctions up to organs damage and failure [10]. In addition, infection-induced nutritional deficiencies appear to diminish immune responsiveness and impair antimicrobial defensive mechanisms, thus making the host more susceptible to secondary or opportunistic infections [9]. Considering these metabolic alterations and the risk of hyper- or hypo-metabolism [10], nutritional interventions may be crucial to improving the response of the host [11]. The aim of this narrative review is to discuss in view of the novel insights: (1) the pros and cons of a tight glycemic control; (2) the nutritional problems and strategies in the different phases of acute stress response; and (3) the reliability and limitations of the available tools to assess energy expenditure.

2. Methods—Literature Search Strategy

Electronic databases (Pubmed, Medline, Embase, Google Scholar, Knowledge Finder) were used to locate and appraise relevant studies. We carried out the search to identify the articles published in English on (i) glycemic control, (ii) nutritional interventions in the different acute stress phases, (iii) the role of the parenteral nutrition and of the immunity-enhancers nutrients and, finally, (iv) the usefulness of predictive equations and indirect calorimetry use in critically ill children. Relevant articles published from January 2007 to March 2017 were identified using the following groups of key terms: "metabolic response" AND "critical illness" AND "infections"; "hyperglycemia" AND "critically ill children" OR "nutritional status" OR "nutrition" OR "nutritional sciences" AND "critical care"; OR "hospitalized children" OR "child, hospitalized" "insulin therapy" AND "critically ill children"; "energy expenditure" AND "critically ill children"; "nutrition" AND "critically ill children" OR "PICU". RCT and largest non-RCT studies were considered. In the eligible studies, we focused on data on infective events (or inflammation status), need of mechanical ventilation, length of PICU stay, and mortality rate (number of deaths) for the different nutritional approaches.

3. Results

3.1. Hyperglycemia and Glycemic Control in Stress Conditions

Stress hyperglycemia remains an unsolved medical condition. It is usually defined as blood glucose level >11.1 mmol/L [12]. Its incidence is very high, ranging between 56% and 86% among patients requiring intensive care [12]. Mechanical ventilation, vasopressor/inotropic infusion, continuous renal replacement therapy, infections, and long lengths of stay are the main risk factors of hyperglycemia that are commonly associated with worse outcomes [1].

Several trials have investigated the consequences of glycemic alterations and of their correction, providing inconclusive results. The effect of targeting age-adjusted normoglycemia with insulin infusion was investigated in a large prospective, randomized, controlled study including 700 critically-ill pediatric patients. Subjects were randomly assigned to intensive insulin treatment (to target blood glucose concentrations of 2.8–4.4 mmol/L in infants and 3.9–5.6 mmol/L in children) or conventional insulin infusion to prevent blood glucose from exceeding 11.9 mmol/L. A significantly shorter PICU stay was found in the intensively treated group (5.5 days vs. 6.2 days, $p = 0.017$) [13]. However, among several neurocognitive outcomes analyzed in a four-year follow-up of these children, only motor coordination ($p \leq 0.03$) and cognitive flexibility ($p = 0.02$) were worse in conventional insulin infusion group [14]. Another study on 97 patients with a median age of two years demonstrated that hyperglycemia was associated with higher morbidity (e.g., need of mechanical ventilation at 30 days) in meningococcal sepsis [15]. Other investigators found significantly reduced number of complications (e.g., incidence of sepsis) with lower (6.7–7.2 mmol/L) compared to higher (8.3–8.9 mmol/L) glucose targets in pediatric patients with burns [16].

On the contrary, a large multicenter trial, including more than 1300 critically-ill children randomly assigned to conventional glycemic control (glucose below 12 mmol/L) or tight glycemic control (4–7 mmol/L), found that the tight glycemic control had no significant effect on need of mechanical ventilation and mortality [17]. In this study, however, the upper limit of tight glycemic control was higher than in previous RCT, reducing the number of patients undergoing insulin treatment. In "The Heart and Lung Failure–Pediatric Insulin Titration (HALF-PINT)" trial (Clinical Trials ID: NCT01565941), glycemic control targeted to blood levels between 2.8–4.4 mmol/L in infants and 3.3–5.6 mmol/L in children was neither associated with lower length of PICU stay nor with mortality, as compared to levels between 8.3 and 10 mmol/L [18].

It may also be considered that tight glycemic control is difficult to achieve and, very often, intensive insulin therapy inevitably increases the risk of hypoglycemic episodes [19]. In turn, if severe and prolonged, low glucose levels may also cause major adverse events. A study investigating the effects of hypoglycemia found that critically-ill infants spending more than 50% of the time with lower (4.4–6.1 mmol/L) glucose levels had more frequent complications (e.g., renal failure) than those with higher (>11.1 mml/L) glucose levels [20]. However, in the Leuven pediatric study, although hypoglycemia was more common in children on intensive insulin therapy and patients developing hypoglycemia had a higher risk of death, this association was not significant and it could be explained by duration of PICU stay [13,14]. Other similar studies showed contrasting results on the association between hypoglycemia episodes and neurological consequences [21,22].

Randomized clinical trials on glycemic control and nutrition in critically-ill children are reported in Table 1. As a result of this section, the targets of glycemic control are still debated. Since normoglycemia is mostly associated with favorable outcomes, in terms of hospital stay and mortality, both hyperglycemia and hypoglycemia should be adequately prevented or promptly managed. However, the adoption of a tight glucose range for critically-ill patients is debated [10]. Indeed, currently, there is no recommendation available on hyperglycemia management provided from pediatric international societies. A "common sense approach" suggests keeping blood glucose between 7.8 and 10.0 mmol/L [23].

3.2. The Acute Phase: Metabolic Steps and Nutritional Implications

In acute stress conditions, circulating glucose and glycogen stores are rapidly depleted. Hence, hepatic gluconeogenesis, fatty acid beta-oxidation, and ketogenesis become the primary source of energy. At a further stage, the energy necessary for the increased gluconeogenesis is provided from either lactate or proteins and AAs [24]. These physiologic mechanisms are mediated by the insulin release switch-off followed by a production of glucagon, cortisol, and epinephrine, and activation of the sympathetic activity. They aim at providing sufficient energy for body metabolism and, among them, ketone bodies mainly supply central nervous energy requirements. During severe infections increased levels of inflammatory cytokines (e.g., IL1 and TNF), ACTH, and growth hormone tend to amplify these pathways and enhance protein breakdown [25–27]. Although these mechanisms are well known, management of ketosis in acute stress conditions is still challenging. On the one hand, ketone bodies are organic acids and, therefore, consume bicarbonates leading to blood acidosis and may cause malaise, nausea, and vomiting [25,28]. On the other hand, the amount of fatty acids resulting from lipolysis may even exceed energy requirements and glucose supplementation may rapidly lead to hyperglycemia and hepatic steatosis [24,25]. Yet, no randomized control trial has been conducted so far on different strategies to manage ketosis in critically-ill children.

Failing to provide adequate amounts of nutrients in the acute phase of stress response also results in exacerbation of existing nutritional problems in children. Malnutrition and infection may indeed interact, reinforcing each other even in milder stage of disease [29–31]. Furthermore, restricted nutritional support may stimulate autophagy, a survival mechanism by which cells break down their own damaged components to recycle intracellular nutrients and generate energy during starvation [32–34]. However, the acute stress response is a complex condition where hypercatabolism and muscular tissue consumption often cannot be reversed even with increased provision of nutrients ("futile cycle of nutrients") [35]. On the other hand, overfeeding, i.e., a caloric intake/resting energy expenditure (REE) ratio >110% or >120% [36], is associated with increased morbidity (e.g., delayed ventilator weaning), prolonged hospitalization, and a higher mortality [37]. Overfeeding may also inhibit autophagy, leading to an increased risk of cell death and organ dysfunction [35]. Being the lower and upper limits of individual energy requirements unknown in critically-ill children and may largely vary in the acute phase of stress conditions, a characteristic paradigm of under- and overfeeding is the unpredictable combination of metabolic and feeding patterns [38].

3.3. Stable and Recovery Phases

During the stable phase, both an early normalization of the catabolic counter-regulatory hormone levels and an increased effect of anabolic hormones occur, however, proteins continue to be wasted while fat stores remain relatively intact [39]. During this phase, the risk of muscle atrophy remains high, especially if this condition lasts several weeks. On the contrary, in the recovery phase, protein synthesis exceeds protein break down. Nutrition in both these phases should slowly increase to allow recovery and growth [40]. A recent systematic review and a single-center study in mechanically-ventilated children calculated a minimum intake of, respectively, 57 and 58 kcal/kg/day to achieve a positive nitrogen balance [41,42]. In both studies, a protein intake of 1.5 g/kg/day was suggested to attain nitrogen balance. In these studies, however, no difference was made between the stable and recovery phases.

3.4. Nutrition: Method of Administration and Immunity-Enhancers Nutrients

It is generally accepted that enteral nutrition is advised in the stable phase and mostly in the recovery phase. In the acute phase, it may also be of benefit, but its composition and timing of administration should be cautiously considered [43].

A large RCT, including 1440 patients from three different PICUs, found that early (within 24 h) parenteral supplementation of AAs was associated with a higher rate of infections and longer PICU stay [44]. Yet, the heterogeneity of the population and the different glycemic control strategies were likely to be biased across the participating centers [43,44]. On the contrary, in a retrospective study of more than 5100 critically-ill children, early enteral nutrition, over the first 48 h of admission, was associated with a lower mortality rate in those with a PICU length of stay at least 96 h [45].

Many strategies to optimize enteral nutrition have been developed recently. It has been suggested that an immune-enhancing formula (i.e., giving patients perioperative nutritional supplements with immunonutritional additives) might improve the general and metabolic conditions in adults with infection [5]. In particular, the importance of AAs, dietary nucleotides, and lipids in modulating immune function has been recognized. For instance, the arginine plasma concentrations are strongly related to the severity of systemic inflammation, being especially low during the acute phase of critical illness [46]. Dietary supplementation with arginine might have positive effects on immune function and reparative collagen synthesis [47]. The role of glutamine supplementation is controversial: experimental work proposed various mechanisms of action, but none of the randomized studies in early life showed any effect on mortality and only a few showed some effect on inflammatory response, organ function, and a trend for infection control [47,48]. Briassoulis et al., in a blinded, randomized, controlled trial, compared nitrogen balance, biochemical indices, antioxidant catalysts, and clinical outcomes in critically-ill children given an immune-enhancing formula or conventional early enteral nutrition [49]. In their cohort, immunonutrition had a favorable effect on some biochemical indices (e.g., natremia) and antioxidant catalysts. However, the mortality rate did not differ between the two groups [49]. A further single-center, randomized, blinded controlled trial in 38 children with septic shock, performed by the same group, compared the effect of early enteral feeding using immune-enhancing with non-immune-enhancing formulas on cytokines. The study showed that immune-enhancing nutrition can interfere with the production of interleukin-6, but no evidence was found regarding the impact on the short-term outcome [50]. Finally, no clinical effect was provided by the immune-enhancing diet in a study including 40 ventilated children with severe head injury [51].

Overall, the present section suggests that, besides the role of the appropriate timing of the parenteral and enteral nutrition, some specific nutrients, particularly AAs, may contribute to an improvement of the immune response. Newer formulations of enteral or parenteral mixtures of AAs meeting the individual needs of different critically-ill populations should be tested [10].

3.5. Energy Expenditure Assessment

In critically-ill children, measured or calculated REE has been proposed to estimate energy intake requirements in children [2]. The five most commonly-used REE prediction formulas are: the World Health Organization (WHO) formula, the Harris–Benedict formula, the Schofield formula based on weight, the Schofield formula based on weight and height, and the Oxford formula [25]. These formulas have been validated in healthy children. However, they were found inaccurate in critically-ill children [38,40,52]. In a prospective cohort study, performed in a PICU setting, standard equations overestimated the energy expenditure and an 83% incidence of overfeeding with cumulative energy excess of up to 8000 kcal/week was observed [11]. Another prospective study found that measured energy expenditure (MEE) during critical illness was much lower than the energy expenditure predicted by formulas [53]. Further studies came to similar conclusions [54,55].

IC has been proposed as a reference method to measure REE. This technique is used to measure the rate of energy production and substrate oxidation in children, both in clinical and research settings. The recent clinical guidelines of the American Society for Parenteral and Enteral Nutrition (ASPEN) for nutritional support of the critically-ill child, suggest that IC measurements should be obtained in patients with suspected metabolic alterations or malnutrition [6]. A study including 150 patients found that 72% of PICU patients were candidates for IC accordingly to the ASPEN guidelines [56]. Particularly, authors suggested prioritizing performing IC in patients <2 years of age, malnourished (underweight/overweight) on admission, or with a PICU stay of >5 days [56].

In recent years, the reliability of ventilator-derived carbon dioxide production (VCO_2) equations was investigated. These simplified methods measure the VCO_2 derived from measurements of exhaled gas volume and carbon dioxide (CO_2) concentrations and seem to be a promising tool when IC is not available or applicable [57]. Recent studies have demonstrated that this technique is a promising option for the determination of energy requirements in children on mechanical ventilation [30]. However, since the variability of RQ (respiratory quotient) influences the accuracy of the $EEVCO_2$ calculation (EE from CO_2 measurements), and many of these approaches assume that the RQ value is a fixed value, the validity of this technique as an alternative to IC is questionable in some circumstances [57].

Strides have been made to build new, compact metabolic monitors to measure REE in PICU and to validate them. However, again, there is a wide range of conditions that may compromise their accuracy. For instance, metabolic monitors' errors were shown to be significantly affected by oxygen concentration and minute ventilation and when used during inhaled anesthesia [58].

4. Conclusions

For critically-ill children, the role of nutrition is evolving from a simple supportive function to the possibility of an effective co-adjuvant therapy. However, the variability of metabolic responses to stress requires testing the hypothesis of an individualized approach to nutrition in critically-ill children, since available data are mainly derived from healthy individuals [59]. Additionally, the possible "intermediary" role of the microbiome should be investigated in future studies in PICUs [59]. Unfortunately, as previously mentioned, in stress conditions, the futile cycle of nutrients may make most interventions that are effective in healthy subjects useless. Indeed, inconclusive data are available in most important issues of critically-ill child nutrition (such as the glycemic control) and no international recommendation exists for the management of many common problems in their day-to-day care [10]. The determination of individual macronutrients needed in the various phases of stress are still more hypothetical than evidence-based. Finally, problems with the current predictive equations and lack of availability of IC are likely to result in continued under- or overfeeding in many critically-ill children, with the associated morbidity [60]. New efforts are urgently needed to develop individualized nutrition strategies and evaluate their effect on relevant health outcomes.

Table 1. Randomized clinical trials on glycemic control and nutrition in critically-ill children.

Authors, years	Country	Groups	Primary Outcomes	Results
Vlasselaers et al. [13]	Belgium	Tight glycemic control (interventional group) $n = 700$, conventional glycemic control (control group) $n = 351$, age = 0–16 years. Statistical power = 80%	Effect of tight glycemic control on duration of PICU stay and inflammation	PICU stay was shorter (5.5 vs. 6.2 days, $p = 0.017$) and C-reactive protein change after 5 days lower (−9.8 mg/L vs. 9.0 mg/L, $p = 0.007$) in interventional vs. control group
Mesotten et al. [14]	Belgium	Tight glycemic control (interventional group) $n = 222$, conventional glycemic control (control group) $n = 234$, age ≤ 16 years. Statistical power = 80%	Effect of tight glycemic control on long-term follow-up of neuro-cognitive-outcomes	No significant difference between the two groups was observed
Jeschke et al. [16]	USA	Tight glycemic control (interventional group) $n = 60$, conventional glycemic control (control group) $n = 159$, age = 0–16 years. The study was underpowered	Effect of tight glycemic control on infectious events	Sepsis was less frequent ($p < 0.05$) in interventional than in control group (8.2% and 22.6% of patients, respectively)
Macrae et al. [17]	England	Tight glycemic control (interventional group) $n = 694$, conventional glycemic control (control group) $n = 675$, age = 0–16 years. Statistical power = 80%	Effect of tight glycemic control on days alive and free from and free from mechanical ventilation at 30 days after enrollment	No significant difference between the two groups was observed
Agus et al. [18]	USA	Tight glycemic control (interventional group) $n = 349$, conventional glycemic control (control group) $n = 360$, age = 2 weeks–17 years. Statistical power = 80%	Effect of tight glycemic control on length of PICU stay	No significant difference between the two groups was observed
Agus et al. [19]	USA	Tight glycemic control (interventional group) $n = 490$, conventional glycemic control (control group) $n = 490$, age = 0–36 months. Statistical power = 80%	Effect of tight glycemic control on mortality, length of PICU stay, and infectious events	No significant difference between the two groups was observed
Sadhwani et al. [21]	USA	Tight glycemic control (interventional group) $n = 121$, conventional glycemic control (control group) $n = 116$, age = 0–36 months. Statistical power not reported	Effect of tight glycemic control on neurodevelopment follow-up	No significant difference between the two groups was observed
Vanhorebee et al. [22]	Belgium	Tight glycemic control (interventional group) $n = 349$, conventional glycemic control (control group) $n = 351$, age = 0–16 years. Statistical power = 80%	Effect of tight glycemic control on neurological injury biomarkers	No significant difference between the two groups was observed
Vanhorebee et al. [44]	Belgium, Netherlands, Canada	Early parenteral nutrition (interventional group) $n = 723$, late parenteral nutrition (control group) $n = 717$, age = 0–17 years. Statistical power = 70%	Effect of macronutrients supplementation timing on infections, need of mechanical ventilation, and length of PICU stay	The early provision of amino-acids, and not glucose or lipids, was associated with worse outcomes
Briassoulis et al. [49]	Greece	Immunonutrition (interventional group), $n = 25$. Conventional enteral nutrition (control group) $n = 25$, age = 8–9.2 years. Statistical power not reported	Effect of immunonutrition on biochemical nutritional markers and hard outcomes (mortality, length of PICU stay, and need of mechanical ventilation)	Immunonutrition had a favorable effect on few nutritional biochemical markers, but not on hard outcomes

Table 1. *Cont.*

Briassoulis et al. [50]	Greece	Immunonutrition (interventional group), $n = 15$, conventional enteral nutrition (control group) $n = 15$, age = 6.5–7.9 years. Statistical power not reported	Effect of immunonutrition on interleukins in septic children	IL-6 levels were lower (11.8 vs. 38.3 pg/mL, $p < 0.001$) and IL-8 higher (65.4 vs. 21 pg/mL, $p < 0.03$) in interventional group compared with control group
Briassoulis et al. [51]	Greece	Immunonutrition (interventional group), $n = 20$, conventional enteral nutrition (control group) $n = 20$, age = 6–10.5 years. Statistical power not reported	Effect of immunonutrition on biochemical nutritional markers and hard outcomes (mortality, length of PICU stay, and need of mechanical ventilation) in severe head injury patients	Except for IL-8 levels and nitrogen balance, no difference was observed between the two groups

PICU, pediatric intensive care unit.

Acknowledgments: The authors thank Alec Villa for the linguistic revision and all the members of the Pediatric Unit for their kindly support.

Author Contributions: C.A. conceived the review. V.D.C. searched the existing literature and selected the included studies. V.D.C., G.P.M., A.M., and V.D. wrote the manuscript under the supervision of C.A., and M.S. and E.C. contributed to revising the manuscript and gave a significant contribution in their area of expertise. All authors read and approved the final manuscript.

Conflicts of Interest: The authors declare no conflict of interest.

References

1. Chwals, W.J. Hyperglycemia management strategy in the pediatric intensive care setting. *Pediatr. Crit. Care Med.* **2008**, *9*, 656–658. [CrossRef] [PubMed]

2. Joosten, K.F.; Kerklaan, D.; Verbruggen, S.C. Nutritional support and the role of the stress response in critically ill children. *Curr. Opin. Clin. Nutr. Metab. Care* **2016**, *19*, 226–233. [CrossRef] [PubMed]

3. Beisel, W.R. Metabolic effects of infection. *Prog. Food Nutr. Sci.* **1984**, *8*, 43–75. [PubMed]

4. Wilson, B.; Typpo, K. Nutrition: A Primary Therapy in Pediatric Acute Respiratory Distress Syndrome. *Front. Pediatr.* **2016**, *4*, 108. [CrossRef] [PubMed]

5. Powanda, M.C.; Beisel, W.R. Metabolic effects of infection on protein and energy status. *J. Nutr.* **2003**, *133*, 322S–327S. [PubMed]

6. Mehta, N.M.; Compher, C.; A.S.P.E.N. Board of Directors. ASPEN Clinical Guidelines: Nutrition support of the critically ill child. *JPEN J. Parenter. Enter. Nutr.* **2009**, *33*, 260–276. [CrossRef] [PubMed]

7. Keller, K.L.; Adise, S. Variation in the Ability to Taste Bitter Thiourea Compounds: Implications for Food Acceptance, Dietary Intake, and Obesity Risk in Children. *Annu. Rev. Nutr.* **2016**, *36*, 157–182. [CrossRef] [PubMed]

8. Preiser, J.C.; Ichai, C.; Orban, J.C.; Groeneveld, A.J. Metabolic response to the stress of critical illness. *Br. J. Anaesth.* **2014**, *113*, 945–954. [CrossRef] [PubMed]

9. De Cosmi, V.; Mehta, N.M.; Boccazzi, A.; Milani, G.P.; Esposito, S.; Bedogni, G.; Agostoni, C. Nutritional status, metabolic state and nutrient intake in children with bronchiolitis. *Int. J. Food Sci. Nutr.* **2017**, *68*, 378–383. [CrossRef] [PubMed]

10. Tavladaki, T.; Spanaki, A.M.; Dimitriou, H.; Briassoulis, G. Alterations in metabolic patterns in critically ill patients-is there need of action? *Eur. J. Clin. Nutr.* **2017**, *71*, 431–433. [CrossRef] [PubMed]

11. Mehta, N.M.; Bechard, L.J.; Dolan, M.; Ariagno, K.; Jiang, H.; Duggan, C. Energy imbalance and the risk of overfeeding in critically ill children. *Pediatr. Crit. Care Med.* **2011**, *12*, 398. [CrossRef] [PubMed]

12. Van den Berghe, G.; Wouters, P.; Weekers, F.; Verwaest, C.; Bruyninckx, F.; Schetz, M.; Bouillon, R. Intensive insulin therapy in critically ill patients. *N. Engl. J. Med.* **2001**, *345*, 1359–1367. [CrossRef] [PubMed]

13. Vlasselaers, D.; Milants, I.; Desmet, L.; Wouters, P.J.; Vanhorebeek, I.; van den Heuvel, I.; Mesotten, D.; Casaer, M.P.; Meyfroidt, G.; Ingels, C.; et al. Intensive insulin therapy for patients in pediatric intensive care: A prospective, randomised controlled study. *Lancet* **2009**, *373*, 547–556. [CrossRef]

14. Mesotten, D.; Gielen, M.; Sterken, C.; Claessens, K.; Hermans, G.; Vlasselaers, D.; Lemiere, J.; Lagae, L.; Gewillig, M.; Eyskens, B.; et al. Neurocognitive development of children 4 years after critical illness and treatment with tight glucose control: A randomized controlled trial. *JAMA* **2012**, *308*, 1641–1650. [CrossRef] [PubMed]

15. Day, K.M.; Haub, N.; Betts, H.; Inwald, D.P. Hyperglycemia is associated with morbidity in critically ill children with meningococcal sepsis. *Pediatr. Crit. Care Med.* **2008**, *9*, 636–640. [CrossRef] [PubMed]

16. Jeschke, M.G.; Kulp, G.A.; Kraft, R.; Finnerty, C.C.; Mlcak, R.; Lee, J.O.; Herndon, D.N. Intensive insulin therapy in severely burned pediatric patients: A prospective randomized trial. *Am. J. Respir. Crit. Care Med.* **2010**, *182*, 351–359. [CrossRef] [PubMed]

17. Macrae, D.; Grieve, R.; Allen, E.; Sadique, Z.; Morris, K.; Pappachan, J.; Parslow, R.; Tasker, R.C.; Elbourne, D.A. Randomized trial of hyperglycemic control in pediatric intensive care. *N. Engl. J. Med.* **2014**, *370*, 107–118. [CrossRef] [PubMed]

18. Agus, M.S.; Wypij, D.; Hirshberg, E.L.; Srinivasan, V.; Faustino, E.V.; Luckett, P.M.; Nadkarni, V.M. Tight Glycemic Control in Critically Ill Children. *N. Engl. J. Med.* **2017**, *376*, 729–741. [CrossRef] [PubMed]

19. Agus, M.S.; Steil, G.M.; Wypij, D.; Costello, J.M.; Laussen, P.C.; Langer, M.; Ohye, R.G. Tight glycemic control versus standard care after pediatric cardiac surgery. *N. Engl. J. Med.* **2012**, *367*, 1208–1219. [CrossRef] [PubMed]

20. Rossano, J.W.; Taylor, M.D.; Smith, E.B.; Fraser, C.D.; McKenzie, E.D.; Price, J.F.; Mott, A.R. Glycemic profile in infants who have undergone the arterial switch operation: Hyperglycemia is not associated with adverse events. *J. Thorac. Cardiovasc. Surg.* **2008**, *135*, 739–745. [CrossRef] [PubMed]

21. Sadhwani, A.; Asaro, L.A.; Goldberg, C.; Ware, J.; Butcher, J.; Gaies, M.; Smith, C.; Alexander, J.L.; Wypij, D.; Agus, M.S. Impact of Tight Glycemic Control on Neurodevelopmental Outcomes at 1 Year of Age for Children with Congenital Heart Disease: A Randomized Controlled Trial. *J. Pediatr.* **2016**, *174*, 193–198. [CrossRef] [PubMed]

22. Vanhorebeek, I.; Gielen, M.; Boussemaere, M.; Wouters, P.J.; Grandas, F.G.; Mesotten, D.; Van den Berghe, G. Glucose dysregulation and neurological injury biomarkers in critically ill children. *J. Clin. Endocrinol. Metab.* **2010**, *95*, 4669–4679. [CrossRef] [PubMed]

23. Briassoulis, G. Are Early Parenteral Nutrition and Intensive Insulin Therapy What Critically Ill Children Need? *Pediatr. Crit. Care Med.* **2014**, *15*, 371–372. [CrossRef] [PubMed]

24. Şimşek, T.; Şimşek, H.U.; Cantürk, N.Z. Response to trauma and metabolic changes: Posttraumatic metabolism. *Ulus. Cerrahi Derg.* **2014**, *30*, 153–159. [PubMed]

25. Felts, P.W. Ketoacidosis. *Med. Clin. N. Am.* **1983**, *67*, 831–843. [CrossRef]

26. Lang, T.F.; Hussain, K. Pediatric hypoglycemia. *Adv. Clin. Chem.* **2014**, *63*, 211–245. [PubMed]

27. Dennhardt, N.; Beck, C.; Huber, D.; Nickel, K.; Sander, B.; Witt, L.H.; Boethig, D.; Sümpelmann, R. Impact of preoperative fasting times on blood glucose concentration, ketone bodies and acid-base balance in children younger than 36 months: A prospective observational study. *J. Anaesthesiol.* **2015**, *32*, 857–861. [CrossRef] [PubMed]

28. Van Veen, M.R.; van Hasselt, P.M.; de Sain-van der Velden, M.G.; Verhoeven, N.; Hofstede, F.C.; de Koning, T.J.; Visser, G. Metabolic profiles in children during fasting. *Pediatrics* **2011**, *127*, 1021–1027. [CrossRef] [PubMed]

29. Martindale, R.G.; Warren, M.; Diamond, S.; Kiraly, L. Nutritional Therapy for Critically Ill Patients. *Nestle Nutr. Inst. Workshop Ser.* **2015**, *82*, 103–116. [PubMed]

30. Dornelles, C.T.; Piva, J.P.; Marostica, P.J. Nutritional status, breastfeeding, and evolution of Infants with acute viral bronchiolitis. *J. Health Popul. Nutr.* **2007**, *25*, 336–343. [PubMed]

31. Hulst, J.; Joosten, K.; Zimmermann, L.; Hop, W.; van Buuren, S.; Büller, H.; van Goudoever, J. Malnutrition in critically ill children: From admission to 6 months after discharge. *Clin. Nutr.* **2004**, *23*, 223–232. [CrossRef]

32. Levine, B.; Mizushima, N.; Virgin, H.W. Autophagy in immunity and inflammation. *Nature* **2011**, *469*, 323. [CrossRef] [PubMed]

33. Casaer, M.P.; Wilmer, A.; Hermans, G.; Wouters, P.J.; Mesotten, D.; Van den Berghe, G. Role of disease and macronutrient dose in the randomized controlled EPaNIC trial: A post hoc analysis. *Am. J. Respir. Crit. Care Med.* **2013**, *187*, 247–255. [CrossRef] [PubMed]

34. Hermans, G.; Casaer, M.P.; Clerckx, B.; Güiza, F.; Vanhullebusch, T.; Derde, S.; Meersseman, P.; Derese, I.; Mesotten, D.; Wouters, P.J.; et al. Effect of tolerating macronutrient deficit on the development of intensive-care unit acquired weakness: A subanalysis of the EPaNIC trial. *Lancet Respir. Med.* **2013**, *1*, 621–629. [CrossRef]

35. Puthucheary, Z.A.; Rawal, J.; McPhail, M.; Connolly, B.; Ratnayake, G.; Chan, P.; Hopkinson, N.S.; Phadke, R.; Dew, T.; Sidhu, P.S.; et al. Acute skeletal muscle wasting in critical illness. *JAMA* **2013**, *310*, 1591–1600. [CrossRef] [PubMed]

36. Kerklaan, D.; Hulst, J.M.; Verhoeven, J.J.; Verbruggen, S.C.; Joosten, K.F. Use of Indirect Calorimetry to Detect Overfeeding in Critically Ill Children: Finding the Appropriate Definition. *J. Pediatr. Gastroenterol. Nutr.* **2016**, *63*, 445–450. [CrossRef] [PubMed]

37. Agostoni, C.; Edefonti, A.; Calderini, E.; Fossali, E.; Colombo, C.; Battezzati, A.; Bertoli, S.; Milani, G.; Bisogno, A.; Perrone, M.; et al. Accuracy of Prediction Formula for the Assessment of Resting Energy Expenditure in Hospitalized Children. *J. Pediatr. Gastroenterol. Nutr.* **2016**, *63*, 708–712. [CrossRef] [PubMed]

38. Briassoulis, G.; Briassouli, E.; Tavladaki, T.; Ilia, S.; Fitrolaki, D.M.; Spanaki, A.M. Unpredictable combination of metabolic and feeding patterns in malnourished critically ill children: The malnutrition–energy assessment question. *Intensive Care Med.* **2014**, *40*, 120–122. [CrossRef] [PubMed]

39. Boonen, E.; Van den Berghe, G. Endocrine responses to critical illness: Novel insights and therapeutic implications. *J. Clin. Endocrinol. Metab.* **2014**, *99*, 1569–1582. [CrossRef] [PubMed]

40. Briassoulis, G.; Venkataraman, S.; Thompson, A.E. Energy expenditure in critically ill children. *Crit. Care. Med.* **2000**, *28*, 1166–1172. [CrossRef] [PubMed]

41. Jotterand Chaparro, C.; Laure Depeyre, J.; Longchamp, D.; Perez, H.M.; Taffé, P.; Cotting, J. How much protein and energy are needed to equilibrate nitrogen and energy balances in ventilated critically ill children? *Clin. Nutr.* **2015**, *35*, 460–467. [CrossRef] [PubMed]

42. Mehta, N.M.; Bechard, L.J.; Zurakowski, D.; Duggan, C.P.; Heyland, D.K. Adequate enteral protein intake is inversely associated with 60-d mortality in critically ill children: A multicenter, prospective, cohort study. *Am. J. Clin. Nutr.* **2015**, *102*, 199–206. [CrossRef] [PubMed]

43. Fivez, T.; Kerklaan, D.; Verbruggen, S.; Vanhorebeek, I.; Verstraete, S.; Tibboel, D.; Guerra, G.G.; Wouters, P.J.; Joffe, A.; Joosten, K.; et al. Impact of withholding early parenteral nutrition completing enteral nutrition in pediatric critically ill patients (PEPaNIC trial): Study protocol for a randomized controlled trial. *Trials* **2015**, *16*, 202. [CrossRef] [PubMed]

44. Vanhorebeek, I.; Verbruggen, S.; Casaer, M.P.; Gunst, J.; Wouters, P.J.; Hanot, J.; Guerra, G.G.; Vlasselaers, D.; Joosten, K.; Van den Berghe, G. Effect of early supplemental parenteral nutrition in the paediatric ICU: A preplanned observational study of post-randomisation treatments in the PEPaNIC trial. *Lancet Respir. Med.* **2017**, *5*, 475–483. [CrossRef]

45. Mikhailov, T.A.; Kuhn, E.M.; Manzi, J.; Christensen, M.; Collins, M.; Brown, A.M.; Dechert, R.; Scanlon, M.C.; Wakeham, M.K.; Goday, P.S. Early enteral nutrition is associated with lower mortality in critically ill children. *JPEN J. Parenter. Enteral Nutr.* **2014**, *38*, 459–466. [CrossRef] [PubMed]

46. Van Waardenburg, D.A.; de Betue, C.T.; Luiking, Y.C.; Engel, M.; Deutz, N.E. Plasma arginine and citrulline concentrations in critically ill children: Strong relation with inflammation. *Am. J. Clin. Nutr.* **2007**, *86*, 1438–1444. [PubMed]

47. Pérez de la Cruz, A.J.; Abilés, J.; Pérez Abud, R. Perspectives in the design and development of new products for enteral nutrition. *Nutr. Hosp.* **2006**, *21*, 98–108. [PubMed]

48. Briassouli, E.; Briassoulis, G. Glutamine randomized studies in early life: The unsolved riddle of experimental and clinical studies. *Clin. Dev. Immunol.* **2012**, *2012*, 749189. [CrossRef] [PubMed]

49. Briassoulis, G.; Filippou, O.; Hatzi, E.; Papassotiriou, I.; Hatzis, T. Early enteral administration of immunonutrition in critically ill children: Results of a blinded randomized controlled clinical trial. *Nutrition* **2005**, *21*, 799–807. [CrossRef] [PubMed]

50. Briassoulis, G.; Filippou, O.; Kanariou, M.; Hatzis, T. Comparative effects of early randomized immune or non-immune-enhancing enteral nutrition on cytokine production in children with septic shock. *Intensive Care Med.* **2005**, *31*, 851–858. [CrossRef] [PubMed]

51. Briassoulis, G.; Filippou, O.; Kanariou, M.; Papassotiriou, I.; Hatzis, T. Temporal nutritional and inflammatory changes in children with severe head injury fed a regular or an immune-enhancing diet: A randomized, controlled trial. *Pediatr. Crit. Care Med.* **2006**, *7*, 56–62. [CrossRef] [PubMed]

52. Briassoulis, G.; Ilia, S.; Meyer, R. Enteral Nutrition in PICUs: Mission Not Impossible! *Pediatr. Crit. Care Med.* **2016**, *17*, 85–87. [CrossRef] [PubMed]
53. Mehta, N.M.; Smallwood, C.D.; Joosten, K.F.; Hulst, J.M.; Tasker, R.C.; Duggan, C.P. Accuracy of a simplified equation for energy expenditure based on bedside volumetric carbon dioxide elimination measurement-a two-center study. *Clin. Nutr.* **2015**, *34*, 151–155. [CrossRef] [PubMed]
54. Rousing, M.L.; Hahn-Pedersen, M.H.; Andreassen, S.; Pielmeier, U.; Preiser, J.C. Energy expenditure in critically ill patients estimated by population-based equations, indirect calorimetry and CO_2-based indirect calorimetry. *Ann. Intensive Care* **2016**, *6*, 16. [CrossRef] [PubMed]
55. Briassoulis, G.; Venkataraman, S.; Thompson, A. Cytokines and metabolic patterns in pediatric patients with critical illness. *Clin. Dev. Immunol.* **2010**, *2010*, 354047. [CrossRef] [PubMed]
56. Kyle, U.G.; Arriaza, A.; Esposito, M.; Coss-Bu, J.A. Is indirect calorimetry a necessity or a luxury in the pediatric intensive care unit? *JPEN J. Parenter. Enter. Nutr.* **2012**, *36*, 177–182. [CrossRef] [PubMed]
57. Oshima, T.; Graf, S.; Heidegger, C.P.; Genton, L.; Pugin, J.; Pichard, C. Can calculation of energy expenditure based on CO2 measurements replace indirect calorimetry? *Crit. Care Med.* **2017**, *21*, 13. [CrossRef] [PubMed]
58. Briassoulis, G.; Briassoulis, P.; Michaeloudi, E.; Fitrolaki, D.M.; Spanaki, A.M.; Briassouli, E. The effects of endotracheal suctioning on the accuracy of oxygen consumption and carbon dioxide production measurements and pulmonary mechanics calculated by a compact metabolic monitor. *Anesth. Analg.* **2009**, *109*, 873–879. [CrossRef] [PubMed]
59. Zeevi, D.; Korem, T.; Zmora, N.; Israeli, D.; Rothschild, D.; Weinberger, A.; Ben-Yacov, O.; Lador, D.; Avnit-Sagi, T.; Lotan-Pompan, M.; et al. Personalized Nutrition by Prediction of Glycemic Responses. *Cell* **2015**, *163*, 1079–1094. [CrossRef] [PubMed]
60. Meyer, R.; Kulinskaya, E.; Briassoulis, G.; Taylor, R.M.; Cooper, M.; Pathan, N.; Habibi, P. The challenge of developing a new predictive formula to estimate energy requirements in ventilated critically ill children. *Nutr. Clin. Pract.* **2012**, *27*, 669–676. [CrossRef] [PubMed]

nutrients

MDPI

Article

Neuroprotective Role of Atractylenolide-I inan In Vitro and In Vivo Model of Parkinson's Disease

Sandeep More and Dong-Kug Choi *

Department of Biotechnology, College of Biomedical and Health Science, Konkuk University,
Chungju 27478, Korea; sandeepbcp@gmail.com
* Correspondence: choidk@kku.ac.kr; Tel.: +82-43-840-3610

Received: 11 January 2017; Accepted: 20 March 2017; Published: 2 May 2017

Abstract: Parkinson's disease (PD) is an age-related neurological disorder characterized by a loss of dopaminergic neurons within the midbrain. Neuroinflammation has been nominated as one of the key pathogenic features of PD. Recently, the inadequate pharmacotherapy and adverse effects of conventional drugs have spurred the development of unconventional medications in the treatment of PD. The purpose of this study is to investigate the anti-neuroinflammatory mechanisms of Atractylenolide-I (ATR-I) in in vivo and in vitro models of PD. Nitrite assay was measured via Griess reaction in lipopolysaccharide (LPS) stimulated BV-2 cells. mRNA and protein levels were determined by a reverse transcription-polymerase chain reaction (RT-PCR) and immunoblot analysis, respectively. Further, flow cytometry, immunocytochemistry, and immunohistochemistry were employed in BV-2 cells and MPTP-intoxicated C57BL6/J mice. Pre-treatment with ATR-I attenuated the inflammatory response in BV-2 cells by abating the nuclear translocation of nuclear factor-κB (NF-κB) and by inducing heme oxygenase-1 (HO-1). The intraperitoneal administration of ATR-I reversed MPTP-induced behavioral deficits, decreased microglial activation, and conferred protection to dopaminergic neurons in the mouse model of PD. Our experimental reports establish the involvement of multiple benevolent molecular events by ATR-I in MPTP-induced toxicity, which may aid in the development of ATR-I as a new therapeutic agent for the treatment of PD.

Keywords: Atractylenolide-I; astrocyte; microglia; neuroinflammation; Parkinson's disease

1. Introduction

Parkinson's disease (PD) is a progressive, neurodegenerative disorder characterized by the loss of dopaminergic neurons in the substantia nigra pars compacta (SNpc) and microglial malfunction. Accumulating evidence from various scientific studies suggests that neuroinflammation is one of the important factors involved in the selective degeneration of the nigral neurons in PD [1–4]. Extracts from medicinal plants and their secondary metabolites have conventionally been used to treat numerous clinical diseases, including inflammation-associated diseases [5]. Atractylenolide-I (ATR-I) is one of the major bioactive ingredients that have been isolated from the rhizomes of *Atractylodes macrocephala* (*A. macrocephala*). *A. macrocephala* is also popularly known as "*Baizhu*" in traditional Chinese medicine and has long been used for a variety of corresponding medicinal properties [6]. This species of plant belongs to the Asteraceae family, and its distribution has been found in some Asian countries, such as Japan, China, and Korea. ATR-I, chemically a sesquiterpene, is a highly lipophilic volatile lactone. In terms of the activities elicited by ATR-I in a biological system, pharmacological studies have demonstrated its gastrointestinal inhibitory effects [7], anti-oxidant activity [8], and anti-cancer activity [9]. ATR-I has been reported to increase the hunger and mid-arm muscle circumference of patients with cachectic cancer, and it can significantly inhibit serum levels of interleukin-1 (IL-1), tumor necrosis factor-α (TNF-α), and urine proteolysis-inducing factors [10]. Several reports have illustrated that ATR-I can selectively antagonize the toll-like receptor-4 or the membrane-bound glucocorticoid receptor, to elicit an anti-inflammatory effect [11,12]. Apart from the bioactivities reported for ATR-I, a systematic and detailed examination of its anti-inflammatory mechanism has not yet been performed. Additionally, the chronic use of existing symptomatic treatment is often associated with debilitating side effects [13] and none seem to halt the progression of PD. So far, the development of effective anti-neuroinflammatory therapies has been impeded by our limited knowledge of the pathogenesis of PD.

Based on these existing evidences, the purpose of our present study is to unravel the mechanisms responsible for the anti-neuroinflammatory effect of ATR-I in LPS- and MPTP-induced neuroinflammatory PD models in BV-2 cells and C57BL/6J mice, respectively. Our results demonstrate the involvement of multiple molecular events mediated by ATR-I for its anti-neuroinflammatory effect in LPS-induced BV-2 cells and MPTP-induced neurotoxicity. Our results may aid in the development of ATR-I as a new therapeutic agent for controlling inflammation associated with PD.

2. Materials and Methods

2.1. Material

ATR-I (purity $\geq$ 98%) was purchased from BaoJi Herbest Bio-Tech Co., Ltd. (Baoji, China). LPS (*Escherichia coli* 0111:B4, Sigma, St. Louis, MO, USA), MPP$^+$, MPTP, Tween-20, bovine serum albumin, dimethyl sulfoxide (DMSO), sodium nitrite, sulfanilamide, *N*-(1-naphthyl) ethylenediamine dihydrochloride, and zinc protoporphyrin IX (ZnPP-IX) were purchased from Sigma. 2,7-dichlorodihydrofluorescein diacetate (H$_2$DCFDA) was obtained from Invitrogen. The 6-well and 96-well tissue culture plates and 100 mm culture dishes were purchased from Nunc Inc. (Aurora, IL, USA). Fetal bovine serum (FBS) was purchased from PAA Laboratories Inc. (Etobicoke, ON, Canada). Dulbecco's modified Eagle medium (DMEM) and phosphate buffer saline (PBS), as well as other cell culture reagents, were obtained from Gibco/Invitrogen (Carlsbad, CA, USA). The protease-inhibitor cocktail tablets and phosphatase-inhibitor cocktail tablets were supplied by Roche (Indianapolis, IN, USA). Antibodies for NF-κB/p65 and nucleolin were obtained from Santa Cruz Biotechnology (Santa Cruz, CA, USA), and the antibodies for inducible nitric oxide synthase (iNOS), ERK, p-ERK, Akt, p-Akt, PI3K, p-PI3K, the inhibitor of kappa B-alpha (IκB-α), p-IκB-α, and β-actin were supplied by Cell Signaling Technology (Danvers, MA, USA). Antibodies for heme oxygenase (HO-1) and manganese superoxide dismutase (MnSOD) were procured from Enzo Life Sciences (Farmingdale, NY, USA). The following vendors were used for the procurement of other antibodies: macrophage-1 antigen (Mac-1) (AbD Serotec), glial fibrillary acidic protein (GFAP) (Abcam, Cambridge, UK), and tyrosine hydroxylase (TH) (Calbiochem, MA, USA). All of the other chemicals used in this study are of analytical grade and were obtained, unless otherwise noted, from Sigma Chemical Co. (St. Louis, MO, USA).

2.2. BV-2 Cell Culture

The BV-2 immortalized murine microglial cell line [14] was provided by Dr. Kyungho Suk, Kyungbook National University. The method used for culturing the BV-2 cells was followed according to our previously published report [15]. ATR-I was dissolved in DMSO. The control/vehicle group was treated with only DMSO.

2.3. Animals and MPTP Administration

Male C57BL6/J mice (age, 8–9 weeks; weight, 25–28 g) (Samtako Bio Korea, Gyeonggi-do, Korea) were acclimatized for 14 days, prior to the drug treatment. Animal experiments and the experimental procedures were approved by the Institutional Animal Care and Use Committee of Konkuk University. The animals were housed in a controlled environment (23 °C ± 1 °C and 50% ± 5% humidity) and were allowed food and water *ad libitum*. The room's lights were switched on between 8:00 a.m. and 8:00 p.m. One hundred and twenty six animals were randomly divided into the following seven groups, comprised of 18 animals each: Vehicle group, MPTP group, ATR-I (30 mg/kg/10 mL) group, Selegiline (10 mg/kg/10 mL) + MPTP, ATR-I (3 mg/kg/10 mL) + MPTP group, ATR-I (10 mg/kg/10 mL) + MPTP group, and ATR-I (30 mg/kg/10 mL) + MPTP group. As demonstrated in the experimental design (Scheme 1), a 10 mg/kg/10 mL MPTP was administered intraperitoneally (i.p.) on four occasions, at an interval of 1 h [16,17], to all of the animals except for those in the vehicle group and ATR-I (30 mg/kg/10 mL) i.e., per se group. Twelve hours after the last MPTP injection, ATR-I (i.p.) or selegiline (i.p.) was administered once daily, for three and seven days. The animals were sacrificed on day three and day seven. The MPTP was freshly prepared in saline before use. ATR-I was suspended in 0.5% methyl cellulose solution containing 1% Tween 80, before dosing. The control/vehicle group was administered as plain methyl cellulose solution.

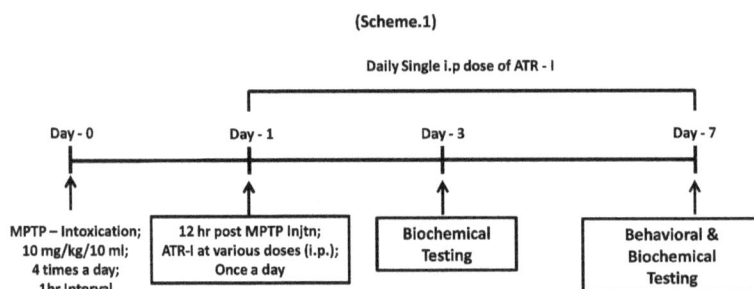

(Scheme.1)

Daily Single i.p dose of ATR - I

Day - 0	Day - 1	Day - 3	Day - 7
MPTP – Intoxication; 10 mg/kg/10 ml; 4 times a day; 1hr Interval	12 hr post MPTP Injtn; ATR-I at various doses (i.p.); Once a day	Biochemical Testing	Behavioral & Biochemical Testing

Scheme 1. Experimental design illustrating the treatment and testing schedule for animal experiments.

2.4. Pole Test

To measure the degree of bradykinesia, a typical symptom of Parkinsonism, a pole test was performed with a slightly modified reported procedure [18]. This test was successively performed three times for each mouse. The experimenter was kept uninformed to the experimental groups.

2.5. Nitrite Measurement Assay

The release of nitric oxide (NO) was measured using the Griess reaction in the cell supernatant [19]. BV-2 cells (2.5×10^4 cells/mL) were seeded in 96-well plates in 200 µL culture medium, pre-treated with three different doses of ATR-I (25 µM, 50 µM, and 100 µM) for 1 h, and were then stimulated with or without LPS (100 ng/mL) for 24 h. The further procedure was carried out in accordance with our previously published data [15]. The results are representative of three independent experiments.

2.6. Immunohistochemistry

The procedure was followed according to our previously published data [20], with slight modifications. The free-floating brain sections from striatum (STR) and SNpc from the mice sacrificed on day three and seven post-MPTP injection were tagged with primary antibodies, including anti-Mac-1 (1:500; AbD Serotec, Oxford, UK) and an anti-TH antibody (1:1500; Calbiochem San Diego, CA, USA), respectively. The samples were visualized by a DAB Peroxidase (HRP) Substrate Kit (Vector Laboratories, CA, USA) following the manufacturers protocol. Fluorescence staining was performed with the primary antibodies of anti-GFAP (1:2000; Abcam), incubated overnight at 4 °C. The sections were then incubated with rabbit antigoat GFAP secondary antibody (1:200) (Alexa Flour 488, Invitrogen, Carlsbad, CA, USA) for 1 h. Stained cells were viewed using a bright field and fluorescence microscope (Carl Zeiss). The quantification of the effects in the brain-tissue sections was performed by using Image J software (Bethesda, MD, USA) to count the number of TH-positive cells in SNpc and STR, respectively. Data were presented as a percentage of the control-group values.

2.7. Isolation of Total RNA and Reverse Transcription-Polymerase Chain Reaction (RT-PCR)

ATR-I at varying concentrations was pre-incubated with BV-2 cells (5.0×105 cells/mL) for 1 h, and was then treated with or without LPS (100 ng/mL) for 6 h. The animal tissues were washed in cold 0.1 M PBS (pH 7.4) and homogenized using a tuberculin syringe with TRIzol, and were subsequently stored at -80 °C until RNA extraction. The total RNA was isolated by extraction with TRIzol (Invitrogen). For RT-PCR, 2.5 µg of the total RNA from each group was reverse transcribed using a First-Strand cDNA Synthesis kit (Invitrogen). The cDNA was then amplified by PCR using specific primers as mentioned previously [15]. The following primers were used for PCR, iNOS: F-5′-CTTGCAAGTCCAAGTCTTGC-3′ and R-5′-GTATGTGTCTGCAGATGTGCTG-3′; TNF-α: F-5′-TTCGAGTGACAAGCCTGTAGC-3′ and R-5′-AGATTGACCTCAGCGCTGAGT-3′; IL-6:   F-5′-GGAGGCTTAATTACACATGTT-3′   and   R-5′-TGATTTCAAAGATGAATTGGAT-3′; IL-1β: F-5′-CATATGAGCTGAAAGCTCTCCA-3′ and R-5′-GACACAGATTCCATGGTGAAGTC-3′; MCP-1:   F-5′-AGATGCAGTTAACGCCCCAC-3′   and   R-5′-GACCCATTCCTTCTTGGGGT-3′; MMP-9:   F-5′-CGCTCATGTACCCGCTGTAT-3′   and   R-5′-TGTCTGCCGGACTCAAAGAC-3′; HO-1:   F-5′-GGATTTGGGGCTGCTGGTTTC-3′   and   R-5′-GCAGTGGCAAAGTGGAGATTG-3′; Mac-1:   F-5′-TCAAGCGGCAGTACAAGGAC-3′   and   R-5′-GCCACACACAGAGCTTGCTT -3′; GFAP: F-5′-TTCCTGTACAGACTTCTCC-3′ and R-5′-CCCTTCAGGACTGCCTTAGT-3′; GAPDH: F-5′-GCAGTGGCAAAGTGGAGATTG-3′ and R-5′-TGCAGGATGCATTGCTGACA-3′. The PCR products were analyzed on 1% agarose gels stained with ethidium bromide, and the bands were visualized by UV light.

2.8. SDS-PAGE and Western Blot Analysis

BV-2 cells (5.0×10^5 cells/mL) were cultured in 6-well plates and were pre-treated for 1 h with respective doses of ATR-I, followed by exposure to LPS (100 ng/mL) for 30 min or 18 h, respectively. The preparation of cell lysates (whole and nuclear), electrophoresis, and the immunoblotting procedures were followed as mentioned previously [15]. For the animal experiment, STR and ventral midbrain (VM) tissues were processed according to our previously published data [21]. PVDF membranes were incubated overnight with anti-iNOS (1:1000), anti-IκB-α (1:1000), anti-p-IκB-α (1:1000), anti-ERK (1:1000), anti-p-ERK (1:1000), anti-Akt (1:1000), anti-p-Akt (1:1000), anti-PI3K (1:1000), anti-p-PI3K (1:1000), anti-HO-1 (1:1000), anti-β-actin (1:2000), anti-MnSOD (1:1000), anti-GFAP (1:50,000), anti-Mac-1 (1:500), anti-TH (1:1000), anti-NF-κB/p65 (1:500), and anti-nucleolin (1:500), followed by a 1 h incubation with horseradish peroxidase-conjugated secondary antibodies (1:1000–5000) (Cell Signaling Technology and Santa Cruz biotechnology). The optical densities of the antibody-specific bands were analyzed with a Luminescent Image Analyzer, LAS-3000 (Fuji, Tokyo, Japan).

2.9. Detection of Intracellular ROS Generation Using a Flow Cytometer

Intracellular reactive oxygen species (ROS) production was measured using the non-fluorescent compound H_2DCFDA, as previously described [22], with slight modifications. H_2DCFDA is a stable nonpolar compound that readily diffuses into cells to produce DCFH. Intracellular ROS in the presence of peroxidase changes DCFH to the highly fluorescent compound DCF; therefore, the amount of ROS is proportional to the fluorescent intensity exhibited by the cells [23]. Briefly, cells (5.0×10^5 cells/well in 6-well plate) were incubated with vehicle (DMSO) or ATR-I (25 µM, 50 µM, and 100 µM) at 37 °C for 1 h, before being stimulated with LPS for another 4 h. Furthermore, cells were treated with 10 µM of H_2DCFDA (dissolved in DMSO) for 45 min at 37 °C. The cells were trypsinized and the fluorescent intensity was analyzed with a FACS Calibur flow-cytometry system.

2.10. Immunofluorescence Assay

BV-2 cells (5×10^5 cells/well) were cultured on sterile cover slips in a 24-well plate and treated with ATR-I (100 µM) for 1 h, followed by LPS (100 ng/mL) for 0.5 h, to detect the intracellular p65 sub-unit of NF-κB. Further procedure for the immunofluorescence assay was followed in accordance with the previously reported method [15].

2.11. Statistical Analyses

Statistical analyses were performed using GraphPad software version 5 (GraphPad, La Jolla, CA, USA). Data are the mean ± standard error of mean (SEM) of at least three independent experiments. Significant differences between the groups were determined using a one-way analysis of variance (ANOVA) followed by Tukey's post-hoc analysis. The p values < 0.05 are considered statistically significant.

3. Results

3.1. ATR-I Suppresses LPS-Induced NO Release in BV-2 Cells

LPS exposure to BV-2 cells from the control group and only ATR-I group did not exhibit any increase in NO levels. Pre-treatment with ATR-I at 25 µM, 50 µM, and 100 µM significantly reduced the NO levels in LPS-stimulated BV-2 cells, in a dose-dependent manner (Figure 1A).

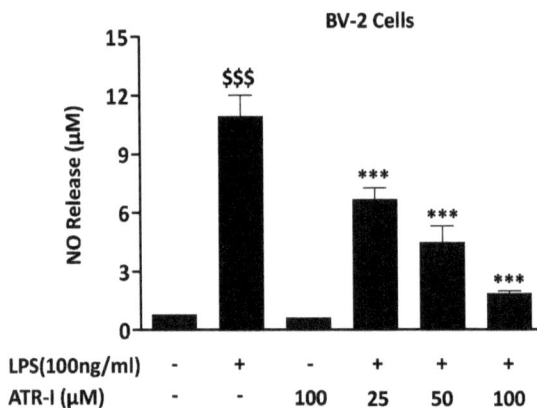

Figure 1. NO release was measured in BV-2 cells ($n = 6$) as mentioned in "Material and Methods" section. $^{\$\$\$}$ $p < 0.001$ vs. vehicle group and *** $p < 0.001$ vs. LPS group.

3.2. ATR-I Diminishes iNOS, TNF-α, IL-6, IL-1β, MCP-1, and MMP-9 mRNA-Expression in LPS-Stimulated BV-2 Cells

The mRNA levels of TNF-α, IL-6, IL-1β, monocyte chemotactic protein-1 (MCP-1), and matrix metallopeptidase-9 (MMP-9) were increased 6 h after LPS treatment (Figure 2A). Further, ATR-I was observed to suppress the LPS-induced mRNA expression of iNOS, TNF-α, IL-6, and IL-1β, in a dose-dependent fashion. However, the inhibitory effects of ATR-I on the mRNA expression of MCP-1 and MMP-9 were not observed to be dose-dependent.

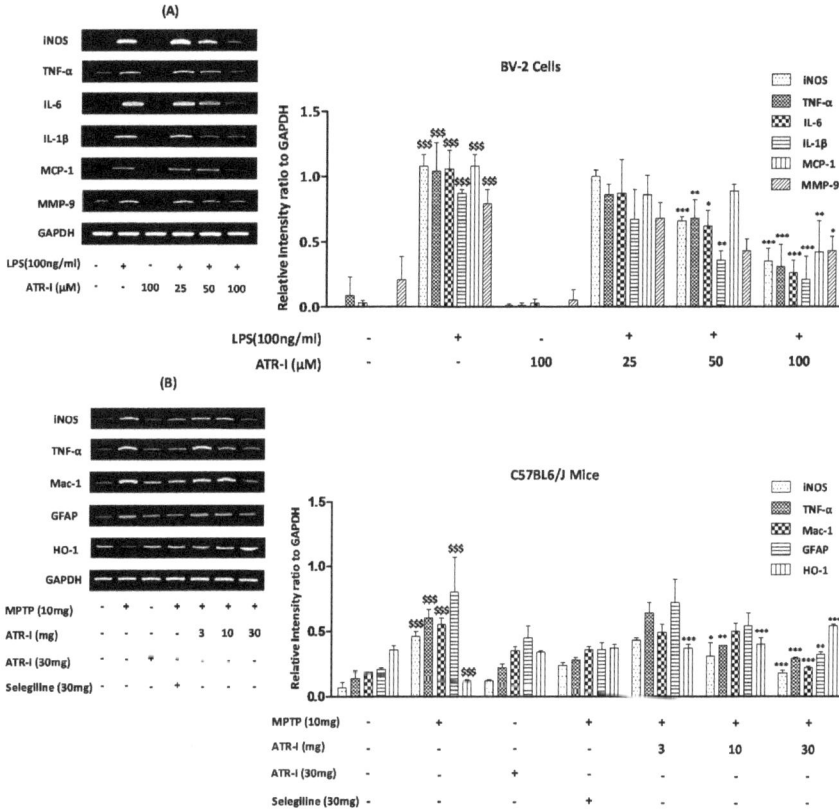

Figure 2. Quantification data are shown in the right panel. (**A**) mRNA levels (n = 6) of iNOS, TNF-α, IL-6, IL-1β, MCP-1, MMP-9 in BV-2 cells and (**B**) mRNA levels (n = 4) of iNOS, TNF-α, Mac-1, GFAP, and HO-1 in mice. GAPDH was used as an internal control. $^{\$\$\$}$ $p < 0.001$ vs. vehicle group; * $p < 0.05$, ** $p < 0.01$, and *** $p < 0.001$ vs. LPS and MPTP-treated group.

3.3. ATR-I Diminishes mRNA-Expression of Microglia-Derived Pro-Inflammatory Mediators and Induces HO-1 Expression in MPTP-Induced Model of Neuroinflammation

MPTP significantly upregulated the mRNA expression of iNOS, TNF-α, GFAP, and Mac-1, and decreased the expression of HO-1 in mice STR. Post-treatment with ATR-I significantly inhibited the mRNA expressions of iNOS, Mac-1, GFAP, and TNF-α in the MPTP-intoxicated mice (Figure 2B). This inhibitory effect of ATR-I was dose-dependent for the mRNA expressions of iNOS, TNF-α, and GFAP. Furthermore, post-treatment with ATR-I significantly and dose-dependently increased the mRNA expression of HO-1 in MPTP-intoxicated mice (Figure 2B).

3.4. ATR-I Diminishes the Microglial and Astroglial Markers in Striatum of MPTP-Intoxicated Mice

Figure 3A depicts the protein content of Mac-1 and GFAP, which was significantly increased in the STR after MPTP exposure, compared to the vehicle group. However, post-treatment with ATR-I significantly decreased the elevated protein levels of Mac-1 and GFAP. On the other hand, ATR-I did not elicit a dose-dependent inhibition of GFAP, as observed in protein blots (Figure 3A). Similarly, results from the staining of the STR revealed that microglial cells from the STR of the vehicle group did not exhibit significant staining. Both the Mac-1 (Figure 3B) (brown) and GFAP (Figure 3B) (green) staining signal significantly increased on day three. Post-treatment with ATR-I mitigated the significantly increased number of Mac-1 and GFAP microglia and astrocytes. Furthermore, these results also synchronized with the immunoblot analysis (Figure 3A), wherein ATR-I effectively abated the Mac-1 and GFAP expressions in the STR of the MPTP-intoxicated mice.

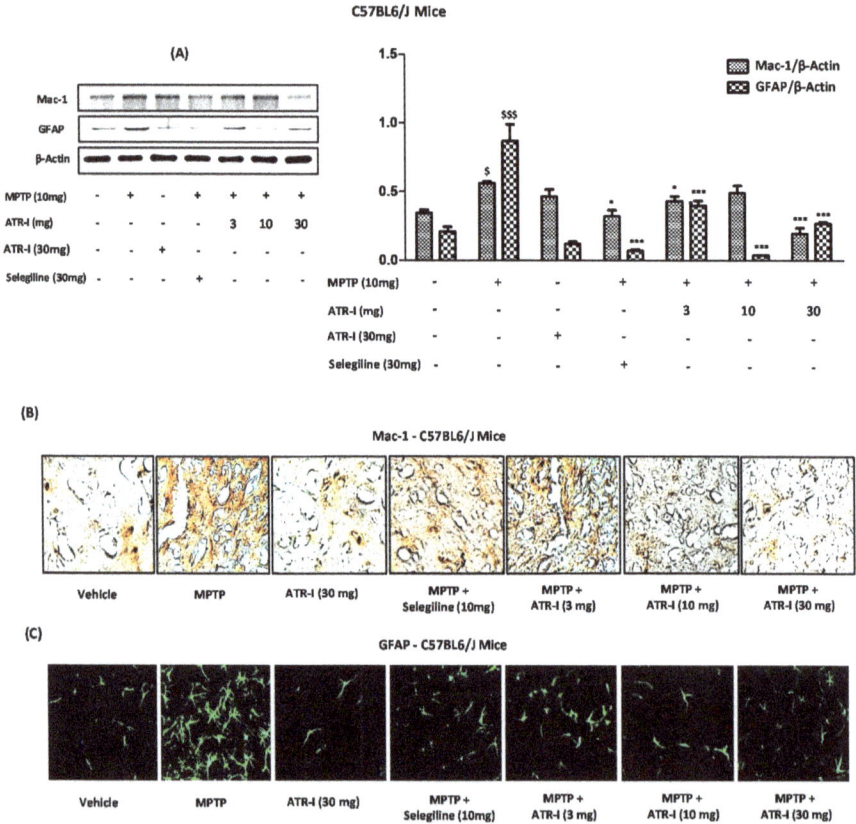

Figure 3. (**A**) Western blot for Mac-1 and GFAP expression in the STR ($n = 3$). $ $p < 0.05$ and $$$ $p < 0.001$ vs. vehicle group; * $p < 0.05$, and *** $p < 0.001$ vs. MPTP-treated group; (**B**) Representative image presenting Mac-1-positive microglia in the STR of MPTP-intoxicated mice ($n = 4$); (**C**) Representative image presenting GFAP-positive astroglia in the STR of MPTP-intoxicated mice ($n = 4$).

3.5. ATR-I Subsides LPS- and MPTP-Induced Increases of iNOS Expression and Modulates LPS-Induced Phosphorylation of ERK, PI3K, and Akt

Exposure to LPS (Figure 4A) and MPTP (Figure 4B) significantly increased the protein expression of iNOS in BV-2 cells and MPTP-intoxicated mice, respectively. However, treatment with ATR-I at different doses reduced the protein expression of iNOS. ATR-I strongly and dose-dependently inhibited ERK phosphorylation (Figure 4D), whereas ATR-I had no effect on p38 and c-Jun N-terminal kinase (JNK). LPS exposure significantly increased PI3K and Akt phosphorylation in contrast to the control. Furthermore, pretreatment with ATR-I dose-dependently and significantly inhibited the phosphorylation levels of PI3K and Akt, respectively (Figure 4D). However, ATR-I treatment alone did not alter the phosphorylation levels of PI3K and Akt in BV-2 cells.

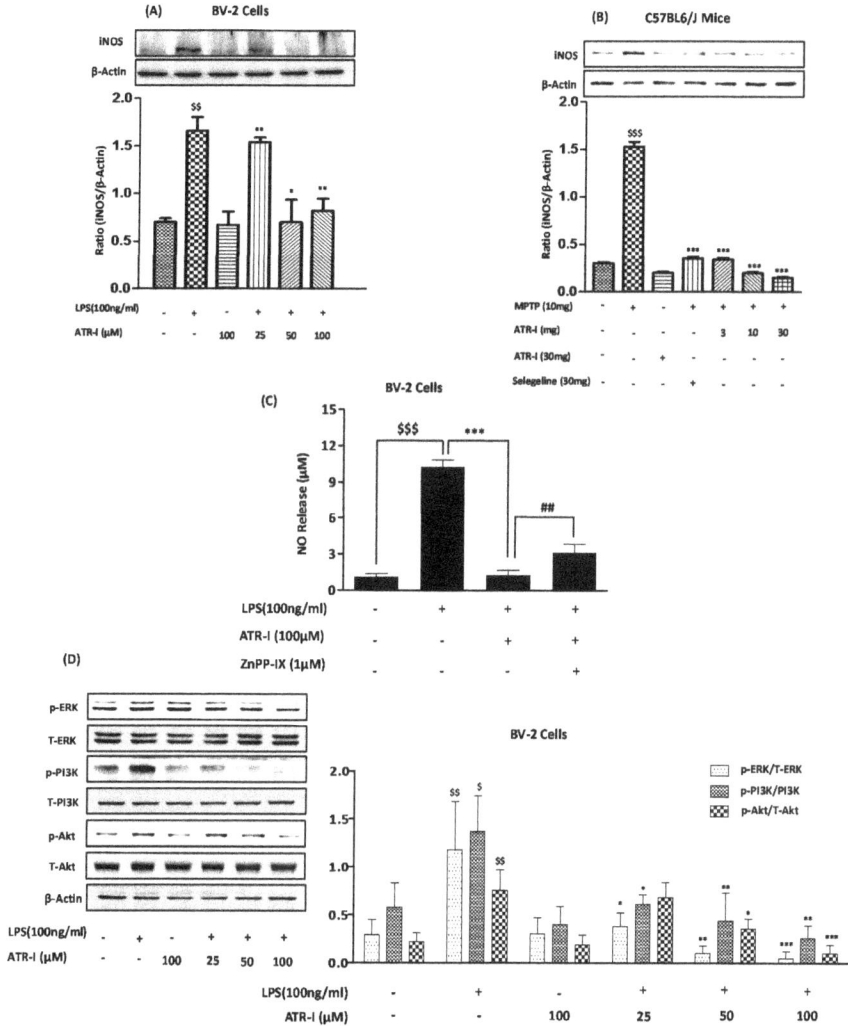

Figure 4. Quantitative protein analysis ($n = 6$) of iNOS (**A,B**) and (**D**) p-ERK-MAPK, p-PI3K, and p-Akt in BV-2 cells. $^{\$\$\$}$ $p < 0.001$, $^{\$\$}$ $p < 0.01$, and $^{\$}$ $p < 0.05$ vs. vehicle group; * $p < 0.05$, ** $p < 0.01$, and *** $p < 0.001$ vs. LPS and MPTP-treated groups. (**C**) NO release was measured in BV-2 cells ($n = 6$) as mentioned in "Material and Methods". $^{\$\$\$}$ $p < 0.001$ vs. vehicle group; *** $p < 0.001$ vs. LPS-treated group; $^{\#\#}$ $p < 0.01$ vs. ATR-I + LPS treated group.

3.6. HO-1 Mediates the Inhibitory Effect of ATR-I on NO Release in LPS-Stimulated BV-2 Cells

ATR-I (100 μM), in combination with LPS, was able to decrease NO release in BV-2 cells (Figure 4C). At 1 μM, ZnPP-IX reversed the inhibition of LPS-stimulated NO production caused by pre-treatment with ATR-I. These data indicate that HO-1 mediates the anti-inflammatory effects of ATR-I in BV-2 cells (Figure 4C).

3.7. ATR-I Ameliorates the Degradation of IκB-α and Inhibits Nuclear Translocation of NF-κB in LPS and MPTP Model of Neuroinflammation

Exposure to LPS (Figure 5A) and MPTP (Figure 5C) significantly increased the protein expressions of p-IκB-α; however, ATR-I significantly reversed the effects of LPS and MPTP on IκB-α phosphorylation, but this effect was not dose-dependent in either of the models. Furthermore, ATR-I significantly inhibited the LPS- and MPTP-induced nuclear translocation of NF-κB/p65, as determined by Western blotting and the immunofluorescence assay (Figure 5B).

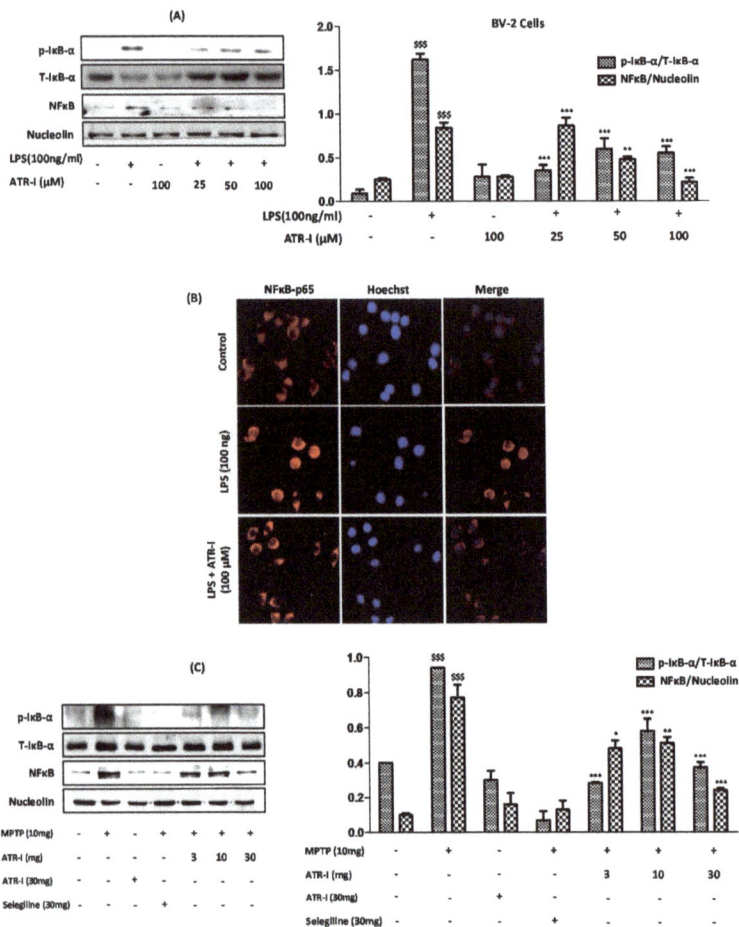

Figure 5. (**A**) Nuclear and whole-cell extracts (*n* = 3) were prepared and analyzed by Western blot to asses (**A**) anti-p-IκB-α and NF-κB/p65; (**B**) The sub-cellular location of the NFκB-p65 subunit was determined by an immunofluorescence assay (*n* = 3). The total nuclear and whole-cell lysates (*n* = 4) of the STR were tagged with (**C**) anti-p-IκB-α and NF-κB/p65. $^{\$\$\$}$ *p* < data compared with the vehicle group; * *p*, ** *p*, *** *p* < data compared with LPS and MPTP-treated groups.

3.8. ATR-I Suppresses the LPS-Induced Production of ROS in BV-2 Cells, While Upregulates the Anti-Oxidant Defense System in the MPTP-Induced Model of Neuroinflammation

As indicated in Figure 6A, the BV-2 cells that were exposed to LPS exhibited a prominent curve shift toward the right, indicating significant ROS generation. Furthermore, the incubation of the BV-2 cells with ATR-I at varying doses significantly reduced the ROS content, as seen by the curve shift towards the left. The lower dose of 25 µM did not result in a significant reduction of ROS content, and instead exhibited a greater extent of ROS release compared to the LPS exposure, which might have been due to a sample variation. The remaining two doses (50 µM and 100 µM), however, had a similar effect, wherein the LPS-induced ROS release was suppressed in the BV-2 cells. In line with these reports, our results from animal experiments indicate that MPTP significantly decreased the protein levels of HO-1 and MnSOD in the STR (Figure 6B) and VMs (Figure 6C) of the mice, as compared to the vehicle group. However, ATR-I was able to reverse the decreased protein levels of HO-I in the STR and VM, in a dose-dependent fashion.

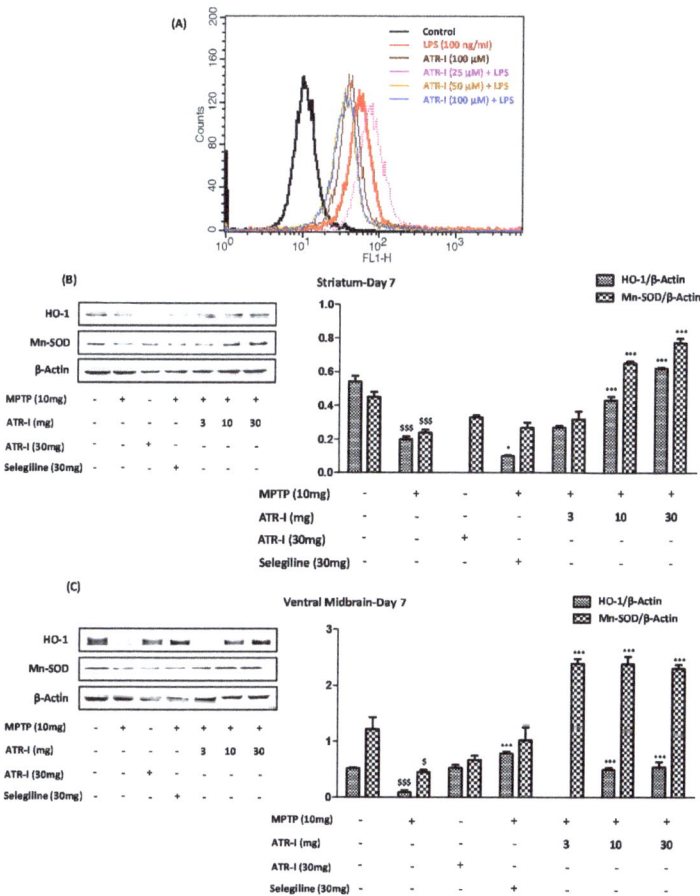

Figure 6. (A) The histogram obtained by fluorometric analysis portrays one representative set of results ($n = 6$) from the three different experiments in BV-2 cells; **(B)** Represents protein levels for HO-1 and MnSOD in STR and **(C)** VM. A densitometric analysis of HO-1 and MnSOD is shown in the right panel. $^\$ p < 0.05$ and $^{\$\$\$} p < 0.001$ vs. vehicle group; $^* p < 0.05$ and $^{***} p < 0.001$ vs. MPTP-treated group.

3.9. ATR-I Mitigates MPTP-Induced Dopaminergic Neurodegeneration and Motor Dysfunction in a Mouse Model of Neuroinflammation

MPTP intoxication significantly decreased the levels of TH expression in the STR and VM (Figure 7A). The MPTP + ATR-I (10 mg/kg, and 30 mg/kg) treatment groups presented a greater extent of TH expression than the MPTP-treated group, in both the STR and VM tissues. However, the TH expression of STR and VM tissues in the 3 mg/kg ATR-I treated mice was lower than those in the MPTP-treated mice. Only treatment with ATR-I did not have an effect on TH expression in the STR, whereas the TH expression in the VM was significantly increased. On the other hand, selegiline was not effective in the reversal of TH expression in the STR; however, selegiline significantly increased TH expression in the VM.

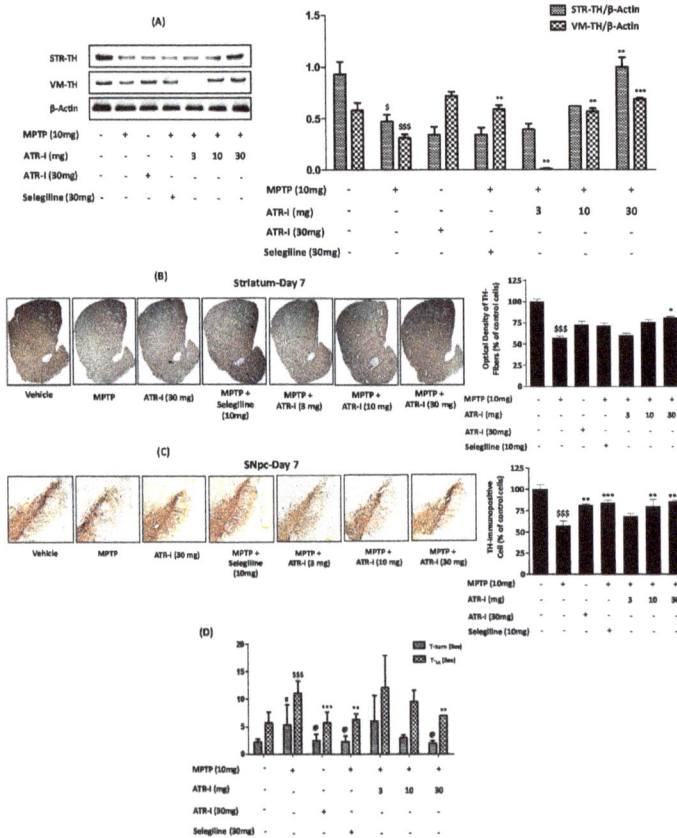

Figure 7. Protective effects of ATR-I against MPTP in STR and VM. Western blot for TH protein expression in STR and VM (**A**); Representative coronal images of TH-positive striatal fibers (**B,C**) SNpc neurons. Densitometric analysis for TH protein expression is provided in the right panel (*n* = 4). $^{\$\$\$}$ *p* < 0.001 and $^{\$}$ *p* < 0.05 vs. vehicle group; * *p* < 0.05, ** *p* < 0.01, and *** *p* < 0.001 vs. MPTP-treated group; (**D**) The Pole test was carried out on day 7 (*n* = 4). $^{\$\$\$}$ *p* < 0.001 and $^{\#}$ *p* < 0.05 vs. vehicle group; $^{@}$ *p* < 0.05, ** *p* < 0.01, and *** *p* < 0.001 vs. MPTP-treated group.

Similarly, as displayed in the representative photographs (Figure 7), numerous TH-positive fibers in STR- and TH-positive neurons in SNpc were observed in the vehicle group. Exposure to MPTP decreased the TH-positive fiber (56.97% ± 4.23%) in the STR (Figure 7B) and the TH-positive cells (57.75% ± 6.89%) in the SNpc, to a significant extent (Figure 7C). ATR-I alone did not modify the TH-positive neurons, as shown by the stain intensities of the STR and SNpc. ATR-I prevented the MPTP-induced loss of TH-positive fibers in the STR (Figure 7B) (60.18% ± 4.50%, 75.78% ± 4.98%, and 80.86% ± 2.56%) and SNpc neurons (Figure 7C) (68.31% ± 6.11%, 80.24% ± 14.22%, and 86.14% ± 1.76%), in a dose-dependent manner.

MPTP intoxication developed bradykinesia, as was confirmed by the turning on the top and the climbing down the pole; however, post-treatment with ATR-I and selegiline significantly decreased the T-turn and T-LA time taken by the mice (Figure 7D).

4. Discussion

The role of sesquiterpene lactones in the inhibition of the central transcription factor NF-κB is well documented [24]. Sesquiterpene lactones indicate the active principle of many of the anti-inflammatory drugs used in traditional medicine. Furthermore, observations have shown that they also possess other bioactivities for hematomas, contusions, sprains, and rheumatic diseases [25]. ATR-I has been documented for a variety of pharmacological uses in eastern Asian countries regarding the treatment of edema, anorexia, splenic asthenia, excessive perspiration, and abnormal fetal movement [26,27]. Moreover, previous findings also advocate the in vivo and in vitro anti-inflammatory activities of ATR-I's [12,28–30].

We used LPS to induce inflammatory pathology in BV-2 cells as they simulate the inflammatory response that occurs in neurodegenerative diseases [31] and MPTP to induce neuroinflammation in C57BL6/J mice, because they replicate a variety of important pathobiochemical features of PD [32,33]. Growing evidence from a variety of studies suggests that proinflammatory factors released from activated microglia play an important role in the initiation and progress of neurodegenerative processes, including PD [34,35]. NO, produced from iNOS, is a proinflammatory factor and nitrosative stress source. [36]. Microglia in PD have been observed to densely grow in the striatum and SNpc with increased expression of proinflammatory mediators, including TNF-α, IL-1β, and IL-6 [37]. Furthermore, during inflammatory conditions, chemokine MCP-1 acts to cause the migration of specific leukocyte sets to inflammation sites, while MMP-9 plays a vital role by degrading the blood-brain barrier [38,39]. Hence, we first investigated the ability of ATR-I to curb the release of NO and proinflammatory mediators. We showed that ATR-I significantly decreased the mRNA expression profile of inflammatory mediators, MCP-1 and MMP-9, in our models. Our data corresponds with a previous report, wherein ATR-I also inhibited NO and TNF-α [40]. Gabriel and coworkers have established that iNOS plays an essential role in the neurotoxic process initiated by MPTP and suggested that inhibitors of iNOS may provide a protective benefit in the treatment of PD [41]. Recent experimental data has designated the dopaminergic system in the midbrain as the most susceptible area following MPTP-mediated inflammatory damage that is caused by an over-activation of astrocytes and microglia [42,43].

Microglia are resident macrophage cells in the brain, that promptly react in response to toxins in their microenvironment and quickly proliferate, become hypertrophic, and persistently increase the expression of a large number of marker molecules, such as Mac-1 and TNF-α [44], and are further transformed to macrophage-like cells in patients with PD [45]. GFAP is a chief intermediate astrocyte filament protein that mediates the interaction of astrocytes with other cells that are essential for the formation and maintenance of myelin. Therefore, GFAP has received much attention as a reliable biomarker in studies of central nervous system diseases [46]. Astrocytes anticipate brain injury via a process termed "reactive gliosis" [47], and the density of GFAP-positive astrocyte is negatively correlated with the reduction of dopaminergic neuron [48]. Therefore, molecules that can modulate microglial and astrocyte activation may potentially exert neuroprotective effects against MPTP. Hence, apart from iNOS, we also chose to study Mac-1 in microglia and GFAP in astroglia, as they are specific markers that represent microglia and astroglia activation [49]. Our experimental data in both in vivo and in vitro studies demonstrated that ATR-I treatment inhibits LPS/MPTP-induced iNOS expression. Also, data from our immunohistochemistry and Western blot studies indicate that the systemic administration of ATR-I diminished the MPTP-induced production of GFAP and Mac-1 in the microglial cells of the STR; however, we did not observe any significant effect of ATR-I on the GFAP and Mac-1 profiles in the microglial cells of the SNpc. Our results are in coherence with earlier reports, wherein the mitigation of microglial inflammation provides neuroprotection in an MPTP model of PD [20].

Recently, it was reported that the activation of the PI3K/Akt pathway is involved in the LPS-induced expression of iNOS [27,50]. Also, an accumulating body of evidence indicates the PI3K/Akt pathway's role in the inflammatory response of LPS, whereby NF-κB is activated through IκB degradation [51,52]. We found that ATR-I significantly and dose-dependently blocked the phosphorylation of PI3K and Akt in response to an LPS challenge. LPS is involved in the stimulation of a variety of intracellular signaling pathways in the microglia that involve MAPKs and NF-κB [53,54]. Furthermore, MAPKs facilitate the gene expression of many inflammatory mediators [55,56].

We found that ERK and Akt are specific targets of ATR-I. Earlier reports suggested the importance of ERK-signaling in the production of pro-inflammatory cytokines, while NF-κB mediates the gene expression of proinflammatory cytokines and additional effector molecules [57]. Pharmacological interference in NF-κB's transcriptional activity may therefore represent an important therapeutic strategy for the mitigation of neuroinflammatory diseases that are linked to oxidative stress. Our data demonstrate that ATR-I inhibits the phosphorylation of the IκB-α protein, thereby disrupting the activation of NF-κB and its nuclear translocation, in response to LPS/MPTP-induced neuroinflammation. A number of recent reports demonstrate that the mitochondrial malfunctioning and increase in ROS leads to neuronal-cell demise via apoptosis [58,59]. Additionally, historical data has also indicated the role of ROS in the activation of NF-κB [60,61]. In the present study, pre-treatment with ATR-I significantly attenuated intracellular ROS generation in BV-2 cells that were exposed to LPS. HO-1 is an anti-inflammatory cytoprotective protein which protects against oxidative stress [62,63]. Recently, several experiments have suggested that HO-1 induction is involved in the inhibition of LPS-induced NO production [64,65]. Our experiments involving ZnPP-IX (specific HO-1 inhibitor) revealed that the induction of HO-1 caused by pre-treatment with ATR-I in LPS-stimulated BV-2 cells, is responsible for abating LPS-induced NO release. Apart from HO-1, MnSOD found exclusively in the mitochondrial matrix [66], plays a vital role in neutralizing ROS [67]. The over-expression of MnSOD is reported to protect cells and attenuate MPTP-induced toxicity [68]. Our present data indicates that the anti-inflammatory effects of ATR-I are, at least in part, due to the inhibition of ROS and the induction of antioxidant enzymes (HO-1 and MnSOD). Our results are also in agreement with previous works, wherein the induction of HO-1 facilitates cellular protection against NO-induced injury [69,70] and MPTP-induced oxidative stress [71–73].

As a rate-limiting enzyme, TH is required for catecholamine synthesis and is a hallmark indicator for evaluating the neuroprotective ability of compounds in PD, since a decrease of TH activity is positively correlated with PD severity [74]. We found ATR-I to restore the population of surviving dopaminergic neurons in the STR and SNpc, in response to MPTP intoxication, thus asserting the neuroprotective effect of ATR-I in the MPTP model of PD. The pole test data revealed the protective benefits of ATR-I for mitigating bradykinesia against MPTP. Our data is in coherence with earlier published data, wherein the maintenance of normal motor behavior is positively correlated with the preservation of TH-positive neurons [20]. Additionally, data concerning the toxicity and efficacy profile of ATR-I in different models of PD would produce more reliable data. This data would further help in sustaining the applicability of ATR-I in clinical studies and also its use in PD.

5. Conclusions

In conclusion, the present study showed significant anti-neuroinflammatory efficacy of ATR-I in the MPTP-induced model of neuroinflammation, through increasing HO-1 and MnSOD levels, along with an increase in the counts of TH-immunoreactive neurons. Additionally, ATR-I exhibits promising efficacy in the LPS-induced model of PD. Taken together, the results indicate that ATR-I may be an effective anti-neuroinflammatory agent to be considered as a lead candidate against PD.

Acknowledgments: This paper was supported by Konkuk University.

Author Contributions: Sandeep More designed, executed, and analyzed the results. Dong-Kug Choi completed the mentoring for designing the experiments. The manuscript was constructed and written by Sandeep More.

Conflicts of Interest: The authors declare no conflict of interest.

References

1. Henchcliffe, C.; Beal, M.F. Mitochondrial biology and oxidative stress in parkinson disease pathogenesis. *Nat. Clin. Pract. Neurol.* **2008**, *4*, 600–609. [CrossRef] [PubMed]
2. Vivekanantham, S.; Shah, S.; Dewji, R.; Dewji, A.; Khatri, C.; Ologunde, R. Neuroinflammation in parkinson's disease: Role in neurodegeneration and tissue repair. *Int. J. Neurosci.* **2015**, *125*, 717–725. [CrossRef] [PubMed]
3. Imaizumi, Y.; Okada, Y.; Akamatsu, W.; Koike, M.; Kuzumaki, N.; Hayakawa, H.; Nihira, T.; Kobayashi, T.; Ohyama, M.; Sato, S.; et al. Mitochondrial dysfunction associated with increased oxidative stress and alpha-synuclein accumulation in park2 ipsc-derived neurons and postmortem brain tissue. *Mol. Brain* **2012**, *5*, 35. [CrossRef] [PubMed]
4. Licker, V.; Turck, N.; Kovari, E.; Burkhardt, K.; Cote, M.; Surini-Demiri, M.; Lobrinus, J.A.; Sanchez, J.C.; Burkhard, P.R. Proteomic analysis of human substantia nigra identifies novel candidates involved in parkinson's disease pathogenesis. *Proteomics* **2014**, *14*, 784–794. [CrossRef] [PubMed]
5. Kao, T.K.; Ou, Y.C.; Lin, S.Y.; Pan, H.C.; Song, P.J.; Raung, S.L.; Lai, C.Y.; Liao, S.L.; Lu, H.C.; Chen, C.J. Luteolin inhibits cytokine expression in endotoxin/cytokine-stimulated microglia. *J. Nutr. Biochem.* **2011**, *22*, 612–624. [CrossRef] [PubMed]
6. Chen, Q.; He, H.; Li, P.; Zhu, J.; Xiong, M. Identification and quantification of atractylenolide i and atractylenolide iii in rhizoma atractylodes macrocephala by liquid chromatography-ion trap mass spectrometry. *Biomed. Chromatogr.* **2013**, *27*, 699–707. [CrossRef] [PubMed]
7. Zhang, Y.; Xu, S.; Lin, Y. gastrointestinal inhibitory effects of sesquiterpene lactones from atractylodes macrocephala. *Zhong Yao Cai* **1999**, *22*, 636–640. [PubMed]
8. Wang, C.C.; Lin, S.Y.; Cheng, H.C.; Hou, W.C. Pro-oxidant and cytotoxic activities of atractylenolide i in human promyeloleukemic hl-60 cells. *Food Chem. Toxicol.* **2006**, *44*, 1308–1315. [CrossRef] [PubMed]
9. Huang, H.L.; Chen, C.C.; Yeh, C.Y.; Huang, R.L. Reactive oxygen species mediation of baizhu-induced apoptosis in human leukemia cells. *J. Ethnopharmacol.* **2005**, *97*, 21–29. [CrossRef] [PubMed]
10. Liu, Y.; Ye, F.; Qiu, G.Q.; Zhang, M.; Wang, R.; He, Q.Y.; Cai, Y. Effects of lactone i from atractylodes macrocephala koidz on cytokines and proteolysis-inducing factors in cachectic cancer patients. *Di Yi Jun Yi Da Xue Xue Bao* **2005**, *25*, 1308–1311. [PubMed]
11. Li, C.; He, L. Establishment of the model of white blood cell membrane chromatography and screening of antagonizing tlr4 receptor component from atractylodes macrocephala koidz. *Sci. China C Life Sci.* **2006**, *49*, 182–189. [CrossRef] [PubMed]
12. Li, C.Q.; He, L.C.; Dong, H.Y.; Jin, J.Q. Screening for the anti-inflammatory activity of fractions and compounds from atractylodes macrocephala koidz. *J. Ethnopharmacol.* **2007**, *114*, 212–217. [CrossRef] [PubMed]
13. Kostic, V.; Przedborski, S.; Flaster, E.; Sternic, N. Early development of levodopa-induced dyskinesias and response fluctuations in young-onset parkinson's disease. *Neurology* **1991**, *41*, 202–205. [CrossRef] [PubMed]
14. Bocchini, V.; Mazzolla, R.; Barluzzi, R.; Blasi, E.; Sick, P.; Kettenmann, H. An immortalized cell line expresses properties of activated microglial cells. *J. Neurosci. Res.* **1992**, *31*, 616–621. [CrossRef] [PubMed]
15. More, S.V.; Park, J.Y.; Kim, B.W.; Kumar, H.; Lim, H.W.; Kang, S.M.; Koppula, S.; Yoon, S.H.; Choi, D.K. Anti-neuroinflammatory activity of a novel cannabinoid derivative by inhibiting the nf-kappab signaling pathway in lipopolysaccharide-induced bv-2 microglial cells. *J. Pharmacol. Sci.* **2013**, *121*, 119–130. [CrossRef] [PubMed]
16. Kohutnicka, M.; Lewandowska, E.; Kurkowska-Jastrzebska, I.; Czlonkowski, A.; Czlonkowska, A. Microglial and astrocytic involvement in a murine model of parkinson's disease induced by 1-methyl-4-phenyl-1,2,3,6-tetrahydropyridine (mptp). *Immunopharmacology* **1998**, *39*, 167–180. [CrossRef]

17. Himeda, T.; Watanabe, Y.; Tounai, H.; Hayakawa, N.; Kato, H.; Araki, T. Time dependent alterations of co-localization of s100beta and gfap in the mptp-treated mice. *J. Neural Transm.* **2006**, *113*, 1887–1894. [CrossRef] [PubMed]

18. Matsuura, K.; Kabuto, H.; Makino, H.; Ogawa, N. Pole test is a useful method for evaluating the mouse movement disorder caused by striatal dopamine depletion. *J. Neurosci. Methods* **1997**, *73*, 45–48. [CrossRef]

19. Green, L.C.; Wagner, D.A.; Glogowski, J.; Skipper, P.L.; Wishnok, J.S.; Tannenbaum, S.R. Analysis of nitrate, nitrite, and [15n]nitrate in biological fluids. *Anal. Biochem.* **1982**, *126*, 131–138. [CrossRef]

20. Kim, B.W.; Koppula, S.; Kumar, H.; Park, J.Y.; Kim, I.W.; More, S.V.; Kim, I.S.; Han, S.D.; Kim, S.K.; Yoon, S.H. A-asarone attenuates microglia-mediated neuroinflammation by inhibiting nf kappa b activation and mitigates mptp-induced behavioral deficits in a mouse model of parkinson's disease. *Neuropharmacology* **2015**, *97*, 46–57. [CrossRef] [PubMed]

21. Kumar, H.; Kim, I.S.; More, S.V.; Kim, B.W.; Bahk, Y.Y.; Choi, D.K. Gastrodin protects apoptotic dopaminergic neurons in a toxin-induced parkinson's disease model. *Evid. Based Complement. Altern. Med.* **2013**, *2013*, 514095. [CrossRef] [PubMed]

22. Bass, D.A.; Parce, J.W.; Dechatelet, L.R.; Szejda, P.; Seeds, M.C.; Thomas, M. Flow cytometric studies of oxidative product formation by neutrophils: A graded response to membrane stimulation. *J. Immunol.* **1983**, *130*, 1910–1917. [PubMed]

23. Eruslanov, E.; Kusmartsev, S. Identification of ros using oxidized dcfda and flow-cytometry. *Methods Mol. Biol.* **2010**, *594*, 57–72. [PubMed]

24. Wagner, S.; Arce, R.; Murillo, R.; Terfloth, L.; Gasteiger, J.; Merfort, I. Neural networks as valuable tools to differentiate between sesquiterpene lactones' inhibitory activity on serotonin release and on nf-kappab. *J. Med. Chem.* **2008**, *51*, 1324–1332. [CrossRef] [PubMed]

25. Willuhn, G. New findings from research on arnica. *Pharm. Unserer Zeit* **1981**, *10*, 1–7. [CrossRef] [PubMed]

26. Commission, C.P. *Pharmacopoeia of the People's Republic of China*; China Medical Science and Technology Press: Beijing, China, 2010; Volume 1.

27. Lin, Y.C.; Kuo, H.C.; Wang, J.S.; Lin, W.W. Regulation of inflammatory response by 3-methyladenine involves the coordinative actions on akt and glycogen synthase kinase 3beta rather than autophagy. *J. Immunol.* **2012**, *189*, 4154–4164. [CrossRef] [PubMed]

28. Endo, K.; Taguchi, T.; Taguchi, F.; Hikino, H.; Yamahara, J.; Fujimura, H. Antiinflammatory principles of atractylodes rhizomes. *Chem. Pharm. Bull. (Tokyo)* **1979**, *27*, 2954–2958. [CrossRef] [PubMed]

29. Park, H.Y.; Lim, H.; Kim, H.P.; Kwon, Y.S. Downregulation of matrix metalloproteinase-13 by the root extract of cyathula officinalis kuan and its constituents in il-1beta-treated chondrocytes. *Planta Med.* **2011**, *77*, 1528–1530. [CrossRef] [PubMed]

30. Sin, K.S.; Kim, H.P.; Lee, W.C.; Pachaly, P. Pharmacological activities of the constituents of atractylodes rhizomes. *Arch. Pharm. Res.* **1989**, *12*, 236–238. [CrossRef]

31. More, S.V.; Kumar, H.; Kim, I.S.; Song, S.Y.; Choi, D.K. Cellular and molecular mediators of neuroinflammation in the pathogenesis of parkinson's disease. *Mediat. Inflamm.* **2013**, *2013*, 952375. [CrossRef] [PubMed]

32. Beal, M.F. Experimental models of parkinson's disease. *Nat. Rev. Neurosci.* **2001**, *2*, 325–334. [CrossRef] [PubMed]

33. Schmidt, N.; Ferger, B. Neurochemical findings in the mptp model of parkinson's disease. *J. Neural Transm.* **2001**, *108*, 1263–1282. [CrossRef] [PubMed]

34. Tansey, M.G.; Goldberg, M.S. Neuroinflammation in parkinson's disease: Its role in neuronal death and implications for therapeutic intervention. *Neurobiol. Dis.* **2010**, *37*, 510–518. [CrossRef] [PubMed]

35. Ha, S.K.; Moon, E.; Kim, S.Y. Chrysin suppresses lps-stimulated proinflammatory responses by blocking nf-kappab and jnk activations in microglia cells. *Neurosci. Lett.* **2010**, *485*, 143–147. [CrossRef] [PubMed]

36. Shahani, N.; Sawa, A. Nitric oxide signaling and nitrosative stress in neurons: Role for s-nitrosylation. *Antioxid. Redox Signal.* **2011**, *14*, 1493–1504. [CrossRef] [PubMed]

37. Hunot, S.; Dugas, N.; Faucheux, B.; Hartmann, A.; Tardieu, M.; Debré, P.; Agid, Y.; Dugas, B.; Hirsch, E.C. Fcεrii/cd23 is expressed in parkinson's disease and induces, in vitro, production of nitric oxide and tumor necrosis factor-α in glial cells. *J. Neurosci.* **1999**, *19*, 3440–3447. [PubMed]

38. Thompson, W.L.; Van Eldik, L.J. Inflammatory cytokines stimulate the chemokines ccl2/mcp-1 and ccl7/mcp-3 through nfkb and mapk dependent pathways in rat astrocytes. *Brain Res.* **2009**, *1287*, 47–57. [CrossRef] [PubMed]

39. Jayasooriya, R.G.; Choi, Y.H.; Kim, G.Y. Glutamine-free condition inhibits lipopolysaccharide-induced invasion of bv2 microglial cells by suppressing of matrix metalloproteinase-9 expression. *Environ. Toxicol. Pharmacol.* **2013**, *36*, 1127–1132. [CrossRef] [PubMed]

40. Li, C.Q.; He, L.C.; Jin, J.Q. Atractylenolide i and atractylenolide iii inhibit lipopolysaccharide-induced tnf-alpha and no production in macrophages. *Phytother. Res.* **2007**, *21*, 347–353. [CrossRef] [PubMed]

41. Liberatore, G.T.; Jackson-Lewis, V.; Vukosavic, S.; Mandir, A.S.; Vila, M.; McAuliffe, W.G.; Dawson, V.L.; Dawson, T.M.; Przedborski, S. Inducible nitric oxide synthase stimulates dopaminergic neurodegeneration in the mptp model of parkinson disease. *Nat. Med.* **1999**, *5*, 1403–1409. [PubMed]

42. Bruck, D.; Wenning, G.K.; Stefanova, N.; Fellner, L. Glia and alpha-synuclein in neurodegeneration: A complex interaction. *Neurobiol. Dis.* **2016**, *85*, 262–274. [CrossRef] [PubMed]

43. Liu, Y.; Hu, J.; Wu, J.; Zhu, C.; Hui, Y.; Han, Y.; Huang, Z.; Ellsworth, K.; Fan, W. Alpha7 nicotinic acetylcholine receptor-mediated neuroprotection against dopaminergic neuron loss in an mptp mouse model via inhibition of astrocyte activation. *J. Neuroinflamm.* **2012**, *9*, 98. [CrossRef] [PubMed]

44. Hunot, S.; Hirsch, E. Neuroinflammatory processes in parkinson's disease. *Ann. Neurol.* **2003**, *53*, S49–S60. [CrossRef] [PubMed]

45. Gerhard, A.; Pavese, N.; Hotton, G.; Turkheimer, F.; Es, M.; Hammers, A.; Eggert, K.; Oertel, W.; Banati, R.B.; Brooks, D.J. In vivo imaging of microglial activation with [11 c](r)-pk11195 pet in idiopathic parkinson's disease. *Neurobiol. Dis.* **2006**, *21*, 404–412. [CrossRef] [PubMed]

46. Kochanek, P.M.; Berger, R.P.; Bayr, H.; Wagner, A.K.; Jenkins, L.W.; Clark, R.S. Biomarkers of primary and evolving damage in traumatic and ischemic brain injury: Diagnosis, prognosis, probing mechanisms, and therapeutic decision making. *Curr. Opin. Crit. Care* **2008**, *14*, 135–141. [CrossRef] [PubMed]

47. Van Eldik, L.J.; Thompson, W.L.; Ranaivo, H.R.; Behanna, H.A.; Watterson, D.M. Glia proinflammatory cytokine upregulation as a therapeutic target for neurodegenerative diseases: Function-based and target-based discovery approaches. *Int. Rev. Neurobiol.* **2007**, *82*, 277–296. [PubMed]

48. McGeer, P.; Itagaki, S.; Boyes, B.; McGeer, E. Reactive microglia are positive for hla-dr in the substantia nigra of parkinson's and alzheimer's disease brains. *Neurology* **1988**, *38*, 1285. [CrossRef] [PubMed]

49. Wu, D.C.; Jackson-Lewis, V.; Vila, M.; Tieu, K.; Teismann, P.; Vadseth, C.; Choi, D.K.; Ischiropoulos, H.; Przedborski, S. Blockade of microglial activation is neuroprotective in the 1-methyl-4-phenyl-1,2,3,6-tetrahydropyridine mouse model of parkinson disease. *J. Neurosci.* **2002**, *22*, 1763–1771. [PubMed]

50. Choi, W.S.; Seo, Y.B.; Shin, P.G.; Kim, W.Y.; Lee, S.Y.; Choi, Y.J.; Kim, G.D. Veratric acid inhibits inos expression through the regulation of pi3k activation and histone acetylation in lps-stimulated raw264.7 cells. *Int. J. Mol. Med.* **2015**, *35*, 202–210. [CrossRef] [PubMed]

51. Wei, J.; Feng, J. Signaling pathways associated with inflammatory bowel disease. *Recent Pat. Inflamm. Allergy Drug Discov.* **2010**, *4*, 105–117. [CrossRef] [PubMed]

52. Park, H.Y.; Han, M.H.; Park, C.; Jin, C.-Y.; Kim, G.-Y.; Choi, I.-W.; Kim, N.D.; Nam, T.-J.; Kwon, T.K.; Choi, Y.H. Anti-inflammatory effects of fucoidan through inhibition of nf-κb, mapk and akt activation in lipopolysaccharide-induced bv2 microglia cells. *Food Chem. Toxicol.* **2011**, *49*, 1745–1752. [CrossRef] [PubMed]

53. Jeng, K.C.; Hou, R.C.; Wang, J.C.; Ping, L.I. Sesamin inhibits lipopolysaccharide-induced cytokine production by suppression of p38 mitogen-activated protein kinase and nuclear factor-kappab. *Immunol. Lett.* **2005**, *97*, 101–106. [CrossRef] [PubMed]

54. Choi, M.M.; Kim, E.A.; Hahn, H.G.; Nam, K.D.; Yang, S.J.; Choi, S.Y.; Kim, T.U.; Cho, S.W.; Huh, J.W. Protective effect of benzothiazole derivative khg21834 on amyloid beta-induced neurotoxicity in pc12 cells and cortical and mesencephalic neurons. *Toxicology* **2007**, *239*, 156–166. [CrossRef] [PubMed]

55. Moron, J.A.; Zakharova, I.; Ferrer, J.V.; Merrill, G.A.; Hope, B.; Lafer, E.M.; Lin, Z.C.; Wang, J.B.; Javitch, J.A.; Galli, A.; et al. Mitogen-activated protein kinase regulates dopamine transporter surface expression and dopamine transport capacity. *J. Neurosci.* **2003**, *23*, 8480–8488. [PubMed]

56. Zhu, C.B.; Blakely, R.D.; Hewlett, W.A. The proinflammatory cytokines interleukin-1beta and tumor necrosis factor-alpha activate serotonin transporters. *Neuropsychopharmacology* **2006**, *31*, 2121–2131. [CrossRef] [PubMed]

57. Miller, A.H.; Maletic, V.; Raison, C.L. Inflammation and its discontents: The role of cytokines in the pathophysiology of major depression. *Biol. Psychiatry* **2009**, *65*, 732–741. [CrossRef] [PubMed]
58. Kong, Y.; Trabucco, S.E.; Zhang, H. Oxidative stress, mitochondrial dysfunction and the mitochondria theory of aging. *Interdiscip. Top. Gerontol.* **2014**, *39*, 86–107. [PubMed]
59. Hwang, O. Role of oxidative stress in parkinson's disease. *Exp. Neurobiol.* **2013**, *22*, 11–17. [CrossRef] [PubMed]
60. Flohe, L.; Brigelius-Flohe, R.; Saliou, C.; Traber, M.G.; Packer, L. Redox regulation of nf-kappa b activation. *Free Radic. Biol. Med.* **1997**, *22*, 1115–1126. [CrossRef]
61. Schoonbroodt, S.; Piette, J. Oxidative stress interference with the nuclear factor-kappa b activation pathways. *Biochem. Pharmacol.* **2000**, *60*, 1075–1083. [CrossRef]
62. Chen, Y.C.; Chen, C.H.; Ko, W.S.; Cheng, C.Y.; Sue, Y.M.; Chen, T.H. Dipyridamole inhibits lipopolysaccharide-induced cyclooxygenase-2 and monocyte chemoattractant protein-1 via heme oxygenase-1-mediated reactive oxygen species reduction in rat mesangial cells. *Eur. J. Pharmacol.* **2011**, *650*, 445–450. [CrossRef] [PubMed]
63. Kim, A.N.; Jeon, W.K.; Lee, J.J.; Kim, B.C. Up-regulation of heme oxygenase-1 expression through camkii-erk1/2-nrf2 signaling mediates the anti-inflammatory effect of bisdemethoxycurcumin in lps-stimulated macrophages. *Free Radic. Biol. Med.* **2010**, *49*, 323–331. [CrossRef] [PubMed]
64. Lin, H.Y.; Shen, S.C.; Chen, Y.C. Anti-inflammatory effect of heme oxygenase 1: Glycosylation and nitric oxide inhibition in macrophages. *J. Cell. Physiol.* **2005**, *202*, 579–590. [CrossRef] [PubMed]
65. Oh, G.S.; Pae, H.O.; Choi, B.M.; Chae, S.C.; Lee, H.S.; Ryu, D.G.; Chung, H.T. 3-hydroxyanthranilic acid, one of metabolites of tryptophan via indoleamine 2,3-dioxygenase pathway, suppresses inducible nitric oxide synthase expression by enhancing heme oxygenase-1 expression. *Biochem. Biophys. Res. Commun.* **2004**, *320*, 1156–1162. [CrossRef] [PubMed]
66. Miriyala, S.; Holley, A.K.; St Clair, D.K. Mitochondrial superoxide dismutase—Signals of distinction. *Anticancer Agents Med. Chem.* **2011**, *11*, 181–190. [CrossRef] [PubMed]
67. Buettner, G.R. Superoxide dismutase in redox biology: The roles of superoxide and hydrogen peroxide. *Anticancer Agents Med. Chem.* **2011**, *11*, 341–346. [CrossRef] [PubMed]
68. Klivenyi, P.; St Clair, D.; Wermer, M.; Yen, H.C.; Oberley, T.; Yang, L.; Flint Beal, M. Manganese superoxide dismutase overexpression attenuates mptp toxicity. *Neurobiol. Dis.* **1998**, *5*, 253–258. [CrossRef] [PubMed]
69. Datta, P.K.; Koukouritaki, S.B.; Hopp, K.A.; Lianos, E.A. Heme oxygenase-1 induction attenuates inducible nitric oxide synthase expression and proteinuria in glomerulonephritis. *J. Am. Soc. Nephrol.* **1999**, *10*, 2540–2550. [PubMed]
70. Mosley, K.; Wembridge, D.E.; Cattell, V.; Cook, H.T. Heme oxygenase is induced in nephrotoxic nephritis and hemin, a stimulator of heme oxygenase synthesis, ameliorates disease. *Kidney Int.* **1998**, *53*, 672–678. [CrossRef] [PubMed]
71. Dal-Cim, T.; Molz, S.; Egea, J.; Parada, E.; Romero, A.; Budni, J.; Martin de Saavedra, M.D.; del Barrio, L.; Tasca, C.I.; Lopez, M.G. Guanosine protects human neuroblastoma sh-sy5y cells against mitochondrial oxidative stress by inducing heme oxigenase-1 via pi3k/akt/gsk-3beta pathway. *Neurochem. Int.* **2012**, *61*, 397–404. [CrossRef] [PubMed]
72. Li, M.H.; Cha, Y.N.; Surh, Y.J. Peroxynitrite induces ho-1 expression via pi3k/akt-dependent activation of nf-e2-related factor 2 in pc12 cells. *Free Radic. Biol. Med.* **2006**, *41*, 1079–1091. [CrossRef] [PubMed]
73. Samoylenko, A.; Dimova, E.Y.; Horbach, T.; Teplyuk, N.; Immenschuh, S.; Kietzmann, T. Opposite expression of the antioxidant heme oxygenase-1 in primary cells and tumor cells: Regulation by interaction of usf-2 and fra-1. *Antioxid. Redox Signal.* **2008**, *10*, 1163–1174. [CrossRef] [PubMed]
74. Kang, K.H.; Liou, H.H.; Hour, M.J.; Liou, H.C.; Fu, W.M. Protection of dopaminergic neurons by 5-lipoxygenase inhibitor. *Neuropharmacology* **2013**, *73*, 380–387. [CrossRef] [PubMed]

nutrients

MDPI

Review

Nutritional Challenges in Duchenne Muscular Dystrophy

Simona Salera [1], Francesca Menni [1], Maurizio Moggio [2], Sophie Guez [1], Monica Sciacco [2] and Susanna Esposito [3],*

[1] Pediatric Highly Intensive Care Unit, Department of Pathophysiology and Transplantation, Università degli Studi di Milano, Fondazione IRCCS Ca' Granda Ospedale Maggiore Policlinico, 20122 Milan, Italy; simona.salera@policlinico.mi.it (S.S.); francescamenni@hotmail.com (F.M.); sguez_2000@yahoo.com (S.G.)

[2] Neuromuscular and Rare Disease Unit, Department of Neuroscience, Foundation IRCCS Ca' Granda Ospedale Maggiore Policlinico, University of Milan, 20122 Milan, Italy; maurizio.moggio@unimi.it (M.M.); monica.sciacco@policlinico.mi.it (M.S.)

[3] Pediatric Clinic, Università degli Studi di Perugia, 06129 Perugia, Italy

* Correspondence: susanna.esposito@unimi.it; Tel.: +39-075-578-4417

Received: 8 February 2017; Accepted: 7 June 2017; Published: 10 June 2017

Abstract: Neuromuscular diseases (NMDs) represent a heterogeneous group of acquired or inherited conditions. Nutritional complications are frequent in NMDs, but they are sometimes underestimated. With the prolongation of survival in patients with NMDs, there are several nutritional aspects that are important to consider, including the deleterious effects of overnutrition on glucose metabolism, mobility, and respiratory and cardiologic functions; the impact of hyponutrition on muscle and ventilatory function; constipation and other gastrointestinal complications; chewing/swallowing difficulties with an increased risk of aspiration that predisposes to infectious diseases and respiratory complications; as well as osteoporosis with an associated increased risk of fractures. The aim of this review is to provide a comprehensive analysis of the nutritional aspects and complications that can start in children with Duchenne muscular dystrophy (DMD) and increase with ageing. These aspects should be considered in the transition from paediatric clinics to adult services. It is shown that appropriate nutritional care can help to improve the quality of life of DMD patients, and a multidisciplinary team is needed to support nutrition challenges in DMD patients. However, studies on the prevalence of overnutrition and undernutrition, gastrointestinal complications, infectious diseases, dysphagia, and reduced bone mass in the different types of NMDs are needed, and appropriate percentiles of weight, height, body mass index, and body composition appear to be extremely important to improve the management of patients with NMD.

Keywords: dysphagia; hyponutrition; neuromuscular disease; nutrition; osteoporosis; overnutrition

1. Introduction

Neuromuscular diseases (NMDs) represent a heterogeneous group of acquired or inherited conditions. They can be classified into four major categories on the basis of their neuroanatomical localization: motor neuron diseases, neuropathies, disorders of the neuromuscular junction, and myopathies [1,2]. The predominant and common symptom across NMDs is hypotonia, which results in muscle weakness, with consequent fatigue, reduced mobility, and diminished physical work capacity. In addition, orthopaedic, cardiac, infectious, and respiratory problems are often present, which result in worsening quality of life for NMD patients and their families [1,2]. The prevention and appropriate treatment of infectious diseases as well as improvement in respiratory and cardiologic assistance, physiotherapy, and other aspects of treatment has recently changed the natural course of NMDs with an increased number of patients living into adulthood [3]. However, despite the improvement in survival, disease progression is only slowed with these therapeutic advances, leading to profound muscle weakness, disability, and complex medical and social needs that worsen with the ageing of patients [4]. Furthermore, positive medical achievements sometimes result in the appearance of treatment-related adverse events (i.e., chronic corticosteroid therapy may cause short stature, delayed puberty, obesity, osteoporosis, infections, and behavioural problems) [5]. At the same time, the appearance of adulthood-related problems (i.e., cardiovascular risk and metabolic syndrome) is a consequence of prolonged life [5].

Nutritional complications are often present in NMDs and worsen with age. Specifically, as the child grows up, there are several nutritional aspects of the disease that are important to consider. Among these aspects, there are the deleterious effects of overnutrition on glucose metabolism, mobility, and respiratory and cardiologic functions; the impact of hyponutrition on muscle and ventilatory function; constipation and other gastrointestinal complications; chewing/swallowing difficulties with an increased risk of aspiration that predisposes to infectious diseases and respiratory complications; as well as osteoporosis. Among these aspects are the deleterious effects of overnutrition, which are associated with an increased risk of fractures [6]. However, appropriate nutritional care can help to improve the quality of life of NMD patients and their families [6,7]. The aim of this review is to provide a comprehensive analysis of the nutritional aspects and complications that can start in children with NMD but increase with ageing. These aspects should be considered in the transition from cohesive paediatric clinics to disjointed adult services. Due to the limited availability of data on other NMDs, this manuscript focuses on Duchenne muscular dystrophy (DMD).

2. Anthropometrics and Growth Charts

Among NMDs, DMD is the only disease for which specific growth charts are available. At birth, weight and length in males with DMD are similar to standard distribution patterns, suggesting that the disease progression is potentially responsible for the differences in height and alterations in body weight [8]. Normal growth charts do not consider the progressive loss of muscle that occurs as muscular dystrophy progresses throughout childhood. Therefore, normal weight results on standard growth charts implies that fat tissue is accumulating [9].

In 1988, Griffiths and Edwards proposed a growth chart for boys with DMD that accounted for progressive muscle loss at a rate of 4% decline per year, thus providing ideal weight guidelines for weight control in the disease [9]. In 2013, West et al. presented a set of growth curves derived from a large cohort of male youths with DMD (i.e., 513 ambulatory males, 2–12 years old) [10]. These curves for weight, height, and body mass index (BMI) demonstrate that DMD males are shorter and tend to be at the extremes of weight and BMI compared with the general male paediatric population [10].

In DMD, despite the absence of steroid use, weight gain seems to increase from ages 7–10 years, and the average weight of ambulatory steroid-naive males with DMD is greater than the averages on the CDC 2000 growth charts [11,12]. This observation suggests that weight gain in DMD is not only a side effect of steroid therapy. It has been found that the greatest risk for overweight/obesity occurs in the preteen to teenage years (9–17.7 years), while undernutrition and weight loss become prevalent at approximately 18 years of age [12,13]. In light of these data, it is recommended that patients should be routinely weighed to identify those individuals with a trend towards excess weight gain or weight loss requiring dietetic evaluation.

Regarding height, several studies have demonstrated that children with DMD are shorter, on average, than the typical male child, regardless of steroid therapy [8,10]. In DMD, height outcome may be predicted by genetics because distal deletions of the DMD gene are more frequently associated with shorter stature [8]. Central mutations are also associated with short stature, but to a lesser degree [8].

Most studies evaluating height in DMD patients have focused on the years prior to loss of ambulation. In a non-ambulatory population it is difficult to obtain reliable and comparable height measurements. Limited range of movement, joint contractures, and scoliosis are additional factors that make for less accurate height measurements. Haapala et al. evaluated the agreement between measured and estimated height in children and young adults with cerebral palsy, and demonstrated that the preferred method when standing height cannot be obtained is to take the sum of individual segmental lengths [14]. It is necessary to standardize the methods for measuring height in non-ambulating patients with DMD or other NMDs in order to disseminate standards and obtain comparable data [15].

Given the noted increases in weight and short stature, it is not surprising that BMI tends to be higher in children with DMD compared with CDC growth trends [10]. DMD patients with body weight and BMI in accordance with standard growth charts may show good nutritional status without a need for dietetic evaluation, but because lean muscle mass can be greatly diminished, they could present with excessive body fat [5,16]. Another aspect to consider is that when the measurement of height is difficult because patients cannot stand up, the calculation of BMI may be approximate, and any error in the measurement of height would lead to an exponential error in the calculation of BMI because height is squared [14].

Overall, the available data documenting an increase in weight, a decrease in stature, and an increase in BMI in patients with DMD should be confirmed in other NMDs, in which it is advisable to carefully use standard growth charts. In addition, due to the limitations reported above, the use of BMI does not seem to be the best way to evaluate the growth of children with NMDs, and body composition should always be considered when interpreting anthropometric measures in these patients.

3. Overnutrition

In DMD and other NMDs, patient overnutrition is multifactorial due to decreased caloric needs associated with decreased physical activity and resting energy expenditure (REE). The excess caloric intake is due to the possible use of medication resulting in increased appetite and caregivers' compassion which causes a lack of caloric restriction (Table 1) [5]. In this group of patients, risks of obesity with ageing are insulin resistance (with increased risk of carbohydrate intolerance and diabetes), dyslipidemia, hypertension, and obstructive sleep apnea [5]. Other complications of being overweight are the acceleration of disease progression due to the exertion of extra force on already weak muscle groups, increased respiratory involvement with worsening pulmonary and cardiac function, and deterioration of skeletal malformations with increased need for orthopaedic surgery [5,17]. In addition, obesity worsens the ability of parents and caregivers to transfer and assist the patient in daily activities when patients lose their independence [15]. Furthermore, as in the general population, being overweight has adverse psychological impacts to further debilitate patients with chronic disease and physical disability [15]. Therefore, obesity can worsen the quality of life of these patients as well as their caregivers.

Table 1. Overnutrition in patients with neuromuscular diseases (NMDs).

Risk Factor
Decreased caloric needs
Decreased resting energy expenditure
Decreased physical activity
Excessive caloric intake due to increased appetite because of medication
Lack of caloric restriction by parents

Chronic treatment with corticosteroids increases the risk of becoming overweight, insulin resistance, and type 2 diabetes mellitus [15]. Corticosteroids may stimulate appetite and food intake and act on liver and fat cell metabolism to promote insulin resistance, hyperglycaemia, and visceral adiposity. Glucose metabolism should be evaluated in the presence of excess weight gain with paired glucose and insulin levels, glycosylated haemoglobin levels, and the oral glucose tolerance test [15].

The mainstays of prevention and treatment of becoming overweight in patients with NMDs is dietary control, as an increase of physical activity obviously has limited practical value. Therefore, dietetic advice can be beneficial to control caloric intake and excess weight gain when implemented prior to the commencement of steroid treatment [15]. A low glycaemic index diet may be useful to control alterations of glucose metabolism, and is based on the avoidance of simple sugars, the consumption of complex carbohydrates that produce relatively small changes in blood glucose, portion control, and increased fibre consumption [15]. Examples of carbohydrate-containing foods with a low glycaemic index include dried beans and legumes, all non-starchy vegetables, most fruits, and many whole-grain breads and cereals. Furthermore, dietetic advice to limit caloric intake is useful, and recommendations include reducing the intake of sugar-containing beverages and calorically-dense foods, paying attention to meals consumed outside the home, increasing the consumption of fruits and vegetables, limiting the addition of oils and fats, and the consumption of only small portions of sweet foods for breakfast and not at the end of meals [18,19]. Finally, behaviour modification techniques include the consumption of meals with family and encouraging patients to eat slowly while recognizing satiety cues [20].

4. Undernutrition

The transition from overnutrition to undernutrition usually occurs with disease progression. Mehta et al. revealed a decline in both weight and BMI Z-scores across a three-year time period in 60 children aged 2–12 years old with spinal muscular atrophy [21]. A significant decline in BMI was noted in 47% of the patients, and the prevalence of severe malnutrition increased from 2% to 17% after a period of three years.

Decreased muscle strength is the main cause of hypoalimentation (Table 2) [22]. Dysphagia, gastrointestinal problems (i.e., constipation, delayed gastric emptying), prolonged mealtime, and dependent feeding are all consequences of muscle weakness [22]. Furthermore, the presence of respiratory failure in the late stage of the disease can cause increased energy requirements [22]. The consequence of hypoalimentation and increased energy requirements is a negative energetic balance and weight loss.

Table 2. Undernutrition in patients with neuromuscular diseases (NMDs).

Main Causes
Decreased muscle strength
Dysphagia
Gastrointestinal problems (i.e., constipation, delayed gastric emptying)
Prolonged mealtime
Dependent feeding
Increased energy requirements because of respiratory failure
Swallowing difficulties

Moreover, a variety of swallowing difficulties are reported in patients with NMDs (i.e., facial weakness, reduced mastication, and poor tongue coordination) [6,12,22]. These problems result in increased mealtime with a consequent decrease in food intake, weight loss or an inability to gain weight; food inhalation into the airways with subsequent breathing problems and recurrent respiratory infections; and increased episodes of choking, coughing, and spluttering while eating resulting in embarrassment and psychological difficulties [6,12,22,23].

Undernutrition can deteriorate respiratory function and blunt immunological responses with increased risk of chest infections and a negative impact on quality of life [22]. Considering all these observations, an integrated multidisciplinary treatment appears to be mandatory to recognize signals that can indicate reduced food intake to avoid a negative impact on nutritional status.

5. Resting Energy Expenditure (REE) and Energy Requirement

Skeletal muscle metabolism is a major determinant of REE, and it is altered by the severe muscle loss that characterizes NMDs. Shimizu-Fujiwara et al. investigated REE in 77 DMD patients aged 10–37 years at various disease stages [13]. REE was significantly lower than the corresponding value in controls. At the advanced stage of the disease, abnormal development and atrophy of the liver—common findings in DMD—can contribute to the decrease in REE. A weak positive correlation between REE and serum levels of rapid turn-over proteins (prealbumin and cholinesterase synthesized by liver) was previously observed [13].

Hankard et al. studied REE in 13 DMD children (aged 8–13 years) and hypothesized that muscle mass loss would decrease REE and therefore contribute to the onset of obesity [24]. Alternatively, Zanardi et al. evaluated nine patients aged 6–12 years, showing that a loss of muscle mass in DMD patients was not associated with a reduction in REE [25].

It has been determined that DMD patients require fewer calories compared to healthy children (80% of the recommended caloric intake for healthy children in ambulatory boys and 70% in non-ambulatory boys) [15]. Therefore, caloric intake should be individualized based on physical activity and the capability of ambulation [15]. However, decreasing caloric intake has the potential to induce a negative energy balance, and in NMDs this condition may increase the loss of lean body mass, which, once lost, does not have the potential to regenerate [17].

Little is known about how energy requirements in NMDs differ from those in DMD. For the particular body composition of NMD patients, predictive energy formulas based on weight may have limited value. Further research is needed to understand energy requirements and to develop specific guidelines for energy prediction in NMD patients. In the interim, overweight prevention is an appropriate treatment for NMD patients; this involves dietary education in the early years after diagnosis and prior to the initiation of corticosteroid therapy. In obese patients, the dietitian should plan a programme for caloric restriction, while considering the devastating effects of excess weight and the potential for muscle loss associated with inducing a negative energy balance.

6. Protein and Fluid Requirement

Little literature is available about specific protein requirements in NMD populations, and it is not possible to draw conclusions. Protein intake should meet the recommended dosage for age because there is no evidence suggesting that NMD patients require additional protein intake.

There is no literature available regarding fluid intake in NMD patients; therefore, no conclusion can be drawn. However, the intake of adequate fluid is recommended to contrast an increased risk of constipation. Fluid calculations based on height and weight are not recommended, because it is often difficult to obtain a true measure of height. It is preferable to use formulas based on weight only (i.e., the Holliday–Segar method), considering individual requirements [5].

7. Gastrointestinal Complications

Complications of the gastrointestinal (GI) tract are relatively frequent in NMDs. Pane et al. evaluated GI involvement in 118 DMD patients with ages ranging between 3 and 35 years, and observed that GI symptoms were reported by 47% of patients [12]. The most common GI complications in NMD patients are delayed gastric emptying, gastroesophageal reflux (GER), and constipation (Table 3) [5]. With increasing survival, gastric and intestinal dilatation related to air swallowing due to ventilator use have also been reported [7].

Table 3. Gastrointestinal complications in patients with neuromuscular diseases (NMDs).

Gastrointestinal Complication	Pathogenesis
Delayed gastric emptying	Altered function of gastric smooth muscle cells
Gastroesophageal reflux	Delayed gastric emptying
Constipation	Immobility Weakness of abdominal wall muscles Inadequate fluid intake

The altered function of gastric smooth muscle cells in NMD patients causes delayed gastric emptying [5]. Borrelli et al. demonstrated that gastric emptying time in patients with DMD was significantly delayed compared with controls and was also worse at follow-up as the disease progresses [26]. Delayed gastric emptying can contribute to GER [5]. In the study by Pane et al., the majority of patients experienced occasional episodes of heartburn, but GER requiring pharmacological treatment was reported in only 4% of patients with DMD [12]. The presence of GER also increases possible risks for aspiration.

Constipation usually occurs in the second decade of life in patients with NMDs, and increases with age [5]. Pane et al. reported that 36% of patients with DMD experienced constipation, and this problem was more frequently reported after 18 years (60% of the patients) [12]. Colon smooth muscle involvement is due to immobility, weakness of abdominal wall muscles, and inadequate fluid intake [5]. These factors are responsible for slower GI transit, increased permanence of stool in the colon for water absorption, and consequent hard, dry stool. Decreased appetite is a complication of constipation, and consequently, it is very important to treat constipation early—especially in malnourished patients [5]. Dietary advice to prevent and/or treat constipation includes adequate fluid intake together with an increase of dietary fibre consumption. Fibre supplementation without satisfactory hydration might worsen symptoms and result in large amounts of hard stool.

8. Dysphagia

Dysphagia is common in NMD patients and affects about one third [27]. Due to the muscle weakness that characterizes all NMDs, oral motor activities are impaired [1]. Furthermore, increased risk of aspiration predisposes to respiratory complications, and this is worsened by co-existing respiratory muscle weakness in patients with compromised airway defence mechanisms and severe coughing [4,23]. Dysphagia can also lead to social and psychological consequences with worsening quality of life that may be associated with loss of satiety, fear of choking, embarrassment, and social isolation secondary to coughing, spluttering, and prolonged feeding times [23].

Table 4 summarizes the main causes of dysphagia. In most NMDs, dysphagia is mainly related to weakness of the oral muscles rather than to the incoordination of sucking, swallowing, and breathing [1]. In progressive NMDs, the majority of children have learned to chew, but masticatory problems may arise due to increasing weakness of masseter and temporal muscles in combination with weak tongue movements, resulting in prolonged mealtimes and feeling of irritation in the throat and choking [1]. Children with non-progressive NMDs (e.g., congenital myopathy) are often not able to feed in the neonatal period, and consequently need to be started on slow tube feeding with oral nutrition to avoid possible pharyngeal dysphagia [1]. In slowly progressive NMDs, feeding and swallowing problems develop insidiously, thus close professional monitoring is needed from an early stage [1].

Table 4. Causes of dysphagia in patients with neuromuscular diseases (NMDs).

Main Causes
Weakness of the oral muscles
Incoordination of sucking and swallowing
Difficulties in breathing

Unintentional weight loss or decline in the expected age-related weight gain can be a sign of dysphagia [7]. Touissant et al. considered an unintentional weight loss greater than 10% in a year to be clinically significant, but it is necessary to consider the initial morphology of patients [27]. Overweight patients have adequate fat reserves, and consequently, in those with good appetite, a diet with high calorie fluids may minimize weight loss [27]. In contrast, underweight patients have limited fat reserves and consequently more intensive nutritional therapy is needed [27].

The primary aim in the treatment of feeding and swallowing disorders is to prevent choking and to avoid aspiration pneumonia [1]. Recent studies have suggested that thickening fluid is probably not appropriate in NMDs [28]. In central forms of dysphagia (i.e., cerebral palsy), poor neurological coordination of the swallowing function is often a risk factor for aspiration during fluid intake. Alternatively, in the neuromuscular form of dysphagia, progressive muscle weakness is the main characteristic, which accompanies solid rather than liquid intake [27,28]. Consequently, it is difficult for children with NMDs to manage thick liquids and solid foods as they experience more problems with post-swallow residues when consuming these substances than when consuming thin liquids. In the presence of mastication problems, softer foods and smaller pieces of food are recommended to ensure that patients have an adequate intake of nutrients (i.e., protein, iron, fibre, and calories) [1].

Poor head control is often present in advanced stages of NMDs, and contributes to worsened feeding and dysphagia [1]. Adaptation of the head posture of children improves the efficiency of swallowing in these cases [1].

Toussaint et al. presented an algorithm to facilitate clinical decisions regarding dysphagia management in patients with DMD [27]. If difficulty in swallowing is present but it occurs without weight loss, the presentation of meals may be modified to allow for easier swallowing by reducing the efforts of chewing and transporting of the bolus [27]. It is advised to stop solid food, to promote pureed meals, and to rinse the throat regularly during and after meals with an appropriate amount of fluid [27]. In the case of unintentional weight loss, a high-caloric diet is proposed (maximal calories in a minimal volume), as well as increasing the caloric and protein density of meals. If the intake of natural food is not adequate, high-energy drinks or powders should be added [27]. A similar approach should be proposed in other NMDs, but specific studies on dysphagia in the various cohorts of patients with NMDs are needed before defining appropriate algorithms for the single conditions.

9. Enteral Nutrition

Enteral nutrition is often required in NMD patients. It is recommended to discuss enteral nutrition with the patient and the family at an early stage of the disease progression, giving enough time for a possible better outcome [17]. Early discussion and an early decision of percutaneous endoscopic gastrostomy (PEG) can also reduce risks associated with anaesthesia if respiratory capacity is not yet compromised [17]. Toussaint et al. recommended PEG when a high caloric diet trial is unsuccessful (i.e., decreased weight even with high calories) [27]. DiVito and Meyers recommend placing a feeding tube if one of these conditions is present: the child aspirates, mealtimes are longer than 30 min, the child is unable to meet nutritional needs, there is weight loss or lack of weight gain for 3 months, there is a decrease of two or more weight or height percentiles [29]. Ramelli et al. reported that PEG was associated with a reduced frequency of chest infections and consequent hospitalization, and it appeared to be more effective in improving weight and height than the use of oral supplementation [30]. The results of Martigne et al. showed improved weight status in many patients with DMD after PEG placement, and likely increased life span and/or quality of life [31]. After PEG placement, if the patient is able to eat with no risk of aspiration, oral feeding can be maintained, but without the need to reach the entire energy requirement [27]. Alternatively, if aspiration is evident, oral feeding is prohibited and the whole volume of calories and nutrients is provided by enteral nutrition [27].

This means that a multidisciplinary team including a dietitian, a gastroenterologist, and a swallowing therapist are needed to: (a) maintain the best nutritional status to prevent both undernutrition and overnutrition; (b) manage GI problems; and (c) monitor and treat dysphagia to prevent aspiration pneumonia and weight loss (Table 5).

Table 5. Main recommendations for the management of nutritional complications in patients with Duchenne muscular dystrophy (DMD).

Main Causes
Identification of a multidisciplinary team including a dietitian, a gastroenterologist, and a swallowing therapist
Maintenance of the best nutritional status to prevent both undernutrition and overnutrition
Management of gastrointestinal problems
Monitoring and treatment of dysphagia to prevent aspiration pneumonia and weight loss

10. Reduced Bone Mass

It is not known whether NMDs directly affect bone, but the necessity of long-term therapy with corticosteroids exposes the patients to unwanted side effects, such as loss of bone mass and an increased risk of fractures [3]. Other risk factors for poor bone health in NMDs include reduced weight-bearing activity and muscle weakness, with a consequent risk of fractures, osteopenia, osteoporosis, scoliosis, bone pain, and poor quality of life [7]. In the study of Bianchi et al., in about two-thirds of DMD cases, calcifediol supplementation, adjustment of dietary calcium intake to the recommended dose, and reduced sodium intake to avoid calciuric effects were able to reduce bone resorption, correct vitamin D deficiency, and increase bone mass [3].

Reaching the recommended calcium intake during childhood and adolescence is necessary to achieve optimal peak bone mass. A calcium-rich diet based on dairy products is usually appreciated by children, and may have the greatest benefit on bone accrual, while calcium supplements are sometimes not well tolerated [3]. Such a diet can become a healthy habit for all patients at risk of low bone mass. The problem of increased fracture risk will become more serious with the prolonged survival of patients.

11. Conclusions

Nutritional complications are very frequent in DMD, but they are sometimes underestimated. However, data collected from DMD patients should be extended to the other NMDs. Studies on the prevalence of overnutrition and undernutrition, GI complications, infectious diseases, dysphagia, and reduced bone mass in all the different types of NMDs are urgently needed. Additionally, appropriate percentiles of weight, height, BMI, and body composition appear to be extremely important to improve a patient's NMD management. Furthermore, problems of drug side-effects on growth and quality of life must be taken in consideration. Based on different muscle involvement and degree of impairment, complications might be different among NMDs. Moreover, gut microbiota could also influence processes including homeostasis, drug pharmacokinetics, and therapeutic response in NMD patients. Meanwhile, appropriate patient management should include dietetic assessment at diagnosis before initiating corticosteroids: (1) when patient is underweight; (2) unintentional weight loss or poor weight gain; (3) patient is overweight or at risk of becoming overweight; (4) major surgery is planned; (5) patient is chronically constipated; or (6) dysphagia is present. In addition, due to the longer life expectancy in NMD patients, nutritional issues and complications related to adult age should be considered. Specifically, increased weight gain together with the inability to exercise can probably increase the risk of developing a cluster of cardiovascular risk factors, as well as metabolic syndrome. Further research on this new adult population with NMDs will enable improved quality of life due to the avoidance of nutritional challenges.

Acknowledgments: This review, including the costs to publish in open access, was supported by a grant from the Italian Ministry of Health (Fondazione IRCCS Ca' Granda Ospedale Maggiore Policlinico Ricerca Corrente 2017 850/02).

Author Contributions: Simona Salera wrote the first draft of the manuscript; Francesca Menni, Maurizio Moggio, Sophie Guez and Monica Sciacco revised the text; Susanna Esposito critically revised the text and made a substantial scientific contributions. All the authors approved the final version of the manuscript.

Conflicts of Interest: The authors declare no conflict of interest.

References

1. Van den Engel-Hoek, L.; de Groot, I.J.M.; de Swart, B.; Erasmus, C.E. Feeding and swallowing disorders in pediatric neuromuscular diseases: An Overview. *J. Neuromuscul. Dis.* **2015**, *2*, 357–369. [CrossRef] [PubMed]

2. Liew, W.K.M.; Kang, P.B. Recent developments in the treatment of Duchenne muscular dystrophy and spinal muscular atrophy. *Ther. Adv. Neurol. Disord.* **2013**, *6*, 147–160. [CrossRef] [PubMed]

3. Bianchi, M.L.; Morandi, L.; Andreucci, E.; Vai, S.; Frasunkiewicz, J.; Cottafava, R. Low bone density and bone metabolism alterations in Duchenne muscular dystrophy: Response to calcium and vitamin D treatment. *Osteoporos. Int.* **2011**, *22*, 529–539. [CrossRef] [PubMed]

4. Archer, S.K.; Garrod, R.; Hart, N.; Miller, S. Dysphagia in Duchenne muscular dystrophy assessed by validated questionnaire. *Int. J. Lang. Commun. Disord.* **2013**, *48*, 240–246. [CrossRef] [PubMed]

5. Davis, J.; Samuels, E.; Mullins, L. Nutrition considerations in Duchenne muscular dystrophy. *Nutr. Clin. Pract.* **2015**, *30*, 511–521. [CrossRef] [PubMed]

6. Moore, G.E.; Lindenmayer, A.W.; McConchie, G.A.; Ryan, M.M.; Davidson, Z.E. Describing nutrition in spinal muscular atrophy: A systematic review. *Neuromuscul. Disord.* **2016**, *26*, 395–404. [CrossRef] [PubMed]

7. Bushby, K.; Finkel, R.; Birnkrant, D.J.; Case, L.E.; Clemens, P.R.; Cripe, L.; Kaul, A.; Kinnett, K.; McDonald, C.; Pandya, S.; et al. Diagnosis and management of Duchenne muscular dystrophy, part 2: Implementation of multidisciplinary care. *Lancet Neurol.* **2010**, *9*, 177–189. [CrossRef]

8. Sarrazin, E.; von der Hagen, M.; Schara, U.; von Au, K.; Kaindl, A.M. Growth and psycomotor development of patients with Duchenne muscular dystrophy. *Eur. J. Pediatr. Neurol.* **2014**, *18*, 38–44. [CrossRef] [PubMed]

9. Griffiths, R.D.; Edwards, R.H.T. A new chart for weight control in Duchenne muscular dystrophy. *Arch. Dis. Child.* **1988**, *63*, 1256–1258. [CrossRef] [PubMed]

10. West, N.A.; Yang, M.L.; Weitzenkamp, D.A.; Andrews, J.; Meaney, F.J.; Oleszek, J.; Miller, L.A.; Matthews, D.; DiGuiseppi, C. Patterns of growth in ambulatory males with Duchenne muscular dystrophy. *J. Pediatr.* **2013**, *163*, 1759–1763. [CrossRef] [PubMed]

11. Poruk, K.E.; Davis, R.H.; Smart, A.L.; Chisum, B.S.; Lasalle, B.A.; Chan, G.M.; Reyna, S.P.; Swoboda, K.J. Observational study of caloric and nutrient intake, bone density, and body composition in infants and children with spinal muscular atrophy type I. *Neuromuscul. Disord.* **2012**, *22*, 966–973. [CrossRef] [PubMed]

12. Pane, M.; Vasta, I.; Messina, S.; Sorleti, D.; Aloysius, A.; Sciarra, F.; Mangiola, F.; Kinali, M.; Ricci, E.; Mercuri, E. Feeding problems and weight gain in Duchenne muscular dystrophy. *Eur. J. Pediatr. Neurol.* **2006**, *10*, 231–236. [CrossRef] [PubMed]

13. Shimizu-Fujiwara, M.; Komaki, H.; Nakagawa, E.; Mori-Yoshimura, M.; Oya, Y.; Fujisaki, T.; Tokita, Y.; Kubota, N.; Shimazaki, R.; Sato, K.; et al. Decreased resting energy expenditure in patients with Duchenne muscular dysptropy. *Brain Dev.* **2012**, *34*, 206–212. [CrossRef] [PubMed]

14. Haapala, H.; Peterson, M.D.; Daunter, A.; Hurvitz, E.A. Agreement between actual height and estimated height using segmental limb lengths for individuals with cerebral palsy. *Am. J. Phys. Med. Rehabil.* **2015**, *94*, 539–546. [CrossRef] [PubMed]

15. Bianchi, M.L.; Biggar, D.; Bushby, K.; Rogol, A.D.; Rutter, M.M.; Tseng, B. Endocrine aspect of Duchenne muscular dystrophy. *Neuromuscul. Disord.* **2011**, *21*, 298–303. [CrossRef] [PubMed]

16. Pessolano, F.A.; Suarez, A.A.; Monteiro, S.G.; Mesa, L.; Dubrovsky, A.; Roncoroni, A.J.; De Vito, E.L. Nutritional assessment of patients with neuromuscular diseases. *Am. J. Phys. Med. Rehabil.* **2003**, *82*, 182–185. [CrossRef] [PubMed]

17. Davidson, Z.E.; Truby, H. A review of nutrition in Duchenne muscular dystrophy. *J. Hum. Nutr. Diet.* **2009**, *22*, 383–393. [CrossRef] [PubMed]

18. CREA—Alimenti e Nutrizione. Linee Guida per una Sana Alimentazione Italiana. Available online: http://nut.entecra.it/648/linee_guida.html (accessed on 30 October 2016).

19. Academy of Nutrition and Dietetics. Position of the Academy of Nutrition and Dietetics: Total diet approach to healthy eating. *J. Acad. Nutr. Diet.* **2013**, *134*, 307–317.

20. Academy of Nutrition and Dietetics. Position of the Academy of Nutrition and Dietetics: Nutrition guidance for healthy children ages 2 to 11 years. *J. Acad. Nutr. Diet.* **2014**, *114*, 1257–1276.

21. Mehta, N.M.; Ndewman, H.; Tarrant, S.; Graham, R.J. Nutritional status and nutrient intake challenges in children with spinal muscular atrophy. *Pediatr. Neurol.* **2016**, *57*, 80–83. [CrossRef] [PubMed]

22. Messina, S.; Pane, M.; De Rose, P.; Vasta, I.; Sorleti, D.; Aloysius, A.; Sciarra, F.; Mangiola, F.; Kinali, M.; Bertini, E.; et al. Feeding problems and malnutrition in spinal muscular atrophy type II. *Neuromuscul. Disord.* **2008**, *18*, 389–393. [CrossRef] [PubMed]

23. Jones, K.; Pitceathly, R.D.S.; Rose, M.R.; McGowan, S.; Hill, M.; Badrising, U.A.; Hughes, T. Interventions for dysphagia in long-term, progressive muscle disease. *Cochrane Database Syst. Rev.* **2016**, *2*, CD004303. [PubMed]

24. Hankard, R.; Gottrand, F.; Turck, D.; Carpentier, A.; Romon, M.; Farriaux, J.P. Resting energy expenditure and energy substrate utilization in children with Duchenne muscular dystrophy. *Pediatr. Res.* **1996**, *40*, 29–33. [CrossRef] [PubMed]

25. Zanardi, M.C.; RiTagliabue, A.; Orcesi, S.; Berardinelli, A.; Uggetti, C.; Pichiecchio, A. Body composition and energy expenditure in Duchenne muscular dystrophy. *Eur. J. Clin. Nutr.* **2003**, *57*, 273–278. [CrossRef] [PubMed]

26. Borrelli, O.; Salvia, G.; Mancini, V.; Santoro, L.; Tagliente, F.; Romeo, E.F.; Cucchiara, S. Evolution of gastric electrical features and gastric emptying in children with Duchenne and Becker muscular dystrophy. *Am. J. Gastroenterol.* **2005**, *100*, 695–702. [CrossRef] [PubMed]

27. Toussaint, M.; Davidson, Z.; Bouvoie, V.; Evenepoel, N.; Haan, J.; Soudon, P. Dysphagia in Duchenne muscular dystrophy: Practical recommendations to guide management. *Disabil. Rehabil.* **2016**, *38*, 2052–2062. [CrossRef] [PubMed]

28. Van den Engel-Hoek, L.; Erasmus, C.E.; van Hulst, K.C.M.; Arvedson, J.C.; de Groot, I.J.; de Swart, B.J. Children with central and peripheral neurologic disorders have distinguishable patterns of dysphagia on videofluoroscopic swallow study. *J. Child Neurol.* **2014**, *29*, 646–653. [CrossRef] [PubMed]

29. DiVito, D.M.T.; Meyers, R. Nutrition assessment of children with neuromuscular disease at the Children's Hospital of Philadelphia. *Top. Clin. Nutr.* **2012**, *27*, 11. [CrossRef]

30. Ramelli, G.P.; Aloysius, A.; King, C.; Davis, T.; Muntoni, F. Gastrostomy placement in paediatric patients with neuromuscular disorders: Indications and outcome. *Dev. Med. Child Neurol.* **2007**, *49*, 367–371. [CrossRef] [PubMed]

31. Martigne, L.; Seguy, D.; Pellegrini, N.; Orlikowski, D.; Cuisset, J.M.; Carpentier, A.; Tiffreau, V.; Guimber, D.; Gottrand, F. Efficacy and tolerance of gastrostomy feeding in Duchenne muscular dystrophy. *Clin. Nutr.* **2010**, *29*, 60–64. [CrossRef] [PubMed]

nutrients

MDPI

Review

Respiratory Tract Infections and the Role of Biologically Active Polysaccharides in Their Management and Prevention

Milos Jesenak [1,*], Ingrid Urbancikova [2,*] and Peter Banovcin [1]

[1] Department of Pediatrics, Comenius University in Bratislava, Jessenius Faculty of Medicine in Martin, University Teaching Hospital, Kollarova 2, 036 59 Martin, Slovakia; galandovedni@gmail.com

[2] Department of Pediatrics, P.J. Safarik University, Faculty of Medicine, Children's Faculty Hospital, Trieda SNP 1, 040 11 Kosice, Slovakia

* Correspondence: jesenak@gmail.com (M.J.); urbancikova@dfnkosice.sk (I.U.); Tel.: +42-143-420-3959 (M.J.); +42-155-235-2882 (I.U.)

Received: 26 April 2017; Accepted: 17 July 2017; Published: 20 July 2017

Abstract: Respiratory tract infections (RTIs) are the most common form of infections in every age category. Recurrent respiratory tract infections (RRTIs), a specific form of RTIs, represent a typical and common problem associated with early childhood, causing high indirect and direct costs on the healthcare system. They are usually the consequence of immature immunity in children and high exposure to various respiratory pathogens. Their rational management should aim at excluding other severe chronic diseases associated with increased morbidity (e.g., primary immunodeficiency syndromes, cystic fibrosis, and ciliary dyskinesia) and at supporting maturity of the mucosal immune system. However, RRTIs can also be observed in adults (e.g., during exhausting and stressful periods, chronic inflammatory diseases, secondary immunodeficiencies, or in elite athletes) and require greater attention. Biologically active polysaccharides (e.g., β-glucans) are one of the most studied natural immunomodulators with a pluripotent mode of action and biological activity. According to many studies, they possess immunomodulatory, anti-inflammatory, and anti-infectious activities and therefore could be suggested as an effective part of treating and preventing RTIs. Based on published studies, the application of β-glucans was proven as a possible therapeutic and preventive approach in managing and preventing recurrent respiratory tract infections in children (especially β-glucans from *Pleurotus ostreatus*), adults (mostly the studies with yeast-derived β-glucans), and in elite athletes (studies with β-glucans from *Pleurotus ostreatus* or yeast).

Keywords: β-glucans; biologically active polysaccharides; immunomodulation; recurrent respiratory tract infections; prevention

---

1. Recurrent Respiratory Tract Infections and Their Management

Respiratory tract infections (RTIs) represent the most common form of infection at every age. Studies have shown that with increasing age, the incidence of respiratory infection declines. However, in specific age (preschool age) or subject groups (e.g., patients with chronic diseases, immunosuppressive therapy, athletes), the frequency of RTIs is so high that diagnosis of recurrent respiratory tract infections (RRTIs) ought to be discussed. Several possible definitions of RRTIs are in the literature, but the most commonly used are those stated by de Martino et al. (2007). Patients with RRTIs should fulfill at least one of the following criteria: ≥6 respiratory infections per year; ≥1 respiratory infection per month involving the upper airways from September to April; or ≥3 respiratory infections per year involving the lower airways [1]. RRTIs are a specific type of respiratory infection with higher frequency compared to the acceptable and expectable number of

RTIs for a particular age group (so-called physiological morbidity). Correct diagnosis of RRTIs can be established after excluding some severe and chronic conditions associated with RRTIs: cystic fibrosis, primary and secondary immunodeficiency syndromes, primary and secondary ciliary dyskinesia syndromes, or congenital anomalies of the respiratory tract. Also, many factors may contribute to higher respiratory morbidity: older siblings of close age and larger families (with increased possibility to get in contact with RTIs), day-care center attendance, prematurity, shortened breast feeding and malnutrition, environmental factors (indoor and outdoor pollution, passive smoking), allergic inflammation (especially in children of risk with allergic, first-degree relatives), chronic focal infections (e.g., adenoid hypertrophy, chronic tonsillitis, sinusitis) or gastroesophageal reflux. In preschoolers, immature mucosal immunity is another important factor [2–4].

The majority of patients with RRTIs do not have any recognizable immunodeficiencies or other pathologies, but some of these subjects may have mild and non-specific deviations in selected immune parameters that are an expression of immature immunity, or transient immune function decline after certain events (e.g., extreme physical activities or post-infectious, stressful, and exhausting periods) [1]. Immunomodulation represents one of the possible and accepted approaches in managing and preventing RRTIs at every age or patient category. Immunomodulation is characterized as a preventive and therapeutic intervention into immune system activity aimed at correcting deviated immune functions. It can either support declined and suppressed immune parameters or normalize the increased and over-acting functions. Many natural and synthetic compounds and medications have immunomodulatory activities, many of them based on traditional medicine. However, only few of them also possess a scientific basis.

2. Biologically Active Polysaccharides as Biological Response Modifiers

Biologically active polysaccharides (BAPs) are one of the most studied natural immunomodulators and due to their confirmed complex mode of action, they can be named as biological responses modifiers. The most important BAPs are β-glucans, which are a heterogeneous group of natural polysaccharides, comprising D-glucose monomers linked by β-glycosidic bonds. Several sources of β-glucans exist: fungi, yeasts, bacteria, algae, and various plants. Studies have shown that immunomodulating potential differs between β-glucans due to their origin, purity, structure, branching level, solubility, and molecular conformation [5–9]. Studies have also shown that the β-glucans can yield similar activities when used in both injectable and oral forms [5].

Studies have shown that β-glucans possess many biological activities. Those receptors on immune and non-immune cell surfaces (e.g., fibroblasts, keratinocytes) which are important in mediation of β-glucan activities include: dectin-1, complement receptor 3, toll-like receptors (TLR), and others [10,11]. Dectin-1 represents probably the most important receptor mediating the biological effects of β-glucans. It is expressed especially on the cells of non-specific immunity, e.g., macrophages, neutrophils, and dendritic cells. It closely collaborates with TLR 2 and 6 (TLR-2/6). Its activation is linked to the several intracellular pathways (e.g., nuclear factor κB or signaling adaptor protein CARD9), which lead to the release of various cytokines. TLR-4 is probably another essential receptor involved in the activation and maturation of dendritic cells after the recognition of β-glucans [11]. Since insoluble β-glucans are not absorbed into the blood, there are several theories or hypotheses trying to explain their mode of action. One possibility is represented by the direct interaction with immune cells in Peyer's patches in small intestine with subsequent immune cell activation [12]. Another possible mechanism could be the ingestion of β-glucans by macrophages followed by the release of fragments into the intercellular microenvironment and activation of other cellular populations. β-glucans can activate innate as well as adaptive immune mechanisms and cells, stimulate the activity of neutrophils and macrophages via surface receptors, make the phagocytosis more effective, support the functions of NK cells, modulate the functions of antigen-presenting cells (e.g., Langerhans cells) and promote antigen presentation, influence the production of cytokines and chemokines, create conditions supporting Th1 lymphocytes, and modulate antibody production [3,13–17].

Table 1. Effect of β-glucans of different origin on the respiratory tract infections and selected parameters in children and adults.

No.	Country of Study	Study Population	Age	Study Design	Main Outcomes	β-Glucan Type (Dose)	Duration of Treatment	Reference
1	Czech and Slovak Republic	215 children with RRTIs	4.7 years	OLS	↓ frequency of RRTIs (positive therapeutic response—≥50% reduction of RRTI frequency—in 71.2% of children ($p < 0.001$)	Pleuran—insoluble β-glucan from *Pleurotus ostreatus* (10 mg/10 kg of body weight)	3 months (& 3 months follow-up)	Jesenak et al., 2010 [18]
2	Spain	151 children with RRTIs	3.0 years	OLS	↓ frequency of RRTIs ($p < 0.001$); ↓ number of otitis media ($p < 0.001$), common cold ($p < 0.001$), tonsillopharyngitis ($p < 0.001$), laryngitis ($p < 0.001$), bronchitis ($p < 0.001$), pneumonia ($p < 0.05$); ↓ number of emergency visits due to respiratory infections ($p < 0.001$); ↓ number of days-off from kindergarten or school ($p < 0.05$); ↓ use of symptomatic therapy ($p < 0.05$)	Pleuran—insoluble β-glucan from *Pleurotus ostreatus* (10 mg/10 kg of body weight)	3 months (& 3 months follow-up)	Sapena Grau et al., 2015 [19]
3	Poland	194 children with RRTIs	3.7 years	OLS	↓ frequency of RRTIs ($p < 0.001$); ↓ number of otitis media ($p < 0.01$), laryngitis ($p < 0.01$), bronchitis ($p < 0.01$), common cold ($p < 0.01$); ↓ number of days-off from kindergarten or school ($p < 0.01$)	Pleuran—insoluble β-glucan from *Pleurotus ostreatus* (10 mg/10 kg of body weight)	3 months (& 3 months follow-up)	Pasnik et al., 2017 [20]
4	Czech and Slovak Republic	175 children with RRTIs	5.6 years	DBPCRT	↓ frequency of RRTIs ($p < 0.05$); ↑ number of healthy children ($p < 0.05$); ↓ number of flu and flu-like diseases ($p < 0.05$); ↓ number of lower respiratory tract infections ($p < 0.05$); immunomodulating effects on antibody production immunomodulating effects on cellular immunity	Pleuran—insoluble β-glucan from *Pleurotus ostreatus* (10 mg/10 kg of body weight)	6 months (& 6 months follow-up)	Jesenak et al., 2013 [3]
5	Slovak Republic	53 adult patients with Crohn's disease	37.0 years	DBPCRT	↓ frequency of accompanying diseases (respiratory tract infections, herpes simplex infections, oral thrush) ($p = 0.019$); Ø effect of Crohn's diseases activity	Pleuran—insoluble β-glucan from *Pleurotus ostreatus* (100 mg/day)	12 months	Batovsky et al., 2015 [21]
6	Germany	162 healthy adults	43.2 years	DBPCRT	↓ number of symptomatic cold episodes ($p = 0.041$); ↓ sleep difficulties caused by cold episodes ($p < 0.028$)	Insoluble yeast β-glucan (900 mg/day)	4 months	Auinger et al., 2013 [22]
7	U.S.A.	77 stressec. adult women	38.0 years	DBPCRT	↓ upper respiratory symptoms ($p < 0.05$); ↑ overall well-being and superior mental/physical energy levels ($p < 0.05$)	Insoluble yeast β-glucan (250 mg/day)	3 months	Talbott et al., 2012 [23]
8	U.S.A.	150 moderately to highly-stressed adults	39.0 years	DBPCRT	↓ upper respiratory tract infection symptoms ($p < 0.05$); ↑ overall well-being and vigor ($p < 0.05$); ↓ fatigue and tension ($p < 0.05$)	Insoluble yeast β-glucan (250 or 500 mg/day)	1 month	Talbott et al., 2010 [24]
9	U.S.A.	40 healthy adults	30.3 years	DBPCRT	Ø differences in the incidence of symptomatic respiratory tract infection; ↑ number of missed day of schollf or work per cold ($p = 0.026$); ↑ quality of life in active group ($p = 0.042$); ↓ average fever score ($p = 0.042$)	Insoluble yeast β-glucan (500 mg/day)	3 months	Feldman et al., 2009 [25]

Table 1. Cont.

No.	Country of Study	Study Population	Age	Study Design	Main Outcomes	β-Glucan Type (Dose)	Duration of Treatment	Reference
10	Germany	94 healthy adults	45.6 years	DBPCRT	Ø differences in the incidence of common cold subjects without incidence of common cold compares to placebo ($p = 0.019$) ↑ number of infections during the most intense season for infection ($p = 0.02$) ↓ of typical common cold symptoms: sore throat and/or difficulty swallowing ($p = 0.034$), hoarseness and/or cough ($p < 0.001$), runny nose ($p < 0.001$)	Insoluble yeast β-glucan (450 mg/day)	7 months	Graubaum et al., 2012 [26]
11	United Kingdom	97 healthy adults	21.0 years	DBPCRT	Ø effect on the incidence of respiratory tract infection ↑ ability to "breathe easily" ($p = 0.049$) Ø effect on chemokines and cytokines production	Insoluble yeast β-glucan (250 mg/day)	3 months	Fuller et al., 2012 [27]
12	Czech Republic	40 children with chronic respiratory problems	10.7 years	DBPCRT	Improvement of mucosal immunity: ↑ lysozyme ($p < 0.05$), ↓ albumin ($p < 0.05$) Improvement in general disease condition	Insoluble yeast β-glucan (100 mg/day)	1 month	Vetvicka et al., 2013 [28]
13	Czech Republic	40 children with chronic respiratory problems	10.7 years	DBPCRT	↑ of salivary immunoglobulins (IgG, IgA, IgM) ($p < 0.05$)	Insoluble yeast β-glucan (100 mg/day)	1 month	Vetvicka et al., 2013 [29]
14	Czech Republic	60 children with chronic respiratory problems	9.7 years	DBPCRT	↓ of salivary lysozyme ($p < 0.05$), calprotectin ($p = 0.015$), albumin ($p < 0.05$)	Insoluble yeast β-glucan (100 mg/day)	1 month	Richter et al., 2014 [30]
15	Czech Republic	56 children with chronic respiratory problems	9.7 years	DBPCRT	↓ of salivary cotinine ($p < 0.05$) and cortisol levels ($p < 0.05$) ↑ of physical endurance ($p < 0.05$)	Insoluble yeast β-glucan (100 mg/day)	1 month	Richter et al., 2014 [31]
16	Czech Republic	40 children with chronic respiratory problems	10.9 years	DBPCRT	↑ of physical endurance ($p < 0.05$)	Insoluble yeast β-glucan (100 mg/day)	1 month	Vetvicka et al., 2013 [32]
17	Czech Republic	77 children with chronic respiratory problems	10.3 years	DBPCRT	Stabilization of the salivary IgA levels	Insoluble yeast β-glucan (100 mg/day)	1 month	Richter et al., 2015 [33]
18	U.S.A.	264 healthy children	3.5 years	DBPCRT	↓ number and duration of acute respiratory infections ($p = 0.007$) ↓ antibiotic use ($p = 0.01$) Immunomodulatory and anti-inflammatory effects	Insoluble yeast β-glucan (26.1 mg/day)	7 months	Li et al., 2014 [34]

DBPCRT—double-blind, placebo-controlled, randomized trial; OLS—open-label study; RRTIs—recurrent respiratory tract infections; ↑—increased/improved, ↓ decreased/worsened, Ø—no effect.

β-glucans are characterized by pluripotent biological properties which can be useful in managing various immune-mediated conditions and infectious diseases, e.g., RRTIs. In terms of managing and preventing RTIs, a number of published reports exist with children or adult subjects in the literature that studied the β-glucans from oyster mushrooms (*Pleurotus ostreatus*), baker's yeast (*Saccharomyces cerevisiae*) and oats (Table 1). In our study we aimed to analyze the possible effect of commonly used insoluble β-glucans in the treatment and prevention of RTIs in children, adults and in athletes. Original articles have been selected for analysis among those published in PubMed and Scopus referenced journals using the following keywords: "respiratory infections", "respiratory tract infections", "recurrent respiratory tract infections", "treatment", "prevention", "β-glucans", "glucans". For the final analysis, only the studies evaluating the effects of orally applied β-glucans on various clinical and laboratory parameters in human subjects with respiratory tract infections were selected.

3. β-Glucans Isolated from *Pleurotus Ostreatus* and Respiratory Tract Infections

The first study analyzing the efficacy and safety of syrup containing a patented complex of biologically active polysaccharides isolated from *Pleurotus ostreatus* was performed in the Czech and Slovak Republics in a group of 215 children, mostly preschoolers. Children were administered a syrup containing pleuran—insoluble β-glucans from *Pleurotus ostreatus*—every morning on an empty stomach for 3 months starting at the beginning of Autumn. A positive therapeutic response (more than 50% reduction of RRTI frequency) was observed in 71.2% of the studied children and the total number of respiratory tract infections declined from 8.9 episodes of RTIs/year to 3.6 episodes/year compared to the previous treatment period ($p < 0.001$) [18].

In another, open-label study from Spain, the effect of the same product on frequency of RTIs and other selected parameters was studied in a group of 151 children. The children were administered the medication for 3 months and were followed-up for 3 additional months. Active treatment decreased the number of the RRTIs from 8.88 ± 3.35 episodes in the previous year to 4.27 ± 2.21 episodes in the study year ($p < 0.001$). Furthermore, the incidence and number of episodes of each type of respiratory tract infection (otitis media, common cold, tonsillopharyngitis, laryngitis, bronchitis, and pneumonia) were significantly reduced. Application of the syrup with pleuran also reduced the number of emergency department visits, use of symptomatic pharmacotherapy, and missed days from kindergarten or school compared to the previous year before treatment. The product showed good or very good tolerability in 90.7% of children and a significant improvement of clinical status was reported by 85.7% of parents [19]. A study of the same design and product was later performed in another group of 194 children in Poland. Generally, supplementation of syrup with pleuran (Imunoglukan P4H® syrup) significantly decreased the total number of RTIs during treatment and the follow-up period compared to the same period of the previous year (4.18 ± 2.132 vs. 8.71 ± 1.89, $p < 0.001$). The syrup demonstrated a significant capacity to prevent and decrease the number of various forms of RTIs (otitis media, laryngitis, bronchitis and common cold) and positively influenced the number of days off from kindergarten or school. As in the previous trials, the product was well tolerated and no serious or adverse events were observed [20].

The positive and preventive effects of pleuran supplementation on respiratory morbidity were also confirmed in two double-blind, placebo-controlled, multicenter randomized trials (DBPCRT). The only published DBPCRT analyzing the preventive effect of β-glucan on RRTIs was performed in a population of 175 children aged 5.65 years. The subjects were randomized into two different treatment groups receiving either syrup containing pleuran and vitamin C or an active placebo—syrup with vitamin C. The children were administered the medication on an empty stomach for 6 months (starting in August to October) and were followed-up for 6 additional months. In the active group, 36% of children did not have any respiratory infections during the treatment period compared to 21% in the placebo group ($p < 0.05$). Active treatment also significantly decreased the frequency of flu and flu-like diseases ($p < 0.05$) as well as the frequency of lower respiratory tract infection ($p < 0.05$). Based on the laboratory results, active treatment showed potential immunomodulatory effects on humoral immunity and supported natural maturation of antibody production. There were no signs of overstimulation in the cellular part of the immune system. Vitamin C was used as an active placebo to demonstrate that the observed laboratory or clinical effects can be attributed to the active substance—pleuran—not to vitamin C, which is also contained in studied Imunoglukan P4H® syrup [3]. Another DBPCRT investigated the preventive effect of pleuran on infectious and non-infectious complications in adult patients with Crohn's disease. All the patients were treated with biological therapy (TNF-α blockers) and were enrolled during the clinical remission of the gastrointestinal disease. These patients usually suffer from recurrent respiratory tract infections due to a complex immunodeficiency of combined origin (immunosuppressant therapy, chronic inflammation, immune dysregulation), each emerging infection having the capacity to alter the disease's stability. Actively treated patients showed decreased frequency of accompanying diseases and emerging infections compared to the placebo arm of the study. Moreover, pleuran was safe and did not worsen the clinical course of Crohn's diseases [21].

4. Yeast and Oat β-Glucans and Respiratory Tract Infections

Other traditionally used β-glucans are isolated from yeast or cereals (mostly from oats). Studies show that β-glucans from various sources (mushrooms, yeast, and cereals) and within one group (e.g., β-glucans from yeasts) could have different immunomodulatory potential and varying clinical effects [7].

Several clinical studies and reports analyze the effects of yeast and cereal β-glucans on RTIs or selected immune parameters. Most of these studies were performed in adult populations. Auinger et al. (2013) investigated the effect of baker's yeast (*Saccharomyces cerevisiae*) on the number of common cold episodes in 162 healthy subjects who were randomized into either active or placebo arms. Supplementation of insoluble yeast β-glucan reduced the number of symptomatic cold infections by 25% compared to placebo ($p = 0.041$) with improvement of sleep difficulties caused by cold episodes ($p = 0.028$). The efficacy of the preparation was positively rated by both patients and physicians [22]. Another authors' group examined the effects of yeast β-glucans in different clinical trials. In a group of 77 stressed, adult women (with moderate level of psychological stress), subjects treated with yeast β-glucan reported fewer upper respiratory symptoms compared to placebo ($p < 0.05$). Active treatment was associated with better overall well-being and superior mental and physical energy levels ($p < 0.05$) [23]. In another study by the same authors, 150 moderately to highly stressed subjects (45 men, 105 women) took either yeast β-glucan (250 mg or 500 mg) or placebo with the aim of influencing the onset of respiratory tract infection and well-being. Two dosing regimens decreased upper respiratory tract infection symptoms ($p < 0.05$) and improved overall health and vigor ($p < 0.05$) with decreased tension and fatigue ($p < 0.05$) compared to placebo. There were no statistically significant differences between two dosing schedules of β-glucan [24].

Some studies have shown no significant differences in the incidence of respiratory tract infection between treatment with β-glucan or placebo. The study with yeast β-glucan in 40 healthy adults did not show any differences in the incidence of symptomatic respiratory tract infection among the study groups (β-glucan versus placebo). However, none of the actively treated subjects missed days from work or school ($p = 0.026$). Moreover, the application of β-glucan improved quality of life ($p = 0.042$) and decreased average fever scores ($p = 0.042$) compared to placebo-treated subjects [25]. Another study also did not observe any differences in the incidence of common cold episodes compared to placebo. However, in the β-glucan-treated patients, a higher number of subjects without incidence of common cold was seen ($p = 0.019$). During the period with the highest incidence of infections, active treatment led to significantly less infection and reduced typical common cold symptoms (the symptoms were less pronounced and subsided faster) ($p = 0.020$) [26]. Fuller at al. (2012) could not discover a statistically significant difference in the number of days with upper respiratory tract infection symptoms when comparing the use of 250 mg of yeast β-glucan to the rice flour-based placebo. The only significant outcome of their study was improved ability to "breathe easily" in the active arm (*p* = 0.049). Additionally, no significant changes in chemokines or cytokines production between the two groups was found [27].

A couple of papers with yeast-derived β-glucans in the management of children with chronic respiratory problems were published by the group of Vetvicka & Richter [28–33]. In the similar cohorts of the children, they evaluated the effect of yeast-derived insoluble β-glucan on different salivary parameters and clinical characteristics. In the first study, they examined salivary inflammatory markers in 40 children with chronic respiratory diseases (recurrent respiratory tract infections, chronic bronchitis, bronchial asthma, respiratory allergies). An oral application over 4 weeks of insoluble yeast β-glucan decreased the concentration of albumin and increased the levels of lysozyme in saliva among the actively treated children, whereas in placebo group, no significant changes were recorded. The authors reported an improvement in general condition regarding chronic respiratory diseases, however, more detailed information cannot be found in this publication [28]. In other two studies they showed positive effect of orally applied β-glucan on the concentration of salivary immunoglobulin A, G and M levels [29] or inflammatory markers (lysozyme, calprotectin, albumin and CRP) in saliva [30]. Some of the results were inconsistent and differed between the publications (e.g., changes in salivary concentration of lysozyme) [28,30]. Concentration of cotinine (a marker of passive smoking exposure) and cortisol decreased in saliva after a 4-week-application of yeast β-glucan compared to placebo. Therefore authors suggested that β-glucan is able to reduce the negative environmental effects on children with chronic respiratory problems [31]. The possible preventive effect of β-glucan in children under physical stress was studied in another two trials of this authors' group. They found significant improvements in physical endurance and exhaled nitric oxide in glucan-treated children [32]. They observed a stabilization of the IgA levels in saliva after a 6-min walking test compared to placebo [33]. Conversely, levels of exhaled nitric oxide decreased in both actively or placebo treated children [32,33]. The preventive effect of β-glucan on RTIs was not evaluated in these studies.

Another interesting approach was studied by Li et al. (2014) in a double-blind, randomized, controlled prospective trial. They analyzed the possible effect of the follow-up formula containing docosahexaenoic acid (DHA), prebiotics PDX/GOS (polydextrose and galacto-oligosaccharides 1:1 ratio), and 8.7 mg yeast β-glucan per dose on respiratory infection compared to children administered an unfortified, cow's milk-based beverage for 28 weeks. Children administered the modified follow-up formula (FUF) had fewer episodes and shorter duration of acute respiratory tract infections, less antibiotic use, and fewer missed days of day-care. The FUF group also had a higher blood concentration of interleukin 10 and white blood cell count at the end of the study. This strategy could be considered a promising tool for improving and supporting maturity of the immune system during early life via feeding. However, whether the observed effect was strictly attributable to only β-glucan or also to the prebiotics or DHA cannot be determined [34].

5. β-Glucans, Recurrent Respiratory Tract Infections, and Sports Medicine

Sports activity has many beneficial health effects at both a physical and mental level. It was shown that short term physical activity has immune activating effects. Conversely, prolonged and exhausting physical activity causes numerous negative changes to immunity. These changes are usually transient, but in the absence of sufficient resting time, immunosuppression becomes more profound and the development of secondary immunodeficiency can be observed. This leads to an increased risk of RTIs with a general negative impact on sports performance. It is possible that supplementing various immunomodulators could minimize post-exercise immunosuppression, improve immune functions, and decrease the rate and severity of RTIs [35]. The characteristics and outcomes of available studies performed with different β-glucans in athletes are summarized in Table 2.

Table 2. Effect of β-glucans of different origin on respiratory tract infections and laboratory parameters in athletes.

No.	Country	Study Population	Age	Study Design	Main Outcomes	β-Glucan Type (Dose)	Duration of Treatment	Reference
1	U.S.A.	60 recreationally active adults	22.5 years	DBPCRT	↑ potential of blood leukocytes to produce IL-2, IL-4, IL-5, IFN-γ ($p < 0.05$) Effect on respiratory morbidity not studied	Insoluble yeast β-glucan (100 mg/day)	20 days (cross-over after 10 days)	Carpenter et al., 2013 [36]
2	Slovak Republic	20 elite athletes	23.3 years	DBPCRT	Prevention of decline in natural killer cell numbers and activity ($p < 0.001$) Effect on respiratory morbidity not studied	Pleuran—insoluble β-glucan from *Pleurotus ostreatus* (100 mg/day)	2 months	Bobovcak et al., 2010 [37]
3	Slovak Republic	50 elite athletes	23.6 years	DBPCRT	↓ incidence of upper respiratory tract infections ($p < 0.001$) ↑ number of natural killer cells ($p < 0.001$) Prevention of decline of phagocytic functions ($p < 0.001$)	Pleuran—insoluble β-glucan from *Pleurotus ostreatus* (200 mg/day)	3 months (& 3 months follow-up)	Bergendiova et al., 2010 [38]
4	U.S.A.	75 marathon runners	36.0 years	DBPCRT	↓ number of upper respiratory tract infection symptoms ($p < 0.05$) ↑ overall health and vigor ($p < 0.05$) ↓ confusion, fatigue, tension, and anger ($p < 0.05$)	Insoluble yeast β-glucan (250 or 500 mg/day)	1 month	Talbott et al., 2009 [39]
5	U.S.A.	182 marathon runners	34.0 years	DBPCRT	↓ number of cold/flu symptom days ($p = 0.026$) ↑ salivary IgA after exercise ($p < 0.05$)	Insoluble yeast β-glucan (250 mg/day)	1 month	McFarlin et al., 2013 [40]
6	U.S.A.	36 trained male cyclists		DBPCRT	Ø effect on incidence of upper respiratory tract infections Ø effect on exercise-induced immune changes	Insoluble oat β-glucan (5.6 g/day)	2 weeks (+ & weeks follow-up)	Nieman et al., 2008 [41]

DBPCRT—double-blind, placebo-controlled, randomized trial; IL—interleukin; IFN—interferon; RRTIs—recurrent respiratory tract infections; ↑—increased/improved, ↓ decreased/worsened, Ø—no effect.

Some studies have only analyzed the effect of different β-glucans on selected laboratory and immune parameters. In a group of 60 recreationally active men and women, 10 days of supplementing with baker's yeast β-glucan resulted in an increase of total (CD14⁺) and pro-inflammatory monocyte (CD14⁺CD16⁺) concentrations. Furthermore, β-glucan boosted lipopolysaccharide-stimulated production of IL-2, IL-4, IL-5, and IFN-γ before and after exercise. Plasma concentration of IL-4, IL-5, and IFN-γ were also greater 2 h after exercise in the β-glucan group compared to the placebo group. Therefore, it can be suggested that β-glucans may modulate immune response and immune reactivity following strenuous exercise [36]. Bobovcak et al. (2010) confirmed that supplementing with pleuran prevented the decline of natural killer cell numbers and activity after the recovery period in elite athletes compared to placebo [37].

Various trials have shown that β-glucan supplementation could decrease RTIs in athletes. Supplementation with insoluble pleuran from *Pleurotus ostreatus* significantly decreased the incidence of upper respiratory tract infection compared to placebo in elite athletes ($p < 0.001$). Active treatment also increased natural killer cell numbers and prevented the decline of phagocytic functions after the physical exertion [38]. In a study with 75 marathon runners, yeast β-glucan was administered for 4 weeks and significantly decreased symptoms of upper respiratory tract infection, confusion, fatigue, anger, and tension, improving overall health and increasing vigor [39]. Similarly, McFarlin et al. (2013) reported a 37% reduction in the number of cold/flu symptom days post-marathon compared to placebo. Interestingly, 2 hours after exercise yeast β-glucan caused a 32% increase in salivary IgA compared to placebo [40]. Only one trial was performed with oat-derived β-glucan. Its application did not prevent or reduce post-exercise-induced immune changes or incidence of upper respiratory tract infection during the treatment and follow-up period [41].

6. Discussion

β-glucans represent a promising group of immunomodulatory substances with pluripotent biological activities and favorable safety profile. Up until now, several mechanisms of action have been supposed and at least partially confirmed, especially through the laboratory and animal studies. However, the exact mechanisms of biological effects in humans are still under investigation. The use of β-glucans in the management and prevention of RTIs have been evaluated in several studies with different patients populations (children, adults, stressed individuals, athletes). Clinical trials showed a potential role of β-glucans in modulation of mucosal and systemic immunity with positive effect on selected inflammatory markers. β-glucans also yielded a positive effect on the parameter of systemic and mucosal humoral immunity what precludes their potential in the management of RTIs. Moreover, an improvement of health status and general well-being was consistently reported. The preventive potential of β-glucans in RRTIs was confirmed especially for insoluble β-glucans isolated from *Pleurotus ostreatus* and until now, only one DBPCRT, which showed a preventive effect in RRTIs in children, was published. The studies with yeast β-glucans reported especially some effects on the incidence of upper respiratory symptoms, but clear preventive effect was not observed. In athletes, the prevention of post-exercise immune suppression and decreased incidence of RTIs was also confirmed.

On the other hand, the published studies had several weaknesses which should be resolved and addressed in the further clinical trials and research. The number of the involved subjects was small in several studies and high inconsistency was noticed in the selection criteria of the patients. Heterogeneity of the studied cohorts caused inconsistent results found in some studies. Detailed information about the particular forms of RTIs treated and prevented by β-glucans is in general missing in the publications. Up to now, an optimal dose, duration and timing of the β-glucans' application has not been clearly defined. Moreover, the applied β-glucans were not characterized in several studies and the purity of the active substance was not reported.

More studies analyzing the preventive effect of β-glucans in the management of RRTIs are needed to confirm the existing data. For the future, the estimation of the optimal effective dose and the duration of the application of β-glucans from particular source should be evaluated. Another important issue which should be resolved is the standardization of the production and extraction with achieving the highest purity of the active substance from the natural β-glucans' sources. Studies could also focus on the possible combinations of the β-glucans with another immune active substances. Recently, intranasal application of β-glucan in combination with resveratrol showed another promising approach for the prevention of upper respiratory tract infections. In a group of 82 children with RRTIs, intranasal application of the mixture of carboxymethyl-β-glucan and resveratrol was able to reduce the number of days with nasal obstruction, rhinorrhea, sneezing, cough, fever, medication use, medical visits and school absence compared to saline isotonic solution. Therefore, this mode of β-glucans' application should be addressed in the further studies [42].

7. Conclusions

Respiratory tract infections represent an important health-care problem and their rational management and effective prevention could have many direct and indirect benefits. Recurrent respiratory tract infections are a special form of RTI typical in children and some specific patient groups. RRTIs have high direct and indirect economic costs and increasing antibiotic resistance is today a serious and emerging problem. Immunomodulation, therefore, represents an interesting approach how to decrease the use of antibiotics and alleviate the economic impact of RTIs. Based on published studies, evidence supports the preventive use of β-glucans in managing RRTIs. Whereas in children, especially β-glucans from *Pleurotus ostreatus* have proven to be effective and safe, most studies performed with adults were especially with yeast-derived β-glucans. Some data demonstrates the efficacy of β-glucans in preventing RTIs in elite athletes. Preventive application of β-glucans may decrease the frequency of various forms of respiratory tract infection, support protective immune mechanisms, and possibly yield other beneficial effects (increased well-being, decreased missed days from school or work, decreased use of other symptomatic or antibiotic therapy).

Acknowledgments: The study was supported by project VEGA 1/0252/14 and the project Center of Experimental and Clinical Respirology (ITMS 26220120004), co-funded from EU sources.

Author Contributions: M.J. performed literature research and wrote the paper; I.U. performed a literature analysis and constructed the summary tables; P.B. controlled literature review and reviewed the paper.

Conflicts of Interest: The authors declare no conflict of interest.

References

1. De Martino, M.; Balloti, S. The child with recurrent respiratory infections: Normal or not? *Pediatr. Allergy Immunol.* **2007**, *18* (Suppl. 18), 13–18. [CrossRef] [PubMed]
2. Ciprandi, G.; Tosca, M.A.; Fasce, L. Allergic children have more numerous and severe respiratory infections than non-allergic children. *Pediatr. Allergy Immunol.* **2009**, *17*, 389–391. [CrossRef] [PubMed]
3. Jesenak, M.; Majtan, J.; Rennerova, Z.; Kyselovic, J.; Banovcin, P.; Hrubisko, M. Immunomodulatory effect of pleuran (β-glucan from *Pleurotus ostreatus*) in children with recurrent respiratory tract infections. *Int. Immunopharmacol.* **2013**, *15*, 395–399. [CrossRef] [PubMed]
4. Schaad, U.B.; Esposito, S.; Razi, C.H. Diagnosis and management of recurrent respiratory tract infections in children: A practical guide. *Arch. Pediatr. Infect. Dis.* **2016**, *4*, e31039. [CrossRef]
5. Vetvicka, V.; Vetvickova, J. A comparison of injected and orally administered β-glucans. *J. Am. Nutraceut. Assoc.* **2008**, *11*, 1–8.
6. Bohn, J.A.; BeMiller, J.N. (1–3)-β-D-glucans as biological response modifiers: A review of structure-functional activity relationships. *Carbohydr. Polym.* **1995**, *28*, 3–14. [CrossRef]
7. Brown, G.D.; Gordon, S. Immune recognition of fungal β-glucans. *Cell. Microbiol.* **2005**, *7*, 471–479. [CrossRef] [PubMed]
8. Stier, H.; Ebbeskotte, V.; Gruenwald, J. Immune-modulatory effects of dietary yeast beta-1,3/1,6-D-glucan. *Nutr. J.* **2014**, *13*, 38. [CrossRef] [PubMed]

9. Volman, J.J.; Ramakers, J.D.; Plat, J. Dietary modulation of immune function by β-glucans. *Physiol. Behav.* **2008**, *94*, 276–284. [CrossRef] [PubMed]
10. Legentil, L.; Paris, F.; Ballet, C.; Trouvelot, S.; Daire, X.; Vetvicka, V.; Ferrieres, V. Molecular interactions of (1→3)-β-D-glucans with their receptors. *Molecules* **2015**, *20*, 9745–9766. [CrossRef] [PubMed]
11. Chan, G.C.F.; Chan, W.K.; Sze, D.M.Y. The effects of β-glucan on human immune and cancer cells. *J. Hematol. Oncol.* **2009**, *2*, 25. [CrossRef] [PubMed]
12. Spriet, I.; Desmet, S.; Willems, L.; Lagrou, K. No interference of the 1,3-β-D-glucan containing nutritional supplement ImunixX with the 1,3-β-D-glucan serum test. *Mycoses* **2010**, *54*, e352–e353. [CrossRef] [PubMed]
13. Haladova, E.; Mojzisova, J.; Smrco, P.; Ondrejkova, A.; Vojtek, B.; Prokes, M.; Petrovova, E. Immunomodulatory effect of glucan on specific and nonspecific immunity after vaccination in puppies. *Acta Vet. Hung.* **2011**, *59*, 77–86. [CrossRef] [PubMed]
14. Lee, J.G.; Kim, Y.S.; Lee, Y.J.; Ahn, H.Y.; Kim, M.; Kim, M.; Cho, M.J.; Cho, Y.; Lee, J.H. Effect of immune-enhancing enteral nutrition enriched with or without beta-glucan on immunomodulation in critically ill patients. *Nutrients* **2016**, *8*, 336. [CrossRef] [PubMed]
15. Oloke, J.K.; Adebayo, E.A. Effectiveness of immunotherapies from oyster mushroom (*Pleurotus ostreatus*) in the management of immunocompromised patients. *Int. J. Immunol.* **2015**, *3*, 8–20.
16. Rop, O.; Mlcek, J.; Jurikova, T. Beta-glucans in higher fungi and their health effects. *Nutr. Rev.* **2009**, *67*, 624–631. [CrossRef] [PubMed]
17. Vetvicka, V.; Vetvickova, J. Physiological effects of different types of β-glucan. *Biomed. Pap. Med. Fac. Univ. Palacky Olomouc. Czech. Rep.* **2007**, *151*, 225–231. [CrossRef]
18. Jesenak, M.; Sanislo, L.; Kuniakova, R.; Rennerova, Z.; Buchanec, J.; Banovcin, P. Imunoglukan P4H® in the prevention of recurrent respiratory infections in childhood. *Cesk Pediatr.* **2010**, *73*, 639–647.
19. Sapena Grau, J.; Pico Sirvent, L.; Morera Ingles, M.; Rivero Urgell, M. Beta-glucans from *Pleurotus mostreatus* for prevention of recurrent respiratory tract infections. *Acta Pediatr. Esp.* **2015**, *73*, 186–193.
20. Pasnik, J.; Slemp, A.; Cywinska-Bernas, A.; Zeman, K.; Jesenak, M. Preventive effect of pleuran (β-glucan isolated from *Pleurotus ostreatus*) in children with recurrent respiratory tract infections—Open-label prospective study. *Curr. Ped. Res.* **2017**, *21*, 99–104.
21. Batovsky, M.; Zamborsky, T.; Khaled, R.; Desatova, B.; Kadleckova, B. Beta-(1,3/1,6)-D-glucan helps to decrease opportunistic infections in Crohn's disease patients treated with biological therapy. *Arch. Clin. Gastroenterol.* **2015**, *1*, 005–008.
22. Auinger, A.; Riede, L.; Bothe, G.; Busch, R.; Gruenwald, J. Yeast (1,3)-(1,6)-beta-glucan helps to maintain the body's defence against pathogens: A double-blind, randomized, placebo-controlled, multicentric study in healthy subjects. *Eur. J. Nutr.* **2013**, *52*, 1913–1918. [CrossRef] [PubMed]
23. Talbott, S.M.; Talbott, J.A. Baker's yeast beta-glucan supplement reduces upper respiratory symptoms and improved mood state in stressed women. *J. Am. Coll. Nutr.* **2012**, *31*, 295–300. [CrossRef] [PubMed]
24. Talbott, S.; Talbott, J. Beta 1,3/1,6 glucan decreases upper respiratory tract infection symptoms and improves psychological well-being in moderate to highly-stressed subjects. *Agro Food Ind. Hi-Tech* **2010**, *21*, 21–24.
25. Feldman, S.; Schwartz, H.I.; Kalman, D.S.; Mayers, A.; Kohrman, H.M.; Clemens, R.; Krieger, D.R. Randomized phase II clinical trials of Wellmune WGP® for immune support during cold and flu season. *J. Appl. Res.* **2009**, *9*, 30–42.
26. Graubaum, H.J.; Busch, R.; Stier, H.; Gruenwald, J. A double-blind, randomized, placebo-controlled nutritional study using an insoluble yeast beta-glucan to improve the immune defence system. *Food Nutr. Sci.* **2012**, *3*, 738–746. [CrossRef]
27. Fuller, R.; Butt, H.; Noakes, P.S.; Kenyon, J.; Yam, T.S.; Calder, P.C. Influence of yeast-derived 1,3/1,6 glucopolysaccharide on circulating cytokines and chemokines with respect to upper respiratory tract infections. *Nutrition* **2012**, *28*, 665–669. [CrossRef] [PubMed]
28. Vetvicka, V.; Richter, J.; Svozil, V.; Rajnohova Dobiasova, L.; Kral, V. Placebo-driven clinical trials of yeast-derived β-(1,3) glucan in children with chronic respiratory problems. *Ann. Transl. Med.* **2013**, *1*, 26. [PubMed]
29. Vetvicka, V.; Richter, J.; Svozil, V.; Rajnohova Dobiasova, K.; Kral, V. Placebo-driven clinical trials of transfer point glucan #300 in children with chronic respiratory problems: Antibody production. *Am. J. Immunol.* **2013**, *9*, 43–47.

30. Richter, J.; Svozil, V.; Kral, V.; Rajnohova Dobiasova, L.; Stiborova, I.; Vetvicka, V. Clinical trials of yeast-derived β-(1,3) glucan in children: Effects on innate immunity. *Ann. Transl. Med.* **2014**, *2*, 15. [PubMed]

31. Richter, J.; Kral, V.; Svozil, V.; Rajnohova Dobiasova, L.; Pohorska, J.; Stiborova, I.; Vetvicka, V. Effects of transfer point glucan #300 supplementation on children exposed to passive smoking-placebo-driven double-blind clinical trials. *J. Nutr. Health Sci.* **2014**, *1*, 1–8.

32. Vetvicka, V.; Richter, J.; Svozil, V.; Rajnohova Dobiasova, L.; Kral, V. Placebo-driven clinical trials of transfer point glucan #300 in children with chronic respiratory problems: III. Clinical findings. *Am. J. Immunol.* **2013**, *9*, 88–93.

33. Richter, J.; Svozil, V.; Kral, V.; Rajnohova Dobiasova, L.; Vetvicka, V. β-glucan affects mucosal immunity in children with chronic respiratory problems under physical stress: Clinical trials. *Ann. Transl. Med.* **2015**, *3*, 52. [PubMed]

34. Li, F.; Jin, X.; Liu, B.; Zhuang, W.; Scalabrin, D. Follow-up formula consumption in 3- to 4-year-olds and respiratory infections: An RCT. *Pediatrics* **2014**, *133*, e1533–e1540. [CrossRef] [PubMed]

35. Majtan, J. Pleuran (β-glucan from *Pleurotus ostreatus*): An effective nutritional supplement against upper respiratory tract infections? *Med. Sport Sci.* **2013**, *59*, 57–61.

36. Carpenter, K.C.; Breslin, W.L.; Davidson, T.; Adams, A.; McFarlin, B.K. Baker's yeast β-glucan supplementation increases monocytes and cytokines post-exercise: Implication for infection risk? *Br. J. Nutr.* **2013**, *109*, 478–486. [CrossRef] [PubMed]

37. Bobovcak, M.; Kuniakova, R.; Gabriz, J.; Majtan, J. Effect of pleuran (β-glucan from *Pleurotus ostreatus*) supplementation on cellular immune response after intensive exercise in elite athletes. *Appl. Physiol. Nutr. Metab.* **2010**, *35*, 755–762. [CrossRef] [PubMed]

38. Bergendiova, K.; Tibeks, E.; Majtan, J. Pleuran (β-glucan from *Pleurotus ostreatus*) supplementation, cellular immune response and respiratory tract infections in athletes. *Eur. J. Appl. Physiol.* **2010**, *111*, 2033–2040. [CrossRef] [PubMed]

39. Talbott, S.; Talbott, J. Effect of BETA 1,3/1,6 glucan on upper respiratory tract infection symptoms and mood state in marathon athletes. *J. Sport Sci. Med.* **2009**, *8*, 509–515.

40. McFarlin, B.K.; Carpenter, K.C.; Davidson, T.; McFarlin, M.A. Baker's yeast beta glucan supplementation increases salivary IgA and decreases cold/flu symptomatic days after intense exercise. *J. Diet. Suppl.* **2013**, *10*, 171–183. [CrossRef] [PubMed]

41. Nieman, D.C.; Henson, D.A.; McMahon, M.; Wrieden, J.L.; Davis, J.M.; Murphy, E.A.; Gross, S.J.; McAnulty, L.S.; Dumke, C.L. β-glucan, immune function, and upper respiratory tract infections in athletes. *Med. Sci. Sports Exerc.* **2008**, *40*, 1463–1471. [CrossRef] [PubMed]

42. Varricchio, A.M.; Capasso, M.; della Volpe, A.; Malafronte, L.; Mansi, N.; Varricchio, A.; Ciprandi, G. Resveratrol plus carboxymethyl-β-glucan in children with recurrent respiratory infections: A preliminary and real-life experience. *Ital. J. Pediatr.* **2014**, *40*, 93. [CrossRef] [PubMed]

nutrients

MDPI

Article

Induced Aberrant Organisms with Novel Ability to Protect Intestinal Integrity from Inflammation in an Animal Model

Helieh S. Oz

Department of Physiology, Internal Medicine, College of Medicine, University of Kentucky Medical Center, Lexington, KY 40536-0298, USA; hoz2@email.uky.edu

Received: 9 June 2017; Accepted: 3 August 2017; Published: 11 August 2017

Abstract: Robust and balanced gut microbiota are required to support health and growth. Overgrowth of gut microbial or pathogens can change ecosystem balance, and compromise gut integrity to initiate gastrointestinal (GI) complications. There is no safe and effective modality against coccidiosis. Antibiotic additives routinely fed to food animals to protect against infection, are entered into the food chain, contaminate food products and pass to the consumers. Hypothesis: induced aberrant organisms possess distinct ultrastructure and are tolerated by immunodeficient-animals yet are non-pathogenic, but immunogenic in various strains of chicks to act as a preventive (vaccine) and eliminating the needs for antibiotic additives. **Methods:** cyclophosphamide-immunodeficient and immune-intact-chicks were inoculated with induced aberrant or normal *Coccidal*-organisms. Immune-intact-chicks were immunized with escalating-doses of organisms. **Results:** Aberrant organisms showed distinct ultrastructure with 8-free-sporozoites which lacked sporocysts walls and veils. Immunodeficient-chicks inoculated with normal-organisms developed severe GI complications but tolerated aberrant-organisms ($p < 0.001$) while they had no detectable antibodies. Naïve-animals challenged with a pathogenic-dose showed GI complications, bloody diarrhea, severe lesions and weight loss. Immune-intact-animals immunized with aberrant forms were protected against high dose normal-pathogenic-challenge infection and gained more weight compared to those immunized with normal-organisms ($p < 0.05$). **Conclusions:** Aberrant organisms possess a distinct ultrastructure and are tolerated in immunodeficient-chicks, yet provide novel immune-protection against pathogenic challenges including diarrhea, malnutrition and weight loss in immune-intact-animals to warrant further investigations toward vaccine production.

Keywords: infection; gut inflammation; immunosuppression; poultry; coccidiosis; aberrant organisms

1. Introduction

Robust and balanced gut microbiota are required to support health and growth. Application of certain medications and chemicals including antibiotics and immunmodulators can alter this delicate balance in digestive tracts and promote the state of disease [1]. Antibiotics are designed and required modality to abolish infectious pathogenic elements and to cure patients. Yet, antibiotics underlay a dual conundrum, as they can destroy the normal gut microbiome and promote growth of pathogenic organisms to cause dysbiosis. Further, altered permeability damages the gut mucosa and prone individuals to diarrhea and gastroenteritis.

Cyclophoshamide (CY) is an alkylating agent and a potent immunosuppressant compound used in humans against autoimmune diseases and cancer chemotherapy. CY affects both T and B cells; however, B cells have a slower rate of recovery [2,3]. CY is metabolized in hepatic cells to yield active substances and to exert cytotoxic effects on certain immune cells [4]. The cytotoxic effect is due to drug cross-link DNA, which can immediately destroy affected cells, render them susceptible to cell death during mitosis, or permit normal cellular activities if DNA repair occur. Indeed, CY-chemotherapy disrupts the gut epithelial barrier, and causes the gut to bypass certain bacteria. Bacteria gather in lymphoid tissue just outside the gut and spur generation of T helper 1 and T helper 17 cells that migrate to the tumor vicinity and destroy tumor cells [5]. In birds, immunoglobulin synthesis and antibody production depend upon the integrity of bursa of fabricius lymphoid organ which is located on the dorsal side of rectal "cloaca". CY-treatment on the first three days after birth is reported to severely depress antibody production as animals fail to produce antibodies against antigenic challenges up to 7–11 weeks, similar to combined bursectomized and in ova X-irradiation [6]. In addition, CY causes significant decreases in lymphoid population in thymus [2]. The splenic structure and preferral lymphoid organs lost germinal and plasma cells. The variation in the action and the outcome reported in different studies are due to the doses, the length and the time for initiation of CY-administration.

Coccidia are highly host specific transmissible Apicomplexan organisms which mainly attack gut mucosa and compromise the immune system to trigger gastrointestinal inflammation, infectious diarrhea, loss of function and morbidity in humans and animals. In contrast, *Toxoplasma*, another member of the family Apicomplexan, is a ubiquitous organism which infects every organ and cell in animals and humans to cause toxoplasmosis [7]. Coccidiosis is one of the most important communicable pathogenic diseases in the food animals industry. Additionally, *Coccidia* predispose infected animals to other pathogens like *Clostridia* and more severe necrotic enteritis [8]. There is no safe and effective therapeutic modality or vaccine to protect against the infection [9].

Antibiotic additives are routinely fed to poultry and livestock as a common practice to protect against the infection and weight loss. These additives contaminate egg, meat, bone and milk products which are transferred into the food chain and consumed with predicted complications. For instance, quinolones are commonly prescribed in patients while used in diets for the poultry industry to increase weight gain and growth [10,11]. Quinolone residues are detected in 50% of eggs at higher concentrations above the limits for edible tissues established by the regulatory agencies, including United States Department of Agriculture [11]. Animal products contaminated with antibiotic residues create a great concern about possible side effects in consumers such as allergies [12] and potential for antibiotic-resistant microbials.

Coccidiosis causes great economic loss and morbidity by reduction in food intake, weight gain and egg production, and also affects the value of meat quality by decreasing feed conversion, maldigestion and malabsorption to lead in mortality [3]. The annual cost of coccidiosis in poultry production has been estimated at $800 million in the USA [13,14], chiefly for anticoccidial drugs which are commonly used to control the infection and to improve weight gain. The constant addition of medications in food animal diets has been a profitable and effective tool against the disease outbreak, but there are drawbacks including development of drug resistance and potential health side effects in consumers. The live vaccine, Coccivac, is a mixture of seven species of poultry *Coccidia* which has been utilized for over five decades in the USA [15]. The animals recover the infection from vaccines and develop immunity which lasts days to weeks. The disadvantages of this vaccine include: poor feed conversion to weight gain; several weeks are required to develop a solid immunity; possibility of spreading infection; difficulties in administering the vaccine and managing the animals. Other possible vaccines in experimental stages are attenuated strains including serially transferred *Eimeria tenella* (*E. tenella*) into chorio-allantoic chick embryos [16].

Ever since its discovery (Fantham and Porter 1900), Eimeria [17] have been described as organisms (oocysts) with four sporocysts, each containing two sporozoites. By utilizing purification procedures, aberrant forms of different Eimerias were induced which matured to contain 8-free-sporozoites with no protective sporocyst walls, as were confirmed by light microscopy [18]. The aberrant organisms proved to be less pathogenic than the normal form in inbred Leghorn-chicks, but similarly immunogenic. The hypothesis of this investigation was: induced aberrant organisms possess a distinct ultrastructure and are tolerated by immunodeficient-animals, yet are non-pathogenic but immunogenic in various strains of chicks to act as preventive (vaccine) and eliminating the need for antibiotic additives. This investigation reports the ultrastructural formation of these novel organisms and further compares their pathogenecity to normal forms in immunodeficient and susceptible animals. In addition, immunogenicity of these aberrant forms is examined and compared to normal organisms utilizing two diverse inbred chicks, Rhode Island Red and New Hampshire strains.

2. Materials and Methods

2.1. Ethical Guidelines for the Use of Animals

This investigation was conducted according to the guidelines of the National Institute of Health and the International and the American Pain Associations. All animal procedures were approved by the University of Kentucky Institutional Animal Care and Use Committee (IACUC) and Institutional Biosafety Committee.

2.2. Animals

One-day-old (neonate) specific pathogen-free inbred Rhode Island Red and inbred New Hampshire chicks were obtained from the Poultry Science, and kept in *Coccidia* and pathogen-free rooms, in disinfected wire-floor cages and provided feed and water *ad libitum*. Each bird was tagged with a leg band and weighed prior, during and at the end of each study. Weight gain/loss was calculated by subtracting the initial weight before inoculation from the weight obtained following challenge and before termination.

Cyclophosphamide Immunosuppressed Birds

Immunodeficient birds were utilized in order to investigate fine pathological differences between the normal or aberrant organisms. Neonate (1-day-old) Rhode Island Red chicks were intramuscularly (IM) administered with 4 mg (100 mg/kg) cyclophosphamide (CY, Sigma Chemical Co., St. Louis, MO, USA) for 4 consecutive days, in order to suppress B and T cells and immune responses [3]. T cells may partially recover after one month; however, B cells' suppression can last for several months.

2.3. Preparation of Organisms (Oocysts)

Mature organisms (sporulated oocysts) of *E. tenella* originally were obtained from Eli Lilly and Co. (Indianapolis, IN, USA). Fresh cultures were prepared from the ceca of donor birds one week after oral inoculation. The contents were homogenized and the immature organisms were separated from debris by sieve and centrifuged at $400 \times g$ and sediments added into 2.5% aqueous potassium dichromate to obtain mature normal organisms. In order to induce aberrant organisms, the homogenate containing immature (unsporulated) organisms were cleaned in a solution of 2.5% sodium hypochlorite, and rinsed with distilled water ($2\times$). The homogenates were centrifuged for 10 min at $400 \times g$ in a saturated salt solution ($9{:}1{:}v.v$), then rinsed in distilled water and cultured in 2.5% aqueous potassium dichromate solution to enhance maturation. Organisms were declared mature when at least 90% had reached maturation (sporulated). The organisms were enumerated using a hemocytometer and diluted into PBS according to the required numbers per each experiment before organisms gavaged directly into crop. Normal control animals received sham (PBS) treatment by gavage.

At 2–6 weeks of age, animals were gavage inoculated via crop respectively with any dose of 240, 300, 450, 2400, or 24,000 mature organisms of either normal or aberrant strains (Scheme 1). Each experimental animal received one dose at a time and up to three doses with 2 weeks interval according the experimental design. Infected animals (New Hampshire, Rhode Island Red) were kept in isolation wired cages into separate rooms and provided with food and water *ad lib*. Cages were daily cleaned and disinfected.

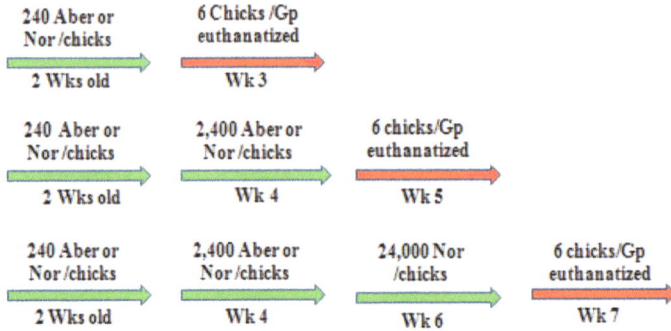

Scheme 1. Schematic timeline immunization for 2 strains of chicks (New Hampshire and Rod Island Red) with aberrant (Aber) or normal (Nor) organisms and challenge with highly pathogenic dose (24,000) of normal organisms.

2.4. Blood and Tissue Collection

One week after the last challenge dose of inoculum, animals were weighed, and blood was collected into syringes from the wing basilic vein using 25 gauge needles. Animals were humanely euthanatized by cervical dislocation immediately followed by cardiac puncture and tissues collected. Blood was centrifuged and serum aliquots were stored at $-20\ ^{\circ}C$ until used. Intestines were immediately removed and examined and a portion of cecal tissues fixed in buffer formalin solution. Slide smears were prepared from the rest of cecal tissues to detect the organisms.

Histopathological Scoring

Histopathological slides were prepared from 6 animals/each group and sections stained with Hematoxylin and Eosin. Slides were numerical-labeled (coded) at the pathology lab during the automated staining process. Every field on each pathological slide was thoroughly examined for changes and possible damage. Cecal lesions were scored from 0 to 5 according to the severity of infection [3,18].

0 = Normal mucosa, and negative cecal smears.

1 = No detectable pathology but organisms detected in the smears.

2 = Scattered petechia on mucosa, organisms present on the smear, normal cecal contents.

3 = Focal inflamed and thickened mucosa, some hemorrhage in lumen.

4 = Multifocal inflammation and thickened mucosa, extensive hemorrhage in the lumen, with little or no fecal contents, weight loss.

5 = Severe inflammation and necrosis, enlarged ceca with blood or sloughed off mucosa, moribund or dead birds.

Slides were decoded after completion of pathological evaluation.

2.5. Maturation of Organisms (Oocysts Sporulation)

Maturation was determined via microscopic examination. Representative samples were examined and percentage of sporulated organisms determined every 4 h. Maturation was achieved when 90% or more of organisms had completed developmental stages.

2.6. Enzyme Linked Immosorbent Assay (ELISA)

The indirect ELISA technique was performed using a soluble oocyst antigen of E. tenella according to [3,18,19]. Each serum sample was heat inactivated at 56 $^{\circ}C$ for 30 min before use and diluted in two-fold increments. Peroxides conjugated γ chain rabbit anti-chicken IgG (Cappel Laboratories, West

Chester, PA, USA) was dispensed into each micro-well coated with 11 µg of antigen, and incubated with o-phenylenediamine (Eastman Kodak Co., Rochester, NY, USA). Optical densities were read at 490 nm on ELISA reader (DYnatech Lab Inc., Alexandria, VA, USA).

2.7. Experimental Design

In the first experiment, 2-weeks-old and 3-weeks-old inbred CY-immunodeficient-animals were inoculated with 300 and 450 organisms, respectively, of either normal or aberrant form. To compare the immunogenic efficacy, 3-weeks-old inbred New Hampshire or Rhode Island Red animals received 240 and 2400 oocysts of normal or aberrant organisms at a two-week interval. Then, animals were challenged with a high pathogenic dose (24,000) of normal organisms. One week after each inoculation, six animals from each group were weighed and bled, then euthanized to collect samples for pathological investigations.

2.8. Frozen Sections

Organisms were centrifuged in 20% (*w/v*) bovine serum albumin (BSA) and 15% (*w/v*) sucrose. The pellets were fixed with 2 drops/mL of 25% glutaraldehyde. Sections were quick frozen in liquid nitrogen and blocks were sectioned at the 18 µm setting on cryostat. Organisms were embedded with mounting agent to protect from shattering during the sectioning procedure. The sections were stained with Hematoxylin and Eosin.

2.9. Electron Micrographs

The ultrastructural modifications of aberrant and normal organisms were investigated during maturation. Therefore, organisms were embedded in cross-linked BSA media and frozen in liquid nitrogen prior to cryostat sectioning at 16 µm. These sections were directly immersed in Carnovsky fixative, re-embedded in BSA, post fixed in osmium tetroxide and embedded in osmium tetroxide and finally embedded in Spurr's Plastic. The blocks were sectioned with KLB ultramicrotome 111 with a diamond knife. The sections were stained with uranyl acetate and lead citrate and examined with Ziess EM Transmission Electron Microscope (TEM) and micrographs were developed.

2.10. Statistical Analysis

Parameters were analyzed with Prism 6 software (Graph Pad Software, Inc., La Jolla, CA, USA). Data are expressed as mean ± SEM unless otherwise stated. One hundred organisms were monitored each time from 3 different samples. Maturation of organisms was calculated when a minimum of 90% of organisms formed infective sporozoites. Weight gain/loss was calculated by subtracting the initial weight before inoculation from the weight obtained following each challenge inoculation and before termination. Statistical analysis was performed utilizing a two-way analysis of variance (ANOVA) followed by Bonferroni multiple comparison post hoc test for comparisons. Statistical significance was set at $p \leq 0.05$.

3. Results

3.1. Maturation and Ultrastructure

Ultrastructural modifications during developmental maturation were investigated to compare aberrant and normal organisms utilizing transmission electron micrographs. Normal organism development was completed after 34 h of maturation (sporulation) when they contained 4 fully formed sporocysts each to shield 2 infective sporozoites within. Sporocysts walls and sporocysts veil ultrastructures visible by TEM (Figure 1A,B) were required to sustain their infectivity, as will be presented. In contrast, aberrant organism maturation was more sluggish and about 23% longer than normal organisms (44 h vs. 34 h).

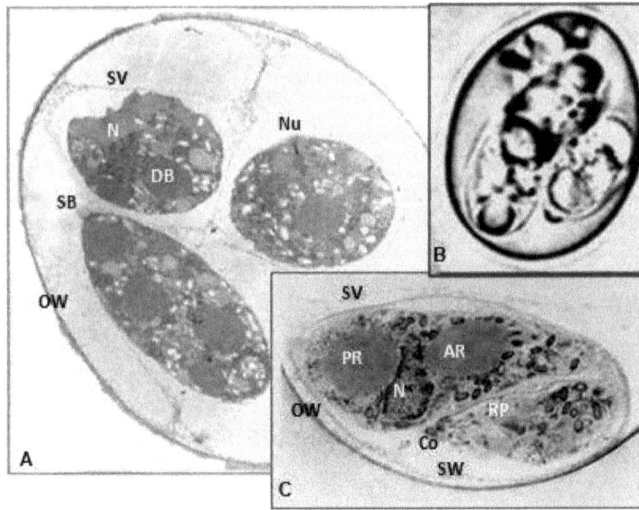

Figure 1. (**A**,**C**) Transmission electron micrographs (TEM) from mature (**A**,**C**) and light micrograph (**B**) from normal organism representative (*E. tenella*). (**A**) Mature normal organism (oocyst) representative 34 h of developmental stage (sporulation) demonstrating 3 of 4 sporocysts, oocyst wall (OW), Stieda body (SB), sporocysts wall (SW), sporocysts veil (SV) dividing the organism into a honey-combed chamber, nucleolus (Nu) ×7000. (**B**) A normal organism under light microscopy with 3 of 4 sporocysts ×800. (**C**) Enlarged micrograph from mature sporocyst contains 2 mature sporozoites within the sporocyst's wall 34 h of developmental stage. Each sporozoite contains refractile bodies ×13,000. Note Nucleus (N), Dense body (DB), Anterior (AR) and posterior refractile bodies (PR), Rhoptry (Rh), nucleolus (Nu), Sprocyst veil (SW), Oocyst wall (OW).

Aberrant organisms had free naïve sporozoites within the oocysts confinement without any protective sporocysts walls and veils (Figure 2A–C) dissimilar to normal ones. Further, aberrant organisms had fragile outer membranes (oocyst wall) which disrupted in occasions during TEM process for ultrastructure investigation. In addition, aberrant organisms contained 2 masses of rudimentary organelles as residuum of sporocysts (Figure 1C). However, normal and aberrant organisms both had an equal number of (8) banana-shape-sporozoites. Aberrant sporozoites similarly contained nucleus, rhoptries, conoidal piercing organelles, and polar rings and fibrillate ultrastructures required for invasion and lodging into gut epithelial cells. Otherwise, aberrant organisms appeared to have similar microstructures to normal ones in shape (round to oval), measurements (25 to 35 μm by 22.5 to 27.5 μm) and the size of each banana-shaped free sporozoite (12.5–14 μm by 2.5 μm). Over 90% of immature organisms processed to mature into the aberrant form, compared to 95% of normal organisms. This study proved the specific protective effects of the sporocysts provided for the longevity and infectivity of the sporozoites.

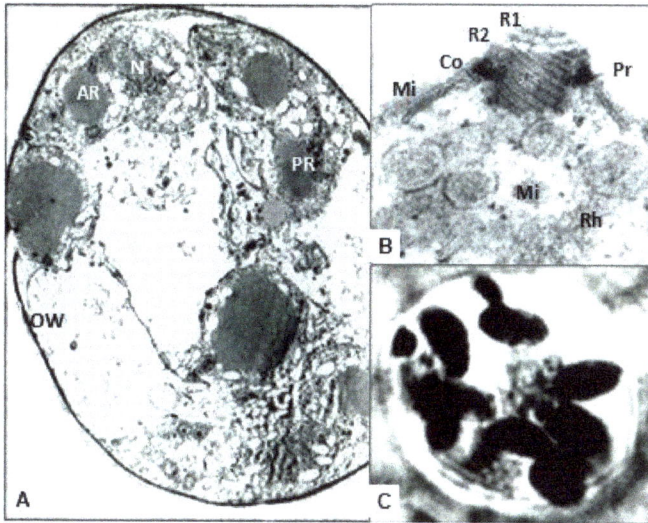

Figure 2. (A–C) Mature aberrant organism. (**A**) Transmission electron micrograph (TEM) from mature aberrant organism representative contains free sporozoites. Micrograph demonstrates section through a mature aberrant organism 44 h after development. Unlike normal, aberrant organisms lack sporocysts, sporocysts wall and veil and stieda body as evidenced. ×7000. (**B**) High magnification of the anterior end from free sporozoite showing conoid (Co), piercing organelles, preconoidal rings (R), polar ring (Pr), and microneme (Mi). ×11,600. (**C**) Cryostat section stained with Hematoxylin and Eosin of mature aberrant organism demonstrates 8 free sporozoites, polar granule and 2 masses of oocyst residuum and lack of sporocyst structure. Light microscopy ×790.

3.2. Immunodeficiency and Comparative Pathogenicity

An induced immunodeficient model was utilized to compare the fine detailed pathogenecity of aberrant and normal organisms. Neonate chicks were exposed to cyclophosphamide treatment for 4 consecutive days, compared to the sham-treatment in immune-intact animals. CY-treated-immunodeficient animals had extensive stunt development as appeared with bare skin and patchy feather growth. Respectively, at 2 or 3 weeks-old, CY-immunodeficient-juvenile chicks were gavage inoculated with low doses of mature (2 weeks with 300 or 3 weeks with 450) normal or aberrant organisms. Animals treated with normal organisms became moribund and developed severe GI inflammatory responses and bloody diarrhea, and gained significantly less weight (Figure 3, Table 1). In contrast, CY-immunodeficient-animals tolerated aberrant organisms with significantly more weight gain and pathological lesions ($p < 0.001$). As was expected, CY-immunodeficient animals developed no detectable (protective) antibodies in contrast to immune-intact animals. In gross pathology, CY-immunodeficient animals had rudimentary thymus, splenic and bursa of Fabricius structures (data not shown) due to the cytotoxic effect of CY on B and T cell formation. Yet, immunodeficient animals treated with normal organisms presented significantly severe pathological damages to the gut than those treated with aberrant forms. These data support that aberrant organisms are significantly less invasive and less pathogenic than normal forms (Figure 3, Table 1). All normal animals were protected against challenge doses with no visible intestinal lesions or organisms detected from their intestinal smears. In contrast, all CY-immunodeficient animals were unprotected against low doses of challenge infection (data not shown).

Table 1. Effects of low doses of aberrant and normal organisms in juvenile immunodeficient or immune-intact Rhode Island Red animals.

Gp	Week Old	CY Treated	Organism/	Weight Gain	Ab Titer
1	2	+	300 N	60 ± 3.4	0
2	2	+	300 A	57 ± 5.7	0
3	3	+	450 N	87 ± 7.2 [b]	0
4	3	+	450 A	106 ± 7	0
5	3	−N *	450 N	95 ± 1.4	50 ± 1
6	3	−N *	0	190 ± 8	0

Cyclophosphamide (CY) immunodeficient (+) or immune-intact (−) chicks were inoculated with a low dose of 300 (2 weeks old) or 450 (3 weeks old) normal or aberrant organisms. Two weeks and three weeks old (CY) immunodeficient animals were chosen when lacking both T and B cells. Data is shown as mean ± SEM. Gp = Group, N = Normal organisms, A = Aberrant organism, N * = Normal Immune-intact birds. Mean Body Weight gain/g. N = 6/group. [b] $p < 0.05$.

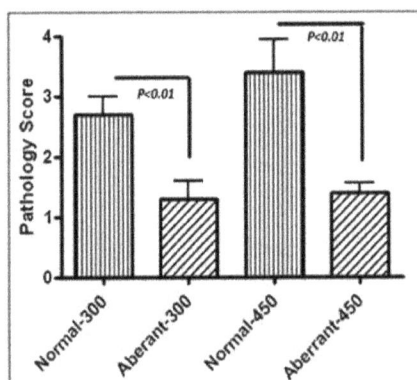

Figure 3. Pathological scores from Cylophosphamide (CY)-immunodeficient animals inoculated with 300 (at 2 weeks of age) or 450 (at 3 weeks of age) of either mature (sporulated) aberrant or normal organisms. Samples are collected 1 week after initiation of infection by oral gavage. Statistically significant pathological damages were detected in animals treated with normal organisms compared to aberrant forms ($p < 0.01$). As expected no antibody production was detected in sera from these CY-immunodeficient animals.

3.3. Organisms and Protective Immunogenicity

There is a strain variation in immune response against organisms amongst chicks. Therefore, 2 different inbred strains, Rhode Island Red and New Hampshire immune-intact animals were utilized for this experiment in order to compare protective immune response against infection. Further, day-old chicks lack immunity but consume their egg yolk which is rich in maternal immunity and become protected against pathogens during the first week of their life. However, in the second week as maternal immunity weans off, chicks become sensitive to infections and start to develop their active immunity. Therefore, two-weeks-old chicks were used to initiate the following investigations as these animals are sensitive to infections yet capable of launching an immune response against pathogens. Two-weeks-old animals were immunized with serial inoculations of aberrant or normal forms (240 and 2400) with 2 weeks interval. All animals were challenged with a high pathogenic dose of normal form (24,000). Those animals immunized with aberrant organisms gained weight more and developed lower cecal pathological lesions than those immunized with normal organisms. These animals were protected against the challenge infection (Tables 2–4). Naïve unimmunized animals challenged with a pathogenic dose (24,000 normal organisms) lost weight and developed severe bloody diarrhea and significantly more extensive inflammatory pathological lesions than immunized animals (unimmunized 3.2–4.1 vs. immunized 0–1) (Table 4).

Table 2. Immune-intact animals were immunized with first dose of normal or aberrant organisms at week 1.

Gp	Birds Strain	Strain	Organism/	Weight Gain	Lesion	Ab Titer
1	New Hampshire	A	240	104 ± 5	1 ± 0.1 [b]	0
2	"	N	240	85 ± 7	1.3 ± 01	0
3	"	C	0	123 ± 7 *	0	0
4	Rhode Island Red	A	240	114 ± 3	1 ± 0.1	0
5	"	N	240	95 ± 2	1 ± 0.1	0
6	"	C	0	190 ± 3 *	0	0

Six birds/group (Gp) were inoculated with Normal (N) or Aberrant (A) organisms (week 1). Data is shown as mean $\pm$ SEM. Normal control (C) birds received (0) sham inoculation. Lesion scores. Mean Body Weight gain/g. N = 6/group. [b] $p < 0.05$. * $p < 0.01$.

Table 3. Animals immunized with second dose of normal or aberrant organisms. Six/group animals which were immunized with 240 and 2400 Normal (N) or Aberrant (A) organisms (week 4) with 2-weeks interval. Control (C) birds received (0) sham inoculation.

Gp	Bird Strain	Strain	Organism/	Weight Gain	Lesion	Ab Titer
1	New Hampshire	A	2400	133.4 ± 6	1 ± 0.1	58.0 ± 2 [b]
2	"	N	2400	133.4 ± 3	1 ± 0.1	50.0 ± 1
3	"	C	0	247.4 ± 2	0	0
4	Rhode Island Red	A	2400	152.3 ± 1 *	1 ± 0.2 [b]	25.0 ± 2 *
5	"	N	2400	104.9 ± 2	1.4 ± 0.1	17.0 ± 1
6	"	C	0	218.9 ± 5	0	0

Data is shown as mean $\pm$ SEM. Lesion scores. Mean Body Weight gain/g. N = 6/group. [b] $p < 0.05$; * $p < 0.01$.

Table 4. Immunized animals with two different doses of normal or aberrant were challenged at 6 weeks of age with high numbers of infective normal organisms.

GP	Bird Strain	Strain	Organism/	Weight Gain	Lesion	Ab Titer
1	New Hampshire	A	24,000	100 ± 2 *	0 [b]	69 ± 4 [b]
2	"	N	24,000	50 ± 3	0.35 ± 0.1	63 ± 2
3	"	N *	24,000	-45 ± 2 [#]	3.2 ± 0.5	50 ± 3
4	"	C	0	247 ± 5	0	0
5	Rhode Island Red	A	24,000	72 ± 3	0.76 ± 0.1 [b]	44 ± 6 [b]
6	"	N	24,000	64 ± 3	1 ± 0.2	81 ± 2
7	"	N *	24,000	$-80 + 4$ [#]	4.1 ± 0.3	ND
8	"	C	0	219 ± 7	0	0

Six birds/group were immunized with 240 and 2400 Normal (N) or Aberrant (A) organisms (weeks 2 and 4) with 2-weeks interval. Control (C) animals received (0) sham inoculation. Infected control group (N *) received no previous immunization before final challenged inoculum. Lesion scores. Mean Body Weight gain/g significant. ND = Not done. [b] $p < 0.05$; * $p < 0.01$; [#] $p < 0.001$.

In general, Rhode Island Red animals were found to be more susceptible to disease, as they developed more severe lesions and significantly less weight gain compared to New Hampshire animals (Tables 1–3). In addition, naïve Rhode Island Red animals developed significantly more severe pathological lesions ($4.1 + 0.3$ vs. $3.2 + 0.5$ $p < 0.05$) and lost weight more ($p < 0.01$) than naïve New Hampshire animals (Table 4).

Animals immunized with aberrant forms were protected against the challenge infection, gained more weight, developed significantly less or no cecal lesions and more antibodies (except Rhode Island Red), compared to those immunized with normal organisms ($p < 0.05$). Rhode Island Red animals immunized with aberrant organisms developed significantly less lesions compared to those immunized with normal organisms, yet released significantly less Ab than those immunized with normal strains. However, the reason behind this discrepancy is not known. Uninfected control animals remained anti-coccidal antibody negative. There was a minor difference between the lesion scores from immunized animals but it did not reach significance. These investigations proved that the sporocysts are required to preserve infectivity in these organisms. Overall, aberrant organisms were less pathogenic in immunodeficient animals yet protective in immunocompetent animals against the challenge infection.

4. Discussion

Coccidiosis causes severe gut inflammation, diarrhea, malabsorption, malnutrition and weight loss. The common preventive practice against coccidiosis in poultry and livestock includes application of antibiotic supplementation into daily diets which contaminates eggs, milk, and meat production. Antibiotics enter the food cycle and are consumed by humans with possible allergies, antibiotic resistance, and other yet unknown side effects. Antibiotic additives include tetracyclines, as commonly used in poultry, and are deposited and remain in bones and meat products. They become a potential human health risk regardless of monitoring appropriate antibiotic withdrawal times before reaching the market [20]. Further, sulfonamide use in the form of anticoccidial additives in food animals has encountered the emergence of drug-resistant strain infections [21]. The Center for Disease Control and Prevention (CDC) has speculated that annually over 2,000,000 sicknesses and 23,000 mortalities are due to microbial resistance to antibiotics in the USA. Recent consumer awareness regarding the antibiotic residue polluting food animal products and antibiotic resistant microbials has created trepidation and specific recommendations to diminish overuse of antibiotics as growth promoters in livestock in the USA [22] and worldwide. However, these recommendations may not be effective or applicable without surrogate implementations such as safe and effective as well as feasible vaccine availability against economically important coccidiosis in the food animal industry. The focus of this investigation was on discovery and establishment of proof-of-concept in basic methodology, for development of aberrant organisms to lead as candidate vaccines in coccidiosis.

Cyclophosphamide is a potent immune-modulating agent discovered in 1959 and ever since has been used in autoimmune diseases and as a chemotherapeutic in cancer. CY is metabolized to active form 4-hydroxy-CY, which spontaneously breaks down to reactive intermediates, phosphoramide mustard and acrolein [23]. Phosphoramide mustard possesses an alkylating property which mediates the anti-proliferative and cytotoxic actions of CY and cross links DNA with mitotic B and T cells. In birds, immunoglobulin synthesis and antibody formation are regulated by Bursa of Fabricius, and CY destroys bursal B cells as well as thymus T cell populations/production [2,3]. After four consecutive days of treating neonate chicks with a cytotoxic agent, CY-treated juvenile animals visibly lacked immune processing organs, thymus and bursa and bared rudimentary splenic structures. CY-immunodeficient animals had pervasive growth stunting and weighed significantly less than immune-intact animals (60%). Therefore, these animals were found to be more susceptible to infections when challenged with a lower dose of organisms and developed more severe lesions than immune-intact animals. Additionally, immunodeficient animals had no detectable antibody titers in their sera even after being exposed with invasive organisms. Immunodeficient animals were challenged with organisms at two or three weeks of age when they lack both active B-cells and T-cells; while recovery of T-cells usually occurs after four weeks and B-cells after 7–12 weeks of age [2,3]. Further, coccidiosis is an immunosuppressive disease which weakens the immune system specifically in juvenile animals. Intestinal pathological scores caused by normal organisms were significantly more severe than those inoculated with aberrant organisms in both experiments. Yet, these animals endured the aberrant organisms proving the concept that altered organisms were significantly of low-pathogenicity even in immunodeficient animals. CY may cause GI effects; however, no attempts were taken to eliminate CY possible side effects on GI tracts except practicing extreme hygiene in animals. It is plausible that some of the pathological findings in these immunodeficient animals could have been provoked by CY more than direct effects from coccidial infection. Nevertheless, CY was similarly used in all immunodeficient animals. Therefore, differences obtained between various groups of immunodeficient animals can be mainly attributed to the pathogenic strength of normal versus aberrant organisms than direct effects from CY on GI tracts.

Strains of chicks with phenotypic differences may express various immunological/pathologic responses and resistance/susceptibility to diseases. In this investigation, two inbred strains, Rhode Island Red and New Hampshire, were chosen in order to compare immune responses against coccidial infection. Differences were detected in weight gain/loss. Naïve Rhode Island Red animals developed more severe lesions compared with New Hampshire animals. New Hampshire animals were significantly more robust and gained weight more and better tolerated challenge infections compared to the Rhode Island Red animals. As expected, uninfected control animals remained anti-coccidial antibody negative, while protective antibody titers were detected in immunized animals with aberrant forms. Yet, both strains became effectively and better protected with aberrant forms against ultimate infectious challenge than those immunized with normal forms. Aberrant organisms provoke immunity without pathogenicity. The pathogenicity after challenge was significantly lower in both groups of animals immunized with aberrant organisms. These included lack of bloody diarrhea and severe weight loss. In addition, the level of Ab production was significantly higher in the New Hampshire group immunized with aberrant, but not in Rhode Island Red animals. It is not clear whether the discrepancy in Ab production is due to immune response differences in animal strains or if other factors may have influenced the results. It is conceivable that higher titers of Ab production may not be protective in some strains of animals as seen in severe cases of infections and autoimmune diseases with unproductive surges of Ab production. These investigations proved that the sporocysts are required to preserve infectivity of the organisms. Overall, both animal strains tolerated better immunization with aberrant forms, gained more weight during immunization and when challenged with high dose pathogenic form compared to those immunized with normal forms, as presented in the results section.

This technique may be applicable to be used in other Apicomplexan organisms. For instance, *Toxoplasma* is another related member of Apicomplexan organisms with severe acute and chronic complications as well as feto-maternal syndrome in humans and animals. *Toxoplasma* has an intra intestinal (coccidian forms) exclusive in cats as a definitive host and in extra intestinal stages is cosmopolitan as it infects intermediate hosts (all animals and man), as well as every single organ and cell in the body [7,24–26]. It is conceivable that the same modifications may be applicable for oocysts of *Toxoplasma* to form aberrant organisms for a safe vaccine to immunize cats against toxoplasmosis as a preventive measure to protect humans and animals against disease.

A century after the original discovery of poultry coccidiosis (*Eimeria*), a unique method of altering infectivity of organisms has been achieved, as presented in this investigation. Previous studies had either relied on genetic variation and selection in chicks [27], or altered organisms by exposing them to irradiation [28]. The viable vaccination with Coccivac, which is a mixture of different pathogenic *Eimerias* in chicks, has been used for over 60 years (Edgar 1952) in the USA [15] with several side effects. The pathogenic live vaccine promotes weight loss and poor feed conversion and extends the required time to develop solid protective immunity response to encounter the disease, the possibility of leading to new infections and difficulties in preparation, administration and husbandry of the flocks. Other experimental attempts for vaccine production include attenuated strains and serial passages of *E. tenella* into chorio-allantoic chick embryo [16] which were proved to be ineffective. Our current report is supported by previous findings related to altered organisms from two different pathogenic strains, *E. tenella* and *E. necatrix*, using two different strains of inbred White Leghorn chicks [18]. In this study, the inbred New Hampshire response was in accord with White Leghorns immunized with aberrant organisms with complete protection from challenge infection and diarrhea. This study further exposed the ultrastructures of organisms and proved specific roles for sporocysts walls and veils to sustain pathogenicity and longevity of the infective sporozoites.

5. Conclusions

Overall, aberrant organisms were significantly less pathogenic in immunodeficient animals compared to the normal organisms, yet protective in immunocompetent animals against severe challenges. As proof of concept, lack of protective ultrastructures (sporocysts walls and veils), reduced pathogenicity, including diarrhea and weight loss, and preserved immunogenicity seems appropriate to indicate a potential use for the aberrant organisms to warrant further trials for possible attenuated vaccine production in future investigations.

Acknowledgments: This study was partly presented at Digestive Disease Week 2016, San Diago, CA, USA. (Gastroenterology 2016; 150(4):S895. DOI:10.1016/S0016-5085(16)33025-6). The investigation was supported partially by the National Institutes of Health NCCAM-AT1490 (HO).

Conflicts of Interest: Author declares to have no commercial or associative interest that represents a conflict of interest in connection with this investigation.

Abbreviations

CY, Cyclophosphamide; **TEM**, transmission electron microscope; *E.*, *Eimeria t*; **H&E**, Hematoxylin and Eosin; **ELISA**, Indirect Enzyme Linked Immunosorbant Assay.

References

1. Pamer, E.G. Resurrecting the intestinal microbiota to combat antibiotic-resistant pathogens. *Science* **2016**, *352*, 535–538. [CrossRef] [PubMed]
2. Misra, R.R.; Bloom, S.E. Roles of dosage, pharmacokinetics, and cellular sensitivity to damage in the selective toxicity of cyclophosphamide towards B and T cells in development. *Toxicology* **1991**, *66*, 239–256. [CrossRef]
3. Oz, H.S.; Markham, R.J.; Bemrick, W.J.; Stromberg, B.E. Enzyme-linked immunosorbent assay and indirect haemagglutination techniques for measurement of antibody responses to *Eimeria tenella* in experimentally infected chickens. *J. Parasitol.* **1984**, *70*, 859–863. [CrossRef]
4. Colvin, M.; Padgett, C.A.; Fenselau, C. A biologically active metabolite of cyclophosphamide. *Cancer Res.* **1973**, *33*, 915–918. [PubMed]
5. Atanasova, R.; Angoulvant, A.; Tefit, M.; Gay, F.; Guitard, J.; Mazier, D.; Fairhead, C.; Hennequin, C. A mouse model for Candida glabrata hematogenous disseminated infection starting from the gut: Evaluation of strains with different adhesion properties. *PLoS ONE* **2013**, *8*, e69664. [CrossRef] [PubMed]
6. Linna, T.J.; Frommel, D.; Good, R.A. Effects of early cyclophosphamide treatment on the development of lymphoid organs and immunological functions in the chickens. *Int. Arch. Allergy Appl. Immunol.* **1972**, *42*, 20–39. [CrossRef] [PubMed]
7. Oz, H.S. Fetal and Maternal Toxoplasmosis. In *Recent Advances in Toxoplasmosis Research*; Lee, C.M., Ed.; Nova Science Publication: New York, NY, USA, 2014; Chapter 1; pp. 1–33. Available online: http://www. Novapublishers.com (accessed on 1 July 2014).
8. Stanley, D.; Wu, S.B.; Rodgers, N.; Swick, R.A.; Moore, R.J. Differential responses of cecal microbiota to fishmeal, *Eimeria* and *Clostridium perfringens* in a necrotic enteritis challenge model in chickens. *PLoS ONE* **2014**, *9*, e104739. [CrossRef] [PubMed]
9. Wunderlich, F.; Al-Quraishy, S.; Steinbrenner, H.; Sies, H.; Dkhil, M.A. Towards identifying novel anti-Eimeria agents: Trace elements, vitamins, and plant-based natural products. *Parasitol. Res.* **2014**, *113*, 3547–3556. [CrossRef] [PubMed]
10. Cheng, G.; Dong, X.; Wang, Y.; Peng, D.; Wang, X.; Hao, H. Development of a novel genetically modified bioluminescent-bacteria-based assay for detection of fluoroquinolones in animal-derived foods. *Anal. Bioanal. Chem.* **2014**, *406*, 7899–7910. [CrossRef] [PubMed]
11. Moscoso, S.; de los Santos, F.S.; Andino, A.G.; Diaz-Sanchez, S.; Hanning, I. Detection of quinolones in commercial eggs obtained from farms in the Espaíllat Province in the Dominican Republic. *J. Food Prot.* **2015**, *78*, 214–217. [CrossRef] [PubMed]

12. Du Toit, G.; Tsakok, T.; Lack, S.; Lack, G. Prevention of food allergy. *J. Allergy Clin. Immunol.* **2016**, *137*, 998–1010. [CrossRef] [PubMed]
13. Sharman, P.A.; Smith, N.C.; Wallach, M.G.; Katrib, M. Chasing the golden egg: Vaccination against poultry coccidiosis. *Parasite Immunol.* **2010**, *32*, 590–598. [CrossRef] [PubMed]
14. Johnson, J.; Reid, W.M.; Jeffers, T.K. Practical immunization of chickens against coccidiosis using an attenuated strain of *Eimeria tenella*. *Poult. Sci.* **1979**, *58*, 37–41. [CrossRef] [PubMed]
15. Williams, R.B. Fifty years of anticoccidial vaccines for poultry (1952–2002). *Avian Dis.* **2002**, *46*, 775–802. [CrossRef]
16. Shirley, M.W.; McDonald, V.; Ballingall, S. Eimeria spp. from the chicken: From merozoites to oocysts in embryonated eggs. *Parasitology* **1981**, *83 Pt 2*, 259–267. [CrossRef] [PubMed]
17. Fantham, H.B.; Porter, A. Minute animal parasites. *Science* **1914**, *40*, 814. [CrossRef] [PubMed]
18. Oz, H.S. Selective Induced Altered Coccidians to Immunize and Prevent Enteritis. *Gastroenterol. Res. Pract.* **2016**, *2016*. [CrossRef] [PubMed]
19. Oz, H.S.; Stromberg, B.E.; Bemrick, W.J. Enzyme-linked immunosorbent assay to detect antibody response against *Eimeria bovis* and *Eimeria zurnii* in calves. *J. Parasitol.* **1986**, *72*, 780–781. [CrossRef] [PubMed]
20. Odore, R.; De Marco, M.; Gasco, L.; Rotolo, L.; Meucci, V.; Palatucci, A.T.; Rubino, V.; Ruggiero, G.; Canello, S.; Guidetti, G.; et al. Cytotoxic effects of oxytetracycline residues in the bones of broiler chickens following therapeutic oral administration of a water formulation. *Poult. Sci.* **2015**, *94*, 1979–1985. [CrossRef] [PubMed]
21. McDougald, L.R.; Fitz-Coy, S.H. Protozoal infections. In *Diseases of Poultry*; Saif, Y.M., Ed.; Blackwell Publishing: Ames, IA, USA, 2009; p. 1352.
22. Diaz-Sanchez, S.; D'Souza, D.; Biswas, D.; Hanning, I. Botanical alternatives to antibiotics for use in organic poultry production. *Poult. Sci.* **2015**, *94*, 1419–1430. [CrossRef] [PubMed]
23. Kawabata, T.T.; Chapman, M.Y.; Kim, D.H.; Stevens, W.D.; Holsapple, M.P. Mechanisms of in vitro immunosuppression by hepatocyte-generated cyclophosphamide metabolites and 4-hydroperoxycyclophosphamide. *Biochem. Pharmacol.* **1990**, *40*, 927–935. [CrossRef]
24. Oz, H.S. Toxoplasmosis Complications and Novel Therapeutic Synergism Combination of Diclazuril plus Atovaquone. *Front. Microbiol.* **2014**, *5*, 1–9. [CrossRef] [PubMed]
25. Chemoh, W.; Sawangjaroen, N.; Nissapatorn, V.; Sermwittayawong, N. Molecular investigation on the occurrence of *Toxoplasma gondii* oocysts in cat feces using TOX-element and ITS-1 region targets. *Vet. J.* **2016**, *215*, 118–122. [CrossRef] [PubMed]
26. Oz, H.S. Feto-Maternal and Pediatric Toxoplasmosis. *J. Pediatr. Infect. Dis.* **2017**, *12*, 1–15. ISSN 1305-7707.
27. Jeffers, T.K. Genetic transfer of anticoccidial drug resistance in *Eimeria tenella*. *J. Parasitol.* **1974**, *60*, 900–904. [CrossRef] [PubMed]
28. Cox, A.B.; Duncan, S.; Levy, C.K. *Eimeria falciformis*: Effects of 60 Co irradiation on infectivity and immunogenicity of sporulated oocysts. *J. Parasitol.* **1977**, *63*, 927–929. [CrossRef] [PubMed]

MDPI AG

St. Alban-Anlage 66

4052 Basel, Switzerland

Tel. +41 61 683 77 34

Fax +41 61 302 89 18

http://www.mdpi.com

Nutrients Editorial Office

E-mail: nutrients@mdpi.com

http://www.mdpi.com/journal/nutrients

www.ingramcontent.com/pod-product-compliance
Lightning Source LLC
Chambersburg PA
CBHW051710210326
41597CB00032B/5431